Human Chorionic Gonadotropin (hCG)

Human Chorionic Gonadotropin (hCG)

Second Edition

Laurence A. Cole
USA hCG Reference Service,
Angel Fire, NM, USA

Stephen A. Butler
MAP Diagnostics,
BioPark, Broadwater Road,
Welwyn Garden City,
Hertfordshire, UK

ELSEVIER

AMSTERDAM • BOSTON • HEIDELBERG • LONDON • NEW YORK • OXFORD
PARIS • SAN DIEGO • SAN FRANCISCO • SINGAPORE • SYDNEY • TOKYO

Elsevier
Radarweg 29, PO Box 211, 1000 AE Amsterdam, Netherlands
The Boulevard, Langford Lane, Kidlington, Oxford OX5 1GB, UK
225 Wyman Street, Waltham, MA 02451, USA

First edition 2010
Second edition 2015

ISBN: 978-0-12-800749-5

British Library Cataloguing-in-Publication Data
A catalogue record for this book is available from the British Library

Library of Congress Cataloging-in-Publication Data
A catalog record for this book is available from the Library of Congress

For Information on all Elsevier Publishing publications
visit our website at http://store.elsevier.com/

Image credit: Nicholas Gibbons & Stephen A. Butler

Contents

List of Contributors

Beata B. Burczynska Biomedical Sciences, Middlesex University, The Burroughs, Hendon, London, UK

Stephen A. Butler MAP Diagnostics, BioPark, Broadwater Road, Welwyn Garden City, Hertfordshire, UK

Francis W. Byrn Obstetrics and Gynecology, University of New Mexico, Albuquerque, NM, USA

Laurence A. Cole USA hCG Reference Service, Angel Fire, NM, USA

Robert O. Hussa Medical College of Wisconsin, Milwaukee, WI, Sunnyvale, CA, USA

Ray K. Iles MAP Diagnostics, BioPark, Broadwater Road, Welwyn Garden City, Hertfordshire, UK

Ervin E. Jones Genetics and IVF Institute, Fairfax, VA, USA

Akira Kobata The Noguchi Institute, Tokyo, Japan

C.V. Rao Herbert Wertheim College of Medicine, Florida International University, Miami, FL, USA

Ulf-Håkan Stenman Helsinki University Central Hospital, Helsinki, Finland

About the Editors

Laurence A. Cole, PhD, and Stephen A. Butler, PhD, wrote this book together.

Laurence A. Cole, PhD

In 1971, Larry Cole began academic life studying medicine in England. A stroke in 1974 put him in a coma for 3 months and left him with amnesia and severe brain damage. He was forced to abandon his career in medicine as a result of this brain damage incident. After spending 1 year in the hospital, he moved to Israel in 1976, then to the United States in 1977. He somehow was fortunate and able to "program" the other side of his brain. He taught himself to learn and memorize science and medicine again and, in 1978, Cole attended the Medical College of Wisconsin, Milwaukee, to work on a PhD in Biochemistry. In 1981, Larry obtained his PhD while studying cancer cell hCG with Robert Hussa, PhD. During this time, Larry also worked with Roland Pattillo, MD, on gestational trophoblastic disease cases. In the same year, he attended the first international symposium on gestational trophoblastic diseases and has attended all 15 consecutive biannual symposia since that time.

In 1983, Larry completed a postdoctoral fellowship with Raymond Ruddon, PhD, at the University of Michigan, Ann Arbor, where he continued to specialize in hCG. His experiences working with Dr. Ruddon are what inspired him to study hCG carbohydrate structure. After his postdoctoral fellowship, Larry joined the faculty of the University of Michigan.

In 1984, Larry took a position at Yale University, where he spent 15 years as part of the Obstetrics and Gynecology faculty. He slowly advanced at Yale University from Assistant Professor to Associate Professor, then to Full Professor. While at Yale University, Larry's research purely focused on hCG, investigating hCG and gestational trophoblastic disease, hCG as a tumor marker, and oligosaccharides and hCG. It was at Yale University that he studied the structure of choriocarcinoma hCG, which ultimately led to his discovery of hyperglycosylated hCG and to new diagnostic protocols in gestational trophoblastic diseases.

In 1999, Larry moved to the University of New Mexico, Albuquerque, as a tenured Full Professor. While at the University of New Mexico, he started the US government CLIA-endorsed program called the USA hCG Reference Service, which globally advises scientists who research hCG, physicians who treat gestational trophoblastic disease, and patients with persistent low levels of hCG. With the Reference Service came his endowment by appreciative hCG patients with false-positive test results. In 2004, he became the Howard and Friedman Distinguished Professor of Obstetrics and Gynecology.

Today, 32 years after receiving his PhD, Dr. Cole still specializes in hCG research. With the discovery of how hCG variants modulate cancer, Larry's research interest now centers on cancer biology and treatment. In 1987, Larry's PhD advisor, Robert Hussa, wrote the first book on hCG. It is with great dignity that Larry follows in the footsteps of his advisor and writes this second version of the second specialized book on hCG in 2013.

Stephen A. Butler, PhD

Stephen Butler, PhD, was a Reader in Biomedical Diagnostics at the Centre for Investigative and Diagnostic Oncology in the School of Science and Technology, Middlesex University, London. In 2014, he left higher education to co-found a diagnostic company specializing in prenatal screening called MAP Diagnostics.

Stephen's first exposure to hCG came as an undergraduate in the Williamson Laboratory for Molecular Oncology founded by Ray Iles in the Department of Reproductive Physiology Department under Tim Chard. Stephen ran hCGβ RIAs and IRMAs for studies into Downs Syndrome and cancer along with reverse transcriptase studies on *CGB* gene expression. After graduating, Stephen returned to Ray's laboratory for his PhD on the structure/function of hCGβ in epithelial cancer and then a fellowship at Yale with Larry Cole. Stephen moved with Larry to New Mexico and helped set up the hCG reference service at a time when phantom hCG was making the laboratory very busy. Stephen returned to the UK after 2½ years. During this time in the USA, Stephen and Larry shared their experiences on hCG (its variants and their structures and functions) and combined their collective evidence to formulate many of the cancer hypotheses which led to the work described in this book. This broadened the whole hCG cancer field and is a central focus of this second edition book.

We thank Camille Sapienz for editing the grammar, punctuation, and diction of this book. We also thank all the other authors who contributed essential chapters to this book: Beata Burcznska; Francis Byrn, MD; Robert Hussa, PhD; Ray Iles; Ervin Jones, MD; Akira Kobata, PhD; C.V. Rao, PhD; and Ulf Stenman, PhD. Once again, we say thank you to everybody involved.

Laurence A. Cole PhD
Stephen A. Butler PhD

Preface

This book was written by twenty-first century scientists in honor of the forefathers and pioneers of hCG research. Without them, we never would have been able to discover the wondrous molecules that we know today. I honor and dedicate this book to those who discovered hCG and to those who first found evidence for an hCG-like molecule. In 1912, Bernhard Aschner at University of Vienna stimulated the genital tract of guinea pigs with injections of water-soluble extracts of human placenta [1]. In 1913, Otfried Fellner, a gynecologist in Vienna, induced ovulation in immature rabbits with saline extracts of human placenta [2]. In 1919, Toyoichi Hirose of the Osaka Medical College stimulated ovulation and normal luteal function in immature rabbits by repeated injection of human placental tissue [3]. All of these works show that there is a clear hormonal link between the placenta and the uterus [1–3]. In 1927, Selamar Aschheim of the University of Berlin in Germany and Bernhard Zondek of the Berlin-Spandau Hospital in Germany demonstrated that pregnant women produce a gonad-stimulating substance [4].They showed that injecting this substance subcutaneously into intact immature female mice produced follicular maturation, luteinization, and hemorrhaging into the ovarian stroma. These findings were confirmed by others [5,6] and the first hCG pregnancy tests were born [4–6].

Around this time, the name human chorionic gonadotropin (hCG) was conceived. Chorion comes from the Latin *chordata*, meaning afterbirth; gonadotropin is used because the hormone is a gonad tropic molecule acting on the ovaries and promoting steroid production. I also dedicate this book to those 21 people who I consider to be the pioneers of hCG research, without whom there would be little to write about:

Akira Kobata, PhD	Tokyo University, Japan.
Bruce Nisula, MD	National Institutes of Health, Bethesda, MD.
Robert Norman, MD	University of Adelaide, Australia.
Bruce Weintraub, MD	National Institutes of Health, Bethesda, MD.
Glenn Braunstein, MD	Cedars-Sinai Medical, Los Angeles, CA.
Irving Boime, PhD	Washington University, St Louis, MO.
Jean-Michel Bidart, PhD	Institut Gustave-Roussy, Villejuif, France.
John Fiddes, PhD	Cold Spring Laboratory, Cold Spring Harbor, NY.
Judith Vaitukaitis, MD	National Institutes of Health, Bethesda, MD.
Klauss Mann, MD	University of Essen, Germany.
Mario Ascoli, PhD	University of Iowa, Iowa City, IA.
Om Bahl, PhD	University at Buffalo, Buffalo, NY.
Robert Canfield, MD	Columbia-Presbyterian Hospital, New York, NY.
Steven Birken, PhD	National Institutes of Health, Bethesda, MD.

Timothy Chard, MD	St Bartholomew's Hospital, London, UK.
Maria Dufau, MD, PhD	National Institutes of Health, Bethesda, MD.
Ulf Stenman, PhD	Helsinki University, Helsinki, Finland.
William Odell, MD, PhD	Utah Medical Center, Salt Lake City, UT.
C.V. Rao, PhD	University of Louisville, Louisville, KY.
Ray Ruddon, MD, PhD	University of Michigan, Ann Arbor, MI.
Francis Morgan, PhD	College of Physicians and Surgeons, Columbia University, New York.

Laurence A. Cole, PhD	USA hCG Reference Service, Angel Fire, NM, USA
Stephen A. Butler, PhD	MAP Diagnostics, BioPark, Broadwater Road, Welwyn Garden City, Hertfordshire, UK

References

[1] Aschner B. Ueber die function der hypophyse. Pflug Arch Gest Physiol 1912;146: 1–147.

[2] Fellner OO. Experimentelle untersuchungen uber die wirkung von gewebsextrakten aus der plazenta und den weiblichen sexualorganen auf das genital. Arch Gynakol 1913;100:641.

[3] Hirose T. Experimentalle histologische studie zur genese corpus luteum. Mitt Med Fakultd Univ ZU 1919;23:63–70.

[4] Aschheim S, Zondek B. Das Hormon des hypophysenvorderlappens: testobjekt zum Nachweis des hormons. Klin Wochenschr 1927;6:248–52.

[5] Zondek B, Aschheim S. The Zondek–Ascheimpregnancy test. Can Med Assoc J 1930;22:251–3.

[6] Friedman MH, Lapham MEA. Simple, rapid procedure for the laboratory diagnosis of early pregnancies. Am J Obstet Gynecol 1931;21:405–10.

Abbreviations

αACT α-subunit activator element
ActD actinomycin D
Ala alanine
Arg arginine
Asn asparagine
ATF1 activating transcription factor 1
BEP chemotherapy regimen: bleomycin, etoposide, and cisplatin
BSO bilateral salpingo-oophorectomy
cAMP cyclic adenosine monophosphate
CG chorionic gonadotropin
CGB chorionic gonadotropin beta gene (s)
CG-H hyperglycosylated chorionic gonadotropin
CHO Chinese hamster ovary
CKGF cystine-knot growth factor
CNS central nervous system
COH controlled ovarian hyperstimulation
COS controlled ovarian stimulation
CRE cAMP response element
CREBP cAMP response element binding protein
CTP human chorionic gonadotropin β-subunit C-terminal peptide
Cys cysteine
DSA Datura stramonium agglutinin
DSD downstream domain
EDTA ethylene diamine tetra-acetic acid
EGF epidermal growth factor
EMA-CO chemotherapy regimen: etoposide, methotrexate, and actinomycin D, alternating with cyclophosphamide and onvocin (vincristine)
EMA-EP chemotherapy regimen: etoposide, methotrexate, and actinomycin D, alternating with etoposide and cisplatin
FDA Food and Drug Administration (USA)
FIGO Federation International of Gynecologic Oncology
free α free α-subunit of human chorionic gonadotropin
free β free β-subunit of human chorionic gonadotropin
free β-βCTP free β-subunit C-terminal peptide
FSH follicle-stimulating hormone
Fuc fucose
FUT fucosyltransferase
Gal galactose
GalNAc *N*-acetylgalactosamine
GATA DNA sequence and DNA coding
GlcNAc *N*-acetylglucosamine

Glu glutamic acid
Gly glycine
GnRH gonadotropin-releasing hormone
GnT *N*-acetylglucosaminyltransferase
GTD gestational trophoblastic disease
GTN gestational trophoblastic neoplasm
hCG human chorionic gonadotropin
hCGα free α-subunit of human chorionic gonadotropin
hCGβ free β-subunit of human chorionic gonadotropin
hCG-βCTP human chorionic gonadotropin β-subunit C-terminal peptide
hCG-H hyperglycosylated human chorionic gonadotropin
hCG-Hβ hyperglycosylated human chorionic gonadotropin free β -subunit
hCGp pituitary hCG
His histamine
hLH human luteinizing hormone
hMG human menopausal gonadotropins
ICE chemotherapy regimen: ifosamide, carboplatin, and etoposide
IFCC International Federation of Clinical Chemistry
Ile Isoleucine
IRP international reference preparation
IRR international reference reagent
IS International Standard
ISOBM International Association of Biological Markers
IU international units
IUI intrauterine insemination
IU/l international units per liter
IVF *in vitro* fertilization
Leu leucine
LIF leukemia inhibitory factor
LH luteinizing hormone
LUF luteinized unruptured follicle syndrome
Lys lysine
Man mannose
MCSF macrophage colony stimulating factor
Met methionine
MoM multiples of the median
MCW Medical College of Wisconsin
mIU milli-international units
mIU/ml milli-international units per milliliter
ml milliliter
MMP metalloproteinase
Mtx methotrexate
NCBI The National Center for Biotechnology Information
NeuAc *N*-acetylneuraminic acid or sialic acid
NGF nerve growth factor
ng/ml nanograms per millilter
NIH National Institutes of Health
Ob/Gyn obstetrics and gynecology
OHSS ovarian hyperstimulation syndrome

OI ovulation induction
Oop oophorectomy
OTC over-the-counter
PAPP-A pregnancy-associated plasma protein-A
PCOS polycystic ovary syndrome
PCR polymerase chain reaction
PDGFB platelet-derived growth factor B
pg/ml picograms per milliliter
PIH pregnancy-induced hypertension
PKA phosphokinase A
PlGF placental growth factor
POC point-of-care
Pro proline
rhCG recombinant human chorionic gonadotropin
RIA radioimmunoassay
RR reference reagent
RT-PCR reverse-transcription polymerase chain reaction
SAB spontaneous abortion or miscarriage
Ser serine
SP1 selective promoter factor 1
ST3Gal sialyltransferase-3-galactose
ST6Gal sialyltransferase-6-galactose
TAH transabdominal hysterectomy
TGFβ transforming growth factor β
Thr threonine
TJA-I Trichosanthes japonica agglutinin-I
TJA-II Trichosanthes japonica agglutinin-II
Trp tryptophan
TSE tissue/trophoblast-specific element
TSH thyroid-stimulating hormone
TVH transvaginal hysterectomy
Tyr tyrosine
URE upstream regulatory element
USD upstream domain
Val valine
VEGF vascular endothelial growth factor

List of Tables

List of Figures

Part A

Introduction

The expanding world of hCG

1

Robert O. Hussa
Medical College of Wisconsin, Milwaukee, WI, Sunnyvale, CA, USA

More than four decades ago, in 1969, the door to the world of hCG opened for me when I joined Roland Pattillo' s research team in the Department of Obstetrics/Gynecology at the Medical College of Wisconsin (MCW) in Milwaukee. Pattillo established the hCG-secreting BeWo choriocarcinoma cell line in 1966 in George Gey's laboratory at Johns Hopkins University [1]. Pattillo literally carried the cell line in a culture flask filled with culture fluid in his shirt pocket during his move from Baltimore to Milwaukee. Pattillo, along with Eleanor Delfs and Richard Mattingly, comprised the trophoblast disease clinical team that migrated from Johns Hopkins University to Milwaukee. Mattingly was chairman of the Department of Obstetrics/Gynecology at MCW and editor-in-chief of *Obstetrics & Gynecology*. From my desk in the laboratory (the room adjoining Pattillo's cell culture laboratory), I would observe Eleanor Delfs personally perform hCG extractions on serum samples from her hydatidiform mole and choriocarcinoma patients before injecting the samples into rats, per her uterine weight bioassay, which was published the year I was born [2].

The emphasis of the Department of Obstetrics/Gynecology at MCW on treatment of patients with trophoblast disease and the study of hCG in patients, as well as in hCG-secreting cell lines, led to the need to quantitatively measure hCG in large numbers of samples. Pattillo's thriving cell culture laboratory soon established a second hCG-secreting trophoblast cell line, JAr [3]. In those years, the glycoprotein hormone field was filled with excitement, such as that engendered by the revelation that hCG, LH, FSH, and TSH all contained a common α-subunit and a hormone-specific β-subunit [4–6]. In addition, the technology of the new radioimmunoassays (RIAs) was making great strides, particularly with the β-subunit-specific RIA created by Vaitukaitis and colleagues in Griff Ross's laboratory at the National Institutes of Health [7]. Our laboratory also evolved in the 1970s and 1980s and began to use RIAs to measure hCG and its α- and β-subunits in our research.

Another significant body of research emanated out of laboratories such as that of Om P. Bahl, where the carbohydrate structure of hCG was initially characterized utilizing digestion with specific exoglycosidases [8]. More excitement for our laboratory came when Pattillo's crackerjack tissue culture staff established the hCG β-secreting CaSki and DoT cell lines from cervical carcinoma [9,10]. It needs to be emphasized that Roland Pattillo has provided a great service to science by willingly providing his trophoblastic and nontrophoblastic cell lines to all requesters over four decades.

Human Chorionic Gonadotropin (hCG). DOI: http://dx.doi.org/10.1016/B978-0-12-800749-5.00001-8

It was in this 1978 setting that a young biochemistry student at MCW, Laurence A. Cole (alias: Larry), began his PhD research in my laboratory. Laurence actually occupied the same desk that I had used in my first years in the laboratory at MCW. Laurence, a United Kingdom native, who greeted me each morning with a friendly "Good morning, sire," was a hard-working, prodigiously productive, and insightful researcher who occasionally found the time to bring special treats for the laboratory, such as fresh scones with clotted cream flown in from London or some of his home-brewed beer.

Larry's dissertation on the characterization of the ectopic hCG β-subunit secreted by the CaSki and DoT cervical carcinoma cell lines [11] led to his first publications [12–15] and established his early fascination with the oligosaccharide structure of hCG. This experience undoubtedly led to Larry Cole's great contributions toward a better understanding of the many variant forms of hCG in pregnancy and cancer, as described in this book. It is indeed fitting that Larry has taken on the monumental task—akin to that of herding cats—of gathering the work of so many experts in the various areas of hCG (most of whom have been contributing to the field for decades) and compiling it in this comprehensive compendium of reviews of all aspects of hCG, from its molecular and oligosaccharide structure and biosynthesis to its applications as a clinical marker and use as a therapeutic agent.

A Web of Science search of this book's areas of emphasis was conducted at the end of 2009 (Table 1.1). Despite the fact that no effort was made to cull through the articles and citations to eliminate redundant and irrelevant publications, the results give an excellent broad perspective of how the world of hCG has expanded over time. The number of publications in a given area is an indicator of the amount of activity in that area. Thus, the number of times a reference is cited by peers should correlate with the publications the fact that experts value these as the most significant.

In the years from 1975 to 1985 and 1986 to 2009, there were more publications and citations of clinical applications of hCG than any other category, followed by receptor activity and biological functions of hCG. In contrast, there were no publications or citations before 1986 of hyperglycosylated hCG, degradation products of hCG, or diet/sports/HIV, and there were only two publications devoted to hCG standards and reference preparations. The lack of publications in these areas of emphasis underscores the timeliness and relevance of the reviews for each of these areas in this book.

After the 1980 reviews [28,29], various aspects of hCG have been reviewed more recently, including three-dimensional structure [17,30], gene expression and cloning [31,32], biosynthesis [33], glycosylation [16], immunochemistry [34], clinical measurement [35–38], receptors [39,40], and infertility [41]. The recently emerging category that has received the most attention (as measured by the 2698 publications since 1985, compared with no previous publications) is that of diet, sports, and HIV. Similarly, the Varki review of the biological roles of oligosaccharides was cited in an astonishing 2975 references [22]; the next most cited reference in hCG literature (738 citations) was that regarding placental implantation (by Cross et al.) [23]; the third most cited publication (602 citations) was of a publication regarding the crystal structure of hCG (by Lapthorn et al.) [20].

Table 1.1 Web of Science hCG search, 1975–2009

Area of hCG research	Articles 1975–1985	Articles 1986–2009	Citations 1975–1985	Citations 1986–2009	Most cited article	Times cited
Biosynthesis	19	448	691	10,286	[16]	274
Glycosylation	9	270	242	7452	[17]	327
Hyperglycosylated hCG	0	9	0	158	[18]	94
Genes, mRNA	11	1677	712	37,713	[19]	551
Structure	9	680	385	16,913	[20]	602
Degradation products	0	9	0	177	[21]	99
Receptor, biological functions	88	3210	2030	74,845	[22]	2975
Clinical applications, infertility, pregnancy, cancer	265	6753	4032	99,162	[23]	738
Diet, sports, HIV	0	154	0	2698	[24]	310
Vaccine, cancer	56	1475	921	23,784	[25]	556
Antibodies	80	1014	1217	20,081	[26]	363
Assays	36	1195	370	19,380	[26]	363
Standards	2	72	66	1122	[27]	78

The yearly trend of publications and citations provides interesting information regarding the dynamics of activity in each of the areas of emphasis described in this book. In the areas of hCG biosynthesis, hCG genes/mRNA, hCG structure, and hCG receptor/biological functions, there was a jump in research activity in 1991–1994, with a corresponding increase in the number of citations starting in 1992. In the area of hCG clinical applications, the publishing activity has remained relatively constant (260–400 articles per year), whereas the number of citations has increased dramatically in linear fashion since 1991 (Table 1.1).Very similar trends were observed for the search areas of hCG assays/tests and vaccines/cancer (not shown). The newly emerging emphasis areas of diet/sports/HIV reveal sporadic publishing activity starting in 1991 and a surge of citations beginning in approximately 1996.

I hope the tapestry of hCG research trends from over the decades will provide an appropriate background for readers and guide them toward a favorite area of interest within the evolving world of hCG as covered in this comprehensive book. As mentioned, several of these topics are being reviewed for the first time; others are continuously updated with new information emanating from the rapidly expanding field of hCG research. I am grateful and proud to have been part of this body of investigation since 1969.

References

[1] Pattillo RA, Gey GO. The establishment of a cell line of human hormone-synthesizing trophoblastic cells *in vitro*. Cancer Res 1968;28:1231–6.

[2] Delfs E. An assay method for human chorionic gonadotropin. Endocrinology 1941;28:196–202.

[3] Pattillo RA, Ruckert A, Hussa R, Bernstein R, Delfs E. The JAr cell line—continuous human multi-hormone production and controls. In Vitro 1971;6:398–9 (Abstract).

[4] Swaminathan N, Bahl OP. Dissociation and recombination of the subunits of human chorionic gonadotropin. Biochem Biophys Res Commun 1970;40:422–7.

[5] Pierce JG. Eli Lilly lecture: the subunits of pituitary thyrotropin—their relationships to other glycoprotein hormones. Endocrinol 1971;89:1331–44.

[6] Bahl OP, Carlsen RB, Bellisario R, Swaminathan N. Human chorionic gonadotropin: amino acid synthesis of the α and β subunits. Biochem Biophys Res Commun 1972;48:416–22.

[7] Vaitukaitis JL, Braunstein GD, Ross GT. A radioimmunoassay which specifically measures human chorionic gonadotropin. Am J Obstet Gynecol 1972;111:751–8.

[8] Bahl OP, Marz L, Kessler MJ. Isolation and characterization of N- and O-glycosidic carbohydrate units of human chorionic gonadotropin. Biochem Biophys Res Commun 1978;84:667–76.

[9] Pattillo RA, Hussa RO, Story MT, Ruckert ACF, Shalaby MR, Mattingly RF. Tumor antigen and human chorionic gonadotropin in CaSki cells: a new epidermoid cervical cancer cell line. Science 1977;196:1456–8.

[10] Pattillo RA, Hussa RO, Ruckert ACF, Story MT, Mattingly RF. hCG-beta production by epidermoid carcinoma of the cervix. Endocrinology (Suppl) 1977;100:179.

[11] Cole LA. Studies with ectopic hCG β: synthesis, glycosylation, isolation and characterization. PhD Dissertation, Milwaukee: Medical College of Wisconsin; 1982.

[12] Cole LA, Hussa RO, Rao CV. Discordant synthesis and secretion of human chorionic gonadotropin and subunits by cervical carcinoma cells. Cancer Res 1981;41: 1615–19.

[13] Cole LA, Hussa RO. Use of glycosidase digested human chorionic gonadotropin and β-subunit to explain the partial binding of ectopic glycoprotein hormones to Con A. Endocrinology 1981;109:2276–8.

[14] Story MT, Cole LA, Hussa RO. A procedure using immobilized antibody for the isolation of the β-subunit of human chorionic gonadotropin from the culture fluid of a β-subunit-secreting nontrophoblastic cell line. J Clin Endocrinol Metab 1981;53: 1090–5.

[15] Cole LA, Birken S, Sutphen S, Hussa RO, Pattillo RA. Absence of the COOH-terminal peptide on ectopic human chorionic gonadotropin β-subunit (hCG β). Endocrinology 1982;110:2198–200.

[16] Van den Steen P, Rudd PM, Dwek RA, Opdenakker P. Concepts and principles of O-linked glycosylation. Crit Rev Biochem Mol Biol 1998;33:151–208.

[17] Wu H, Lustbader JW, Liu Y, Canfield RE, Hendrickson WA. Structure of human chorionic gonadotropin at 2.6 å resolution from MAD analysis of the selenomethionyl protein. Structure 1994;2:545–58.

[18] Elliott MM, Kardana A, Lustbader JW, Cole LA. Carbohydrate and peptide structure of the α- and β-subunits of human chorionic gonadotropin from normal and aberrant pregnancy and choriocarcinoma. Endocrine 1997;7:15–32.

[19] Kumar TR, Wang Y, Lu N, Matzuk MM. Follicle stimulating hormone is required for ovarian follicle maturation but not male fertility. Nat Genet 1997;15:201–4.

[20] Lapthorn AJ, Harris DC, Littlejohn A, Lustbader JW, Canfield RE, Machin KJ, et al. Crystal structure of human chorionic gonadotropin. Nature 1994;369:455–61.

[21] Rotmensch S, Cole LA. False diagnosis and needless therapy of presumed malignant disease in women with false-positive human chorionic gonadotropin concentrations. Lancet 2000;355:712–15.

[22] Varki A. Biological roles of oligosaccharides—all of the theories are correct. Glycobiology 1993;3:97–130.

[23] Cross JC, Werb Z, Fisher SJ. Implantation and the placenta—key pieces of the development puzzle. Science 1994;266:1508–18.

[24] Hansen JE, Lund O, Tolstrup N, Gooley AA, Williams KL, Brunak S. NetOglyc: prediction of mucin type *O*-glycosylation sites based on sequence context and surface accessibility. Glycoconj J 1998;15:115–30.

[25] Mead GM, Stenning SP, Cook P, International Germ Cell Cancer Consensus Group. International germ cell consensus classification: a prognostic factor-based staging system for metastatic germ cell cancers. J Clin Oncol 1997;15:594–603.

[26] Jayasena SD. Aptamers: an emerging class of molecules that rival antibodies in diagnostics. Clin Chem 1999;45:1628–50.

[27] Cole LA, Kardana A. Discordant results in human chorionic gonadotropin assays. Clin Chem 1992;38:263–70.

[28] Hussa RO. Biosynthesis of human chorionicgonadotropin. Endocr Rev 1980;1:268–94.

[29] Hussa RO. The clinical marker hCG. New York, NY: Praeger Publishers; 1987.

[30] Lustbader JW, Yarmush DL, Birken S, Puett D, Canfield RE. The application of chemical studies of human chorionic gonadotropin to visualize its three-dimensional structure. Endocr Rev 1993;14:291–311.

[31] Habener JF. Molecular cloning of hormone genes. Clifton, NJ: The Humana Press; 1987.
[32] Jameson JL, Hollenberg AN. Regulation of chorionic gonadotropin gene expression. Endocr Rev 1993;14:203–321.
[33] Merz WE. Biosynthesis of human chorionic gonadotropin: a review. Eur J Endocr 1996;135:269–84.
[34] O'Connor JF, Birken S, Lustbader JW, Krichevsky A, Chen Y, Canfield RE. Recent advances in the chemistry and immunochemistry of human chorionic gonadotropin: impact on clinical measurements. Endocr Rev 1994;15:650–83.
[35] Chard T. Pregnancy tests: a review. Hum Reprod 1992;7:701–10.
[36] Duffy MJ. Clinical uses of tumor markers: a critical review. Crit Rev Clin Lab Sci 2001;38:225–62.
[37] Reis FM, D'Antona D, Petraglia F. Predictive value of hormone measurements in maternal and fetal complications of pregnancy. Endocr Rev 2002;23:230–57.
[38] Stenman UH, Alfthan H, Hotakainen K. Review. Human chorionic gonadotropin in cancer. Clin Biochem 2004;37:549–61.
[39] Ascoli M, Fanelli F, Segaloff DL. The lutropin/choriogonadotropin receptor, a 2002 perspective. Endocr Rev 2002;23:141–74.
[40] Kleinan G, Krause G. Thyrotropin and homologous glycoprotein hormone receptors: structural and functional aspects of extracellular signaling mechanisms. Endocr Rev 2009;30:133–51.
[41] Macklon NS, Stouffer RL, Giudice LC, Fauser BCJM. The science behind 25 years of ovarian stimulation for *in vitro* fertilization. Endocr Rev 2006;27:170–207.

History and introduction to human chorionic gonadotropin, a group of five independent growth factors

2

Laurence A. Cole
USA hCG Reference Service, Angel Fire, NM, USA

2.1 History

It is difficult to say who was the first discoverer of the hormone that we today call human chorionic gonadotropin (hCG). In 1912, Bernhard Aschner stimulated the genital tract of guinea pigs with injections of water-soluble extracts of human placenta [1]. This was followed in 1913 by Otto Fellner inducing ovulation in immature rabbits with saline extracts of human placenta [2]. In 1919, Hirose stimulated ovulation and normal luteal function in immature rabbits by repeated injection of human placental tissue [3]. These resulted in findings that were the first discoveries of an hCG-like hormone. At approximately this time, the name human chorionic gonadotropin was conceived: *Chorion* is Latin for "placenta" and the hormone is produced by the placenta, and *chorionic gonadotropin* because the hormone is tropic, acting on the female gonad tissue (ovaries), and promoting steroid-induced actions.

2.2 The pregnancy test

In 1927, Aschheim and Zondek demonstrated that the blood and urine of pregnant women contained a gonad-stimulating substance [4]. They showed that injecting this substance subcutaneously into intact immature female mice produced follicular maturation, luteinization, and hemorrhaging into the ovarian stroma. These findings were confirmed by others [5,6] and the first hCG/pregnancy tests were born [4–6]. These early tests primarily used urine to promote ovulation in mice and rabbits [4–10] and were commonly referred to as the "rabbit" or Friedman test (Table 2.1).

Over the next four decades, bioassays like the rabbit test were the only practical way to detect pregnancy or measure hCG [4–10]. In 1960, we saw the first antibody-based pregnancy test. The first antibody-based tests examined hemagglutination inhibition and latex agglutination [11,12]. These were insensitive slide tests that detected hCG at a concentration of 1000 mIU/ml or greater (Table 2.1).

Human Chorionic Gonadotropin (hCG). DOI: http://dx.doi.org/10.1016/B978-0-12-800749-5.00002-X

Table 2.1 History of the laboratory pregnancy test (the hCG and hCGβ assay)

Year	Test description	Sensitivity mIU/ml	Test time	Reference described
1927	Urine promotes mice corpora lutea	5000	5 days	Ascheim and Zondek [4]
1931	Urine hCG promotes rabbit ovulation	5000	1.5 days	Friedman and Lapham [6]
1934	Urine hCG promotes toad ovulation	3000	18 h	Shapiro and Zwarenstein [7]
1941	Serum hCG raises rat uterine weight	3000	3 days	Delfs [8]
1943	Urine causes hyperemia in rat ovary	400	2 h	Kupperman et al. [9]
1948	Urine hCG promotes toad sperm ejection	500	2 h	Galli and Mainini [10]
1960	Blood hCG inhibits hemagglutination	1000	1 day	Wide and Gemzell [11]
1962	Serum hCG promoted latex agglutination	3000	2 min	Wide [12]
1964	Serum hCG radioimmunoassay	5	4 h	Paul and Odell [13]
1972	Serum hCGβ radioimmunoassay	5	4 h	Vaitukaitis et al. [14]
1980	Serum enzyme immunometric assay	5	30 min	Batzer [15]
1984	Chemiluminescent immunometric assay	1	30 min	Barnard et al. [16]

In 1964, the competitive hCG radioimmunoassay (RIA) was invented [13,14,17–20] and revolutionized pregnancy testing. At last a test was available that could measure hCG as low as 5 mIU/ml, and it could measure pregnancy close to the day of a missed period. The invention of the RIA led to readily available pregnancy testing/hCG measurement at clinical laboratories throughout the world [13,14,17–20].

The initial RIAs were problematic because they used an antibody against hCG dimer and detected both hCG and luteinizing hormone (LH). The problem was that the α-subunit of hCG was identical to the α-subunit of LH, and the β-subunit of hCG was 80% homologous with the β-subunit of LH. Hence, the early hCG dimer RIA detected both hCG and LH and, thus, could only demonstrate pregnancy and exclude LH by showing a continual increase in hormone levels.

In 1973, Vaitukaitis et al. introduced the hCGβ test [14], an RIA pregnancy test using an antibody against the β-subunit of hCG. The hCGβ test was the first hCG-specific RIA. Unlike its predecessor, which detected both hCG and LH, the hCGβ test measured hCG alone and did not detect LH [14]. This was an important distinction because LH has no relationship to pregnancy and need not be measured when testing for pregnancy. The hCGβ test RIA became the world standard for the next 20 years. Even today, in the age of immunometric assays, both physicians and textbooks still describe hCG tests as hCGβ tests.

The discovery of monoclonal antibodies in 1975 was paramount to the development of the modern immunometric hCG tests [21]. Modern two- or three-antibody immunometric hCG assays were developed in 1981. These assays relied on one or two antibodies to capture hCG by binding one site, and a tracer antibody to bind and mark hCG by binding a distant site. With these assays came the concept of antibody enzyme labeling and high-sensitivity fluorimetric and spectrometric detection [15,22–26]. The advent of chemiluminescent and europium labeling, automation, and sensitive detection led to the rapid high-sensitivity hCG tests that are used today.

Dual-antibody immunometric technologies are the principle of most modern point-of-care rapid pregnancy tests in the physician's office and in home or over-the-counter rapid pregnancy tests [26]. In these tests a positive result is indicated by a line formed in the plastic window by the immobilized antibody–hCG dye antibody complex.

2.3 Introduction to hCG

In this book we examine every aspect of hCG. After introducing hCG (Part A), we look at hCG synthesis, secretion, structure, and action on receptors (Part B). Then, we examine hCG continuous dissociation and degradation (Part D). Biological functions of hCG and related molecules are then considered: first pregnancy (Part D), then normal pituitary (Part E), then evolution (Part F), gestational trophoblastic disease (Part G), and finally cancer (Part H). Clinical applications of hCG therapy are then considered (Part I). Then, hCG assays and antibodies are carefully examined (Part J), as are test applications and standards (Part K). Finally, some general methods are suggested (Part L), and the book finishes with some general comments and discussions about the future (Part M).

To start this introduction to hCG, I take the bold initiative of calling hCG a key group of molecules and biochemical marvel, because hCG and hyperglycosylated hCG and their free β-subunits are the most acidic proteins known to humans (isoelectric point = pI = 3.2–3.5). They are also the most glycosylated glycoproteins known to humans (30–42% sugar) [27]. Their acidity makes them master ultrapotent long growth factors with an exceptionally long circulating half-life [27]. hCG variants act as both hormones on the hCG/LH receptor and as autocrines or cytokines on the type II transforming growth factor β (TGFβ) receptor [28–30]. This inerrant dual receptor action also makes them a molecular marvel.

hCG is a also a key molecule because it exists in human blood or urine as at least 18 different isomers or degradation products (Table 2.2), including six biologically active independent molecules and six molecules sharing a common amino

Table 2.2 **hCG isomers and independent variants**

Molecule	Production	Detected	Biological action
hCG	Syncytiotrophoblast cells	Serum and urine	Hormone, acts on LH/hCG receptor
Nicked hCG	Cleaved by elastase	Serum and urine	
Nicked hCG −CTP	Cleaved by elastase	Serum and urine	
Sulfated hCG	Gonadotrope cells	Serum and urine	Hormone, acts on LH/hCG receptor
Nicked sulfated hCG	Cleaved by elastase	Serum and urine	
Hyperglycosylated hCG	Cytotrophoblast cells	Serum and urine	Autocrine, acts on TGFβ-II receptor
Nicked hCG-H	Cleaved by elastase	Serum and urine	Autocrine, acts on TGFβ-II receptor
Nicked hCG-H −CTP	Cleaved by elastase	Serum and urine	?
hCG free β-subunit	Syncytiotrophoblast cells	Serum and urine	Autocrine, acts on TGFβ-II receptor
Nicked hCGβ	Cleaved by elastase	Serum and urine	Autocrine, acts on TGFβ-II receptor
Nicked hCGβ −CTP	Cleaved by elastase	Serum and urine	?
β-core-fragment	Renal degradation product	Urine only	
hCGβ-H	Cytotrophoblast cells	Serum and urine	Autocrine, acts on TGFβ-II receptor
Nicked hCGβ-H	Cleaved by elastase	Serum and urine	Autocrine, acts on TGFβ-II receptor
Nicked hCGβ-H −CTP	Cleaved by elastase	Serum and urine	?
β-core-fragment-H	Renal degradation product	Urine only	
Free α-subunit	Syncytiotrophoblast cells	Serum and urine	
Sulfated free α-subunit	Gonadotrope cells	Serum and urine	

The term −CTP indicated missing the β-subunit C-terminal peptide. Syncytiotrophoblast cells are the multinuclear cells of the placenta, cytotrophoblast cells are the root cells of the placenta and gonadotrope cells are part of the anterior pituitary gland. hCG-H is hyperglycosylated hCG, hCGβ is hCG free β-subunit and hCGβ-H is hyperglycosylated hCG free β-subunit.

acid sequence with placental hCG [27] (Table 2.3). Table 2.3 compares the five independent hCG variants with separate biological functions. The other thing that makes hCG a key molecule is that it controls so many widely varying events in human life. It controls the evolution of humans and the development of the human brain [37,38]. It controls human life in pregnancy and multiple events and

Table 2.3 Properties of 6 biologically active independent variants of hCG

Parameter	hCG	Sulfated hCG	hCG-H	Fetal hCG	hCGβ	hCGß-H
Action	Endocrine	Endocrine	Autocrine	Endocrine	Autocrine	Autocrine
Site of action	LH/hCG	LH/hCG	TGFß	LH/hCG	TGFß	TGFß
Total molecular weight	37,180	36,150	42,800	?		27,600
Amino acids α-subunit	92	92	92	?	0	0
Amino acids ß-subunit	145	145	145	?	145	145
Peptide molecular weight	26,200	26,200	26,200	?	16,000	16,000
O-linked sugar units	4	4	4	?	4	4
N-linked sugar units	4	4	4	?	2	2
Molecular weight sugars	10,980	9950	16,600	?	6200	11,600
Percentage sugars	30%	28%	39%	?	28%	42%
pI, acidic peak	3.5	?	3.2	?	3.5	3.2
Metabolic clearance rate	36 h	20 h	?	?	?	?

Amino acid content, molecular weight and sugar contents determined from published structures as determined by Elliott and colleagues. for hCG and hyperglycosylated hCG [31], Birken and colleagues for sulfated pituitary hCG [32] and Valmu et al. for hyperglycosylated hCGß [33]. The molecular weight of common hCG dimer amino acid backbone is that as determined by Morgan and colleagues [34]. Molecular weight of N- and O-linked sugar side chains is added to these values. Isoelectric points are those published by Sutton [35] and metabolic clearance rate are those established by Wehmann and Nisula [36].

complications leading to childbirth [39–41]. On the other side of existence, it can drive human death or is a principal promoter of human cancers and their malignancy [28–30,42–44]. This is why I call it a key molecule: it is the master of both human life and death.

This book describes the biological function of all six independent hCG variant molecules. Table 2.3 compares the biological properties of the six biologically active independent hCG variant molecules. They are called independent because they each are produced by independent body systems or have completely independent bodily functions. Hyperglycosylated hCG, hCG free β-subunit, and hyperglycosylated hCG free β-subunit are autocrines that act on a TGFβ type II receptor [30], whereas hCG, sulfated hCG, and fetal hCG (made by fetus, structure unknown [45]) are hormones that act on a LH/hCG receptor.

References

[1] Aschner B. Ueber die function der hypophyse. Pflug Arch Gest Physiol 1912;146: 1–147.

[2] Fellner OO. Experimentelle untersuchungen uber die wirkung von gewebsextrakten aus der plazenta und den weiblichen sexualorganen auf das genital. Arch Gynakol 1913; 100:641.

[3] Hirose T. Experimentalle histologische studie zur genese corpus luteum. Mitt Med Fakultd Univ ZU 1919;23:63–70.

[4] Aschheim S, Zondek B. Das Hormon des hypophysenvorderlappens: testobjekt zum Nachweis des hormons. Klin Wochenschr 1927;6:248–52.

[5] Zondek B, Aschheim S. The zondek-ascheim pregnancy test. Can Med Assoc J 1930;22:251–3.

[6] Friedman MH, Lapham ME. A simple, rapid procedure for the laboratory diagnosis of early pregnancies. Am J Obstet Gynecol 1931;21:405–10.

[7] Shapiro HA, Zwarenstein H. Metabolic changes associated with endocrine activity and the reproductive cycle in Xenopus laevis: IV. The effects of injection of ovarian and pituitary extracts on the serum calcium in normal, ovariectomised and hypophysectomised toads. J Exp Biol 1934;11:267–72.

[8] Delfs E. An assay method for human chorionic gonadotropin. Endocrinology 1941;28:196–202.

[9] Kupperman HS, Greenblatt RB, Noback CR. A two and six-hour pregnancy test. J Clin Endo crinol 1943;3:548–50.

[10] Galli-Mainini C. Pregnancy test using the male batrachia. J Am Med Assoc 1948;138: 121–5.

[11] Wide L, Gemzell CA. An immunological pregnancy test. Acta Endocrinol 1960;35: 261–7.

[12] Wide L. An immunological method for the assay of human chorionic gonadotropin. Acta Endocrinol (Kobenhavn) 1962;4:1–111.

[13] Paul WE, Odell WD. Radiation inactivation of the immunological and biological activities of human chorionic gonadotropin. Nature 1964;203:979–80.

[14] Vaitukaitis JL, Braunstein GD, Ross GT. A radioimmunoassay which specifically measures human chorionic gonadotropin in the presence of human luteinizing hormone. Am J Obstet Gynecol 1972;113:751–8.
[15] Batzer FR. Hormonal evaluation of early pregnancy. Fertil Steril 1980;34:1–13.
[16] Barnard GJ, Kim JB, Brockelbank JL, Collins WP, Gaier B, Kohen F. Management of choriogonadotropin by chemiluminescense immunoassay and immunochemiluminometric assay: 1. Use of isoluminol derivatives. Clin Chem 1984;130:538–41.
[17] Wilde CE, Orr A, Bagshaw K. A radioimmunoassay for human chorionic gonadotropin. Nature 1965;205:191–2.
[18] Lunenfeld B, Eshkol A. Immunology of human chorionic gonadotropin (HCG). Vit Hor 1967;25:137–90.
[19] Aono T, Goldstein DP, Taymor ML, Dolch K. A radioimmunoassay method for human pituitary luteinizing hormone (LH) and human chorionic gonadotropin (HCG) using 125-I-labeled LH. Am Journal Obstet Gynecol 1967;98:996–1001.
[20] Rushworth AG, Orr AH, Bagshawe KD. The concentration of HCG in the plasma and spinal fluid of patients with trophoblastic tumours in the central nervous system. Br J Cancer 1968;22:253–7.
[21] Kohler G, Milstein C. Continuous cultures of fused cells secreting antibody of predefined specificity. Nature 1975;256:495–7.
[22] Maggio ET, Nakamura RM. Biomedical advances: 1. Clinical assays employing enzymes in innovative ways. Ligand Rev 1981;3:16–24.
[23] Ishikawa E, Kawai T, Miyai K. Enzyme immunoassay. Tokyo: Igaku-Shoin; 1981:1–280.
[24] Hussa RO, Hudson En. A two-site immunometric assay in evaluation of low levels of serum hCG. Am Clin Prod Rev 1984;3:12–17.
[25] Sekiya T, Furuhashi Y, Goto S, Kaseki S, Tomoda Y, Kato K. Sandwich-type enzyme immunoassay for human chorionic gonadotropin. J Endocrinol Invest 1981;4:275–9.
[26] Joshi UM, Roy R, Sheth AR, Shah HP. A simple and sensitive color test for the detection of human chorionic gonadotropin. Obstet Gynecol 1981;57:252–4.
[27] Cole LA. hCG, the wonder of today's science. Reprod Biol Endocrinol 2012;10:1–18.
[28] Butler SA, Ikram MS, Mathieu S, Iles RK. The increase in bladder carcinoma cell population induced by the free beta subunit of hCG is a result of an anti-apoptosis effect and not cell proliferation. Brit J Cancer 2000;82:1553–6.
[29] Berndt S, Blacher S, Munuat C, Detilleux J, Evain-Brion D, Noel A, et al. Hyperglycosylated human chorionic gonadotropin stimulates angiogenesis through TGF-ß receptor activation. FASEB J 2013; Abstract 12-213686.
[30] Cole LA, Butler SA. Hyperglycosylated hCG hCGß and hyperglycosylated hCGß: interchangeable cancer promoters. Molec Cellul Endocrinol 2012;349:232–8.
[31] Elliott MM, Kardana A, Lustbader JW, Cole LA. Carbohydrate and Peptide structure of the α- and β-subunits of human chorionic gonadotropin from normal and aberrant pregnancy and choriocarcinoma. Endocrine 1997;7:15–32.
[32] Birken S, Maydelman Y, Gawinowicz MA, Pound A, Liu Y, Hartree AS. Isolation and characterization of human pituitary chorionic gonadotropin. Endocrinology 1996;137:1402–11.
[33] Valmu L, Alfthan H, Hotakainen K, Birken S, Stenman UH. Site-specific glycan analysis of human chorionic gonadotropin beta-subunit from malignancies and pregnancy by liquid chromatography – electrospray mass spectrometry. Glycobiology 2006;16:1207–18.
[34] Morgan FJ, Birken S, Canfield RE. The amino acid sequence of human chorionic gonadotropin. J Biol Chem 1975;250:5247–58.

[35] Sutton JM. Charge variants in serum and urine hCG. Clin Chem Acta 2004;341:199–203.
[36] Wehmann RE, Nisula BC. Metabolic and renal clearance rates of purified human chorionic gonadotropin. J Clin Invest 1981;68:184–94.
[37] Cole LA, Khanlian SA, Kohorn EI. Evolution of the human brain, chorionic gonadotropin and hemochorial implantation of the placenta: insights into origins of pregnancy failures, preeclampsia and choriocarcinoma. J Reprod Med 2008;53:449–557.
[38] Cole LA. hCG and hyperglycosylated hCG in the establishment and evolution of hemochorial placentation. J Reprod Immunol 2009;82:111–17.
[39] Strott CA, Yoshimi T, Ross GT, Lipsett MB. Ovarian physiology: relationship between plasma LH and steroidogenesis by the follicle and corpus luteum; effect of HCG. J Clin Endocrinol Metab 1969;29:1157–67.
[40] Pierce JG, Parsons TF. Glycoprotein hormones: structure and function. Ann Rev Biochem 1981;50:65–95.
[41] Rao CV. Differential properties of human chorionic gonadotropin and human luteinizing hormone binding to plasma membranes of bovine corpora luteal. Acta Endocrinol 1979;90:696–710.
[42] Carter WB, Sekharem M, Coppola D. Purified hCG induces apoptosis in breast cancer. Breast Cancer Res Treatm 2006;100:S243–4.
[43] Jankowska A, Gunderson SI, Andrusiewicz N, Burczynska B, Szczerba A, Jarmolowski A, et al. Reduction of human chorionic gonadotropin beta subunit expression by modified U1 snRNA caused apoptosis in cervical cancer cells. Molec Cancer 2008;7:26–9.
[44] Li D, Wen X, Ghali L, Al-Shalabi FM, Docherty SM, Purkis P, et al. hCG beta expression by cervical squamous carcinoma—*in vivo* histological association with tumour invasion and apoptosis. Histopathology 2008;53:147–55.
[45] Goldsmith PC, McGregor WG, Raymoure WJ, Kuhn RW, Jaffe RB. Cellular localization of chorionic gonadotropin in human fetal kidney and liver. J Clin Endocrinol Metab 1983;57:54–61.

Part B

Synthesis, Structure and Function

The molecular genetics of hCG

3

Stephen A. Butler[1], Beata B. Burczynska[2] and Ray K. Iles[1]
[1]MAP Diagnostics, BioPark, Broadwater Road, Welwyn Garden City, Hertfordshire, UK,
[2]Biomedical Sciences, Middlesex University, The Burroughs, Hendon, London, UK

We know that hCG is a complicated molecule. It is not surprising, then, that the transcription and regulation of hCG gene expression are also complicated. In the first complete text on hCG [1], a short paragraph was dedicated to the complex gene cluster of hCGβ, which was first fully described only 5 years previously [2] and mapped only 1 year before [3]. Since then, the advances in molecular biology have expanded the field but, despite these advances, many questions posed in the early 1980s still remain unanswered. However, the most important question is, exactly how is hCG gene expression controlled?

3.1 The LH/hCG gene cluster

hCG α-subunit (hCGα) gene expression is fairly straightforward. Common to all of the glycoprotein hormones are luteinizing hormone (LH), follicle-stimulating hormone (FSH), thyroid-stimulating hormone (TSH), and hCG. The amino acid sequence of the α-subunit is encoded by a single copy gene (*CGA*) on chromosome 6 [4]. The exact location is currently reported as 6q12-q21, the genomic location of which is found at 6: 87795216–87804865 bp according to National Center for Biotechnology Information (NCBI) GenBank.

The β-subunit of hCG (hCGβ), however, is encoded in a region on chromosome 19 at 19q13.32 by a cluster of 6 *CGB* paralogs originally designated *CGB1*, *CGB2*, *CGB3*, *CGB5*, *CGB7*, and *CGB8*; these are adjacent to the *CGB4* gene encoding the hLH β-subunit [2,5–8]. According to NCBI GenBank, the *CGB* cluster is found between the genomic locations 19:49519237 bp (LHB) and 49561603 bp (*CGB7*). The exact locations and transcriptional orientation of the *CGB* genes within the cluster are shown in Figure 3.1. Sequencing studies have since shown that genes *CGB7* and *CGB3* also have allelic variants *CGB6* and *CGB9*, respectively [9,10]. *CGB3* is

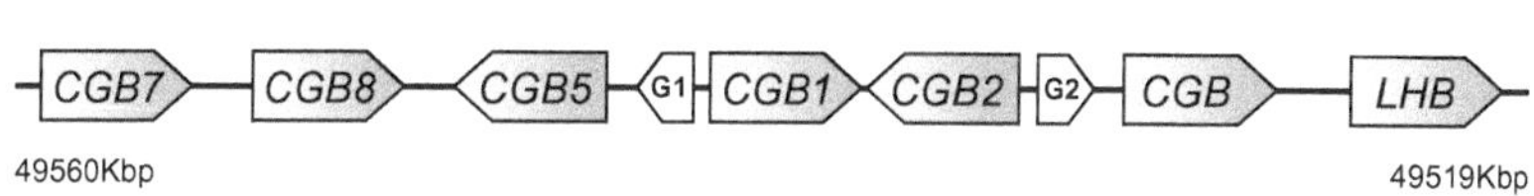

Figure 3.1 A diagrammatical representation of the arrangement of genes in the *LHB/CGB* gene cluster on chromosome 19q13.32, where the exact positions are indicted in Kbp. The figure indicates the current gene assignments within the transcription orientation arrows, and G1 and G2 indicate the position of the snaRs.

Human Chorionic Gonadotropin (hCG). DOI: http://dx.doi.org/10.1016/B978-0-12-800749-5.00003-1

now simply reported as *CGB* and was probably the first true *CGB* gene in the cluster; it arises adjacent to the LHβ gene (*CGB4*). Two types of genes were proposed based on their sequence and function: *CGB3*, *CGB5*, *CGB7*, *CGB8*, and *CGB1* or *CGB2*. *CGB1* and *CGB2* genes evolved in the cluster as a result of a 736-bp/724-bp DNA fragment insertion, which resulted in the formation of a new 5′-upstream region with an alternative start codon and a new exon 1. This insertion also caused a single base-pair frameshift in the open reading frame (ORF) for exons 2 and 3 and, as a consequence, may result in a novel protein (132aa). This hypothetical protein product of *CGB1* and *CGB2* genes would be completely different from the hCGβ product after translation of *CGB*, *CGB4*, *CGB5*, *CGB7*, and *CGB8* genes.

Bioinformatics searches have led to the discovery of small noncoding RNAs family genes (*snaR)* that are present within the proximal end of *CGB1* and *CGB2* promoters (*snaR-G1* and *snaR-G2*, respectively; see Figure 3.1) [11]. The snaRs (*snaR-A*, *snaR-C*, and *snaR-G2*) have been found to be most abundant in human testes but have been minimally detected in the brain, placenta, and other human tissues. *snaR-A* is highly expressed in many immortalized cell lines (293 cells, HeLa, HepG2, LNCap, HT-1080, MCF-7) and may indicate the function of these small noncoding RNAs in cellular transformation and immortalization [11,12].

The small noncoding RNAs type G (*SnaRs-G*) are located within the promoter region of *CGB1* and *CGB2*. They can be found 86 bp upstream of the transcriptional start sites of the *CGB1/2* genes and are in opposite orientation to *CGB*, *CGB5*, *CGB6*, *CGB7*, and *CGB8* (Figure 3.1). The genomic coding for *snaR-G* overlap with predicted binding sides for *CGB* transcription factors are essential for hCG gene expression and, therefore, fetal implantation and early placental development [11,13]. The location of *snaRs* suggests an additional regulatory function of *CGB1/2* gene expression through an epigenetic mechanism [11].

This unique cluster was initially thought to have been a result of an evolutionary change after a duplication of the α-subunit gene [14,15], based on the high degree of structural homology in peptide sequences, which indicates that hCGβ and hCGα were orthologs and shared a common ancestral gene. This was soon dismissed, in part, after the development of full sequence data, which showed large discrepancies in alignment between intronic and exonic regions [1]. Although not a recent duplication, newer data suggest that the similarities extend much further to early evolutionary divergence, because regions involving cystine knot formation appear to be highly conserved in both hCG subunits as well as in other cystine knot growth factors [16]. However, it is accepted that the first *CGB* gene arose as a result of a duplication event from *LHB*, and that this occurred at some point in primate divergence approximately 35–55 million years ago [17]. With this event, a single point mutation occurred in the duplicated gene toward the carboxyl end (corresponding to amino acid 114) of the DNA coding sequence. This point mutation resulted in frameshift (Gln > Arg) and readthrough of the termination codon into the untranslated region adjacent to the terminal *LHB* exon [14]. This effectively gave rise to what we refer to as the hCG β-subunit carboxy terminal peptide (CTP) and serves as the largest sequence difference between the hCG β-subunit and LH β-subunit in the mature proteins. Additional duplication events in the *CGB/LH* cluster lead to a variable number of copies of *CGB* genes between primate species; there are three copies in rhesus macaque, six in humans, and approximately 50 copies in gorilla [18]. The pattern of an increased number of genes within the cluster, along with the evolution of primate species, is compelling to read [18] and is expressed in terms of clinical significance later in this section (see Chapter 17).

The current published human sequences for each of the six *CGB* genes are now readily available at the NCBI resources. Fiddes and coworkers and Boime and coworkers together described the genetics of both hCG subunits in a series of publications from the late 1970s to the mid 1980s, revealing the loci and cluster maps and establishing the foundations for the evolution of the genes that we are now exploring [3–5,14,15,19–23].

3.2 Control of hCG gene expression: hCGα

CGA gene expression was reported in 1979 (see Figure 3.2) [19] and was shown to be translated in placental and tumor tissue soon after [20,21]. Online versions of the gene sequences and their transcripts can now be readily obtained from websites such as that of NCBI and Ensembl from NM_00735.3. The ORF of *CGA* variants and corresponding amino acid sequences are shown in Figure 3.3.

The α-subunit mRNA is 10-times more abundant in trophoblast tissue than in the β-subunit. Thus, the regulatory mechanisms resulting in the α-subunit and β-subunit gene expression are likely to be different. The pathways involved in the upregulation of hCG genes are not fully understood; however, cAMP has clearly been shown to increase both α-subunit and β-subunit unit gene expression in placental and choriocarcinoma cells [24–26]. All of the glycoprotein hormone genes contain 5′ cAMP response element (CRE) enhancer sequences that act together with specific placental or pituitary promoters [24]. The promoter region for the α-subunit gene contains two identical neighboring CREs found between −146 and −111, which are responsible for the binding of CRE binding proteins (CREBPs) and other members of the B-Zip transcription factor family [27–32].

At position −180 to −151, an upstream regulatory element (URE) is located and overlapped by several additional binding protein regions [28,30,31,33–36]. Initially designated URE-1 and URE-2 [30], they were later identified as α-subunit activator element (αACT) and tissue/trophoblast-specific element (TSE) [35]. Although these are essentially the same regions, the mapping experiments conducted showed differences in the precise location of each region. URE-1/ αACT is located at −161 to −142 and binds one of the members of the ubiquitous GATA family (a family of transcriptional factors) of DNA-binding proteins. URE-2/TSE is located at −182 to −159 and binds the TSE-binding protein (TSEBP). In addition, there are two further regulatory sequences: the downstream domain (DSD) located at −172 to −151 and the upstream domain (USD) at −177 to −156 [30]. These regions overlap each other while at the same time overlapping the URE-1/αACT sites. This could indicate that these regions are activated

1
TCATTGGACGGAATTTCCTGTTGATCCCAGGGCTTAGATGCAGGTGGAAACACTCTGCT
GGTATAAAAGCAGGTGAGGACTTCATTAACT***GCAGTTACTG***

101
AGAACTCATAAGACGAAGCTAAAATCCCTCTTCGGATCCACAGTCAACCGCCC
TGAACACATCCTGCAAAAAGCCCAGAGAAAGGTAATATGAATGAAAT

201
AATTTTGGGGGACTTTAATTGAGGAGTAAGATATTTGAGAATA //6.4Kb

234
TTTTTTTTTTTTTTTTTTGCCATGTCTGTCTGCAG***GAGCGCC*****ATGGATTACTACAGAA**
AATATGCAGCTATCTTTCTGGTCACATTGTCGGTGTTTCTGC

334
ATGTTCTCCATTCCGCTCCTGATGTGCAGGGTGCGTGACCAAATTTGTGGTTCAAGTAA
TAAGGACAACACACATTT //1.7Kb

411
TTCTTTTTGAGTCTTTTTTGGATATTTTACTCTGCCTTTTTTTTTCCCTGATAG**ATTGCCC**
AGAATGCACGCTACAGGAAAACCCATTCTTCTCCCAGCC

511
GGGTGCCCCAATACTTCAGTGCATGGGCTGCTGCTTCTCTAGAGCATATCCCAC
TCCACTAAGGTCCAAGAAGACGATGTTGGTCCAAAAGAACGTCACC

611
TCAGAGTCCACTTGCTGTGTAGCTAAATCATATAACAGGGTAAGAACCTCAAGAT
CCCCAGAAGCTTT //1.4Kb

679
ATAATATGTTTTTTTTTCCTTCCCCTTTAG**GTCACAGTAATGGGGGGTTTCAAAGTG**
GAGAACCACACGGCGTGCCACTGCAGTACTTGTTATTATCACA

779
AATCT***TAA***ATGTTTTACCAAGTGCTGTCTTGATGACTGCTGATTTTCTGGAATGGAAAA
TTAAGTTGTTTAGTGTTTATGGCTTTGTGAGATAAAACTCT

879
CCTTTTCCTTACCATACCACTTTGACACGCTTCAAGGATATACTGCAGCTTTACTGCCTT
CCTCCTTATCCTACAGTACAATCAGCAGTCTAGTTCTTTT

979
CATTTGGAATGAATACAGCATTAAGCTTGTTCCACTGCAAATAAAGCCTTTTAAATCATCA
TTCAATCACTGAATTATCATTTTTCTTCAAAGTAAG

Figure 3.2 Original published gene sequence of CG α indicating introns and exons (in bold black). The amino acids coded for by the italicized (bold) regions are removed during posttranslational modifications; therefore, remaining regions code for only the mature protein. Large portions of intronic regions were omitted from the published sequence, indicated by //. Data adapted from Ref. [4].

by at least two different binding proteins that may be specifically expressed by either pituitary or placental tissue. Although distinct base-pair regions have been established in murine and equine models, human gene expression studies have failed to confirm this. These observations are summarized in Figure 3.4.

```
  1 acactctgct ggtataaaag caggtgagga cttcattaac tgcagttact gagaactcat
 61 aagacgaagc taaaatccct cttcggatcc acagtcaacc gccctgaaca catcctgcaa
121 aaagcccaga gaaaggagcg ccatggatta ctacagaaaa tatgcagcta tctttctggt
181 cacattgtcg gtgtttctgc atgttctcca ttccgctcct gatgtgcagg agacagggtt
241 tcaccatgtt gcccaggctg ctctcaaact cctgagctca agcaatccac ccactaaggc
301 ctcccaaagt gctaggatta cagattgccc agaatgcacg ctacaggaaa acccattctt
361 ctcccagccg ggtgccccaa tacttcagtg catgggctgc tgcttctcta gagcatatcc
421 cactccacta aggtccaaga agacgatgtt ggtccaaaag aacgtcacct cagagtccac
481 ttgctgtgta gctaaatcat ataacagggt cacagtaatg gggggtttca aagtggagaa
541 ccacacggcg tgccactgca gtacttgtta ttatcacaaa tcttaaatgt tttaccaagt
601 gctgtcttga tgactgctga ttttctggaa tggaaaatta agttgtttag tgtttatggc
661 tttgtgagat aaaactctcc ttttccttac cataccactt tgacacgctt caaggatata
721 ctgcagcttt actgccttcc tccttatcct acagtacaat cagcagtcta gttcttttca
781 tttggaatga atacagcatt tagcttgttc cactgcaaat aaagcctttt aaatcatcat
841 tcaaaaaaaa aaaaaaaaaa a
```

Figure 3.3 ORFs of human *CGA* adapted from NCBI reference sequence NM_001252383.1. Entire mRNA sequence with the ORF highlighted in gray. Exons are from 1–135, 136–230, 231–323, 324–508, and 509–844.

In studies of human villous trophoblasts, it has been established that the entire region to 365 bp upstream is sufficient to control all cell-specific transcription of the *CGA* gene. It was further confirmed that the CREs are required for transcription following luciferese reporter assays [37]. Interestingly, it was also shown that activating transcription factor 1 (ATF1) was strongly implicated in CRE binding and, to a lesser extent, CREB-1; however, ATF2, ATF3, ATF4, and CREB-2 do not appear to be involved [37]. According to an additional finding in human studies, CREBP binding appears to be dependent on URE binding [36].

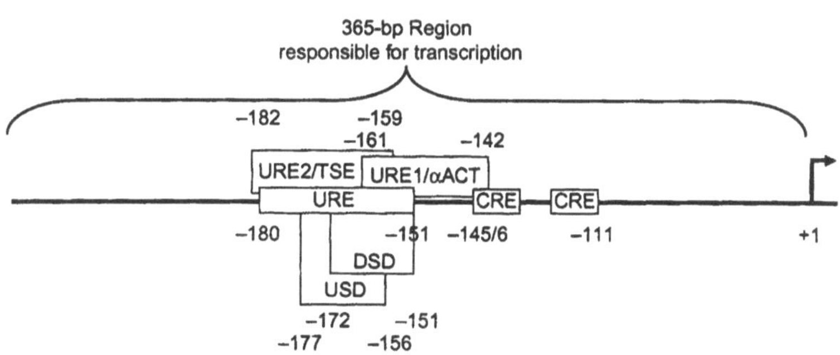

Figure 3.4 A diagrammatic representation of the *CGA* promoter sequence based on the work of Knofler et al. [37], with significant additions from others [24–36,38]. In addition to the CREs, the URE is overlapped by several additional binding protein regions; such an overlap could indicate that these regions are activated by at least two different tissue-specific binding proteins.

3.3 Control of hCG gene expression: hCGβ

The control of *CGB* gene expression is independent from that of *CGA*; when hCGα continues to be produced throughout gestation, hCGβ diminishes and, therefore, appears to regulate the overall hCG concentration in the blood [39]. In light of these early observations, it follows that expression of hCGβ must also be under very tight control and that there are factors influencing *CGB* expression that are likely to be distinct from those regulating expression of *CGA*. The expression of hCGβ can be detected in cultures of villous cytotrophoblasts [39,40], and the production increases over time. However, it has been noted that there is a delay of more than 12 h in hCGβ expression, which appears to be consistent with cytotrophoblast fusion to syncytiotrophoblast tissue (at which point maximum hCG production is initiated) [40]. Again, this highlights a distinct role for the expression of hCGβ as governor of overall hCG production, even during very early gestation.

Early studies of transfected murine cells *in vitro* showed that *CGB5*, *CGB3* (now *CGB*), and *CGB8* are transcriptionally active, in a descending order of magnitude [41]. Real-time PCR analysis of placental tissue mRNA showed that genes *CGB7* and either *CGB1* or *CGB2* were also transcribed, but at a lower level. Bo and Boime concluded that at least five of the six *CGB* genes are expressed *in vivo* [9]. More recently, placental tissue from normal and failing pregnancies was examined by RT-PCR; in the majority of samples, *CGB7* was shown to be inactive, as were *CGB1* and *CGB2*. This study concluded that expression of *CGB8* was, in fact, dominant, and *CGB5* and *CGB* followed in equal proportion. This was noted to change slightly during the third trimester, when *CGB8* and *CGB5* were shown to be co-dominantly expressed and *CGB* was expressed at lower levels [42].

CGB1 and *CGB2* have long been regarded as pseudogenes [2], and there is some disagreement regarding the functional nature of the proteins produced by these genes. The message arising from either gene was noted to be shorter than those from the other genes (a result of alternative splicing) [9]. Transcription of *CGB1* and *CGB2* results in at least four major splice variants (see Figure 3.5), only one of which represents a product that may translate into a functional hCGβ protein, when and if the complete intronic sequence is removed during splicing [9,43]. The differences in the transcribed mRNA and translated protein products are compared in Figures 3.6 and 3.7. In more recent studies, *CGB1* splice variants were again detected, but not those of *CGB2*. These genes (*CGB1* and *CGB2*) appear to be upregulated to some extent during the first trimester of pregnancy [42] and are upregulated in normal pregnancy and ectopic pregnancies but not in miscarriages, suggesting a potential role in implantation [42]. We discuss this further in Chapter 23, where we explore the potential implications of these genes in cancer, where we now see a potential role for *CGB2* expression in many common epithelial cancers [44]. A hypothetical protein of *CGB1* and *CGB2* has been postulated as a 132-amino-acid peptide arising from an alternate viable ATG start codon located 47 bp upstream from the standard initiation site [12]. All possible splice variants are shown in Figure 3.5. This same study also describes small noncoding RNA family genes (snaR), which are only present within the proximal end of *CGB1* and *CGB2* promoters (as indicated in Figure 3.1) and may be implicated in the regulation of the expression of *CGB1* and *CGB2* [12].

The variations in reports surrounding *CGB* paralog expression indicate that tissue-specific promoters regulate gene expression within the *CGB* cluster from different cell lineages. Unlike the *CGA* gene, identifying promoter sequence regions in

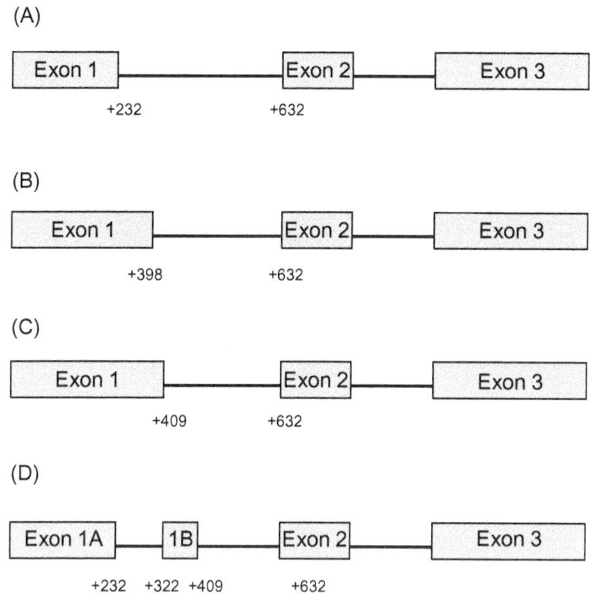

Figure 3.5 *CGB* splice variants including both empirical and hypothetical protein sequences of splice variants from genes *CGB1* and *CGB2*. (A) With fully spliced intron 1, splice donor site not truncated and possible ATG1 or ATG2 ORFs. From ATG1, the product for gene *CGB1* would be a 132aa protein with a molecular mass of 14.0 kDa, or from ATG2 a 155aa protein with a molecular mass of 16.7 kDa; for gene *CGB2* from ATG1, the predicted product would be a 132aa protein with a molecular mass of14 kDa, or from ATG2 a 163aa protein with a molecular mass of 17.4 kDa. (B) Splice variant *CGB2* + 166 bp (from intron 1) from ATG1 may give a 60aa protein with molecular mass of 6.9 kDa. (C) Splice variant *CGB1* + 176 bp. (D) Novel hypothetical splice variant with additional deletion within exon 1, to give exon 1A and exon 1B.

CGB genes has been extremely difficult because of the absence of known consensus sequences (like the CAAT or the TATA box) within 200-bp upstream of the cap site. Initial studies indicated that expression was higher using transfected constructs (including upstream sequences of −279 bp), and basal promoter regions were identified between −78 and −40 [41,45]. A region rich in cytosine and guanine repeats (−187 to −38) is also believed to be involved in basal transcription. Subsequently, a basal transcription factor binding region between −37 and +104 was identified [46]. Later studies indicated that there is a large cAMP-responsive region upstream between −311 and −200 bp on the *CGB* gene; this is required in its entirety for gene expression [47]. Sequences within this region (−311 to −202) appear to be required for basal transcription and, according to Steiger and coworkers, lies between −305 and −279 [35].

Within this region, there are multiple binding sites, including two CREBP binding domains (as seen in the α-gene promoter region). Some binding sites within this region appear to be more important than others. The tandem CRE 1

```
CGB1      ------ATGTCAAAG AGGCTGCTGCTGTTGCTGCTGCTGAGCATGGGCGGGACATGGGCA
CGB2      ------ATGTCAAAG GGGCTGCTGCTGTTGCTGCTGCTGAGCATGGGCGGGACATGGGCA
CGB7      ATGGAGATGTTCCAG GGGCTGCTGCTGTTGCTGCTGCTGAGCATGGGCGGGACATGGGCA
CGB       ATGGAGATGTTCCAG GGGCTGCTGCTGTTGCTGCTGCTGAGCATGGGCGGGACATGGGCA
CGB5      ATGGAGATGTTCCAG GGGCTGCTGCTGTTGCTGCTGCTGAGCATGGGCGGGACATGGGCA
CGB8      ATGGAGATGTTCCAG GGGCTGCTGCTGTTGCTGCTGCTGAGCATGGGCGGGACATGGGCA
                ****  ..** .********************************************

CGB1      TCCAAGGAGCCGCTTCGGCCACGGTGCCGCCCCATCAATGCCACCCTGGCTGTGGAGAAG
CGB2      TCCAAGGAGCCGCTTCGGCCACGGTGCCGCCCCATCAATGCCACCCTGGCTGTGGAGAAG
CGB7      TCCAGGGAGATGCTTCGGCCACGGTGCCGCCCCATCAATGCCACCCTGGCTGTGGAGAAG
CGB       TCCAAGGAGCCGCTTCGGCCACGGTGCCGCCCCATCAATGCCACCCTGGCTGTGGAGAAG
CGB5      TCCAAGGAGCCGCTTCGGCCACGGTGCCGCCCCATCAATGCCACCCTGGCTGTGGAGAAG
CGB8      TCCAAGGAGCCGCTTCGGCCACGGTGCCGCCCCATCAATGCCACCCTGGCTGTGGAGAAG
          ****.****. *************************************************

CGB1      GAGGGCTGCCCCGTGTGCATCACCGTCAACACCACCATCTGTGCCGGCTACTGCCCCACC
CGB2      GAGGGCTGCCCCGTGTGCATCACCGTCAACACCACCATCTGTGCCGGCTACTGCCCCACC
CGB7      GAGGGCTGCCCCGTGTGCATCACCGTCAACACCACCATCTGTGCCGGCTACTGCCCCACC
CGB       GAGGGCTGCCCCGTGTGCATCACCGTCAACACCACCATCTGTGCCGGCTACTGCCCCACC
CGB5      GAGGGCTGCCCCGTGTGCATCACCGTCAACACCACCATCTGTGCCGGCTACTGCCCCACC
CGB8      GAGGGCTGCCCCGTGTGCATCACCGTCAACACCACCATCTGTGCCGGCTACTGCCCCACC
          ************************************************************

CGB1      ATG ACCCGCGTGCTGCAGGGGGTCCTGCCGGCCCTGCCTCAGGTGGTGTGCAACTACCGC
CGB2      ATG ACCCGCGTGCTGCAGGGGGTCCTGCCGGCCCTGCCTCAGGTGGTGTGCAACTACCGC
CGB7      ATG ACCCGCGTGCTGCAGGGGGTCCTGCCGGCCCTGCCTCAGGTGGTGTGCAACTACCGC
CGB       ATG ACCCGCGTGCTGCAGGGGGTCCTGCCGGCCCTGCCTCAGGTGGTGTGCAACTACCGC
CGB5      ATG ACCCGCGTGCTGCAGGGGGTCCTGCCGGCCCTGCCTCAGGTGGTGTGCAACTACCGC
CGB8      ATG ACCCGCGTGCTGCAGGGGGTCCTGCCGGCCCTGCCTCAGGTGGTGTGCAACTACCGC
          *** ********************************************************

CGB1      GATGTGCGCTTCGAGTCCATCCGGCTCCCTGGCTGCCCGCGCGGCGTGAACCCCGTGGTC
CGB2      GATGTGCGCTTCGAGTCCATCCGGCTCCCTGGCTGCCCGCGCGGCGTGAACCCCGTGGTC
CGB7      GATGTGCGCTTCGAGTCCATCCGGCTCCCTGGCTGCCCGCGCGGCGTGAACCCCGTGGTC
CGB       GATGTGCGCTTCGAGTCCATCCGGCTCCCTGGCTGCCCGCGCGGCGTGAACCCCGTGGTC
CGB5      GATGTGCGCTTCGAGTCCATCCGGCTCCCTGGCTGCCCGCGCGGCGTGAACCCCGTGGTC
CGB8      GATGTGCGCTTCGAGTCCATCCGGCTCCCTGGCTGCCCGCGCGGCGTGAACCCCGTGGTC
          ************************************************************

CGB1      TCCTACGCCGTGGCTCTCAGCTGTCAATGTGCACTCTGCCGCCGCAGCACCACTGACTGC
CGB2      TCCTACGCCGTGGCTCTCAGCTGTCAATGTGCACTCTGCCGCCGCAGCACCACTGACTGC
CGB7      TCCTACGCCGTGGCTCTCAGCTGTCAATGTGCACTCTGCCGCCGCAGCACCACTGACTGC
CGB       TCCTACGCCGTGGCTCTCAGCTGTCAATGTGCACTCTGCCGCCGCAGCACCACTGACTGC
CGB5      TCCTACGCCGTGGCTCTCAGCTGTCAATGTGCACTCTGCCGCCGCAGCACCACTGACTGC
CGB8      TCCTACGCCGTGGCTCTCAGCTGTCAATGTGCACTCTGCCGCCGCAGCACCACTGACTGC
          ************************************************************

CGB1      GGGGGTCCCAAGGACCACCCCTTGACCTGTGATGACCCCCGCTTCCAGGACTCCTCTTCC
CGB2      GGGGGTCCCAAGGACCACCCCTTGACCTGTGATGACCCCCGCTTCCAGGCCTCCTCTTCC
CGB7      GGGGGTCCCAAGGACCACCCCTTGACCTGTGATGACCCCCGCTTCCAGGCCTCCTCTTCC
CGB       GGGGGTCCCAAGGACCACCCCTTGACCTGTGATGACCCCCGCTTCCAGGACTCCTCTTCC
CGB5      GGGGGTCCCAAGGACCACCCCTTGACCTGTGATGACCCCCGCTTCCAGGACTCCTCTTCC
CGB8      GGGGGTCCCAAGGACCACCCCTTGACCTGTGATGACCCCCGCTTCCAGGACTCCTCTTCC
          *************************************************.**********

CGB1      TCAAAGGCCCCTCCCCCCAGCCTTCCAAGTCCATCCCGTCTCCCGGGGCCCTAG------
CGB2      TCAAAGGCCCCTCCCCCCAGCCTTCCAAGCCCATCCCGACTCCCGGGGCCCTCAGACACC
CGB7      TCAAAGGCCCCTCCCCCCAGCCTTCCAAGTCCATCCCGACTCCCGGGGCCCTCAGACACC
CGB       TCAAAGGCCCCTCCCCCCAGCCTTCCAAGCCCATCCCGACTCCCGGGGCCCTCGGACACC
CGB5      TCAAAGGCCCCTCCCCCCAGCCTTCCAAGTCCATCCCGACTCCCGGGGCCCTCGGACACC
CGB8      TCAAAGGCCCCTCCCCCCAGCCTTCCAAGTCCATCCCGACTCCCGGGGCCCTCGGACACC
          ***************************** ********:*************..

CGB1      ------------------
CGB2      CCGATCCTCCCACAATAA
CGB7      CCGATCCTCCCACAATAA
CGB       CCGATCCTCCCACAATAA
CGB5      CCGATCCTCCCACAATAA
CGB8      CCGATCCTCCCACAATAA
```

Figure 3.6 ORFs of *CGB* genes (*CGB1*, *CGB2*, *CGB7*, *CGB*, *CGB5*, and *CGB8*) showing multiple sequence alignments based on sequences adopted from NCBI database. The sequences are variable in length and show exons including mRNA from start codon to stop codon only. The asterisks indicate where *CGB* gene sequence differences can be seen and spaces indicate exon boundaries.

```
CGB8    MEMFQGLLLLLLLSMGGTWASKEPLRPRCRPINATLAVEKEGCPVCITVNTTICAGYCPT    60
CGB5    MEMFQGLLLLLLLSMGGTWASKEPLRPRCRPINATLAVEKEGCPVCITVNTTICAGYCPT    60
CGB     MEMFQGLLLLLLLSMGGTWASKEPLRPRCRPINATLAVEKEGCPVCITVNTTICAGYCPT    60
CGB7    MEMFQGLLLLLLLSMGGTWASREMLRPRCRPINATLAVEKEGCPVCITVNTTICAGYCPT    60
CGB1    --MSKRLLLLLLLSMGGTWASKEPLRPRCRPINATLAVEKEGCPVCITVNTTICAGYCPT    58
CGB2    --MSKGLLLLLLLSMGGTWASKEPLRPRCRPINATLAVEKEGCPVCITVNTTICAGYCPT    58
          * : ***************:* ************************************

CGB8    MTRVLQGVLPALPQVVCNYRDVRFESIRLPGCPRGVNPVVSYAVALSCQCALCRRSTTDC    120
CGB5    MTRVLQGVLPALPQVVCNYRDVRFESIRLPGCPRGVNPVVSYAVALSCQCALCRRSTTDC    120
CGB     MTRVLQGVLPALPQVVCNYRDVRFESIRLPGCPRGVNPVVSYAVALSCQCALCRRSTTDC    120
CGB7    MTRVLQGVLPALPQVVCNYRDVRFESIRLPGCPRGVNPVVSYAVALSCQCALCRRSTTDC    120
CGB1    MTRVLQGVLPALPQVVCNYRDVRFESIRLPGCPRGVNPVVSYAVALSCQCALCRRSTTDC    118
CGB2    MTRVLQGVLPALPQVVCNYRDVRFESIRLPGCPRGVNPVVSYAVALSCQCALCRRSTTDC    118
        ************************************************************

CGB8    GGPKDHPLTCDDPRFQDSSSSKAPPPSLPSPSRLPGPSDTPILPQ    165
CGB5    GGPKDHPLTCDDPRFQDSSSSKAPPPSLPSPSRLPGPSDTPILPQ    165
CGB     GGPKDHPLTCDDPRFQDSSSSKAPPPSLPSPSRLPGPSDTPILPQ    165
CGB7    GGPKDHPLTCDDPRFQASSSSKAPPPSLPSPSRLPGPSDTPILPQ    165
CGB1    GGPKDHPLTCDDPRFQDSSSSKAPPPSLPSPSR------------    151
CGB2    GGPKDHPLTCDDPRFQASSSSKAPPPSLPSPSRLPGPSDTPILPQ    163
        **************** ****************
```

Figure 3.7 Multiple sequence alignments of *CGB* expressed proteins from all genes (*CGB8*, *CGB5*, *CGB*, *CGB7*, *CGB1*, and *CGB2*) showing sequence differences.

(−299 to −289) and CRE 2 (−240 to −219) regions both bind c-Jun (a downregulator of hCGβ expression) [45]; −305 to −249 binds Oct-3/4, which has been shown to silence hCG expression by more than 90% [48]. A TSE site has also been identified. This site contains a sequence similar to the α-subunit gene promoter at −301 to −275. It also has an additional two sequences within this essential region that have also been shown to bind TSEBPs, suggesting coordination between hCGβ and hCGα expression [35]. Sites −311 to −274 and −250 to −200 (which do not bind CREBPs) are involved in binding activating protein 2 (AP2), an upregulator of hCGβ expression [49,50]. These observations are summarized in Figure 3.8.

We now know that this complex region spanning 118 bp actually has multiple AP2 and selective promoter factor 1 (SP1) binding regions [50] in addition to c-Jun and Oct-3/4 transcription factor binding regions. A complex model proposes that CGB transcription is essentially activated through SP1 and AP2 binding, which can be enhanced by cAMP. An upstream binding site for SP1/AP2 (−318 to −279) preferentially binds SP1, which blocks AP2 binding to the neighboring site. SP1 and AP2 can also bind to the site located at −220 to −188. SP3 suppresses basal transcription through inhibition of SP1 [52,53]. c-Jun or Oct-3/4 can potentially silence *CGB* transcription activation almost completely [47,50,51]. The most recent studies suggest that the model may be even more complex, with a fine interplay between SP1/3 and AP2(α) [54]. The model appears to confirm a role for SP3 in suppressing SP1 basal transcription in cytotrophoblasts while it is being suppressed in syncytial tissue, thus allowing for the known elevation in *CGB* transcription at this juncture in gestation.

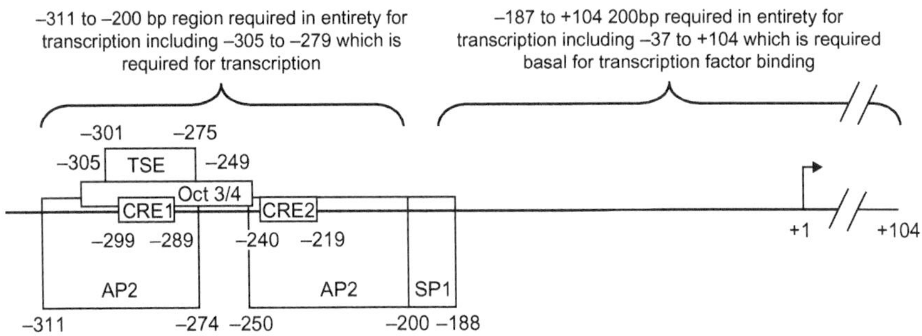

Figure 3.8 Promoter sequence of *CGB5*, *CGB*. Transcription is controlled by the SP2 and SP1 binding site between −311 and −188; this is affected positively by cAMP. Schematic drawing. Data from Ref. [45–53].

3.4 Summary

In conclusion, we have come far in hCG molecular genetics, but definitive answers still elude us. This is certainly attributable, in no small part, to the similarity in the sequences of each *CGB* paralog in the LHβ/hCGβ gene cluster. However, because of more recent work, the holes in the molecular puzzle surrounding hCGβ gene expression are starting to be completed. Clues from the primate evolution of *CGB* genes and insights into the control of gene expression are certain to reveal the importance and significance of each gene in the cluster. Subsequently, the effect of any resulting functional protein, after translation and posttranslational modification, can be explained.

References

[1] Hussa RO. The clinical marker hCG. New York, NY: Praeger Publishers; 1987.

[2] Boorstein WR, Vamvakopoulos NS, Fiddes JC. hCG β-subunit is encoded by at least eight genes arranged in tandem and inverted in pairs. Nature 1982;300:419–22.

[3] Policastro PF, Daniels-McQueen S, Carle G, Boime I. A map of the hCG β-LHβ gene cluster. J Biol Chem 1986;261:5907–16.

[4] Fiddes JC, Goodman HM. The gene encoding the common alpha subunit of the four glycoprotein hormones. J Mol Appl Genet 1981;1:3–18.

[5] Policastro P, Ovitt CE, Hoshina M, Fukuoka H, Boothby MR, Boime I. The beta subunit of human chorionic gonadotropin is encoded by multiple genes. J Biol Chem 1983; 258:11492–9.

[6] Julier C, Weil D, Couillin P, Cote JC, Nguyen VC, Foubert C, et al. The beta chorionic gonadotropin betaluteinising gene cluster maps to human chromosome 19. Hum Genet 1984;67:174–7.

[7] Graham MY, Otani T, Boime I, Olsen MV, Carle GF, Chaplin DD. Cosmid mapping of the human chorionic gonadotropin beta subunit genes by field inversion electrophoresis. Nucleic Acids Res 1987;15:4437–48.

[8] Jameson JL, Lindell CM, Habener JM. Gonadotropin and thyrotropin and subunit gene expression in normal and neoplastic tissues characterised by using specific ribonucleic acid hybridisation probes. J Clin Endocrinol Metab 1987;64:319–26.
[9] Bo M, Boime I. Identification of the transcriptionally active genes of the chorionic gonadotropin β gene *in vivo*. . J Biol Chem 1992;267:3179–84.
[10] Bidart JM, Baudin E, Troalen F, Bellet D, Schlumberger M. Eutopic and ectopic production of glycoprotein hormones alpha and beta subunits. Ann Endocrinol 1997; 58:125–8.
[11] Parrott AM, Mathews MB. Novel rapidly evolving hominid RNAs bind nuclear factor 90 and display tissue-restricted distribution. Nucleic Acids Res. 2007;35(18):6249–58.
[12] Parrott AM, Sriram G, Liu Y, Mathews MB. Expression of type II chorionic gonadotropin genes supports a role in the male reproductive system. Mol Cell Biol 2011;31 (2):287–99.
[13] Hallast P, Rull K, Laan M. The evolution and genomic landscape of CGB1 and CGB2 genes. Mol Cell Endocrinol. 2007;260,262:2–11.
[14] Fiddes JC, Goodman HM. The cDNA for the β-subunit of hCG suggests evolution of a gene by readthrough into the 3′ untranslated region. Nature 1980;286:684–7.
[15] Fiddes JC, Talmage K. Structure, expression and evolution of the genes for the human glycoprotein hormones. Recent Prog Horm Res 1984;40:43–74.
[16] Vitt UA, Hsu SY, Hsueh AJ. Evolution and classification of cystine knot-containing hormones and related extracellular signaling molecules. Mol Endocrinol 2001;15: 681–94.
[17] Maston GA, Ruvolo M. Chorionic gonadotropin has a recent origin within primates and an evolutionary history of selection. Mol Biol Evol 2002;19:320–35.
[18] Hallast P, Saarela J, Palotie A, Laan M. High divergence in primate-specific duplicated regions: human and chimpanzee chorionic gonadotropin beta genes. BMC Evol Biol 2008;8:195.
[19] Fiddes JC, Goodman HM. Isolation, cloning and sequence analysis of the cDNA for the alpha-subunit of human chorionic gonadotropin. Nature 1979;281:351–6.
[20] Boothby M, Kukowska J, Boime I. Imbalanced synthesis of human chorionic gonadotropin alpha and beta subunits reflects the steady state levels of the corresponding mRNAs. J Biol Chem 1983;258:9250–3.
[21] Boothby M, Ruddon RW, Anderson C, McWilliams D, Boime I. A single gonadotropin alpha-subunit gene in normal tissue and tumor-derived cell llines. J Biol Chem 1981;256:5121–7.
[22] Talmadge K, Boorstein WR, Fiddes JC. The human genome contains seven genes for the beta subunit of chorionic gonadotropin but only one gene for the beta subunit of LH. DNA 1983;2:281–9.
[23] Talmadge K, Vamvakopoulos NC, Fiddes JC. Evolution of the genes for the beta subunits of human chorionic gonadotropin and luteinizing hormone. Nature 1984;307: 37–40.
[24] Burnside J, Nagelberg SB, Lippman SS, Weintraub BD. Differential regulation of hCG alpha and beta subunit mRNAs in JEG-3 choriocarcinoma cells by 8-bromo-cAMP. J Biol Chem 1985;260:12705–9.
[25] Chin WW, Gharib SD. Organization and expression of gonadotropin genes. Adv Exp Med Biol 1986;205:245–65.
[26] Ringler GE, Kao LC, Miller WL, Strauss III JF. Effects of 8-bromo-cAMP on expression of endocrine functions by cultured human trophoblast cells: regulation of specific mRNAs. Mol Cell Endocrinol 1989;61:13–21.

[27] Silver BJ, Bokar JA, Virgin JB, Vallen EA, Milsted A, Nilson JH. Cyclic AMP regulation of the human glycoprotein hormone alpha-subunit gene is mediated by an 18-basepair element. Proc Natl Acad Sci USA 1987;84:2198–202.
[28] Delegeane AM, Ferland LH, Mellon PL. Tissue-specific enhancer of the human glycoprotein hormone alpha-subunit gene: dependence on cyclic AMP-inducible elements. Mol Cell Biol 1987;7:3994–4002.
[29] Hoeffler JP, Meyer TE, Yun Y, Jameson JL, Habener JF. Cyclic AMP-responsive DNA-binding protein: structure based on a cloned placental cDNA. Science 1988;242: 1430–3.
[30] Jameson JL, Albanese C, Habener JF. Distinct adjacent protein-binding domains in the glycoprotein hormone alpha gene interact independently with a cAMP-responsive enhancer. J Biol Chem 1989;264:16190–6.
[31] Jameson JL, Powers AC, Gallagher GD, Habener JF. Enhancer and promoter element interactions dictate cyclic adenosine monophosphate mediated and cell-specific expression of the glycoprotein hormone alpha-gene. Mol Endocrinol 1989;3:763–72.
[32] Habener JF. Cyclic AMP response element binding proteins: a cornucopia of transcription factors. Mol Endocrinol 1990;4:1087–94.
[33] Bokar JA, Keri RA, Farmerie TA, Fenstermaker RA, Andersen B, Hamernik DL, et al. Expression of the glycoprotein hormone alpha-subunit gene in the placenta requires a functional cyclic AMP response element, whereas a different cis-acting element mediates pituitary-specific expression. Mol Cell Biol 1989;9:5113–22.
[34] Fenstermaker RA, Farmerie TA, Clay CM, Hamernik DL, Nilson JH. Different combinations of regulatory elements may account for expression of the glycoprotein hormone alpha-subunit gene in primate and horse placenta. Mol Endocrinol 1990;4:1480–7.
[35] Steiger DJ, Buscher M, Hecht JH, Mellon PL. Coordinate control of the alpha- and beta-subunit genes of human chorionic gonadotropin by trophoblast-specific element-binding protein. Mol Endocrinol 1993;7:1579–88.
[36] Pittman RH, Clay CM, Farmerie TA, Nilson JH. Functional analysis of the placenta-specific enhancer of the human glycoprotein hormone alpha subunit gene: emergence of a new element. J Biol Chem 1994;269:19360–8.
[37] Knofler M, Saleh L, Strohmer H, Husslein P, Wolschek MF. Cyclic AMP and differentiation-dependent regulation of the proximal αHCG gene promoter in term villous trophoblasts. Mol Hum Reprod 1999;5:573–80.
[38] Balfour NJ, Franklyn JA, Gurr JA, Sheappard MC. Multiple DNA elements determine basal and thyroid hormone regulated expression of the human glycoproteinα subunit gene in pituitary cells. J Mol Endocrinol 1990;4:187–90.
[39] Braunstein GD, Rasor JL, Engvall E, Wade ME. Interrelationships of human chorionic gonadotropin, human placental lactogen, and pregnancy-specific β 1-glycoprotein throughout normal human gestation. Am J Obstet Gynecol 1980;138:1205–13.
[40] Kato Y, Braunstein GD. Discordant secretion of placental protein hormones in differentiating trophoblasts *in vitro*. J Clin Endocrinol Metab 1989;68:814–20.
[41] Otani T, Otani F, Bo M, Chaplin D, Boime I. Promoter sequences in the chorionic gonadotropin β subunit gene. In: Mochizuki M, Hussa R, editors. Placental protein hormones. Amsterdam: Elsevier; 1988.
[42] Rull K, Laan M. Expression of β-subunit of human chorionic gonadotropin genes during the normal and failed pregnancy. Hum Reprod 2005;20:3360–8.
[43] Dirnhofer P, Hermann M, Hittmair A, Hoermann R, Kapelari K, Berger P. Expression of the human chorionic gonadotropin-β gene cluster in human pituitaries and alternative use of exon 1. J Clin Endocrinol Metab 1996;81:4212–17.

[44] Burczynska BB, Kobrouly L, Butler SA, Naase M, Iles RK. Novel insights into the expression of CGB1 & 2 genes by epithelial cancer cell lines secreting ectopic free hCGβ. Anticancer Res 2014;34(5):2239–48.
[45] Otani T, Boime I, Mochizuki M. Studies on the transcriptional regulatory elements of the hCG beta gene cluster using a choriocarcinoma cell line which expresses hCG beta eutopically. Nippon Sanka Fujinka Gakkai Zasshi 1989;41:1885–90.
[46] Hollenberg AN, Pestell RG, Albanese C, Boers ME, Jameson JL. Multiple promoter elements in the human chorionic gonadotropin beta subunit genes distinguish their expression from the luteinizing hormone beta gene. Mol Cell Endocrinol 1994;106: 111–19.
[47] Albanese C, Kay TW, Troccoli NM, Jameson JL. Novel cyclic adenosine 3′,5′-monophosphate response element in the human chorionic gonadotropin beta-subunit gene. Mol Endocrinol 1991;5:693–702.
[48] Liu L, Roberts RM. Silencing of the gene for the β subunit of human chorionic gonadotropin by the embryonic transcription factor Oct-3/4. J Biol Chem 1996;271:16683–9.
[49] Johnson W, Albanese C, Handwerger S, Williams T, Pestell RG, Jameson L. Regulation of the human chorionic gonadotropin α- and β-subunit promoters by AP-2. J Biol Chem 1997;272:15405–12.
[50] LiCalsi C, Christophe S, Steger DJ, Buescher M, Fischer W, Mellon PL. AP-2 family members regulate basal and cAMP-induced expression of human chorionic gonadotropin. Nucleic Acids Res 2000;28:1036–43.
[51] Pestell RG, Hollenberg AN, Albanese C, Jameson JL. c-Jun represses transcription of the human chorionic gonadotropin alpha and beta genes through distinct types of CREs. J Biol Chem 1994;269:31090–6.
[52] Johnson W, Jameson JL. AP-2 (activating protein 2) and Sp1 (selective promoter factor 1) regulatory elements play distinct roles in the control of basal activity and cyclic adenosine 3′,5′-monophosphate responsiveness of the human chorionic gonadotropin-β promoter. Mol Endocrinol 1999;13:1963–75.
[53] Krikun G, Schatz F, Mackman N, Guller S, Demopoulos R, Lockwood CJ. Regulation of tissue factor gene expression in human endometrium by transcription factors Sp1 and Sp3. Mol Endocrinol 2000;14:393–400.
[54] Knofler M, Saleh L, Bauer S, Galos B, Rotheneder H, Husslein P, et al. Transcriptional regulation of the human chorionic gonadotropin β gene during villous trophoblast differentiation. Endocrinology 2004;145:1685–94.

Structure, synthesis, and secretion of hCG and hyperglycosylated hCG

4

Laurence A. Cole[1] *and Stephen A. Butler*[2]
[1]USA hCG Reference Service, Angel Fire, NM, USA, [2]MAP Diagnostics, BioPark, Broadwater Road, Welwyn Garden City, Hertfordshire, UK

hCG is a hormone composed of an α-subunit and a β-subunit that are joined by noncovalent hydrophobic and ionic interactions. Hyperglycosylated hCG is an autocrine sharing the amino acid composition of hCG. The molecular weight of hCG is 37,180 and that of hyperglycosylated hCG is 42,800. They are unusual glycoproteins in that 30–39% of the molecular weight (30% in hCG and 39% in hyperglycosylated hCG) are derived from the sugar side-chains. The entire story of hCG and hyperglycosylated hCG has been one of ever-changing understanding. The amino acid sequence of their subunits, for instance, was first proposed by Carlsen et al. [1] and by Bellisario et al. in 1973 [2]; it was then corrected by Morgan et al. in 1975 [3]. Similarly, the structure of the N-linked sugar side-chains was first proposed by Bahl in 1969 [4], modified by Endo et al. in 1979 [5], and finalized by Elliott et al. in 1997 [6]. The structure of the O-linked oligosaccharides was proposed by Kessler et al. in 1979 [7], modified by Cole et al. in 1985 [8], and finalized by Valmu et al. in 2006 [9]. The position of the disulfide bonds on hCG was proposed by Giudice and Pierce in 1979 [10], modified by Mise and Bahl in 1981 [11], and finalized by Lapthorn et al. in 1994 [12].

The hormonal function of hCG has always been considered as gonadotropic or promoting progesterone production. Today, the function of hCG is still described as promoting progesterone, but we now know that hCG has many other important functions besides this. From the time of implantation, hCG takes over the promotion of progesterone production from luteinizing hormone (LH) by acting on the LH/hCG receptor in ovarian corpus luteal cells. This continues for approximately 4 weeks. After that time, the syncytiotrophoblast cells of the placenta take over progesterone production. We now know that hCG plays key roles in promoting the fusion and differentiation of cytotrophoblast to syncytiotrophoblast cells [13], as well as in promoting angiogenesis in the maternal vasculature [11,14–16], but it also has other roles in placental–fetal development [17,18]. The full biological function of hCG is still being slowly resolved (see Chapters 12–25). We examine the structure, synthesis, and function of the hormone hCG in detail.

4.1 Amino acid sequence of hCG

The amino acid sequence of hCG was firmly established by Morgan et al. in 1975 [3]. They discovered that hCG has an α-subunit consisting of 92 individual amino

Human Chorionic Gonadotropin (hCG). DOI: http://dx.doi.org/10.1016/B978-0-12-800749-5.00004-3

acids and a β-subunit consisting of 145 amino acids. The hCG α-subunit sequence is identical to the sequences of the α-subunits of LH, follicle-stimulating hormone (FSH), and thyroid-stimulating hormone (TSH)—a family of molecules called *glycoprotein hormones*. In contrast, the β-subunit sequence is unique to hCG. Morgan et al. [3] found significant problems with Bahl et al.'s original amino acid sequence published in 1973 [1,2], including three erroneous amino acids on the β-subunit and others on the α-subunit. The corrected amino acid sequence (see Figure 4.1) shows

α-subunit of hCG

1 11
ala-pro-asp-val-gln-asp-cys-pro-glu-cys-thr-leu-gln-glu-asp-pro-phe-phe-ser-gln-

21 31
pro-gly-ala-pro-ile-leu-gln-cys-met-gly-cys-cys-phe-ser-arg-ala-tyr-pro-thr-pro-

41 51 52
leu-arg-ser-lys-lys-thr-met-leu-val-gln-lys-asn-val-thr-ser-glu-ser-thr-cys-cys-
N

61 71 78
val-ala-lys-ser-tyr-asn-arg-val-thr-val-met-gly-gly-phe-lys-val-glu-asn-his-thr-

N
81 91 92
ala-cys-his-cys-ser-thr-cys-tyr-tyr-his-lys-ser

β-subunit of hCG

1 11 13
ser-lys-glu-pro-leu-arg-pro-arg-cys-arg-pro-ile-asn-ala-thr-leu-ala-val-glu-lys-
N

21 30 31
glu-gly-cys-pro-val-cys-ile-thr-val-asn-thr-thr-ile-cys-ala-gly-tyr-cys-pro-thr-
N

41 51
met-thr-arg-val-leu-gln-gly-val-leu-pro-ala-leu-pro-gln-val-val-cys-asn-tyr-arg

61 71
asp-val-arg-phe-glu-ser-ile-arg-leu-pro-gly-cys-pro-arg-gly-val-asn-pro-val-val-

81 91
ser-tyr-ala-val-ala-leu-ser-cys-gln-cys-ala-leu-cys-arg-arg-ser-thr-thr-asp-cys-

101 111
gly-gly-pro-lys-asp-his-pro-leu-thr-cys-asp-asp-pro-arg-phe-gln-asp-ser-ser-ser-

121 127 132 138
ser-lys-ala-pro-pro-pro-ser-leu-pro-ser-pro-ser-arg-leu-pro-gly-pro-ser-asp-thr-
O **O** **O** **O**

141 145
pro-ile-leu-pro-gln

Figure 4.1 The amino acid sequence of hCG subunits as shown by Morgan et al. [3]. The symbols N and O mark the sites of the Asn-linked N-linked and Ser-linked O-linked oligosaccharides on hCG subunits.

two N-linked oligosaccharides on the hCG α-subunit at residues 52 and 78, two N-linked oligosaccharides on the hCG β-subunit at residues 13 and 30, and four O-linked oligosaccharides on the C-terminal peptide of the hCG β-subunit at residues 121, 127, 132, and 138 [3].

The hCG peptide sequence has 18 acidic and 24 basic amino acids. As a result, the balance of the amino acid charge is on the basic side in sequence. The basic isoelectric point, however, is made acidic by the addition of eight oligosaccharides, each terminating with one or two sialic acid residues. The sialic acid makes hCG an acidic molecule with peaks reaching an isoelectric point of pI 3.5 (hCG) or 3.2 (hyperglycosylated hCG) [19]. The acidity repels circulating hCG from binding the renal glomerular basement membrane. This gives hCG and hyperglycosylated hCG a long (36 h) circulating half-life in blood compared with the 0.43-h circulating half-life of its nonacidic sister hormone, leutinizing hormone (LH) [20]. hCG completely lacks the aromatic amino acid tryptophan. The α-subunit has eight aromatic amino acids and the β-subunit has just five, making hCG and its subunits deficient in aromatic amino acids and difficult to detect by absorbance at 280 nm (A_{280} extinction coefficient $1.41 \times 10^4\ M^{-1}\ cm^{-1}$ [21]). The sequence of the C-terminal extension on the β-subunit of hCG (the site of the four O-linked oligosaccharides) is like a proteoglycan sequence, a polymer of Ser and Pro. The C-terminal sequence formed between residues 118 and 145 consists of 27 amino acids, including 8 Ser and 8 Pro residues. The high Pro content and the four O-linked oligosaccharides on the C-terminal peptide prevent this region from significant folding. Examining the amino acid sequence of hCG subunits (Figure 4.1), both hCG and hyperglycosylated hCG (hCG-H), which are two independent molecules, share this common amino acid sequence [6]. However, they are independent molecules with completely different oligosaccharide structures [6] and separate biological functions (see Part D and E).

4.2 Carbohydrate structure of hCG

The structure of the O-linked and N-linked oligosaccharides attached to hCG is also a story of refinement. Originally, the N-linked and O-linked structures were determined using classical glycosidase digestion methods [5–8]. In recent years, oligosaccharides at specific attachment sites and their effect on molecular weight have been examined by specialized HPLC and mass spectrometry methods [6,9]. Glycosidase digestion and specialized HPLC studies of hCG, or pregnancy hCG, show that the average pregnancy β-subunit contains 86% trisaccharide/tetrasaccharide-type O-linked oligosaccharides and 15% hexasaccharide/pentasaccharide-type structures (Figure 4.2) [5,6]. Sialic acid studies show that the average O-linked oligosaccharides contain 1.25 pmol/pmol sialic acid, which explains the trisaccharide/tetrasaccharide and hexasaccharide/pentasaccharide balance (Figure 4.2). Recent mass spectrometry studies show that all hCG molecules contain one hexasaccharide/pentasaccharide structure at residue 121, regardless of the source of hCG [9].

(A) Simple

Trisaccharide — NeuAc α2,3 Gal β1,3 GalNAc -O- Ser

Tetrasaccharide —
NeuAc
α2,6
NeuAc α2,3 Gal β1,3 GalNAc -O- Ser

Bisaccharide — NeuAc α2,3 GalNAc -O- Ser

(B) Complex

Hexasaccharide —
NeuAc α2,3 Gal β1,3 GlcNAc
β1,6
NeuAc α2,3 Gal β1,3 GalNAc -O- Ser

Pentasaccharide —
Gal β1,3 GlcNAc
β1,6
NeuAc α2,3 Gal β1,3 GalNAc -O- Ser

Figure 4.2 O-linked oligosaccharides on hCG and hyperglycosylated hCG. (A) Simple and (B) complex.

However, if this were true, then the average pregnancy hCG would contain 25% hexasaccharide/pentasaccharide, rather than the 15% indicated by glycosidase and specialized HPLC studies [6].

When we examined the N-linked oligosaccharides on the α-subunit, we found that the average pregnancy hCG contains 49% structure SM (monoantennary), 37% structure NN, 7.3% structure NNF, 4.5% structure NNNF, and 2.4% structure NNN (Figure 4.3) [6]. Examination of the N-linked oligosaccharides attached to the β-subunit shows that the average pregnancy hCG contains 47% structure NNF, 31% structure NN, 12.6% structure NNNF, 5.4% structure NM, and 3.8% structure NNN (Figure 4.3). As such, the pregnancy α-subunit contains 6.9% triantennary and the β-subunit contains 16% triantennary oligosaccharides.

4.3 hCG primary structure

The primary structure of hCG comprises the linear amino acid sequence β-subunit of 145 residues [1]. This is larger than the amino acid sequence of LHβ by 30 amino acids. As a result, hCG β-subunit has a unique, serine-rich, 30-amino-acid carboxy-terminal extension. This came from read-through into the 3′ untranslated region of the LH β gene [22]. Within this primary sequence, 12 cysteine residues provide the thiol groups for the six disulfide bridges that form during initial folding. The serines within the carboxy-terminal extension also provide the anchors for the four O-linked glycosylations (Figure 4.1). A high degree of sequence homology has

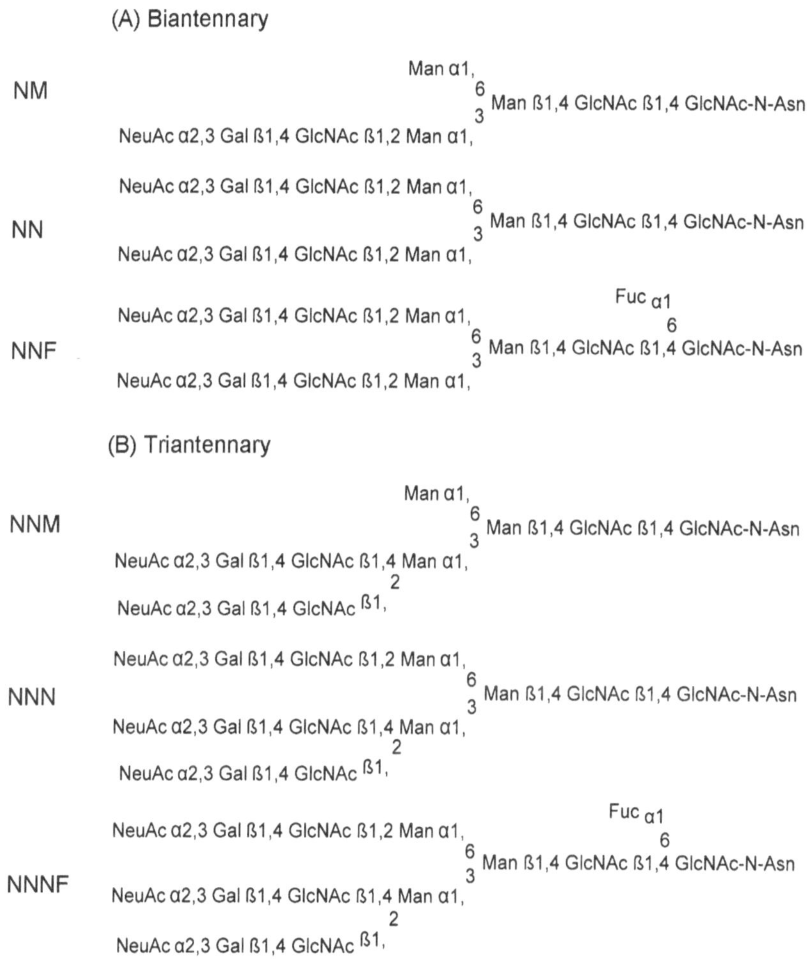

Figure 4.3 N-linked oligosaccharides on hCG and hCG-H. (A) Biantennary and (B) triantennary.

been maintained between the first 115 amino acids of hCG β and the other three β-subunits of the glycoprotein hormone family (LH β, 82%; TSH, 46%; FSH β, 36%) (Figure 4.1). In each case, the most conserved regions occur around the common 12 cysteine residues [13].

4.4 hCG secondary structure

The exact nature of the molecular folding patterns within hCG remained unresolved for 20 years after the initial hCG sequence was revealed. Many attempts at crystallization were impeded by the high degree of complex glycosylations found on the molecule. Although two studies successfully deglycosylated hCG and obtained viable crystals, X-ray diffraction was not performed to an adequate resolution [14,15]. Initial studies of the secondary folding patterns of molecular hCG were reviewed by Pierce and Parsons [13]. It was estimated that in the intact holohormone, only 5−8% of the molecule consisted of α-helices, and that a much higher percentage (25−40%) was accounted for by β-pleated regions.

Using methods similar to those of [12,14] grew hCG crystals from ammonium sulfate solutions after deglycosylation with anhydrous hydrofluoric acid. Electron density maps gave rise to tracings of the α-subunits and β-subunits, including amino acids 5−89 of hCG α and 2−111 of hCG β. The remaining 34 amino acids of the hCG β carboxy-terminal extension were not visible and therefore were believed to adopt a random conformation [12]. It was later revealed that there are three distinct areas of β-pleated structures in each subunit: two β-sheet bulges (62−65 and 79−82 in the α-subunit; 59−62 and 85−88 in the β-subunit) and a loop of double-stranded β-sheet-like structure that, in hCG α, is punctuated with two helix turns involving residues 38−46.

4.5 hCG tertiary structure

Many attempts at cysteine pairing have been made since hCG was first sequenced. Although there was an agreement on three of the disulfide pairs within hCG β (Cys 23−72, Cys 26−110, and Cys 93−100) and three disulfide pairs within hCG α (Cys 11−35, Cys 32−64, and Cys 63−91) [11,16], other possible hCG α and hCG β disulfide bonds remained contested. Reduction and alkylation studies [11,17], receptor binding studies [18,19], antigenic mapping studies [20], and cysteine mutagenic studies [21] all failed to adequately confirm the exact disulfide pairing. It was only when the crystal structure was established in 1994 that the complete disulfide linkages were revealed [12].

As found in crystal X-ray defraction studies, the Cys−Cys assignments of Mise and Bahl [11] contained four correctly assigned disulfides on the β-subunit. The six disulfide pairings for hCG β were established as Cys 9−57, Cys 23−72, Cys 26−110, Cys 34−88, Cys 38−90, and Cys 93−100; the five for hCG α were established as Cys 7−31, Cys 10−60, Cys 28−82, Cys 32−84, and Cys 59−87 [12]. Three of the cystine bonds formed a central cluster that has been termed a "cystine knot" structure [12,23]. This structure provides the molecule with its characteristic folding pattern and stabilizes the formation of a single, long, loop-like structure set opposite to two similarly shaped hairpin loops.

The positions of the three cystines in the cystine knot are almost identical in both subunits. Two disulfide bridges (linking residues 34−88 and 38−90 on the β-subunit and 28−82 and 32−84 on the α-subunit) join the antiparallel strands of the peptide chain, forming a central loop through which the third disulfide passes (linking residues 9−57 on the β-subunit and 10−60 on the α-subunit) (Figure 4.4).

The long loop of the α-subunit maintains its shape as a result of the main chain hydrogen bonds of the β-sheet and α-helix-like structures mentioned previously. Studies by Lustbader et al. [15] show that three disulfide linkages are quickly added to the β-subunit before synthesis is complete in the endoplasmic reticulum. These disulfide linkages are β9−57, β34−88, and β38−90. Three more disulfide linkages are later added to the β-subunit at β23−72, β93−100, and β26−110.

Within the β-subunit, however, the corresponding structure is stabilized with side-to-main chain hydrogen bonding from Arg 43 to Pro 50 and leucine (Leu) 52, and from Gln 54 to Met 4 1. In the α-subunit, the remaining two cystines keep the C-terminal and N-terminal sequences away from the central body of the molecule. In the β-subunit, another cystine (Cys 23−72) maintains a tight association between the opposing hydrophobic and hydrophilic regions of the hairpin loops. The β-subunit forms another unique structure, which has been termed the "seat belt," at residues β70−β105. This region bends away from the central knot as a result of the presence of another disulfide (Cys 93−100), which is then attached to the first hairpin loop by the final crosslink (Cys 26−110) (Figure 4.5).

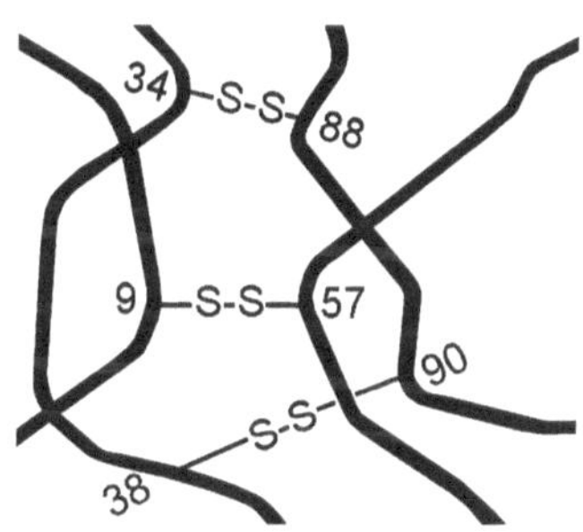

Figure 4.4 The cystine knot structures on hCG β-subunit.

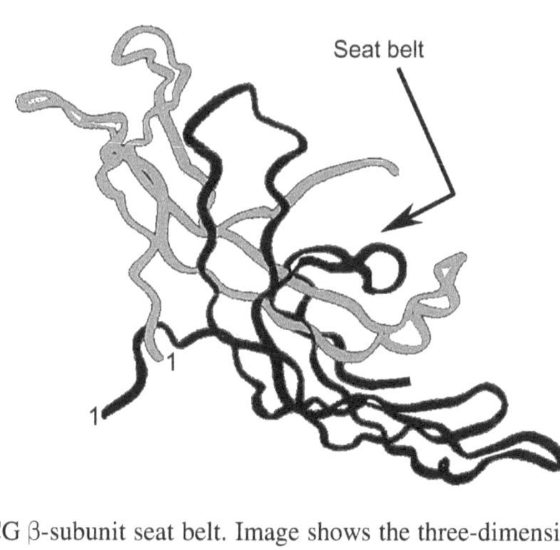

Figure 4.5 The hCG β-subunit seat belt. Image shows the three-dimensional structure of hCG as shown by Lapthorn et al. [12]. Black line is β-subunit and gray line is α-subunit.

4.6 hCG quaternary structure

With regard to subunit interactions and receptor binding, the three-dimensional structural patterns of hCG have been extensively reviewed [24,25], confirmed, and updated [12]. The "cystine knot" structure, and the binding loop regions stabilized by it, account for both structural and functional interactions (Figure 4.5). The subunits come together in a "top-to-tail" orientation and, although the α-subunits and β-subunits dimerize, they are not joined by covalent bonds. Instead, the α-subunit sits in a pocket of the β-subunit and is held in place primarily by the "seat belt" (Figure 4.5). Site-directed mutagenic studies of the conserved cysteine residues have shown that interruption of any β-subunit disulfide bond prevents the formation of an intact hCG heterodimer [21,24]. The explanation for the necessity of the central cystines became apparent after the publication of the crystal structure [12]. It was revealed that the β-sheet segments, held close to the cystine knot in each subunit, came together in the heterodimer to form a short, seven-stranded β-barrel, at the center of which lies β34-36 [12], a sequence considered essential for dimer formation [25] (Figure 4.5).

Although they are not covalently linked, the subunits are held in very close proximity to each other, with a total surface area of 4525 $\AA^2$ [12] or 3860 $\AA^2$ [26]. This tight association is maintained almost entirely through hydrogen bonds, the majority of which are located between the hairpin/long loop interfaces of each subunit. The most important feature in maintaining heterodimer integrity, however, is probably the seat belt (Figure 4.5). A short strand of β-sheet between the inner side of the seat belt and the α-subunit form part of the central barrel, and the Cys 26–110 bridge is believed to "click" into place, thus "hugging" the α-subunit and finalizing α–β dimerization. Chemical modification studies of both subunits indicated that Tyr 41 and Tyr/Phe 69 on the α-subunit and tyrosines or phenylalanines on the β-subunit are conserved in β-subunit regions (e.g., β37, β61, and β84). These are located in areas of subunit interaction and implicate binding. These tyrosines are protected from modification in the intact hCG molecule [13]. Huang and Puett [27] investigated Gln 54, which is conserved in all β-subunits, and noted that lysine (Lys) 54 mutants did not associate with the α-subunit and that Glu 54 only gave rise to a slight reduction in hormonal activity.

4.7 Combination of hCG subunits

Saccuzzo et al. [28] examined the combination of hCG subunits. Their work showed that the rate-limiting step for the combination of α-subunits and β-subunits is the rate of disulfide linkage in the formation of the β-subunit. Approximately 70% of the β-subunit made in a trophoblast cell is degraded and not used for combination. Only 50% of the nondegraded β-subunit is bound as α–β dimer. One explanation for incomplete dimer formation appears to be the biochemical differences between the free and the combined β-subunits that limit combination of the free β-subunit [28]. As explained in Chapter 6 of this book, free α-subunit and free β-subunit are both hyperglycosylated, which may limit subunit combination.

4.8 Synthesis and secretion of hCG

hCG is produced in placental syncytiotrophoblast cells, whereas the variant hyperglycosylated hCG (hCG-H) is made in placental cytotrophoblast cells [29]. Both hCG and hyperglycosylated hCG have the same amino acid sequences [6], indicating that both originate from the same α-subunit and β-subunit genes. hCG α-subunit is encoded by one gene on chromosome 6q21,1-23 [30]. This is a single gene coding for the common hCG, LH, FSH, and TSH α-subunit. In contrast, hCG β-subunit is coded for by as many as eight back-to-back genes on chromosome 19q13.3 [31]. As detailed in Chapter 3, genes 5, 3, and 8 of these 8 isomers are natural promoters with high transcriptional activity. The 5′ flanking regions of the five other genes contain gaps and deletions, making them less commonly transcribed [31].

It is assumed that the β-subunit amino acid sequence (shown in Figure 4.1) derives from genes 5, 3, and 8. Cis and trans gene activators of hCG transcription include activating protein-2, cAMP response element binding, upstream regulatory elements, and cAMP response element [32]. Promoters of hCG expression include glucocorticoids, cAMP, dehydroepiandrosterone, and gonadotropin-releasing hormone (GnRH) [32].

As hCG subunits are translated, dolichol adds a standard N-linked oligosaccharide structure to two asparagine residues on the α−subunit (residues 52 and 78) and to two N-linking sites on the β-subunit (residues 13 and 30). These are further processed after translation of the subunit and initial sugar processing is completed in the endoplasmic reticulum. N-linked oligosaccharide processing continues in the Golgi apparatus before the rapid secretion of hCG in secretory granules [33]. Disulfide bridge formation in hCG subunits, 11 disulfide bridges total, also starts in the endoplasmic reticulum before translation of the subunits is complete. The disulfide bridge formation continues after translation [28]. Addition of all disulfide bridges on the α-subunit and β-subunits is thought to limit complete folding of the structure [28].

Finally, O-linked oligosaccharides are added to the C-terminal peptide of the β-subunit of hCG, thus completing hCG synthesis in the Golgi apparatus [28,33]. hCG is rapidly secreted in unique syncytiotrophoblast secretory granules formed in the Golgi apparatus [34,35]. hCG is not accumulated in cells. It is only rapidly secreted [35]. In fact, it is rapidly secreted in a pulsatile manner [36].

Figure 4.6 shows the hyperglycosylated hCG concentrations and total hCG concentration in serum through the length of a pregnancy, as observed by the US hCG Reference Service. As shown, hCG production reaches a peak concentration at 10 weeks of gestation (since last menstrual bleeding). Hyperglycosylated hCG is first detected at 3−4 weeks of gestation. At this time, hyperglycosylated hCG concentration is much higher than hormone hCG concentration.

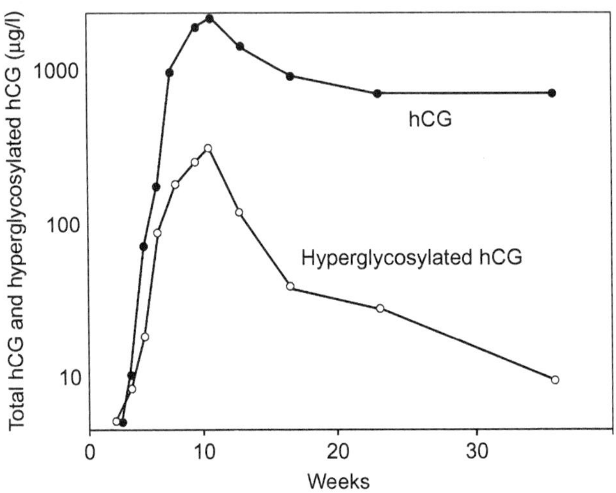

Figure 4.6 Serum hCG and hyperglycosylated hCG during the course of pregnancy.

Similarly, total hCG is approximately twice as high as hCG in urine during the second and third trimesters of pregnancy. This is due to detection of the β-core fragment, which can predominate in urine at this time. Urine hCG results are generally lower than serum hCG results, particularly during the second and third trimesters of pregnancy. hCG production during pregnancy is generally regulated by dissociation and degradation, by differentiation of cytotrophoblast cells, and by hCG-promoted differentiation [37]. In early pregnancy, hCG concentration is initially proportional to the growing mass of cytotrophoblast and syncytiotrophoblast cells (Figure 4.6). Due to hCG-promoted differentiation and the accumulating mass of syncytiotrophoblast cells (the hCG-producing cells), hCG levels increase rapidly during 6–9 weeks of gestation (Figure 4.6). At 10 weeks, hCG reaches a peak. This peak is somewhat artificial or a cell differentiation artifact, because it is due to the increasing production of hCG with increasing syncytiotrophoblast mass, combined with the growing differentiation of syncytiotrophoblast cells with three to four nuclei into terminally differentiated cells or a syncytium with 30–50 nuclei.

These terminal supernucleated cells are poor producers of hCG because they are all cytoplasm-limited. This reduces overall hCG production, starting at the 10-week peak and continuing until term (Figure 4.6). The 10-week peak represents a combination of increasing hCG from syncytiotrophoblast cells and decreasing hCG from those supernucleated cells forming a syncytium.

References

[1] Carlsen RB, Bahl OP, Swaminathan N. Human chorionic gonadotropin: linear amino acid sequence of the β-subunit. J Biol Chem 1973;248:6810–27.

[2] Bellisario R, Carlsen RB, Bahl OP. Human chorionic gonadotropin: linear amino acid sequence of the α-subunit. J Biol Chem 1973;248:6796–809.
[3] Morgan FJ, Birken S, Canfield RE. The amino acid sequence of human chorionic gonadotropin. The alpha subunit and beta subunit. J Biol Chem 1975;250:5247–58.
[4] Bahl OP. Human chorionic gonadotropin: II. Nature of the carbohydrate units. J Biol Chem 1969;244:575–83.
[5] Endo Y, Yamashita K, Tachibana Y, Tojo S, Kobata A. Structures of the asparagine-linked sugar chains of human chorionic gonadotropin. J Biochem 1979;85:669–79.
[6] Elliott MM, Kardana A, Lustbader JW, Cole LA. Carbohydrate and peptide structure of the α- and β-subunits of human chorionic gonadotropin from normal and aberrant pregnancy and choriocarcinoma. Endocrine 1997;7:15–32.
[7] Kessler MJH, Mise T, Ghai RD, Bahl OP. Structure and location of the O-glycosidic carbohydrate units of human chorionic gonadotropin. J Biol Chem 1979;254:7909–14.
[8] Cole LA, Birken S, Perini F. The structures of the serine-linked sugar chains on human chorionic gonadotropin. Biochem Biophys Res Comm 1985;126:333–9.
[9] Valmu L, Alfthan H, Hotakainen K, Birken S, Stenman UH. Site-specific glycan analysis of human chorionic gonadotropin β-subunit from malignancies and pregnancy by liquid chromatography–electrospray mass spectrometry. Glycobiology 2006;16: 1207–18.
[10] Giudice LC, Pierce JG. Studies on the disulfide bonds of glycoprotein hormones. J Biol Chem 1979;254:1164–9.
[11] Mise T, Bahl OP. Assignment of disulfide bonds in the β-subunit of human chorionic gonadotropin. J Biol Chem 1981;256:6587–92.
[12] Lapthorn AJ, Harris DC, Littlejohn A, Lustbader JW, Canfield RE, Machin KJ, et al. Crystal structure of human chorionic gonadotrophin. Nature 1994;369:455–61.
[13] Pierce JG, Parsons TF. Glycoprotein hormones: structure and function. Ann Rev Biochem 1981;50:465–95.
[14] Harris DC, Machin KJ, Evin GM, Morgan FJ, Isaacs NW. Preliminary X-ray diffraction analysis of human chorionic gonadotropin. J Biol Chem 1989;264:6705–9.
[15] Lustbader JW, Birken S, Pileggi NF, Kolks MA, Pollak S, Cuff ME, et al. Crystallization and characterization of human chorionic gonadotropin in chemically deglycosylated and enzymatically desialylated states. Biochemistry 1989;28:9239–43.
[16] Willey KP, Leidenberger F. Functionally distinct agonist and receptor binding regions in hCG. J Biol Chem 1989;264:19716–29.
[17] Cornell JS, Pierce JG. Studies on the disulfide bonds of glycoprotein hormones. Locations in the alpha chain based on partial reductions and formation of 14C-labeled S-carboxymethyl derivatives. J Biol Chem 1974;249:4166–74.
[18] Keutmann HT, Charlesworth MC, Mason KA, Ostrea T, Johnson L, Ryan RJ. A receptor-binding region in human choriogonadotropin/lutropin beta subunit. Proc Natl Acad Sci 1987;84:2038–42.
[19] Ryan RJ, Charlesworth MC, McCormick DJ, Milius RP, Keutmann HT. The glycoprotein hormones: recent studies of structure–function relationships. Fed Am Soc Exp Biol 1988;2:2661–9.
[20] Moyle WR, Matzuk MM, Campbell RK, Cogliani E, Dean-Emig DM, Krichevsky A, et al. Localization of residues that confer antibody binding specificity using human chorionic gonadotropin/luteinizing hormone beta subunit chimeras and mutants. J Biol Chem 1990;265:8511–18.
[21] Suganuma N, Matzuk MM, Boime I. Elimination of disulfide bonds affects assembly and secretion of the hCG β subunit. J Biol Chem 1989;264:19302–7.

[22] Fiddes JC, Talmage K. Structure, expression and evolution of the genes for the human glycoprotein hormones. Recent Prog Hormone Res 1984;40:43–74.
[23] Murray-Rust J, McDonald NQ, Blundell TL, Hosang M, Oefner C, Winkler F, et al. Topological similarities in TGF-P2, PDGF-BB and NGF define a superfamily of polypeptide growth factors. Structure 1993;1:53–159.
[24] Ryan RJ, Keutmann HT, Charlesworth MC, McCormick DJ, Milius RP, Calvo FO, et al. Structure–function relationships of gonadotrophins. Recent Prog Hormone Res 1987; 43:383–417.
[25] Chen F, Puett D. A single amino acid residue replacement in the beta subunit of human chorionic gonadotrophin results in the loss of biological activity. J Mol Endocrinol 1992;8:87–9.
[26] Wu H, Lustbader JW, Liu Y, Canfield RE, Hendrickson WA. Structure of human chorionic gonadotropin at 2.6 Å resolution from MAD analysis of the selenomethionyl protein. Structure 1994;2:545–58.
[27] Huang J, Puett D. On the role of the invariant glutamine at position 54 in the human choriogonadotropin beta subunit. Mol Cell Biochem 1994;136:183–6.
[28] Saccuzzo B, Huth JR, Ruddon RW. Combination of the chorionic gonadotropin free β-subunit with α. Endocrinology 1990;126:384–91.
[29] Xing Y, Williams C, Campbell RK, Cook S, Knoppers M, Addona T, et al. Threading of a glycosylated protein loop through a protein hole: implications for combination of human chorionic gonadotropin subunits. Protein Sci 2001;10:226–35.
[30] Fiddes JC, Goodman HC. The gene encoding the common alpha subunit of four human glycoprotein hormones. J Mol Appl Genet 1981;1:3–10.
[31] Boorstein WR, Vamvakopoulos NC, Fiddes JC. Human chorionic gonadotropin beta-subunit is encoded by at least eight genes arranged in tandem and inverted pairs. Nature 1982;300:419–22.
[32] Becker KL, Bilezikian JP. Endocrinology of trophoblastic tissue. In: Becker KL, editor. Principle and practice of endocrinology. Lipcott W. and Wilkins; 2004. p. 1096–100.
[33] Hannover JA, Elting J, Mintz GR, Lennarz WJ. Temporal aspects of the N-and O-glycosylation of human chorionic gonadotropin. J Biol Chem 1982;257:10172–7.
[34] Morrish DW, Marusyk H, Olivia S. Demonstration of specific secretory granules for human chorionic gonadotropin in placenta. J Histochem Cytochem 1987;35:93–101.
[35] Ruddon RW, Hanson CA, Addison NJ. Synthesis and processing of human chorionic gonadotropin subunits in cultured choriocarcinoma cell. Proc Natl Acad Sci 1979;76: 5143–7.
[36] Odell WD, Griffin J. Pulsatile secretion of human chorionic gonadotropin in normal adults. N Engl J Med 1987;317:1688–91.
[37] Shi QJ, Lei ZM, Rao CV, Lin J. Novel role of human chorionic gonadotropin in differentiation of human cytotrophoblasts. Endocrinology 1993;132:1387–95.

Three-dimensional structures of hCG and hyperglycosylated hCG

5

Laurence A. Cole
USA hCG Reference Service, Angel Fire, NM, USA

5.1 X-ray crystallography

For years, three groups worked meticulously to generate hCG crystals for X-ray crystallography and multi-wavelength anomalous diffraction methods in hopes of determining the fine three-dimensional (3D) structure of hCG [1–4]. Unfortunately, the heterogeneity of the eight carbohydrate side chains on hCG prevented any crystallization. Even desialylization, or removal of the principal charged carbohydrate component, did not allow fine crystal formation [1]. Using an anhydrous hydrofluoric acid treatment, they carefully removed the entire N-linked oligosaccharides from hCG, leaving the deglycosylated protein intact. To remove the four O-linked oligosaccharides, they cleaved the 34-amino-acid C-terminal peptide at β111. It was assumed that the C-terminal peptide was a random, nonfolded component, similar to a semipolymer of proline and serine.

The hCG and hyperglycosylated hCG root structure missing carbodydarte side chains and a ß-subunit C-terminal or just half of the intact hCG or hyperglycosylated hCG molecule was crystallized. X-ray crystallographic analysis and multi-wavelength anomalous diffraction were performed [1–4]. The fine 3D structure of deglycosylated hCG missing the β-subunit C-terminal peptide was determined and disulfide bridges, charge interactions, hydrophobic interactions, and hydrogen bonding of the α–β dimer were all investigated [1–4] (Figure 5.1).

Lapthorn, Lustbader, and Wu [1–4] surely tried their best to resolve this impossible dilemma using half a molecule (Figure 5.1), but it is clear that X-ray crystallographic methods will probably never reveal the actual structure of the real molecules hCG and hyperglycosylated hCG. Glycosylation and a C-terminal peptide greatly affect the final structure.

hCG α-subunit is 92 amino acids and hCG β-subunit is 145 amino acids [5]. Almost 50% of the molecular weight of hCG and hyperglycosylated hCG comes from the four N-linked oligosaccharides and four O-linked oligosaccharides and the β-subunit C-terminal peptide (the items removed) [5]. The crystals used by Lapthorn et al., Lustbader et al., and Wu et al. [1–4] comprised α-subunit residues 5–89 and β-subunit 2–111 with no oligosaccharides. The major problem with this meticulous research is that the crystal was formed with a molecule with a molecular weight of 21,446, no sugars, and no hCG β-subunit C-terminal peptide, or just 58% of hCG's molecular weight of 37,180. We ask, does this crystal structure really

Human Chorionic Gonadotropin (hCG). DOI: http://dx.doi.org/10.1016/B978-0-12-800749-5.00005-5

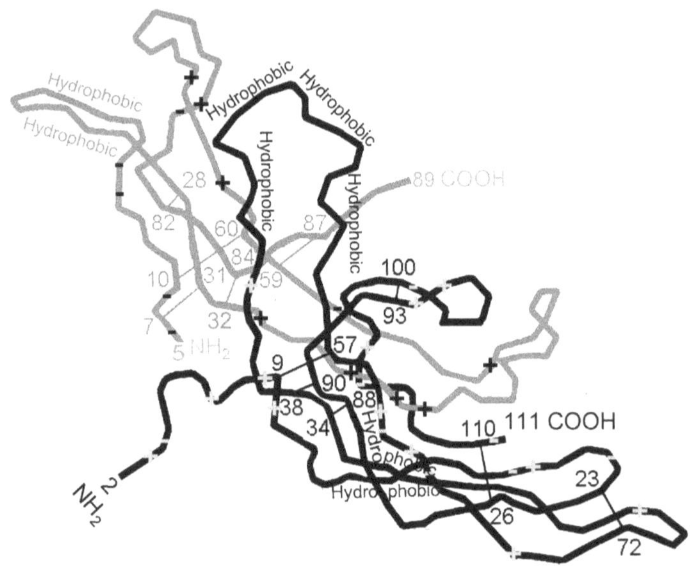

Figure 5.1 The 3D structure of hCG as proposed by Wu et al. [1], Lapthorn et al. [2], and Lustbader et al. [3,4]. Plus and minus signs indicate sites of charged amino acids.

have any relationship to hCG or hyperglycosylated hCG? hCG and hyperglycosylated hCG both share a common amino acid structure of deglycosylated hCG root. However, hCG binds an hCG/LH hormone receptor and hyperglycosylated hCG binds a TGFβ type II receptor. There is clearly a distinct structural element that separates these two molecules that is not examined by this 3D structure. We ask, is the 3D structure a pseudo-impression of either hCG or hyperglycosylated hCG, or anything to having do with hCG or hyperglycosylated hCG?

The deglycosylated hCG missing the β-subunit C-terminal peptide used for the crystal studies was shown to bind the LH/hCG hormone receptor, but with extremely reduced affinity [1–4]. Nobody has tested this molecule against the TGFβ type II receptor in cancer cases or using cytotrophoblast cells. Does this suggest that this half of an hCG molecule is a form of hCG or a variant of hyperglycosylated hCG?

5.2 Refining this model

Adding oligosaccharides increases the molecular weight of hCG and hyperglycosylated hCG by 30–39%, but does this really change its shape? We know that three disulfide bonds are added quickly to the β-subunit of hCG at linkages at β9–57, β34–88, and β38–90. These are the β-subunit cystine-linked inherited disulfide bonds that come from hCG's common evolutionary root with TGFβ [6]. As has been shown, these linkages form the core structure of the hCG β-subunit. They are essential to the formation of α–β dimer structures and to LH/hCG receptor binding [3,7]. We ask again, is regular hCG different from the deglycosylated structure missing the β-subunit C-terminal peptide (Figure 5.1)? Adding charge from sialic acid lowers the isoelectric point of hCG from pI 8.0 to pI 3.5 and extends the circulating half-life from 0.060 to 36 h [8], but does it do anything more?

We conclude that the structure of glycosylated hCG and hyperglycosylated hCG are probably very different from the structures proposed by Wu et al., Lapthorn et al., and Lustbader et al. [1–4]. We studied the proposed structure carefully and examined the effects that the β-subunit C-terminal peptide and the N-linked and O-linked oligosaccharides are likely to have on the final hCG and hyperglycosylated hCG 3D structures.

It is a fact that hCG, desialylated hCG, deglycosylated hCG, and hCG missing the C-terminal peptide and O-linked oligosaccharides all have hCG/LH receptor hormonal activity [3,4,9]. As such, hCG may simply resemble the proposed crystal model with oligosaccharides and C-terminal peptide added (Figure 5.1). The position of the oligosaccharides at α52 has been moved slightly toward the surface of the molecule to accommodate the extreme charge of the oligosaccharide (Figure 5.2).

In modeling hyperglycosylated hCG, the following seven critical observations are considered:

- Critical observation 1: The molecule hCG contains 12 sialic acid residues: approximately three to four on β-subunit N-linked oligosaccharides, five to six on β-subunit O-linked oligosaccharides, and two to three on α-subunit N-linked oligosaccharides. Hyperglycosylated hCG characteristically contains 19 sialic acid residues: five to six on β-subunit N-linked oligosaccharides, six to eight on β-subunit O-linked oligosaccharides, and three to five on α-subunit N-linked oligosaccharides [3].
- Critical observation 2: C1, C2, C3, C5, and C7 preparations of hyperglycosylated hCG dimer dissociate distinctly more rapidly (mean dissociation ½ time 18 h) than regular pregnancy hCG dimer preparations P1, P2, P3, P4, and P5 (mean dissociation ½ time 36 h), suggesting much less α–β subunit interaction or less α–β subunit charge and hydrophobic and hydrogen bond linkages between subunits in hyperglycosylated hCG dimers [1,10].
- Critical observation 3: Antibody B204 radioimmunoassay, specific to the free β-subunit and the β-core fragment of hCG, has zero affinity for pregnancy hCG dimer (preparations P1, P2, P3, P4, and P5) but has limited detection of C1, C2, C3, C5, and C7 choriocarcinoma hyperglycosylated hCG dimer. This suggests that hyperglycosylated hCG has greater exposure to the free β-subunit within its structure [11].
- Critical observation 4: Attachment of an N-linked oligosaccharide at α-subunit residue 52 is critical for hCG biological activity in the hCG/LH receptor [11,12]. The 3D aglyco-C-terminal peptide-less hCG structure is altered somewhat by attachment of this N-linked oligosaccharide [12].
- Critical observation 5: Hyperglycosylated hCG dimer and hyperglycosylated hCG free β-subunit and hCG free β-subunit, but not hCG dimer, function through binding and antagonizing a cancer cell TGFβ-II receptor [4,13]. A structure must be uncovered on these hCG variants that facilitated TGF-β2 binding, and this must be absent on hCG dimer.

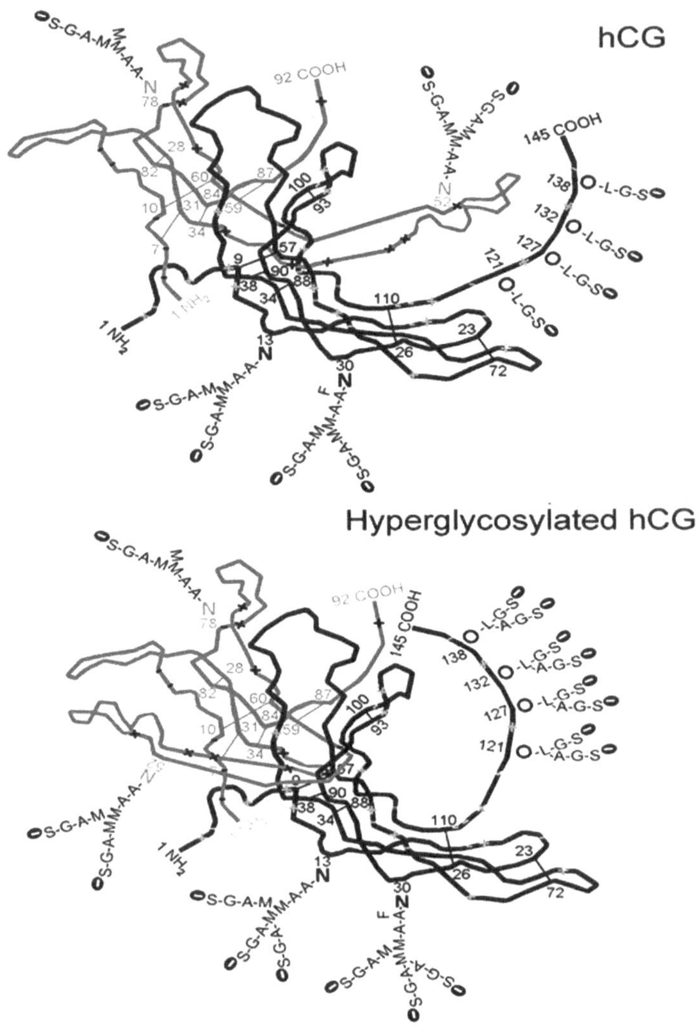

Figure 5.2 Proposed complete intact structures of hCG and hyperglycosylated hCG.

- Critical observation 6: hCG free β-subunit, but not hCG dimer, binds and antagonizes a TGFβ-II receptor, suggesting that α-subunit covers the TGF-β2 binding site [4,13].
- Critical observation 7: The principal structural difference between hCG and hyperglycosylated hCG is on the β-subunit C-terminal peptide is 75% hexasaccharide type O-linked oligosaccharide on hyperglycosylated hCG versus 14% on hCG [13].

Considering all these observations, the reattachment of the β-subunit C-terminal peptide, the part of the molecule that most grossly differentiates hCG and hyperglycosylated hCG (see critical observation 12), probably grossly affects the hCG 3D structure (Figure 5.1). The attachment of the N-linked and O-linked oligosaccharides may also make significant differences.

Ruddon et al. [12] and Beebe et al. [11] have studied the disulfide bridges on Jar and JEG-3 choriocarcinoma hCG, or hyperglycosylated hCG, and have shown that the bridge structure is exactly the same on pregnancy hCG and choriocarcininoma hyperglycosylated hCG. The charged amino acids are marked in the model (Figure 5.1) and are indisputable, as are pertinent hydrophobic sequences. This was the model that was carefully examined, along with the seven critical observations, to suggest a 3D structure for hCG and hyperglycosylated hCG.

The proposed structure for hyperglycosylated hCG considerers the seven critical observations. The structure is based on these observations (Figure 5.2). As shown, the very highly charged C-terminal peptide on hyperglycosylated hCG fold into the core of the hCG structure, balanced by the positive charge of the α-subunit amino acids. This charge movement pushes the similarly charged N-linked oligosaccharide (α52) and the α38–60 loop to move to the other side of the molecule. This movement exposes the β-subunit (critical observation 3) and limits $\alpha - \beta$ subunit interaction (critical observation 2), exposing the cystine knot structure in the core of the β-subunit (critical observation 5) (Figure 5.2).

This is the proposed structure of hyperglycosylated hCG with a TGFβ binding site (Figure 5.2). Research is needed to confirm these structure models for hCG and hyperglycosylated hCG.

References

[1] Wu H, Lustbader JW, Liu Y, Canfield RE, Hendrickson WA. Structure of human chorionic gonadotropin at 2.6 Å resolution from MAD analysis of the selenomethionyl protein. Structure 1994;2:545–58.

[2] Lapthorn AJ, Harris DC, Littlejohn A, Lustbader JW, Canfield RE, Machin KJ. Crystal structure of hCG. Nature 1994;369:455–61.

[3] Lustbader JW, Birken S, Pileggi NF, Kolks MAG, Pollak S, Cuff ME, et al. Crystallization and characterization of human chorionic gonadotropin in chemically deglycosylated and enzymatically desialylated states. Biochemistry 1989;28:9239–43.

[4] Lustbader JW, Wang C, Zhang X, Birken S, Wu H, Brown JM, et al. Human chorionic gonadotropin: progress in determining its tertiary structure. Glycoprotein hormones. New York, NY: Springer-Verlag; 1994. p. 81–102

[5] Morgan FJ, Birken S, Canfield RE. The amino acid sequence of human chorionic gonadotropin. J Biol Chem 1975;250:5247–58.

[6] Vitt UA, Hsu SY, Hsush AJW. Evolution and classification of cytosine kny-containing hormones and related extracellular signaling molecules. Mol Endicrinol 2001;15: 681–94.

[7] Mishra AK, Makal SD, Iyer KS. Disulfide bonds Cys9–Cys57, Cys34–Cys88 and Cys34–Cys90 of the β-subunit of human chorionic gonadotropin are crucial for heterodimer formation with α-subunit: experimental evidence for the conclusions from the crystal structure of hCG. Biochem Biophys Acta 2003;1645:49–55.

[8] Rosa C, Amr S, Birken S, Wehmann R, Nisula B. Effects of desialylation on human chorionic gonadotropin on its metabolic clearance rate in humans. J Clin Endocrinol Metab 1984;59:1215–19.

[9] Ronco AM, Tijmes M, Santibanes JF, Moraga P. Effect of desialylated human chorionic gonadotropin (hCG) on the bioactivity of rat Leydig cell. Cell Biochem Func 1998;16:21–6.
[10] Policastro P, Ovitt CE, Hoshona M, Fukuoka H, Bothby MR, Boime I. The ß-subunit of human chorionic gonadotropin is encoded by multiple genes. J Biol Chem 1983;258: 11492–9.
[11] Beebe JS, Huth JR, Ruddon RW. Combination of the chorionic gonadotropin free beta-subunit with alpha. Endocrinology 1990;126:384–91.
[12] Ruddon RW, Krzesicki RF, Norton SE, Beebe JS, Peters BP, Perini F. Detection of a glycosylated, incompletely folded form of chorionic gonadotropin beta subunit that is a precursor of hormone assembly in trophoblastic cells. J Biol Chem 1987;262: 12533–40.
[13] Elliott MM, Kardana A, Lustbader JW, Cole LA. Carbohydrate and Peptide structure of the α- and β-subunits of human chorionic gonadotropin from normal and aberrant pregnancy and choriocarcinoma. Endocrine 1997;7:15–32.

Structures of hCG free α-subunit and free β-subunit

6

Laurence A. Cole
USA hCG Reference Service, Angel Fire, NM, USA

hCG and hyperglycosylated hCG are dimers composed of an α-subunit and a β-subunit linked together by hydrophobic and ionic interactions in a noncovalent manner. The α-subunit is common to hCG, hyperglycosylated hCG, and other glycoprotein hormones, including luteinizing hormone (LH), follicle-stimulating hormone (FSH), and thyroid-stimulating hormone (TSH). The β-subunit of hCG, in contrast, is unique and an independent molecule that gives each glycoprotein hormone specificity. During pregnancy, the placenta produces limited amounts of each hormone's β-subunit and excess amounts of the common α-subunit. The excess amounts of the α-subunit lead to hCG dimer production, hyperglycosylated hCG dimer production, and the secretion of an excess free α-subunit [1,2]. This excess α-subunit is then converted into a free α-subunit with an alternative carbohydrate structure. Free α-subunit is a glycoprotein hormone-processing intermediate with no known independent functions. It cannot combine with β-subunit [1,2].

In pregnancy serum, free β-subunit is primarily derived from the dissociation of regular hCG and hyperglycosylated hCG. The free β-subunit is detected in serum and urine samples. Excess amounts of β-subunit may be created in hydatidiform moles or choriocarcinoma cells. This excess is secreted as a free β-subunit. Cancer cells retrodifferentiate or are otherwise transformed and express the β-subunit of hCG. This might lead to free β-subunit production by cancer cells [3]. Free β-subunit has been shown to have a biological function separate from that of hCG. In cancer cells, the free β-subunit produced acts as an autocrine on the same cancer cells, promoting cell growth, cell invasion, and cell metastases by antagonizing a TGFβ type 2 receptor [3–7]. In this chapter, we describe free α-subunit and β-subunit structures, their synthesis, and properties in detail.

6.1 Free α-subunit

In the normal synthesis of all glycoprotein hormones (i.e., LH, FSH, TSH, and hCG), the β-subunit is limited and controls hormone synthesis; this results in an excess α-subunit [8,9]. Free α-subunit is made by the gonadotrope and thyrotrope cells of the pituitary gland during the menstrual cycle and by placental cells during pregnancy. All these cells produce excess free α-subunit in addition to the hormone.

The excess α-subunit made during pregnancy is converted into a free α-subunit. The free α-subunit is larger than the α-subunit of hCG because of the extra large

Human Chorionic Gonadotropin (hCG). DOI: http://dx.doi.org/10.1016/B978-0-12-800749-5.00006-7

oligosaccharide side chains (Figure 6.1), giving the appearance of two α-subunit molecules, α-subunit of hCG from dissociation of hCG, and free α-subunit of hCG being freely secreted. However, only one gene codes for all α-subunits. As shown by numerous researchers, the large free α-subunit with large oligosaccharide side chains cannot combine with β-subunit to form an hCG dimer [1,2,10,11].

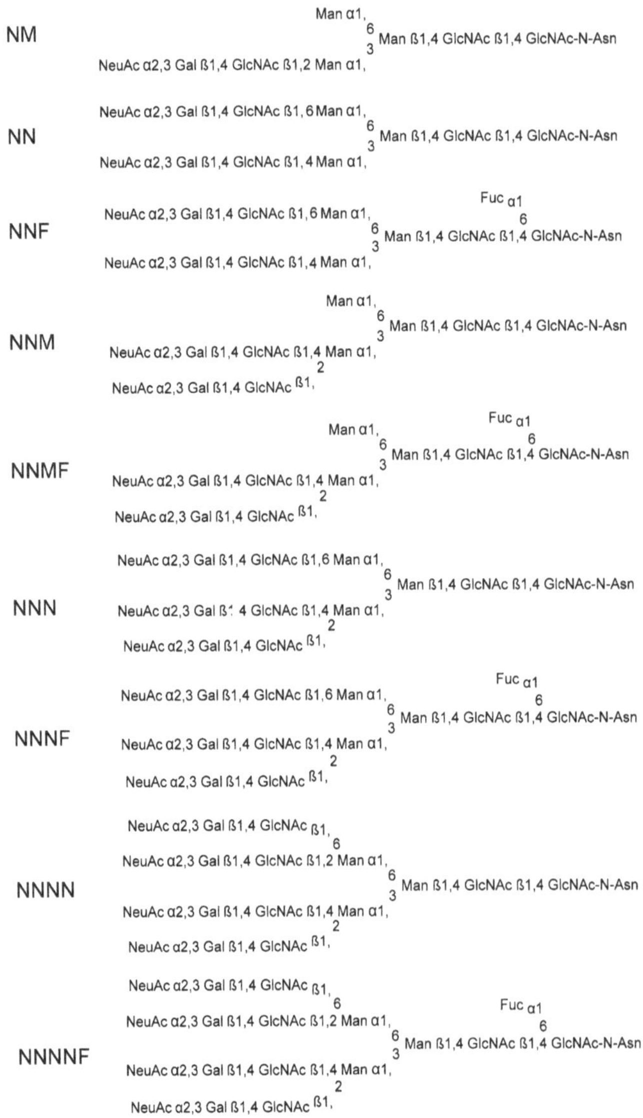

Figure 6.1 The N-linked oligosaccharides attached to the hCG free α-subunit.
Source: Data from Ref. [12,13].

Multiple studies have examined the oligosaccharide structures of free α-subunit but have been limited to lectin and endoglycosidase specificity-based techniques rather than mass spectrometry or carbohydrate sequencing methods. It was demonstrated that after complete removal of large oligosaccharides from free α-subunit by N-glycanase treatment, free α-subunit can combine with β-subunit to form a dimer [11]. This confirms that only the large oligosaccharides side chains on free α-subunit block the combination of subunits [11].

Lens culinaris lectin binds oligosaccharides, except when an N-acetylglucosamine (GlcNAc) is β1,4-linked to the α1,3 mannose (Man) in the core of N-linked oligosaccharides. *Datura stramonium* lectin differentiates a GlcNAc that is β1,6-linked to the α1,6 mannose in the core of N-linked oligosaccharides. Using these lectins [12,13], it has been shown that the β1,6 GlcNac is found on α1,6 mannose and β1,4 GlcNAc is found on the α1,3 mannose of hCG free α-subunit (Figure 6.1). This unusual β1,4 and β1,6 GlcNAc linkage differentiates the structure of the free α-subunit from the structures of the hCG dimer α-subunit and the hyperglycosylated hCG α-subunit (see Chapters 5 and 7).

Although monoantennary and biantennary sugar structures are located on the α-subunit of the hCG dimer, multiple larger oligosaccharides are found on free α-subunits (Chapter 4). Monoantennary (NM) oligosaccharides account for 12% of total free α-subunit structure, biantennary (NN and NNF) oligosaccharides account for 18%, and biantennary terminal mannose structures (NNM and NNMF) account for 21%; triantennary structures (NNN and NNNF) account for 23% and tetraantennary (NNNN and NNNF) oligosaccharide structures account for 8% of total free α-subunit structures (Figure 6.1). Each of these structures can be fucosylated. In the free α-subunit, it is estimated that oligosaccharides are 50% fucosylated. Thus, an equal mixture of fucosylated and nonfucosylated structures are found on free α-subunits [12,13]. The oligosaccharides on the average sample of free α-subunits are mostly a mixture of biantennary, triantennary, and tetraantennary containing an average of 2.7 antennae.

As pregnancy advances, there is a clear shift in the structure of hCG free α-subunits from 60.4% NNN with the GlcNAc β1,4 Man α1,3 structure to only 24.8% [12]. The reverse occurs with the GlcNAc β1,6 mannose α1,6 structure: the proportion surges in the latter part of pregnancy (late second trimester versus early second trimester). As pregnancy advances, the degree of fucosylation of free α-subunits also increases. What these changes mean is uncertain. The free α-subunit concentration in blood and urine starts low in the first trimester of pregnancy and continuously increases as pregnancy advances. In the third trimester of pregnancy, the total hCG concentration equals hCG plus hyperglycosylated hCG plus free β-subunit [14]. As free α-subunit production increases, it appears that its structure changes. The free α-subunit has no known biological function, so the oligosaccharide structural changes have no known biological significance.

Crystals have been made from the deglycosylated free α-subunit. The folding of these crystals has been studied using NMR spectroscopy [15]. These studies have shown that the free α-subunit folding abnormalities are distinct and not found on the α-subunit of the hCG dimer. A structural disorder occurs at residues 33–57, and it changes in the hairpin loops at residues 20–23 and 70–74 [15]. Once again, we note that the free α-subunit has no known biological function. Hence, this difference in folding has no known biological significance.

The free α-subunit can be ectopic, meaning it is produced by nontrophoblastic cancer cells. Ectopic free α-subunit was isolated from a cancer patient's urine [16] and was shown not to combine with β-subunit. The amino acid sequence was then investigated and glutamic acid (Glu) at residue 56 was replaced with alanine (Ala) at residue 56; in other words, an acidic amino acid was replaced with a hydrophobic residue. This probably affected the three-dimensional structure of the free α-subunit. In this case, it is assumed that either a mutation in the single α-subunit gene or an error in amino acid translation occurred. The codons GAA and GAC code for Glu, whereas GCA and GCC code for Ala. It is easy to see how a single point mutation might explain a Glu–Ala switch.

Although the free α-subunit is considered to be a primarily N-glycosylated molecule, it can be O-glycosylated [17,18]. For example, an O-linked oligosaccharide has been located on the free α-subunit produced in choriocarcinoma and on that made by Jar choriocarcinoma cells. The O-linked oligosaccharides detected on hCG have a structure of NeuAc α2,3 Gal β1,3 GalNAc [17,18]. The free α-subunit with an O-linked oligosaccharide is similar in structure to that described. It is concluded that in addition to two N-linked sugar units, the free α-subunit contains an O-linked oligosaccharide located on tryptic peptide 36–42. The tryptic peptide contains a Thr residue at amino acid 39, which is the likely attachment point of the O-linked oligosaccharide.

6.2 Free β-subunit

A free α-subunit is made primarily during pregnancy and is the artifact of excess α-subunit and limited β-subunit production [14,19]. Moving from pregnancy to a complete hydatidiform mole and choriocarcinoma, the β-subunit can be made in excess and leads to the detection of a naturally secreted free β-subunit in serum and urine [20]. As discussed later (Chapter 20), cancer cells ectopically express the hCG free β-subunit and hyperglycosylated hCG free β-subunit, allowing it to be detected in serum samples [21–26]. Acevedo et al. found that the free β-subunit is present in all cell membranes and in every cancer known to humans [25,26]. This is a controversial finding that others have not been able to confirm. The problem is that only 30% of cancers produce high enough levels of free β-subunit to be detected in serum (Chapter 20). In most cancer cases, the free β-subunit is excreted in urine as the terminally degraded product β-core fragment. β-core fragment comprises β-subunit residues 6–40 linked to residues 55–92 by disulfides [27]. β-core fragment has degraded oligosaccharides, and a mannose core structure, Man α1,3 (Man α1,6) Man β1,4GcNAc β1,4GlNAc [27]. Serum free β-subunit reveals 30% of malignancies; tests using urine β-core fragment can detect 55% of malignancies (Chapter 20).

As recently shown [4,28–30], malignancies produce an autocrine free β-subunit that acts on a TGFβ receptor on those same malignancies to inhibit apoptosis, promoting cancer cell growth and cell invasion. Free β-subunit is a biologically active cytokine-like molecule. For 20 years, the free β-subunit on SDS polyacrylamide gels has been observed migrating as a slightly larger molecule than the hCG dimer β-subunit [20,31]. It has been suggested that the molecular weight of the free β-subunit is 24,000, compared with 22,200 for the β-subunit of hCG [31]. In recent years, mass spectrometry has confirmed the molecular weight of the free β-subunit.

Recently, the structure of the free β-subunit has been investigated using the urine of a patient with a complete hydatidiform mole [32]. A complete hydatidiform mole normally produces a pregnancy-like hCG dimer. However, the free β-subunit studied had a hyperglycosylated hCG-like oligosaccharide structure [32]. Like the hyperglycosylated hCG β-subunit, the free β-subunit contained 37% triantennary N-linked structures at one glycosylation site (Asn 13) and 30% triantennary structures at the second N-linked site (Asn 30) (Figure 6.2). The O-linked oligosaccharides were hyperglycosylated hexasaccharide/pentasaccharide in structure: 85% at Ser 121; 86% at Ser 127 and 132; and 79% at Ser 138 [32]. It is inferred that free β-subunit, when secreted freely, can be hyperglycosylated. This is probably the case regardless of whether free β-subunit is produced by a complete hydatidiform mole or cancer cells, and it is different from the dissociated free β-subunit observed in blood in pregnancy.

Valmu et al. [32] speculated that the hyperglycosylation of the β-subunit might block combination with the hCG α-subunit, just as hyperglycosylation blocks free α-subunit combination. Interestingly, Butler et al. [33] have both shown that the isolated β-subunit of pregnancy hCG works as well as the free β-subunit produced by cancer cells to promote cancer cell growth and cell invasion. Hence, hyperglycosylation is not an intrinsic part of the biological activity of the free β-subunit; a receptor binding site is obviously exposed in both molecules. Interestingly, the free β-subunit has been shown to homodimerize, making a β–β dimer [33]. The dimer is made by multiple cancers. The dimer, however, can confuse hCG immunometric assay results when using antibodies to two sites on the β-subunit.

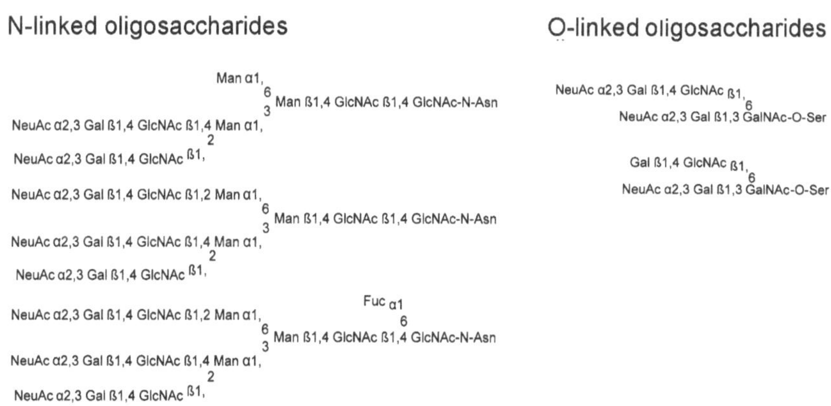

Figure 6.2 Proposed structure of hCG free β-subunit oligosaccharide side chains.
Source: Data from Ref. [32].

References

[1] Posillico EG, Handwerger S, Tyrey L. Human chorionic gonadotropin α-subunit of normal placenta: characterization of synthesis and association with β-subunit. Biol Reprod 1985;32:1101–8.

[2] Beebe JS, Kresinski RF, Norton SE, Perini F, Peters BP, Ruddon RR. Identification and characterization of subpopulations of free α-subunit that vary in their ability to combine with chorionic gonadotropin-β. Endocrinology 1989;124:1613–24.

[3] Muller C, Cole LA. The quagmire of hCG and hCG testing in gynecologic oncology. Gynecol Oncol 2009;112:663–72.

[4] Butler SA, Ikram MS, Mathieu S, Iles RK. The increase in bladder carcinoma cell population induced by the free beta subunit of hCG is a result of an anti-apoptosis effect and not cell proliferation. Br J Cancer 2000;82:1553–6.

[5] Iles RK. Ectopic hCG β expression by epithelial cancer: malignant behavior metastasis and inhibition of tumor cell apoptosis. Mol Cell Endocrinol 2007;260:264–70.

[6] Carter WB, Sekharem M, Coppola D. Human chorionic gonadotropin induces apoptosis in breast cancer. Breast Cancer Res Treat 2006;100:S243–4.

[7] Iles RK. hCGβ expression by cervical squamous carcinoma: *in vivo* histological association with tumour invasion and apoptosis. Histopathologyology 2008;53:147–55.

[8] Kourides IA, Hoffman BJ, Landon MD. Difference in glycosylation between secreted and pituitary free alpha-subunit of the glycoprotein hormones. J Clin Endocrinol Metab 1980;51:1372–7.

[9] Blackman M, Gershengorn A, Weintraub BD. Excess production of free alpha subunits by mouse pituitary thyrotropic tumor cells *in vitro*. Endocrinology 1978;102:499–508.

[10] Prosillo EG, Hadwerger S, Tyrey L. Demonstration of intracellular and secreted forms of large human chorionic gonadotropin alpha subunit in cultures of normal placental tissue. Placenta 1983;4:439–48.

[11] Blithe DL, Iles RK. The role of glycosylation in regulating glycoprotein hormone free α-subunit and free β-subunit combination in the extraembryonic coelomic fluid of early pregnancy. Endocrinology 1995;136:903–10.

[12] Nemansky M, Thotokura NR, Lyons CD, Ya S, Reinhold BB, Reinhold VN, et al. Developmental changes in the glycosylation of glycoprotein hormone free α-subunit during pregnancy. J Biol Chem 1998;273:12068–76.

[13] Blithe DL. Carbohydrate composition of the α-subunit of human chorionic gonadotropin (hCGα) and the free α molecules produced in pregnancy: most free α and some combined hCGα molecules are fucosylated. Endocrinology 1990;126:2788–99.

[14] Cole LA. Immunoassay of hCG, its free subunits and metabolites. Clin Chem 1997;43: 2233–43.

[15] Erbel PJ, Karimi Nejad Y, De Beer T, Boelens R, Kamerling JP, Vliegenthart JF. Solution structure of the alpha-subunit of human chorionic gonadotropin. Eur J Biochem 1999;260:490–8.

[16] Nishimura R, Shin J, Ji I, Russell Middaugh C, Kruggel W, Lewis RV, et al. A single amino acid substitution in an ectopic α-subunit of a human choriogonadotropin. J Biol Chem 1986;261:10475–7.

[17] Cole LA, Perini F, Birken S, Ruddon RW. An oligosaccharide of the O-linked type distinguishes the free from the combined form of hCG α-subunit. Biochem Biophys Res Comm 1984;122:1260–7.

[18] Cole LA. Distribution of O-linked sugar units on hCG and its free α-subunit. Mol Cell Endocrinol 1987;50:45–57.

[19] Cole LA. Human chorionic gonadotropin (hCG), free α (free α), free β (free β) and β-core fragment (β-core). Diagn Endocrinol Metab 1997;15:199–220.
[20] Cole LA, Hartle RJ, Laferla JJ, Ruddon RW. Detection of the free β-subunit of human chorionic gonadotropin (hCG) in cultures of normal and malignant trophoblast cells, pregnancy serum, and sera of patients with choriocarcinoma. Endocrinology 1983;113:1176–8.
[21] Cole LA, Wang Y, Elliott M, Latif M, Chambers JT, Chambers SK, et al. Urinary human chorionic gonadotropin free beta-subunit and beta-core fragment: a new marker for gynecological cancers. Cancer Res 1988;48:1356–60.
[22] Nagelberg SB, Cole LA, Rosen SW. A novel form of ectopic human chorionic gonadotrophin β-subunit in the serum of a woman with epidermoid cancer. J Endocrinol 1985;107:403–8.
[23] Hussa RO, Fein HG, Tattillo RA, Nagelberg SB, Rosen SW, Weintraub BD, et al. A distinctive form of human chorionic gonadotropin β-subunit-like material produced by cervical cancer cells. Cancer Res 1986;46:1948–54.
[24] Cole LA, Nam J-H, Chambers JT, Schwarz PE. Urinary gonadotropin fragment, a new tumor marker. Gynecol Oncol 1990;36:391–4.
[25] Acevedo HF, Tong JY, Hartsock RJ. Human chorionic gonadotropin-beta subunit gene statement in cultured human fetal and cancer cells of different types and origins. Cancer 1995;76:1467–75.
[26] Acevedo HF, Hartstock RJ. Metastatic phenotype correlates with high expression of membrane-associated complete β-human chorionic gonadotropin *in vivo*. Cancer 1996;78:2388–99.
[27] McChesney R, Wilcox AJ, O' Connor JF, Weinberg CR, Baird DB, Schlatterer JP, et al. Intact HCG, free HCG β subunit and HCG β core fragment: longitudinal patterns in urine during early pregnancy. Hum Reprod 2005;20:928–35.
[28] Crawford RA, Iles RK, Carter PG, Caldwell CJ, Shepherd JH, Chard T. The prognostic significance of beta human chorionic gonadotropin and its metabolites in women with cervical carcinoma. J Clin Pathol 1998;51:685–8.
[29] Cole LA, Butler SA. Hyperglycosylated hCG, hCGβ, hyperglycosylated hCGβ: interchangeable cancer promoters. Mol Cell Endocrinol 2012;349:232–8.
[30] Berndt S, Blacher S, Munuat C, Detilleux J, Evain-Brion D, Noel A, et al. Hyperglycosylated human chorionic gonadotropin stimulates angiogenesis through TGF-ß receptor activation. FASEB J 2013; Abstract 12-213686.
[31] Ruddon RW, Hartle RJ, Peters BP, Anderson C, Huot RI, Stromberg K. The biosynthesis and secretion of chorionic gonadotropin subunits by organ cultures of first trimester human placenta. J Biol Chem 1981;256:11389–92.
[32] Valmu L, Alfhan H, Hotakainen K, Birken S, Stenman U-H. Site-specific glycan analysis of human chorionic gonadotropin 11-subunit from malignancies and pregnancy by liquid chromatography–electrospray mass spectrometry. Glycobiology 2006;16:1207–18.
[33] Butler SA, Laidler P, Porter JR, Keiman AT, Chard T, Cowan DA, et al. The β-subunit of hCG exists as a homodimer. J Mol Endocrinol 1999;22:182–92.
[34] Ozturk M, Berkowitz R, Goldstein D, Bellet D, Wands JR. Differential production of human chorionic gonadotropin and free subunit in gestational trophoblastic disease. Am J Obstet Gynecol 1988;158:192–8.
[35] Elliott MM, Kardana A, Lustbader JW, Cole LA. Carbohydrate and peptide structure of the α- and β-subunits of human chorionic gonadotropin from normal and aberrant pregnancy and choriocarcinoma. Endocrine 1997;7:15–32.
[36] Kobata A, Takeuchi M. Structure, pathology and function of the N-linked sugar chains of human chorionic gonadotropin. Biochim Biophys Acta 1990;1455:315–26.

Glycobiology of hCG

7

Akira Kobata
The Noguchi Institute, Tokyo, Japan

7.1 Characteristic features of the sugar chains of glycoproteins

Glycoproteins are the proteins to which various sugar chains are covalently linked. Two major types of sugar chains (N- and O-linked) are found in glycoproteins. N-linked sugar chains contain an *N*-acetylglucosamine (GlcNAc) residue at their reducing termini, which is linked to the amide group of an asparagine (Asn) residue of a polypeptide. O-linked sugar chains contain an *N*-acetylgalactosamine (GalNAc) residue at their reducing termini, which is linked to a serine (Ser) or threonine (Thr) residue of a polypeptide backbone. Accumulation of the structural data of various glycoproteins revealed that N-linked sugar chains include more structural rules than O-linked sugar chains.

All N-linked sugar chains contain the 3 mannose (Man), 2 GlcNAc pentasaccharide: Manα1–6(Manα1–3)Manβ 1–4GlcNAc β1–4GlcNAc as a common core, which is called *trimannosyl core* in this review. Based on the structures and locations of the extra sugar residues added to the trimannosyl core, N-linked sugar chains are further classified into three subgroups: complex type; high-mannose type; and hybrid type (Figure 7.1). Sugar chains classified as the complex type contain no mannosyl residues other than the trimannosyl core. Outer chains with a GlcNAc residue at their reducing terminal are linked to two Man residues of the trimannosyl core. The high-mannose sugar chains contain only α-mannosyl residues in addition to the trimannosyl core. A heptasaccharide with two branching structures, as enclosed by a dotted line in Figure 7.1, is commonly included in this type of sugar chain. Variations are formed by the locations and numbers of up to four Manα1–2 residues linked to the three nonreducing terminal α-mannosyl residues of the common heptasaccharide. Hybrid sugar chains were so named because the oligosaccharides have the structural characteristics of both high-mannose and complex sugar chains. One or two α-mannosyl residues are linked to the Man α1–6 arm of the trimannosyl core, as in the case of the high-mannose type, and the outer chains found in the complex sugar chains are linked to the Man α1–3 arm of the trimannosyl core. Presence or absence of an α-fucosyl residue linked to the C-6 position of the proximal GlcNAc residue and a GlcNAc residue linked to the C-4 position of the β-mannosyl residue of the trimannosyl core (called a bisecting GlcNAc) contribute to the structural variation of the complex and hybrid sugar chains [1].

Among the three subgroups of N-linked sugar chains, the complex type has the largest structural variation. This variation is formed mainly by two structural

Human Chorionic Gonadotropin (hCG). DOI: http://dx.doi.org/10.1016/B978-0-12-800749-5.00007-9

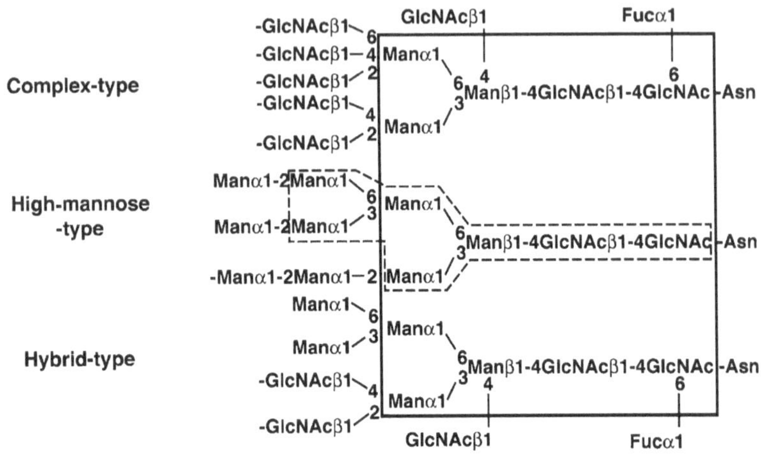

Figure 7.1 Three subgroups of N-linked sugar chains. Structures within the solid line are the pentasaccharide core common to all N-linked sugar chains. The bars in front of the nonreducing terminal monosaccharides indicate that the sugars can be further extended by adding sugars.
Source: Modified from Figure 6 in Ref. [1].

factors. Between one and five outer chains are linked to the trimannosyl core by different linkages (Figure 7.1), resulting in formation of monoantennary, biantennary, triantennary, tetra-antennary, and penta-antennary sugar chains (Figure 7.2).

Two isomeric triantennary sugar chains containing either the GlcNAc β1–4 (GlcNAc β1–2)Manα1–3 group or the GlcNAc β 1–6(GlcNAc β1–2)Manα 1–6 group can be found. These isomeric sugar chains are called 2,4- and 2,6-branched triantennary sugar chains, respectively. Starting from the β-*GlcNAc* residues located at the nonreducing termini of the oligosaccharides in Figure 7.2, various outer chains are formed. Combinations of the antennary with various outer chains can form a large number of different complex sugar chains.

In contrast to N-linked sugar chains, O-linked sugar chains have fewer structural rules. These sugar chains can be categorized into at least four groups according to their core structures (Figure 7.3). In addition, O-linked sugar chains with the GlcNAc β1–6GalNAc core and the GalNAc β1–3GalNAc core are also found in a limited number of glycoproteins.

7.2 Biosynthetic pathways of sugar chains of glycoproteins to form characteristic features

O-linked sugar chains are formed by stepwise addition of monosaccharides to the Ser and Thr residues of polypeptides from nucleotide sugars. In contrast, N-linked sugar chains are formed by a series of complex pathways including lipid-linked

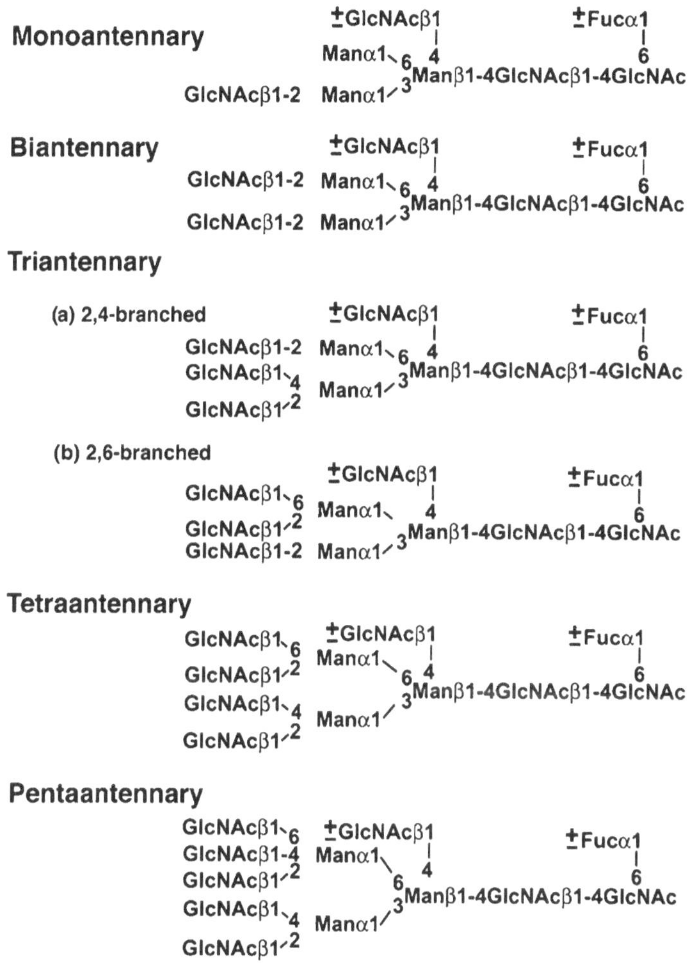

Figure 7.2 Branching of complex sugar chains.

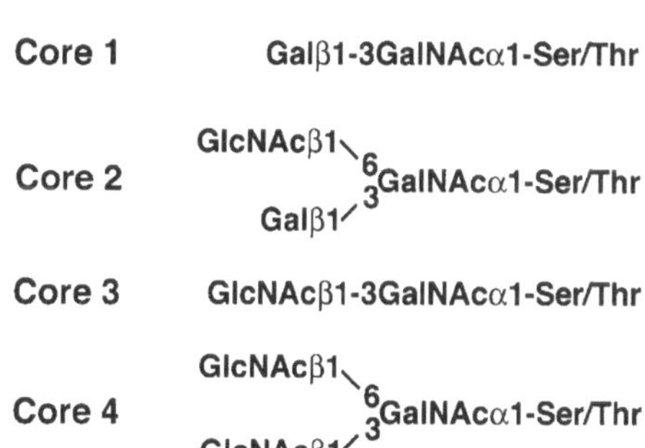

Figure 7.3 Four major core structures found in O-linked sugar chains.

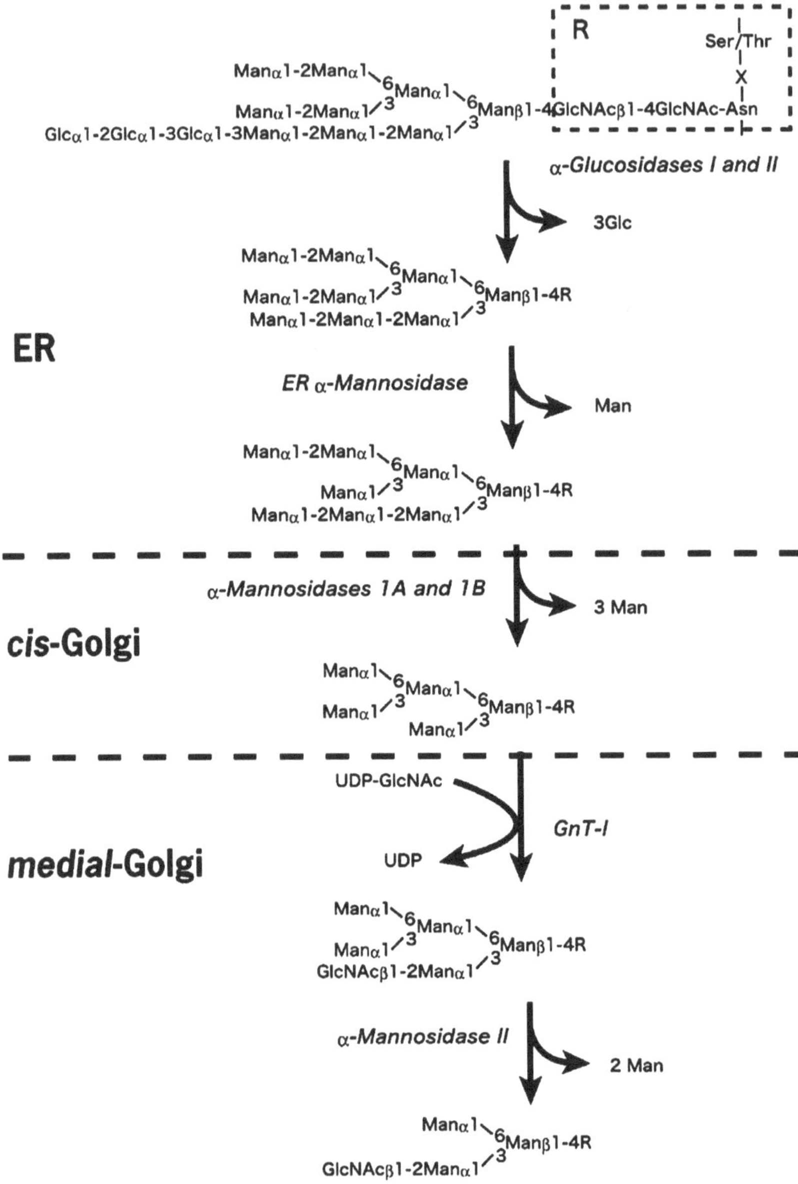

Figure 7.4 Processing pathway in the biosynthesis of N-linked sugar chains.

intermediates [2]. First, Glc_3-Man_9-$GlcNAc_2$-P-P-Dol is formed by a complicated process starting from dolichol phosphate (Dol-P) as an acceptor [3]. The tetradecasaccharide moiety of the lipid derivative is then transferred *en bloc* to the Asn residue of the polypeptide chain, which is translated in the rough endoplasmic reticulum by the catalytic action of an oligosaccharyltransferase complex residing in the endoplasmic membrane [4]. Only the Asn residue in the sequence of Asn-Xaa-Ser/Thr (where Xaa can be any amino acid other than proline) is glycosylated. Accordingly, these tripeptide sequences are called potential glycosylation sites. Asn-Xaa-Cys can be glycosylated equally well as Asn-Xaa-Ser/Thr [5]; however, very few of the tripeptide sequences in natural glycoproteins are glycosylated [6], probably because the SH group of the Cys residue quickly forms an S−S linkage with another Cys residue, and cannot contribute to the ring formation as in the case of an Asn-Xaa-Ser/Thr group.

The completely translated polypeptide with the tetradecasaccharide is then transported to the Golgi apparatus. During this transport, three α-glucosyl residues and at least one Manα1−2 residue are removed by the action of two α-glucosidases and an α-mannosidase residing in the membrane of the endoplasmic reticulum (Figure 7.4). After being translocated to the *cis*-Golgi, the N-linked sugar chain of the polypeptide is converted to Man 5-GlcNAc 2 by the action of Golgi α-mannosidase I, which removes all Manα1−2 residues from the sugar chain (Figure 7.4). A series of high-mannose sugar chains is considered to be the intermediary product of this trimming process. When the glycoprotein is translocated to the *medial*-Golgi, GlcNAc residue is added at the C-2 position of the Manα1−3 arm of the trimannosyl core by the action of *N*-acetylglucosaminyltransferase-I (GnT-I) [7]. Addition of this GlcNAc residue probably changes the steric arrangement of the two α-mannosyl residues linked to the Manα1−6 arm so that they can be removed by Golgi α-mannosidase II [7]. These are the entire features of the processing pathway that forms the prototype of monoantennary complex sugar chains.

Starting from a monoantennary complex sugar chain, a series of prototypes of the complex sugar chains is formed by the action of various GnTs (Figure 7.5). Each β-*N*-acetylglucosamine residue is further elongated by the action of various glycosyltransferases and sulfotransferases. Glycosyltransferases that catalyze these reactions have strict specificities for donor nucleotide sugars and acceptor sugar chain structures.

7.3 The hCG sugar chains from urine of pregnant women and placenta

hCG is a heterodimer composed of α- and β-subunits. Both subunits contain two N-linked sugar chains [8,9], and the β-subunit contains four O-linked sugar chains in addition [9]. Endo et al. investigated several hCG samples purified from pooled urine of healthy pregnant women, and elucidated the complete structures of the N-linked sugar chains of hCG [10]. As shown in Figure 7.6A, five sialylated oligosaccharides were found to occur in all samples. Occurrence of A-3 and A-5 in hCG was also independently reported by Kessler et al. [11]. On sialidase digestion, the five acidic oligosaccharides were converted to the three neutral oligosaccharides shown in Figure 7.6B. Studies of the hCG sample (purified from human placenta) revealed that it contains both sialylated and nonsialylated N-linked sugar chains; however, it

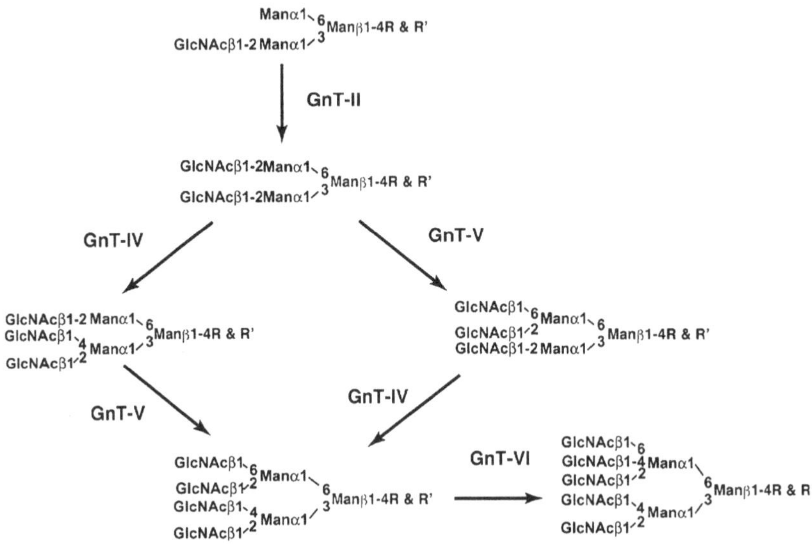

Figure 7.5 Formation of branching structures of complex sugar chains. R and R′ represent the GlcNAc β1−4GlcNAc and the GlcNAc β1−4(Fucα 1−6)GlcNAc groups, respectively.

gave oligosaccharides A, C, and D (shown in Figure 7.6B) in the molar ratio of approximately 1:2:1 after desialylation, as in the case of urinary hCG [12].

hCG can be dissociated into α- and β-subunits by treating with 8 M urea [13]. These subunits are termed hCGα and hCGβ in this chapter. A comparative study of the sugars released from the hCGα and the hCGβ of placental hCG by hydrazinolysis revealed that the oligosaccharides shown in Figure 7.6B are not evenly distributed in the subunits [12]. hCGα contained nearly equal amounts of oligosaccharides A and C, but did not contain D. In contrast, hCGβ contained nearly equal amounts of oligosaccharides C and D, but did not contain A. These results indicated site-specific *N*-glycosylation in hCG; the two N-linked sugar chains of α-subunit are never fucosylated, and one of them remains at the monoantennary stage. In contrast, both N-linked sugar chains of β-subunit are converted to biantennary complex sugars, and one of them is fucosylated.

7.4 Characteristic features of the sugar chains of free α-subunit

A small amount of free-floating α-subunit occurs in the urine of pregnant women and was aptly named free α-subunit. Interestingly, in contrast to the hCGα dissociated from hCG, this free α-subunit cannot bind to hCGβ. A study of the sugar chains of the free α-subunit revealed that it contains only one N-linked sugar chain [14]. A structural study of the oligosaccharides (released from free α-subunit by

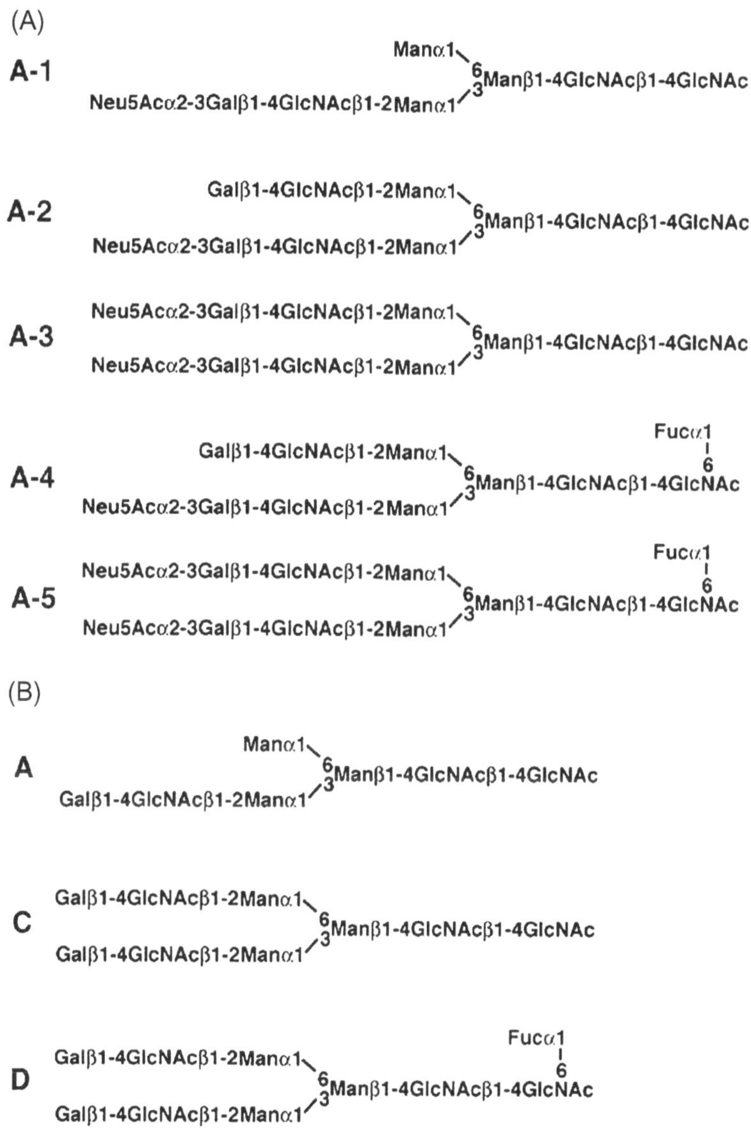

Figure 7.6 Structures of the N-linked sugar chains of hCG purified from the urine of pregnant women (A) and their desialylated forms (B).

hydrazinolysis) indicated that they were sialylated oligosaccharides C and D (Figure 7.6B) in a molar ratio of 91:9 [14].

Based on these findings, we hypothesized the biosynthetic mechanism of the N-linked sugar chains of hCG, as shown in Figure 7.7. Both α- and β-subunits of hCG (produced in the rough endoplasmic reticulum of the trophoblasts of placenta) have two tetradecasaccharides (Glc_3-Man_9-$GlcNAc_2$) at their potential *N*-glycosylation

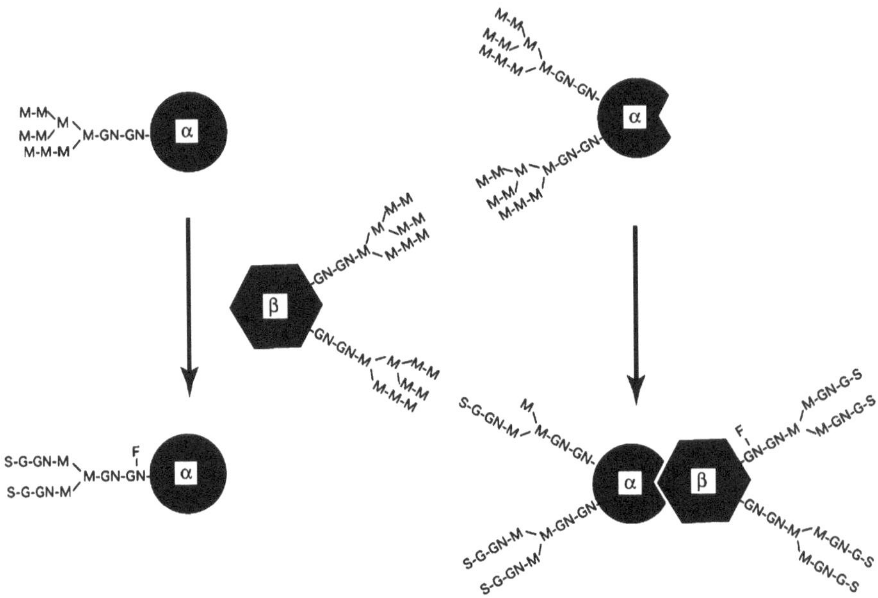

Figure 7.7 Maturation of the N-linked sugar chains of hCG and of free α-subunit. S, sialic acid; G, galactose; M, mannose; F, fucose; GN, *N*-acetylglucosamine.
Source: Data from Ref. [1].

sites as a result of the catalysis of oligosaccharyltransferase complex. These nascent sugar chains are converted to $Man_{8\sim9}$-$GlcNAc_2$ by the action of α-glucosidases and *ER*-α-mannosidase in the endoplasmic reticulum (Figure 7.4). During this early processing, a small portion of the α-subunit fails to accept Glc_3-Man_9-$GlcNAc_2$ at its one potential *N*-glycosylation site. This failure probably occurs because folding of the polypeptide moiety of the α-subunit hides one of the two potential *N*-glycosylation sites of the α-subunit, as found in the case of ovalbumin. Ovalbumin contains two potential *N*-glycosylation sites Asn292-Leu-Thr and Asn311-Leu-Ser in its polypeptide [15]; however, Asn311 is never glycosylated. Denatured ovalbumin can be further *N*-glycosylated by an *in vitro* system, indicating that Asn311 residue might be buried within the folded peptide, before accepting the tetradecasaccharide from the dolichol derivative.

The α-subunit with two Man $_{8\sim9}$-$GlcNAc_2$ will associate with a β-subunit, which also has two $Man_{8\sim9}$-$GlcNAc_2$. The four $Man_{8\sim9}$-$GlcNAc_2$ of the heterodimer will be processed to Manα 1−6 (GlcNAc β1−2Manα1−3)Manβ 1−4GlcNAc β 1−4GlcNAc when the heterodimer reaches the *medial*-Golgi. Maturation of the four hCG N-linked sugar chains to complex sugar chains will then be controlled by the steric effect of the polypeptide moieties of the two subunits. By this mechanism, one N-linked sugar chain of the α-subunit remains in the monoantennary state, whereas the other is converted to a biantennary complex sugar chain. In contrast, the control effect will allow the two N-linked sugar chains of the β-subunit to become a biantennary complex sugar chain, but will inhibit one sugar chain from being fucosylated.

The α-subunit with one $Man_{8\sim9}$-$GlcNAc_2$ cannot combine with a β-subunit; therefore, it reaches the Golgi as a free α-subunit. Because the maturation of its N-linked sugar chain is not controlled by the steric effect of the polypeptide moiety of the β-subunit, it will grow to an oligosaccharide D (Figure 7.6B) by the complete action of the glycosylation machinery of trophoblasts.

The work of Matzuk and Boime turned this hypothesis into a reality [16]. They prepared mutants of the α-subunit gene, in which one of the two *N*-glycosylation sites was eliminated by site-directed mutagenesis. Expression of each mutant gene in CHO cells, together with the gene of wild-type β-subunit of hCG, revealed that removal of the Asn52 *N*-glycosylation site significantly decreased the capacity of the α-subunit to bind with β-subunit; removal of the Asn78 site did not eliminate this binding capacity with β-subunit, but rather caused the mutant subunit to be degraded quickly in the absence of β-subunit. Therefore, the free α-subunit might lack the sugar chain at its Asn52.

7.5 Comparative studies of the N-linked sugar chains of hCG

Trophoblastic diseases are histologically classified as hydatidiform mole, invasive mole, and choriocarcinoma. The prognosis of these diseases differs greatly. Hydatidiform mole is considered a benign lesion, but some of the moles apparently show more malignant characteristics, such as invasion into surrounding tissues and metastasis. Because of these somewhat malignant characteristics, they are distinguished from typical moles by using the name "invasive mole." Choriocarcinoma shows the characteristics of a malignant tumor. A study of the oligosaccharides released from an hCG sample purified from the urine of a patient with choriocarcinoma revealed that none of the oligosaccharides was sialylated, although four moles of oligosaccharides were released from one mole of the hCG sample, as in the case of urinary hCG samples from healthy pregnant women [17]. Further study of the N-linked sugar chains of hCG samples purified from the urine of three additional patients with choriocarcinoma revealed that deletion of sialic acid is not a common phenomenon to choriocarcinoma hCG [18]; however, it was found that the eight oligosaccharides listed in Table 7.1 were included in all choriocarcinoma hCG samples as the neutral portions of the oligosaccharides [17,18].

Study of the N-linked sugar chains of urinary hCG samples purified from patients with hydatidiform mole revealed that these samples contain exactly the same N-linked sugar chains as urinary hCG samples obtained from healthy pregnant women [18]. The sugar chains were highly sialylated, and only oligosaccharides A, C, and D in Table 7.1 were detected in the molar ratio of approximately 1:2:1 after desialylation. Study of the N-linked sugar chains of urinary hCG samples obtained from patients with invasive mole revealed that these chains contain the six sialylated oligosaccharides shown in Table 7.1 [19]. The alteration of the N-linked sugar chains of choriocarcinoma hCG can be induced by the increase and expression of two enzymes.

Table 7.1 **Structures of the desialylated N-linked sugar chains of urinary hCG from pregnant women and from patients with trophoblastic diseases**

	Sugar chain structures	Pregnant women	Hydatidiform mole	Invasive mole	Choriocarcinoma
A	Manα1-6(Galβ1-4GlcNAcβ1-2Manα1-3)Manβ1-4GlcNAcβ1-4GlcNAc	+	+	+	+
B	Manα1-6(Galβ1-4GlcNAcβ1-2Manα1-3)Manβ1-4GlcNAcβ1-4(Fucα1-6)GlcNAc	–	–	+	+
C	Galβ1-4GlcNAcβ1-2Manα1-6(Galβ1-4GlcNAcβ1-2Manα1-3)Manβ1-4GlcNAcβ1-4GlcNAc	+	+	+	+
D	Galβ1-4GlcNAcβ1-2Manα1-6(Galβ1-4GlcNAcβ1-2Manα1-3)Manβ1-4GlcNAcβ1-4(Fucα1-6)GlcNAc	+	+	+	+
E	Manα1-6(Galβ1-4GlcNAcβ1-4(Galβ1-4GlcNAcβ1-2)Manα1-3)Manβ1-4GlcNAcβ1-4GlcNAc	–	–	–	+
F	Manα1-6(Galβ1-4GlcNAcβ1-4(Galβ1-4GlcNAcβ1-2)Manα1-3)Manβ1-4GlcNAcβ1-4(Fucα1-6)GlcNAc	–	–	–	+
G	Galβ1-4GlcNAcβ1-2Manα1-6(Galβ1-4GlcNAcβ1-4(Galβ1-4GlcNAcβ1-2)Manα1-3)Manβ1-4GlcNAcβ1-4GlcNAc	–	–	+	+
H	Galβ1-4GlcNAcβ1-2Manα1-6(Galβ1-4GlcNAcβ1-4(Galβ1-4GlcNAcβ1-2)Manα1-3)Manβ1-4GlcNAcβ1-4(Fucα1-6)GlcNAc	–	–	+	+

The molar ratios of total fucosylated oligosaccharides of the four choriocarcinoma hCG samples reach almost 50% of the total oligosaccharides, which are twice that of normal and hydatidiform mole hCG. This evidence indicates that the level of fucosyltransferase−8 (FUT−8) [20], which forms the Fuc α 1−6GlcNAc group, is increased in choriocarcinoma.

Apparently, oligosaccharides E, F, G, and H are formed by adding the Gal β1−4GlcNAc β1−4 group to oligosaccharides A, B, C, and D in Table 7.1, respectively. Therefore, GnT-IV [21], which is responsible for formation of the GlcNAc β1−4Manα1−3 group, should be ectopically expressed in choriocarcinoma. Although the enzyme is detected in many cells other than trophoblasts, oligosaccharides E and F have not been found in the glycoproteins produced by normal tissues (Figure 7.2). This indicates that GnT-IV in normal cells cannot catalyze pathway 2 in Figure 7.8. This evidence indicates that the ectopically expressed GnT-IV in choriocarcinoma acquired new characteristic to widen its substrate specificity.

Because hCG purified from the urine of invasive mole patients contains oligosaccharides G and H (Table 7.1), ectopic expression of GnT-IV also occurs in this lesion. However, the absence of oligosaccharides E and F in the hCG indicates that the newly expressed GnT-IV did not acquire wider substrate specificity. Based on the sum of this evidence, oligosaccharides E and F were named abnormal biantennary complex sugar chains. Occurrences of these sugar chains were later found in some glycoproteins produced by other tumors, such as γ-glutamyltranspeptidase purified from human hepatoma [22], and carcinoembryonic antigen obtained from liver metastases of primary colon cancers [23].

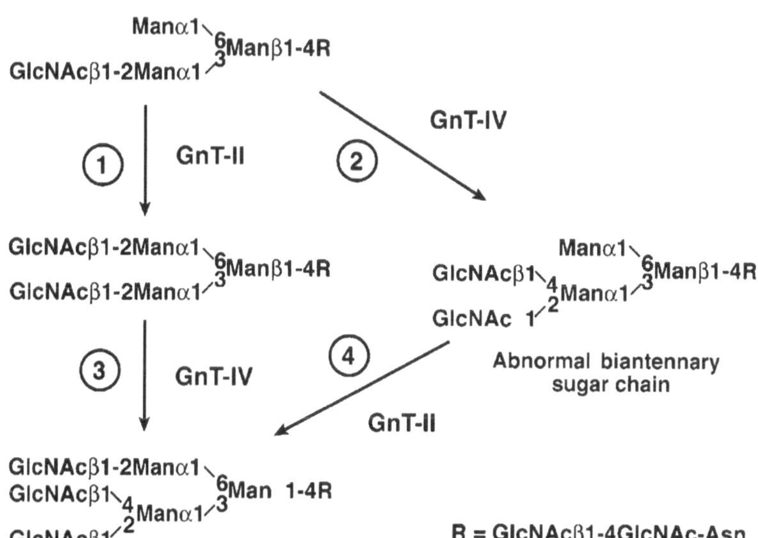

Figure 7.8 Biosynthesis of the abnormal biantennary sugar chains found in choriocarcinoma cells. GnT, *N*-acetylglucosaminyl transferase.

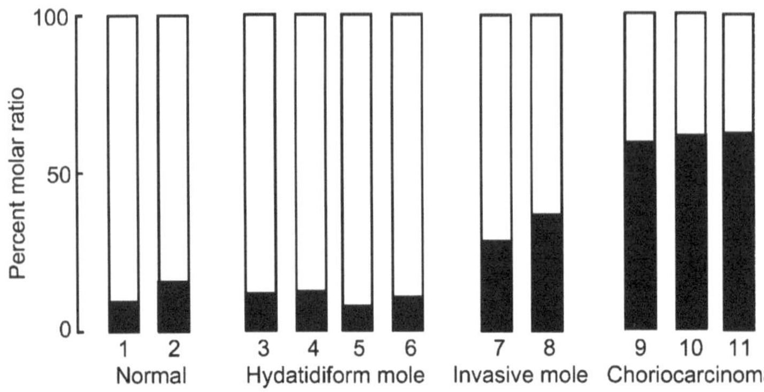

Figure 7.9 Percent molar ratio of the O-linked sugar chains with core 1 and core 2 (shown in white and black, respectively) in various urinary hCG samples.
Source: Data from Ref. [26].

7.6 Alteration induced in the O-linked sugar chains of hCG by malignant transformation of trophoblasts

As already described, the β-subunit of hCG contains four O-linked sugar chains. Structures of the O-linked sugar chains of hCG were elucidated as sialylated core 1 (shown in Figure 7.3) by Kessler et al. [24]. Cole et al. [25] later found that small amounts of sialylated core 2 (shown in Figure 7.3) are included as the O-linked sugar chains of hCG. Amano et al. [26] analyzed the O-linked sugar chains of hCG purified from urine of patients with a variety of trophoblastic diseases and found that tumor-related structural alterations were induced in the O-linked sugar chains as well; however, the alteration is not qualitative, as in the case of N-linked sugar chains, but rather is quantitative. The proportion of oligosaccharides with core 2 prominently increased in choriocarcinoma hCG (Figure 7.9). A moderate but significant increase was also observed in invasive mole hCG, but not in hydatidiform mole hCG.

7.7 Altered expression of GnT-IV in choriocarcinoma cells

Molecular biological studies of GnT-IV in placenta and choriocarcinoma cell lines have revealed some of the enzymatic background of the altered *N*-glycosylation of choriocarcinoma hCG. GnT-IV is expressed in many animal species [21,27–30]; however, GnT-IV activities in these cells are lower than those in other GnTs that are responsible for the branching of complex N-linked sugar chains (shown in Figure 7.5). Oguri et al. successfully purified GnT-IV from bovine small intestine [31]. As reported by Gleeson and Schachter [21], the purified enzyme required the presence of the GlcNAc β1–2Manα1–3 group in the acceptor oligosaccharides. Cloning of the gene encoding this enzyme revealed that the enzyme has a type II membrane-bound protein structure but has no homology with other cloned GnTs [32].

To investigate the enzymatic basis of the altered *N*-glycosylation of the hCG produced in choriocarcinoma cells, human cDNA for GnT-IV was obtained from a human liver cDNA library. Unexpectedly, two active GnT-IV genes with 91% and 64% homology to the bovine GnT-IV gene were obtained [33]. The translation products of these genes were named GnT-IVa and GnT-IVb, respectively. Expressions of the two genes in various organs were strikingly different. The GnT-IVb gene was expressed in all human organs at almost the same level, whereas the GnT-IVa gene was highly expressed in specific organs such as pancreas, thymus, small intestine, and leukocytes [34].

When activities of the glycosyltransferases related to the formation of the abnormal biantennary sugar chains were comparatively investigated in normal placenta and several choriocarcinoma cell lines, the GnT-IV activity was strikingly increased in the cancer cells [35]. GnT-III activity was also increased, although GnT-I, GnT-II, β1,4-galactosyltransferase and α-mannosidase II activities were not increased significantly from those of normal placenta. In addition, GnT-II activities in normal placenta and choriocarcinoma cells were found to be lower than the activities in other cells that do not produce monoantennary sugar chains [35].

Northern blot analysis revealed that the GnT-IVa gene was extraordinarily overexpressed in the cancer cells, whereas the GnT-IVb gene was expressed at the same level as in normal placenta [35]. So far, no difference in the substrate specificities of GnT-IVa and GnT-IVb has been found. It was also found that both enzymes produce the abnormal biantennary structures *in vitro*. The data described thus far indicate that low GnT-II activity and extremely high GnT-IV activity induced by overexpression of the GnT-IVa gene are the enzymatic basis of the formation of the abnormal biantennary sugar chains in choriocarcinoma cells. Although pathways 2 and 3 in Figure 7.8 were strongly enhanced, the monoantennary sugar chain will be mainly converted to the abnormal biantennary sugar chain because pathway 1 was very weak. The abnormal biantennary sugar chain is then derived to the triantennary sugar chain through pathway 4, which should be weak as well; however, these speculations cannot explain the subtle differences detected in the N-linked sugar chains of choriocarcinoma hCG and invasive mole hCG, as already discussed. Therefore, more detailed studies of the kinetic parameters of the enzymes and an investigation of their topology in the Golgi membrane of the cells of the two diseases must be performed in the future.

7.8 Glycosylated hCG as a diagnostic marker of trophoblastic diseases

Differential diagnosis of trophoblastic diseases is clinically important for the proper treatment of patients. Because oligosaccharides G and H (Table 7.1) were detected in both invasive mole hCG and choriocarcinoma hCG, any method that specifically detects the hCG containing the Galβ 1−4GlcNAc β1−4(Galβ 1−4GlcNAc β1−2) Man group in the sugar moieties could be useful in discriminating these patients from pregnant women or patients with hydatidiform mole.

By investigating the behavior of various complex oligosaccharides on several immobilized lectin columns, Yamashita et al. [36] found that these oligosaccharides can be separated into three groups by passing through a *Datura stramonium* agglutinin (DSA)–Sepharose column. The oligosaccharides, which are weakly bound to the column and eluted with buffer in the restricted fraction, all contain the nonsubstituted Galβ 1–4GlcNAc β1–4(Galβ 1–4GlcNAc β 1–2)Man group. The oligosaccharides, which are strongly bound to the column and eluted from the column with the buffer containing a 1% mixture of GlcNAc β1–4 oligomers, have either the nonsubstituted Galβ 1–4GlcNAc β1–6(Galβ 1–4GlcNAc β1–2)Man group or the nonsubstituted Gal β1–4GlcNAc β1–3Galβ 1–4GlcNAc β1 group as their partial structures. The oligosaccharides that contain none of these groups pass through the column without any interaction.

The binding specificity of the DSA – Sepharose column was expected to be useful for discriminating hCG with or without oligosaccharides G and H (Table 7.1), so the behavior of urinary hCG from various trophoblastic diseases on this column was investigated [37]. Almost all hCG in the urine of a pregnant woman passed through the column without interaction. The elution pattern did not change even after the urine was pretreated by sialidase digestion. In contrast, only a portion of hCG in the urine of a patient with choriocarcinoma passed through the column. The remainder was not eluted even with the buffer containing a 1% mixture of GlcNAc β 1–4 oligomers, but it was completely recovered by elution with 0.1 N acetic acid. This unexpectedly strong binding might have occurred because the hCG molecule contains at least two oligosaccharides with the Gal β 1–4GlcNAc β1–4(Galβ 1–4GlcNAc β1–2) Man group. Therefore, the elution step with the buffer containing a 1% mixture of GlcNAc β1–4 oligomers was omitted, and the amounts of hCG in the two fractions obtained by elution with simple buffer and 0.1 N acetic acid were measured to determine the percentage of hCG with the Galβ 1–4GlcNAc β1–4(Galβ 1–4GlcNAc β1–2) Man group in the sugar chains.

As shown in Figure 7.10, the values for normal pregnant women and patients with hydatidiform mole were less than 15%, and the values did not increase after sialidase digestion. The values for patients with invasive mole were also small, but sialidasetreated samples afforded much higher values. The data for patients with choriocarcinoma varied more than others. Some of them behaved very similarly to invasive mole hCG. However, one of the choriocarcinoma hCG samples bound completely to the column without sialidase treatment, indicating that some choriocarcinoma hCG lacked sialic acid residues. Therefore, affinity-column chromatography with use of a DSA – Sepharose can be used effectively to discriminate malignant hCG from nonmalignant hCG in urine samples. By assigning the cutoff value as 20%, desialylated hCG samples of patients with invasive mole and choriocarcinoma can be completely discriminated from those of patients with hydatidiform mole and pregnant women.

Birken et al. produced a monoclonal antibody B152 by using choriocarcinoma hCG as an antigen [38]. Although the exact epitope recognized by the antibody had not been clearly elucidated except for its possible occurrence in the COOH-terminal peptide region of the β-subunit [39], it was recently estimated that the antibody

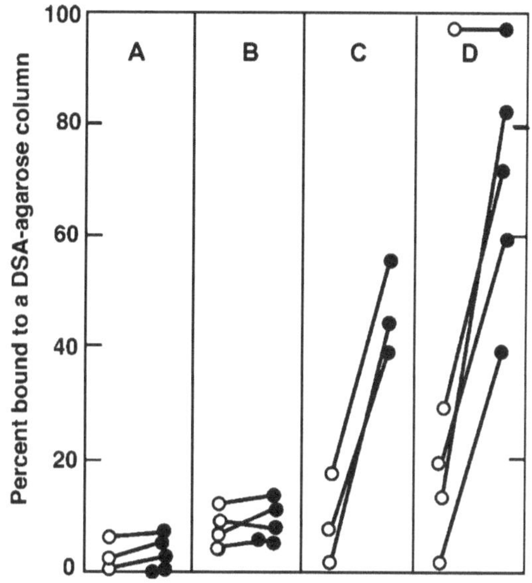

Figure 7.10 Percent molar ratio of urinary hCG bound to a DSA-Sepharose column before (0) and after (e) sialidase digestion. (A) Urine samples from normal pregnant women, (B) those from patients with hydatidiform mole, (C) those from patients with invasive mole, and (D) those from patients with choriocarcinoma.

recognizes core 2 O-linked sugar chains on Ser132 of the β-subunit [40]. Usefulness of this antibody for discriminating various trophoblastic diseases [41,42] and for the diagnosis of several abnormal pregnancies has been described [39,43–48].

7.9 Functional role of the hCG sialic acid residues

Many studies revealed that modification of the hCG N-linked sugar chains alters the hormonal activity of hCG [49–53]. These reports indicated that complete removal of sialic acid residues from hCG enhances its binding to the target cells but reduces its hormonal activity to approximately 50%. Removal of the whole N-linked sugar chains from hCG further increased the binding of hCG to its target cells but almost completely eliminated its hormonal activity. Deglycosylated hCG behaves as an antagonist to native hCG [49]. Calvo and Ryan [54] reported that the glycopeptides mixture (obtained from hCG and hCGα by exhaustive pronase digestion) blocks hCG signal transduction; thus, they suggested that a lectin-like membrane component in addition to hCG receptor might be involved in the signal transduction. Thotakura et al. reported that glycopeptides and oligosaccharides (obtained from other glycoproteins) can also prevent the hormonal action of hCG, and they suggested that a lectin-like polypeptide portion might be included in the hCG receptor itself [55]. After the study of the functional role of the N-linked sugar chains of α-subunit (as introduced in a previous section of this chapter) [16], Matzuk et al. [56,57] further indicated that all four N-linked sugar chains of hCG are important for constructing the correct conformation of the two subunits by comparatively investigating the bioactivities of hCG that lacked one or more of the four N-linked sugar chains by site-directed mutagenesis. The role of N-linked sugar chains in intracellular folding of hCG β was also investigated by Feng et al. [58].

To elucidate the mechanism of suppression of the hormonal activity of hCG by desialylation, Amano et al. [59] investigated the functional role of the sialic acid residues of hCG. It was found that all sialic acid residues of hCG occur as the Neu5Acα2−3Gal group [10]. To find out whether this particular disaccharide group is important in expressing the biological function of hCG, the following experiment was performed. MA-10 cells, a mouse Leydig tumor cell line established by Ascoli [60], produce cyclic AMP (cAMP) in response to the addition of hCG to their culture medium. When hCG was desialylated, its hormonal activity was reduced to approximately 50%, as in the case of the other target cell lines. When the desialylated hCG was resialylated by incubation with CMP-NeuAc and Gal β 1−4GlcNAc: α2−6sialyltransferase, the isomeric hCG containing the Neu5Acα 2−6Gal group thus obtained gave almost the same dose−response curve of cAMP production as the natural hCG [47]. These results indicated that the sialic acid residues of hCG are important for the full expression of their hormonal activity *in vitro*, but the effect is independent of their linkage to the galactose residues.

Nemansky et al. [61] confirmed this finding and provided additional important evidence. They found that the decrease of hormonal activity of hCG caused by desialylation was restored only by the addition of a Sia α2−6 residue, but not a Gal β1−3 residue, to the galactose moiety of the Galβ 1−4GlcNAc β1−2Manα 1−3Man arm of the N-linked sugar chains. Further α6-sialylation of the galactose residue of the Galβ 1−4GlcNAc β1−2Manα1−6Man arm reduced the hormonal activity of hCG, indicating that sialylation of the outer chain on the Manα1−3Man arm, rather than the Manα1−6Man arm, of the N-linked sugar chains of hCG plays an essential role in signal transduction. They also indicated that sialylation of the O-linked sugar chains is not important in LH/hCG receptor signaling.

To elucidate the role of sialic acid residues in the signal transduction of hCG, an *N*-acetylneuraminic acid hexamer, obtained from partial degradation of colominic acid, was added to the reaction mixture of hCG and MA-10 cells. The oligosaccharide did not inhibit the binding of hCG to the surface of the target cells, but cAMP production was reduced to 50% when 2 mM solution of the hexasaccharide was added [62]. These results indicated that the hexasaccharide can only inhibit the interaction of the sialic acid residues of hCG with the specific binding site on the cell surface, but it does not influence the binding of the peptide portion of hCG to the receptor. The possibility that the action of the sialic acid hexamer might be attributable to a nonspecific anionic polymer effect was ruled out because the addition of fucoidin did not show any inhibition of the [^{3}H]hCG binding to the cell surface receptor or cAMP production by hCG.

Presence of the sialic acid binding site on the MA-10 cells was confirmed by using 3′−sialyllactose-conjugated BSA as a probe [62]. Based on the data

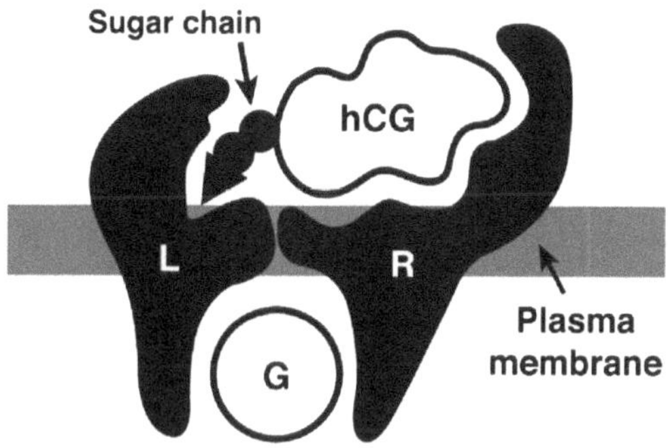

Figure 7.11 Schematic presentation of hCG-receptor complex. G, Gs protein; L, lectin; R, hCG receptor.
Source: Data modified from Ref. [62].

indicating that sialic acid residues bind directly to the cell surface, a model of the hCG-receptor complex was constructed, as shown in Figure 7.11. Dual interaction of the peptide portion and the sialylated N-linked sugar chain of hCG with respective binding sites is essential for signal transduction. However, it is still not clear whether the lectin-like membrane component is a part of the hormone receptor or a part of a different molecule, as shown in Figure 7.11. In connection with this, it is interesting that a region homologous to the soybean lectin was detected in the human hCG receptor [63].

Amano and Kobata [64] attempted to assign the N-linked sugar chains of hCG, which interact with the lectin-like component on MA-10 cells. By investigating which sugar chains are resistant to sialidase digestion in the presence of MA-10 cells, they found that the sialic acid residues linked to A and D (Figure 7.6B) became resistant to sialidase digestion by binding to the target cells. These results indicated that one of each sialylated N-linked sugar chains of hCGα and hCGβ are covered when the hormone binds to the target cells. Therefore, one or both of the sugar chains at these two sites should play a critical role in signal transduction in the hormonal action of hCG.

7.10 Future prospects

Several reports conflict with the data described so far in this review. Weisshaar et al. [65] reconfirmed the finding that site-specific distribution of the N-linked sugar chains occurs at the four *N*-glycosylation sites of hCG. They also found less strict distribution than that supposed by Mizuochi and Kobata 30 years ago (based on data of the sugar pattern analyses of commercial hCGα and hCGβ, which were guaranteed more than 99% pure) [12]. Weisshaar et al. also reported that small amounts of sialylated hybrid sugar chains Manα1−3Manα1−6(Galβ 1−4GlcNAc β1−2Manα1−3)Manβ 1−4GlcNAc β 1−4GlcNAc and Manα1−6(Manα1−3)

Manα1−6 (Galβ 1−4GlcNAc β1−2Manα1−3)Manβ 1−4GlcNAc β1−4GlcNAc are included as the sugar that is linked at Asn52 of hCGα [65]. These sugar chains could be included in the minor peaks of gel-permeation chromatography, which Mizuochi and Kobata neglected to analyze.

Several articles have reported the occurrence of triantennary sugar chains in normal hCG samples. However, these samples were purified from urine samples collected from many pregnant women. Because triantennary sugar chains occur in large amounts in urinary hCG from invasive mole and choriocarcinoma patients, it is possible that the larger sugar chains might have originated from trophoblastic disease hCG and contaminated the pooled urine. Our analyses of hCG samples purified from the urine of several healthy pregnant women never revealed the occurrence of triantennary sugar chains. Furthermore, the behaviors of desialylated hCG samples from urine of healthy pregnant women in a DSA − Sepharose column (see Figure 7.10) clearly indicated that no triantennary sugar chains occur in these samples.

Valmu et al. [40] analyzed the oligosaccharide profiles of the two *N*-glycosylation sites and four *O*-glycosylation sites of β-subunits dissociated from hCG samples purified from the urine of a patient with choriocarcinoma, a patient with invasive mole, and three pregnant women at different stages of fertilization by using liquid chromatography−electrospray mass spectrometry. The reported data provided much useful information. Most of the N-linked sugar chains linked to Asn13 were not fucosylated, whereas those linked to Asn30 were fucosylated. These results extended the information reported by Mizuochi and Kobata [12]; however, Valmu et al. also found triantennary and even small amounts of tetraantennary sugar chains in both sites. Regarding O-linked sugar chains, they reported the occurrence of site-specific distribution of sialylated core 1 and core 2 sugar chains. According to their data, approximately 25% of the O-linked sugar chains of pregnancy hCG contain core 2; this amount is much larger than that in the data shown in Figure 7.9 and by Elliott et al. [66] However, it must be pointed out that mass spectrometric analysis cannot yield quantitative data like the hydrazinolysis-radioactive labeling used in our studies.

Work published by Elliott et al. cannot be overlooked when addressing this problem [66]. By analyzing the carbohydrate structures of hCG samples purified from the urine of normal pregnant women, they also confirmed the occurrence of subunit-specific *N*-glycosylation, indicating that hCGα contains oligosaccharides A and C (Figure 7.6B) in the percent molar ratio of 36.7 and 49.3; hCGβ contains C and D as the major sugar chains. As described, Elliott et al. also reported that hCGα contains small amounts of oligosaccharides D and H (Table 7.1), and hCGβ contains a small amount of monoantennary and triantennary N-linked sugar chains. Because these researchers analyzed individual samples, the problem of trophoblastic-disease hCG contamination in pooled pregnancy urine sample could not be ruled out.

By using lectin affinity column chromatography, Skarulis et al. [67] found that the glycosylation pattern of hCG changes as gestation progresses. Therefore, more detailed structural analysis of the N-linked sugar chains of hCG produced at different stages of gestation must be performed to confirm this evidence.

The crystal structure of hCG was reported by Lapthorn et al. [68]. They indicated that only the sugar chains at Asn52 of hCGα are present at the interface of α- and β-subunits. The other three N-linked sugar chains are located on the outer face of the heterodimer molecule.

Purohit et al. [69] purified hCG samples from the culture media of insect cells transfected with the hCGα gene, which lacked the potential *N*-glycosylation site at Asn52 or Asn78, together with the intact hCG β gene. Their studies of the circular dichroism measurements and dissociation rates of the two mutant hCGs concluded that the absence of sugar chain at Asn52, but not at Asn78, resulted in conformational changes in the mutant. Based on this evidence, Purohit et al. considered that the loss of hormonal activity of hCG (lacking the sugar chain at Asn52) was probably attributable to a conformational change in the heterodimer rather than to the loss of interaction of the Asn52 sugar chain with the lectin on the target cells. A conformational study of the sugar chains of hCG in solution using NMR, however, revealed that the sugar chains at Asn52 appear to extend into solution [70,71]. These data support the dual-receptor theory described in the previous section of this chapter. Thijssen-van Zuylen et al. [72] later reported that the Asn52 linked sugar chain of the hCG is not susceptible to digestion with peptide *N*-glycosidase F, in contrast to that of free hCGα. Therefore, more information is required for discussing the actual conformation of the sugar chains of hCG in solution.

In 1998, Fukushima et al. [73] found that a part of choriocarcinoma patient serum hCG and hCG purified from the culture media of choriocarcinoma cell lines JEG-3 and BeWo binds to an immobilized *Trichosanthes japonica* agglutinin-I (TJA-I) column and an immobilized *Trichosanthes japonica* agglutinin-II (TJA-II) column; serum hCG from pregnant women completely passes through both columns without interaction. The TJA-I column specifically retains the oligosaccharides containing the Siaα2−6Galβ1−4GlcNAc group [74]; the TJA-II column retains the oligosaccharides containing the Fucα1−2Galβ1 group [75]. Based on these findings, Fukushima et al. thoroughly investigated the activities of glycosyltransferases in human placenta, JEG-3 cells, and BeWo cells. The α2−3sialyltransferase activities in the microsomal fractions of placenta and from the two choriocarcinoma cell lines were almost the same; however, the α2−6sialyltransferase activities of the two choriocarcinoma cell lines were much higher than that of placenta. The α1−2 fucosyltransferase activities in the microsomal fractions of placenta, JEG-3, and BeWo were 0.14, 0.60, and 0.29 nmol/h/mg of microsomal protein, respectively. These results indicated that the activities of α2−6sialyltransferase and of α1−2 fucosyltransferase are increased significantly as a result of malignant transformation of trophoblasts.

Measurement of the level of enzyme transcripts by competitive PCR revealed that ST6 Gal−I [76] transcripts in JEG-3 and BeWo cells are approximately 40-times higher than that in placenta; no difference was detected in the level of ST3Gal-III [77] or ST3Gal-IV [78] transcripts among the three samples. By using the same technique, it was confirmed that the levels of the fucosyltransferase I and II transcripts [79], which are responsible for formation of the Fuc α 1−2Gal group in JEG-3 cells and BeWo cells, were more than 20-times higher than in placenta.

Accordingly, the Siaα2−6Galβ1−4GlcNAc group and the Fuc α1−2Galβ 1−4GlcNAc group are expressed on the N-linked sugar chains of hCG by malignant transformation of trophoblasts, and these alterations could be effectively used for the diagnosis of choriocarcinoma.

More recently, Takamatsu et al. [80] reported additional interesting evidence regarding the N-linked sugar chains of hCG produced by malignant cells. As described in Section 7.8, both GnT-III and GnT-IV increased tremendously in choriocarcinoma cell lines [35]. Because bisected complex sugar chains have never been found in the N-linked sugar chains of hCG to date, Takamatsu et al. chose JEG-3 cells, which produce the largest amount of hCG among the choriocarcinoma cell lines, and investigated the structures of the N-linked sugar chains of hCG produced by this cell line. As reported by Mizuochi et al. [17,18], seven oligosaccharides (except for B in Table 7.1) were detected; however, they also found bisected C, D, G, and H as major components, in accordance with the enhanced expression of GnT-III in JEG-3 cells. Accordingly, altered expression of glycosyltransferases in trophoblasts by malignant transformation could be more complicated than discussed in this review.

References

[1] Kobata A. Structures and functions of the sugar chains of glycoproteins. Eur J Biochem 1992;209:483−501.

[2] Kornfeld R, Kornfeld S. Assembly of asparagine-linked oligosaccharides. Annu Rev Biochem 1985;54:631−64.

[3] Hemming FW. The coenzymic role of phosphodolichols. In: Montreuil J, Vliegenthart JFG, Schachter H, editors. Glycoproteins. Amsterdam: Elsevier; 1995. p. 127−44.

[4] Bause E, Hardt B. Oligosaccharyltransferase complex, Ribophorin-I, Ribophorin-II, OST48, and DAD1. In: Taniguchi N, Honke K, Fukuda M, editors. Handbook of glycosyltransferases and related genes. Tokyo: Springer; 2002. p. 577−86.

[5] Ronin C, Granier C, Caseti C, Bouchilloux S, Van Rietschaten J. Synthetic substrates for thyroid oligosaccharide transferase. Effects of peptide chain length and modifications in the Asn-Xaa-Thr-region. Eur J Biochem 1981;118:159−64.

[6] Stenflo J, Fernlund P. Amino acid sequence of the heavy chain of bovine protein C. J Biol Chem 1982;257:12180−90.

[7] Stanley P. *N*-acetylglucosaminyltransferase-I. In: Taniguchi N, Honke K, Fukuda M, editors. Handbook of glycosyltransferases and related genes. Tokyo: Springer; 2002. p. 61−9.

[8] Bellisario R, Carlsen RB, Bahl OP. Human chorionic gonadotropin: linear amino acid sequence of the α subunit. J Biol Chem 1973;248:6797−807.

[9] Carlsen RB, Bahl OP, Swaminathan N. Human chorionic gonadotropin: linear amino acid sequence of the β subunit. J Biol Chem 1973;248:6810−25.

[10] Endo Y, Yamashita K, Tachibana Y, Tojo S, Kobata A. Structures of the asparagine-linked sugar chains of human chorionic gonadotropin. J Biochem 1979;85:669−79.

[11] Kessler MJ, Reddy MS, Shah RH, Bahl OP. Structures of *N*-glycosidic carbohydrate units of human chorionic gonadotropin. J Biol Chem 1979;254:7901−8.

[12] Mizuochi T, Kobata A. Different asparagine-linked sugar chains on the two polypeptide chains of human chorionic gonadotropin. Biochem Biophys Res Commun 1980;97: 772–8.

[13] Swaminathan N, Bahl OP. Dissociation and recombination of the subunits of human chorionic gonadotropin. Biochem Biophys Res Commun 1970;40:422–7.

[14] Kawano T, Endo T, Nishimura R, Mizuochi T, Mochizuki M, Kochibe N, et al. Structural differences found in the sugar chains of eutopic and ectopic free α-subunits of human glycoprotein hormone. Arch Biochem Biophys 1988;267:787–96.

[15] McReynolds L, O'Malley BW, Nisbet AD, Fothergill JE, Givol D, Fields S, et al. Sequence of chicken ovalbumin mRNA. Nature 1978;273:723–8.

[16] Matzuk MM, Boime I. The role of the asparagine-linked oligosaccharides of the α subunit in the secretion and assembly of human chorionic gonadotropin. J Cell Biol 1988; 106:1049–59.

[17] Mizuochi T, Nishimura R, Derappe C, Taniguchi T, Hamamoto T, Mochizuki M, et al. Structures of the asparagine-linked sugar chains of human chorionic gonadotropin produced in choriocarcinoma: appearance of triantennary sugar chains and unique biantennary sugar chains. J Biol Chem 1983;258:14126–9.

[18] Mizuochi T, Nishimura R, Taniguchi T, Utsunomiya T, Mochizuki M, Derappe C, et al. Comparison of carbohydrate structure between human chorionic gonadotropin present in urine of patients with trophoblastic diseases and healthy individuals. Jpn J Cancer Res 1985;76:752–9.

[19] Endo T, Nishimura R, Kawano T, Mochizuki M, Kobata A. Structural differences found in the asparagine-linked sugar chains of human chorionic gonadotropins purified from the urine of patients with invasive mole and with choriocarcinoma. Cancer Res 1987;47:5242–5.

[20] Miyoshi E, Taniguchi N. α 6-Fucosyltransferase (FUT8). In: Taniguchi N, Honke K, Fukuda M, editors. Handbook of glycosyltransferases and related genes. Tokyo: Springer; 2002. p. 259–63.

[21] Gleeson PA, Schachter H. Control of glycoprotein synthesis. J Biol Chem 1983;258: 6162–73.

[22] Yamashita K, Totani K, Kuroki M, Matsuoka Y, Ueda I, Kobata A. Structural studies of the carbohydrate moieties of carcinoembryonic antigens. Cancer Res 1987;47: 3451–9.

[23] Yamashita K, Totani K, Iwaki Y, Takamizawa I, Tateishi N, Higashi T, et al. Comparative study of the sugar chains of γ-glutamyltranspeptidases purified from human hepatocellular carcinoma and from human liver. J Biochem 1989;10:728–35.

[24] Kessler MJ, Mise T, Ghai RD, Bahl OP. Structure and location of the *O*-glycosidic carbohydrate units of human chorionic gonadotropin. J Biol Chem 1979;254:7909–14.

[25] Cole LA, Birken S, Perini F. The structures of the serine-linked sugar chains on human chorionic gonadotropin. Biochem Biophys Res Commun 1985;126:333–9.

[26] Amano J, Nishimura R, Mochizuki M, Kobata A. Comparative study of the mucin-type sugar chains of human chorionic gonadotropin present in the urine of patients with trophoblastic diseases. J Biol Chem 1988;263:1157–65.

[27] Yamashita K, Tachibana T, Ohkura T, Kobata A. Enzymatic basis for the structural changes of asparagine-linked sugar chains of membrane glycoproteins of baby hamster kidney cells induced by polyoma transformation. J Biol Chem 1985;260:3963–9.

[28] Koenderman AH, Wijermans PW, van den Eijnden DH. Changes in the expression of *N*-acetylglucosaminyltransferase III, IV, V associated with the differentiation of HL-60 cells. FEBS Lett 1987;222:42–6.

[29] Nishikawa A, Gu J, Fujii S, Taniguchi N. Determination of *N*-acetylgluco saminyltransferases III, IV and V in normal and hepatoma tissues of rats. Biochim Biophys Acta 1990;1035:313–18.

[30] Ogier-Denis E, Bauvy C, Moutsia R, Aubery M, Codogno P. Increased UDP-GlcNAc: alpha-mannoside beta (1–4) *N*-acetylglucosaminyltransferase activity during chick embryo development. Biochim Biophys Acta 1990;1054:149–53.

[31] Oguri S, Minowa MT, Ihara Y, Taniguchi N, Ikenaga H, Takeuchi M. Purification and characterization of UDP-*N*-acetylglucosamine:alpha1,3-d-mannoside beta1,4-*N*-acetylglucosaminyltransferase (*N*-acetylglucosaminyltransferase-IV) from bovine small intesti. J Biol Chem 1997;272:22721–7.

[32] Minowa MT, Oguri S, Yoshida A, Hara T, Iwamatsu A, Ikenaga H, et al. cDNA cloning and expression of bovine UDP-*N*-acetylglucosamine: alpha1,3-d-mannoside beta1,4-*N*-acetylglucosaminyltransferase IV. J Biol Chem 1998;273:11556–62.

[33] Yoshida A, Minowa MT, Takamatsu S, Hara T, Ikenaga H, Takeuchi M. A novel second isoenzyme of the human UDP-*N*-acetylglucosamine:alpha1,3-d-mannoside beta1,4 *N*-acetylglucosaminyltransferase family: cDNA cloning, expression, and chromosomal assignment. Glycoconjugate J 1998;15:1115–23.

[34] Yoshida A, Minowa MT, Takamatsu S, Hara T, Oguri S, Ikenaga H, et al. Tissue-specific expression and chromosomal mapping of a human UDP-*N*-acetylglucosamine: alpha1,3d-mannoside beta1,4-*N*-acetylglucosaminyltransferase. Glycobiology 1999;9:303–10.

[35] Takamatsu S, Oguri S, Yoshida A, Minowa MT, Nakamura K, Takeuchi M, et al. Unusually high expression of *N*-acetylglucosaminyltransferase-IVa in human choriocarcinoma cell lines: a possible enzymatic basis of the formation of abnormal biantennary sugar chain. Cancer Res 1999;59:3949–53.

[36] Yamashita K, Totani K, Ohkura T, Takasaki S, Goldstein IJ, Kobata A. Carbohydrate binding properties of complex-type oligosaccharides on immobilized *Datura stramonium* lectin. J Biol Chem 1987;262:1602–7.

[37] Endo T, Iino K, Nozawa S, Iizuka A, Kobata A. Immobilized *Datura stramonium* agglutinin column chromatography, a novel method to discriminate the urinary hCG of patients with invasive mole and choriocarcinoma from those of normal pregnant women and patients with hydatidiform mole. Jpn J Cancer Res 1988;79:160–4.

[38] Birken S, Krichevsky A, O'Connor J, Schlatterer J, Cole LA, Kardana A, et al. Development and characterization of antibodies to a nicked and hyperglycosylated form of hCG from a choriocarcinoma patient. Endocrine 1999;10:137–44.

[39] Kovalevskaya G, Birken S, Kakuma T, Ozaki N, Sauer M, Lindheim S, et al. Differential expression of human chorionic gonadotropin (hCG) glycosylation isoforms in failing and continuing pregnancies: preliminary characterization of the hyperglycosylated hCG epitope. J Endocrinol 2002;172:497–506.

[40] Valmu L, Alfthan H, Hotakainen K, Birken S, Stenman UH. Site-specific glycan analysis of human chorionic gonadotropin β-subunit from malignancies and pregnancy by liquid chromatography–electrospray mass spectrometry. Glycobiology 2006;16:1207–18.

[41] Cole LA, Butler SA, Khanlian SA, Giddings A, Muller CY, Seckl MJ, et al. Gestational trophoblastic diseases: 2. Hyperglycosylated hCG as a reliable marker of active neoplasia. Gyn Oncol 2006;102:150–8.

[42] Cole LA, Muller Y. hCG in the management of quiescent and chemorefractory gestational trophoblastic diseases. Gyn Oncol 2010;116:3–9.

[43] Cole LA, Khanlian SA, Sutton JM, Davies S, Stephens N. Hyperglycosylated hCG (invasive trophoblast antigen, ITA), a key antigen for early pregnancy detection. Clin Biochem 2003;36:647–55.

[44] Cole LA, Shahabi S, Oz UA, Bahado-Singh RO, Mahoney MJ. Hyperglycosylated hCG (invasive trophoblast antigen) immunoassay: a new basis for gestational Down syndrome screening. Clin Chem 1999;45:2109–19.
[45] O'Connor JF, Ellish N, Kakuma T, Schlatterer J, Kovalevskaya G. Differential urinary gonadotropin profiles in early pregnancy and early pregnancy loss. Prenat Diagn 1998;18:1232–40.
[46] Sasaki Y, Ladner DG, Cole LA. Hyperglycosylated hCG: the source of pregnancy failures. Fertil Steril 2008;89:1871–86.
[47] Pandian R, Cole LA, Palomaki GE. Second-trimester maternal serum invasive trophoblast antigen: a marker for Down syndrome screening. Clin Chem 2004;50:1433–5.
[48] Bahado-Singh RO, Oz AU, Kingston JM, Shahabi S, Hsu CD, Cole LA. The role of hyperglycosylated hCG in trophoblast invasion and the prediction of subsequent preeclampsia. Prenat Dagn 2002;22:478–81.
[49] Moyle WR, Bahl OP, Maerz L. Role of carbohydrate of human chorionic gonadotropin in the mechanism of hormone action. J Biol Chem 1975;250:9163–9.
[50] Chen H-C, Shimohigashi Y, Dufau ML, Catt KJ. Characterization and biological properties of chemically deglycosylated human chorionic gonadotropin: role of carbohydrate moieties in adenylate cyclase activation. J Biol Chem 1982;257:14446–52.
[51] Manjunath P, Sairam MR. Biochemical, biological, and immunological properties of chemically deglycosylated human choriogonadotropin. J Biol Chem 1982;257:7109–15.
[52] Amano J, Nishimura R, Sato S, Kobata A. Altered glycosylation of human chorionic gonadotropin decreases its hormonal activity as determined by cyclic-adenosine 3′,5′ monophosphate production in MA-10 cells. Glycobiology 1990;1:45–50.
[53] Goverman JM, Parson THF, Pierce JG. Enzymatic degradation of the subunits of chorionic gonadotropin: effects on formation of tertiary structure and biological activity. J Biol Chem 1982;257:15059–64.
[54] Calvo FO, Ryan RJ. Inhibition of adenylyl cyclase activity in rat corpora luteal tissue by glycopeptides of human chorionic gonadotropin and the alpha-subunit of hCG. Biochemistry 1985;24:1953–9.
[55] Thotakura NR, Weintraub BD, Bahl OP. The role of carbohydrate in human choriogo nadotropin (hCG) action: effects of N-linked carbohydrate chains from hCG and other glycoproteins on hormonal activity. Mol Cell Endocrinol 1990;70:263–72.
[56] Matzuk MM, Boime I. Site-specific mutagenesis defines the intracellular role of the asparagine-linked oligosaccharides of chorionic gonadotropin β subunit. J Biol Chem 1988;263:17106–11.
[57] Matzuk MM, Keene JL, Boime I. Site specificity of the chorionic gonadotropin N-linked oligosaccharides in signal transduction. J Biol Chem 1989;264:2409–14.
[58] Feng W, Matzuk MM, Mountjoy K, Bedows E, Ruddon RW, Boime I. The asparagine-linked oligosaccharides of the human chorionic gonadotropin beta subunit facilitate correct disulfide bond pairing. J Biol Chem 1995;270:11851–9.
[59] Amano J, Sato S, Nishimura R, Mochizuki M, Kobata A. Sialic acids, but not their linkage to galactose residues, are required for the full expression of the biological activity of human chorionic gonadotropin. J Biochem 1989;105:339–40.
[60] Ascoli M. Characterization of several clonal lines of cultured leydig tumor cells: gonadotropin receptors and steroidogenic responses. Endocrinology 1981;108:88–95.
[61] Nemansky M, DeLeeuw R, Wijnands RA, van den Eijnden DH. Enzymic remodeling of the N- and O-linked carbohydrate chains of human chorionic gonadotropin: effects on biological activity and receptor binding. Eur J Biochem 1995;227:880–8.

[62] Amano J, Kobata A. Direct interaction of the sialic acid residue of human lutropin and chorionic gonadotropin with target cell is necessary for the full expression of their hormonal action. Arch Biochem Biophys 1993;305:618–21.
[63] McFarland KC, Sprengel R, Phillips HS, Kohler M, Rosenblit N, Nikolics K, et al. Lutropin-choriogonadotropin receptor: an unusual member of the G protein-coupled receptor family. Science 1989;245:494–9.
[64] Amano J, Kobata A. Structures and function of the N-linked sugar chains of glycohormones. In: Biopolymers and bioproducts: structures, function and application, Bangkok: Samakkhisan Public C; 1995. pp. 210–17.
[65] Weisshaar G, Hiyama J, Renwick AGC. Site-specific *N*-glycosylation of human chori onic gonadotropin-structural analysis of glycopeptides by one- and two-dimensional 1H NMR spectroscopy. Glycobiology 1991;1:393–404.
[66] Elliott MM, Kardana A, Lustbader JW, Cole LA. Carbohydrate and peptide structure of alpha- and beta-subunits of human chorionic gonadotropin from normal and aberrant pregnancy and choriocarcinoma. Endocrine 1997;7:15–32.
[67] Skarulis MC, Wehmann RE, Nisula BC, Blithe DL. Glycosylation changes in human chorionic gonadotropin and free alpha subunit as gestation progresses. J Clin Endocrinol Metab 1992;75:91–6.
[68] Lapthorn AJ, Harris DC, Littlejohn A, Lustbader JW, Canfield RE, Machin KJ, et al. Crystal structure of human chorionic gonadotropin. Nature 1994;369:455–61.
[69] Purohit S, Shao K, Balasubramanian SV, Bahl OP. Mutant of human choriogonadotropin lacking *N*-glycosyl chains in the alpha-subunit. 1. Mechanism for the differential action of the N-linked carbohydrates. Biochemistry 1997;36:12355–63.
[70] Weller CT, Lustbader J, Seshadri K, Brown JM, Chadwick CA, Kolthoff CE, et al. Structural and conformational analysis of glycan moieties in situ on isotopically 13C, 15N-enriched recombinant human chorionic gonadotropin. Biochemistry 1996;35: 15–23.
[71] DeBeer T, van Zuylen CW, Leeflang BR, Hard K, Boelens R, Kaptein R, et al. NMR studies of the free alpha subunit of human chorionic gonadotropin: structural influences of *N*-glycosylation and the beta subunit on the conformation of the alpha subunit. Eur J Biochem 1996;241:229–42.
[72] Thijssen-van Zuylen CWEM, de Beer T, Leeflang BR, Boelens R, Kaptein R, Kamerling JP, et al. Mobilities of the inner three core residues and the man(alpha 1–6) branch of the glycan at Asn78 of the alpha-subunit of human chorionic gonadotropin are restricted by the protein. Biochemistry 1998;37:1933–40.
[73] Fukushima K, Hara-Kuge S, Seko A, Ikehara Y, Yamashita K. Elevation of α 2–6 sialyltransferase and α 1–2 fucosyltransferase activities in human choriocarcinoma. Cancer Res 1998;58:4301–6.
[74] Yamashita K, Umetsu K, Suzuki T, Ohkura T. Purification and characterization of a Neu5Acα2–6Galβ 1–4GlcNAc and HSO 3-6Galβ1–4GlcNAc specific lectin in tuberous roots of *Trichosanthes japonica*. Biochemistry 1992;31:11647–50.
[75] Yamashita K, Ohkura T, Umetsu K, Suzuki T. Purification and characterization of a Fucα1–2Galβ1- and GalNAc β1-specific lectin in root tubers of *Trichosanthes japonica*. J Biol Chem 1992;267:25414–22.
[76] Hamamoto T, Tsuji S. ST3Gal-I. In: Taniguchi N, Honke K, Fukuda M, editors. Handbook of glycosyltransferases and related genes. Tokyo: Springer; 2002. p. 295–300.
[77] Kitazume-Kawaguchi S, Tsuji S. ST3Gal-III. In: Taniguchi N, Honke K, Fukuda M, editors. Handbook of glycosyltransferases and related genes. Tokyo: Springer; 2002. p. 279–83.

[78] Kitazume-Kawaguchi S, Tsuji S. In: Taniguchi N, Honke K, Fukuda M, editors. Handbook of glycosyltransferases and related genes. Tokyo: Springer; 2002. p. 284–8.
[79] Oriol R, Mollicone R. α 2-Fucosyltransferases (FUT1, FUT2, and Sec1). In: Taniguchi N, Honke K, Fukuda M, editors. Handbook of glycosyltransferases and related genes. Tokyo: Springer; 2002. p. 205–17.
[80] Takamatsu S, Katsumata T, Inoue N, Watanabe T, Fujibayashi Y, Takeuchi M. Abnormal biantennary sugar chains are expressed in human choriocarcinoma cell line JEG-3. Glycoconj J 2004;20:473–81.
[81] Moremen KW. α-Mannosidase-II. In: Taniguch N, Honke K, Fukuda M, editors. Handbook of glycosyltransferases and related genes. Tokyo: Springer; 2002. p. 600–6.

Detecting hCG and hCG variants using MALDI-ToF-MS

8

Stephen A. Butler and Ray K. Iles
MAP Diagnostics, BioPark, Broadwater Road, Welwyn Garden City, Hertfordshire, UK

For more than 50 years there was one kind of hCG, a gonadotropic hormone produced by the trophoblast cells of the placenta. With developments in immunoassay technology toward the end of the 1970s, the free β-subunit and free α-subunit could be measured independently and it became possible to distinguish them from the heterodimeric hormone hCG. Now, hCG is considered to be a very heterogeneous molecule with variation in peptide sequence and, to a greater extent, variation in glycosylation, both of which are described in detail in the chapters of this book.

The mass of hCG is often reported as approximately 36 kDa, but the mass range of the intact holohormone is possibly between 34 and 40 kDa (Figures 8.1–8.3). As such, hCG is now believed to be a collective term that encompasses a range of

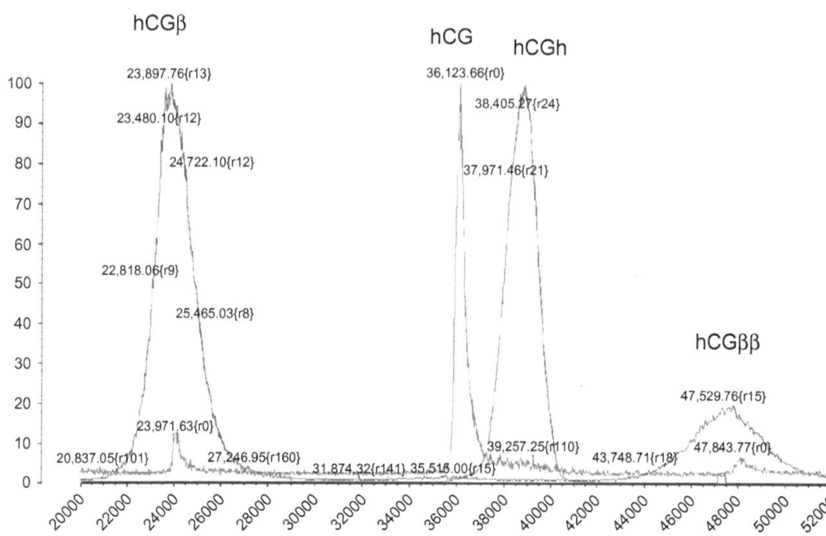

Figure 8.1 MALDI-ToF-MS obtained from the overlay of three spectra from the three hCG standards. Recombinant hCGβ in the peak at 23,897.76 m/z, recombinant hCG in the peak at 36,123.66 m/z, hyperglycosylated hCG (obtained from the collection described in Ref. [1]) in the peak at 38,405.27 m/z, and hCGββ dimer seen in the peak at 47529.76 m/z as described in Ref. [2].
Source: Reproduced and modified from Ref. [3].

Human Chorionic Gonadotropin (hCG). DOI: http://dx.doi.org/10.1016/B978-0-12-800749-5.00008-0

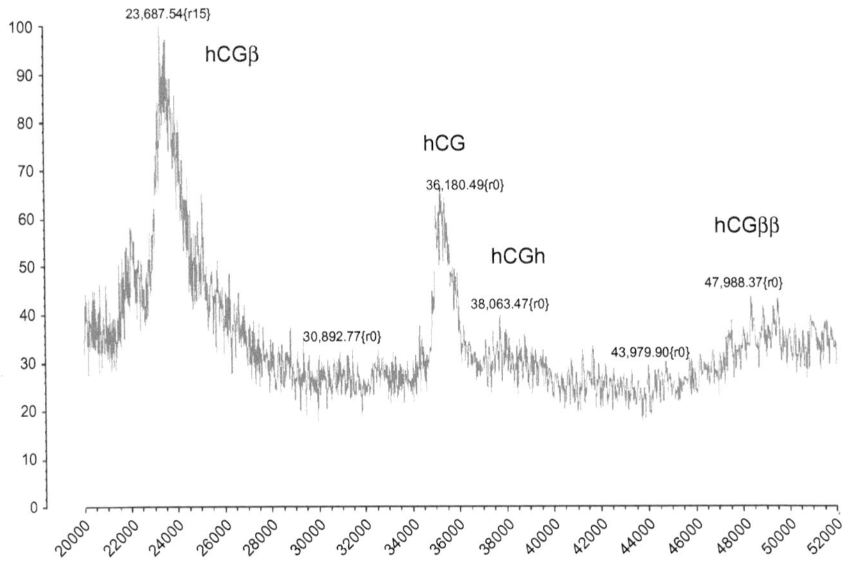

Figure 8.2 MALDI-ToF-MS obtained from the analysis of embryo culture fluid of a blastocyst before embryo transfer. All forms of hCG can be seen: hCGβ at 23,687.54 m/z; hCG at 36,180.49 m/z; hCGh at 38,063.47; and hCGββ dimers at 47,988.37 m/z. Poor spectral quality is attributable to the pmol concentrations of hCG found within the embryo secretome.
Source: Reproduced and modified from Ref. [3].

molecules with essentially the same peptide structure but variable glycosylation. Variants recognized by the IFCC include hCG, hCGβ, hCGα, hCGn, and hCGβcf, which are detailed throughout this book. Other variants that are not formally recognized (because no international standard has ever been produced) include hyperglycosylated hCG, sulfated hCG (or pituitary hCG), familial hCG, and quiescent hCG.

Not surprisingly, this heterogeneity has led to some confusion. Significant efforts were made by both the IFCC and ISOBM working groups regarding hCG standardization to characterize individual variants of hCG based on structure and epitope mapping and to develop standards to formalize the quantification of hCGs. The groups managed to prepare and characterize new standards for the recognized hCG variants (detailed in Chapter 37). To date, the structure of hCG has been characterized by HPLC mass spectrometry and by crystallographic analysis. Using trypsin digestion, peptide mass mapping of hCG and its subunits has also been performed using matrix-assisted laser desorption/ionization time-of-flight mass spectrometry (MALDI-ToF-MS) [4,5].

However, little further progress has been made in recent years and new variants keep being described. The official standards remain those detailed in this chapter. Interestingly, some of these nonstandard variants of hCG have been shown to be of potential clinical use, indicating disorders in pregnancy and cancer; most notable of these is hCGh in early pregnancy or choriocarcinoma. In some cases these multiple

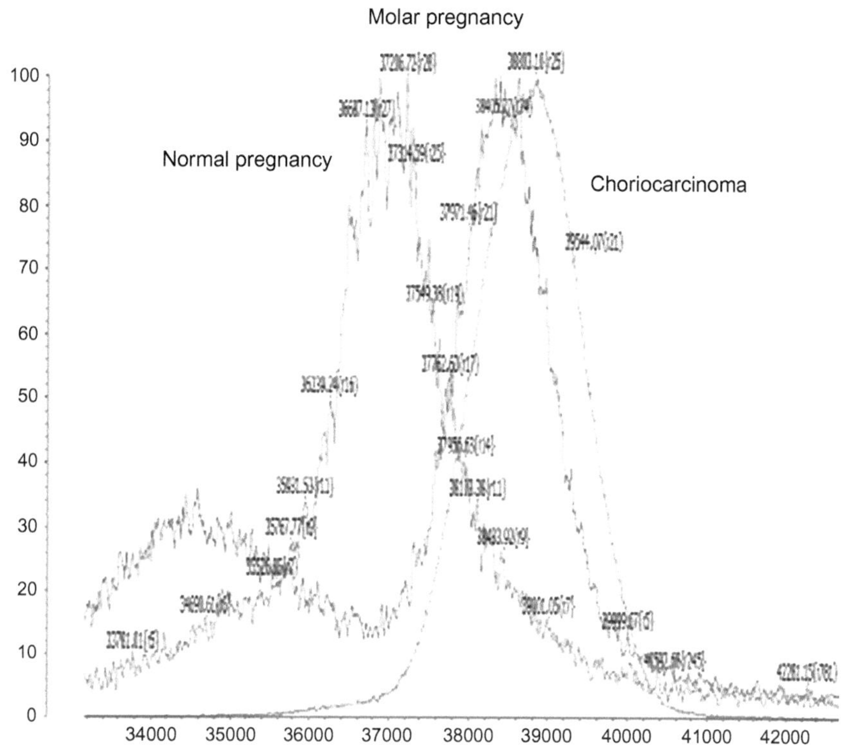

Figure 8.3 MALDI-ToF-MS obtained from the overlay of three spectra from the three samples analyzed in this example: normal, molar, and choriocarcinoma pregnancy urine samples. The neat normal pregnancy urine shows a peak at 36,687 m/z and represents the normal condition. The molar pregnancy urine shows a peak at 38,405 m/z within a range of 37,600–39,200 m/z and the choriocarcinoma urine sample shows a peak at 38,803 m/z, which lies within the range of 37,900–39,600 m/z.

variants have led to problems in management of pregnancy and cancer, with poor assay selection resulting in false-positive results, most notably Phantom hCG, and false-negative results when the Ab pairing is too restrictive to ensure detection of all forms or at least the variants of interest [6]. It has been difficult to quantify specific subtypes when, broadly speaking, the peptides (or at least key epitopes within the peptides) are very similar and developing highly specific antibodies to these variants has been fruitless; B151 for hCGn and B152 for hCGh are the last notable successes. Molecular biology approaches have failed to significantly address the variability and specificity problems associated with antibodies or antibody pairings in protein studies. This is largely because the *CGB* gene cluster is unsurprisingly very complex with little or no sequence differences between the nine genes, pseudogenes, and alleles.

Carbohydrate heterogeneity within hCG has been reported previously, with variable monoantennary, biantennary, and triantennary carbohydrate structures being found in normal and abnormal pregnancies [1,7]. In previous studies we described the purification of hCGβcf from pregnancy urine that was reduced by dithiothreitol (DTT) reduction and analyzed by MALDI-ToF-MS. It was concluded that the mass spectrum of DTT-reduced hCGβcf, although not precisely defining hCGβcf glycosylation, would appear to result in a distinctive fingerprint [8]. Recently, we showed that variability in hCGβcf could now be demonstrated using a direct MALDI-ToF-MS approach [9] and that by analyzing certain m/z regions we could accurately diagnose disorders in pregnancy and cancer.

MALDI-ToF-MS is not a quantitative technique, and this is seen as a critical disadvantage when the raw mass spectra produced cannot be used directly to produce comparative data for diagnostic tests. However, hCG is an abundant molecule in the urine of patients who are pregnant or have cancer, and sample analysis by MALDI presents an opportunity to identify qualitative changes in hCG, which may be indicative of clinical disorders. More than 15 years ago, we first used MALDI-ToF-MS to describe the occurrence of hCGbb homodimers [2]; since then, we have been able to optimize a method to quickly and reliably study mass shifts in hCG and its variants and use this information to identify disorders in pregnancy and cancer.

The research is still in the early stages of development, but it has been possible to identify variation in mass standards of hCG, hCGb, and hCGh (see Figure 8.1) and compare this to the profiles of hCG in culture fluid produced by embryos during assisted reproduction (see Figure 8.2 from [3]). It has also been possible to assess profiles of the hCG produced in normal pregnancy compared with that produced in the urine of patients with a normal gestation, a molar pregnancy, and choriocarcinoma (see Figure 8.3). In some cases, it is possible to see baseline separations between hCG variant forms (Figure 8.1); in other cases, there are clear distinctions in peak masses, even when there is considerable overlap at the baseline (Figure 8.3). Using more complex approaches, we are now able to identify disorders in pregnancy, such as Down syndrome, from a single spot urine test. Inevitably, as MALDI technology advances, the distinctions between masses of hCG variants are likely to improve and may prove useful as a companion diagnostic to more traditional hCG immunoassays by providing more qualitative information to the clinical chemist and clinician.

The authors thank Dr. Jameel Lutoo for his assistance in the analysis and presentation of hCG profiles using MALDI-ToF-MS.

References

[1] Elliott MM, Kardana A, Lustbader JW, Cole LA. Carbohydrate and peptide structure of the alpha- and beta-subunits of human chorionic gonadotropin from normal and aberrant pregnancy and choriocarcinoma. Endocrine 1997;7(1):15–32.

[2] Butler SA, Laidler P, Porter JR, Kicman AT, Chard T, Cowan DA, et al. The beta-subunit of human chorionic gonadotrophin exists as a homodimer. J Mol Endocrinol 1999;22(2):185–92.

[3] Butler SA, Luttoo J, Freire MO, Abban TK, Borrelli PT, Iles RK. Human chorionic gonadotropin (hCG) in the secretome of cultured embryos: hyperglycosylated hCG and

hCG-free beta subunit are potential markers for infertility management and treatment. Reprod Sci 2013;20(9):1038–45.

[4] Laidler P, Cowan DA, Hider RC, Keane A, Kicman AT. Tryptic mapping of human chorionic gonadotropin by matrix-assisted laser desorption/ionization mass spectrometry. Rapid Commun Mass Spectrom 1995;9(11):1021–6.

[5] Valmu L, Alfthan H, Hotakainen K, Birken S, Stenman UH. Site-specific glycan analysis of human chorionic gonadotropin beta-subunit from malignancies and pregnancy by liquid chromatography—electrospray mass spectrometry. Glycobiology 2006;16(12): 1207–18.

[6] Cole LA, Shahabi S, Butler SA, Mitchell H, Newlands ES, Behrman HR, et al. Utility of commonly used commercial hCG immunoassays in the diagnosis and management of trophoblastic diseases. Clin Chem 2001;47(2):308–15.

[7] Stenman UH, Tiitinen A, Alfthan H, Valmu L. The classification, functions and clinical use of different isoforms of HCG. Hum Reprod Update 2006;12(6):769–84.

[8] Jacoby ES, Kicman AT, Laidler P, Iles RK. Determination of the glycoforms of human chorionic gonadotropin beta-core fragment by matrix-assisted laser desorption/ionization time-of-flight mass spectrometry. Clin Chem 2000;46(11):1796–803.

[9] Iles RK, Cole LA, Butler SA. Direct analysis of hCGβcf glycosylation in normal and aberrant pregnancy by matrix-assisted laser desorption/ionization time-of-flight mass spectrometry. Int J Mol Sci 2014;15(6):10067–82.

The hCG/LH hormone receptor

9

Laurence A. Cole[1] *and Stephen A. Butler*[2]
[1]USA hCG Reference Service, Angel Fire, NM, USA, [2]MAP Diagnostics, BioPark, Broadwater Road, Welwyn Garden City, Hertfordshire, UK

hCG binds a common receptor with luteinizing hormone (LH). This receptor is located on ovarian corpus luteal cells for promotion of progesterone production, on myometrial blood vessels for promotion of angiogenesis, on cytotrophoblast and syncytiotrophoblast cells for enhancement of trophoblast cell differentiation, on myometrial muscle cells for suppression of contractions, on trophoblast cells for immunosuppression and macrophage suppression for foreign invading cells, on multiple fetal cell tissues for enhancement of growth, and on testicular Leydig cells for promoting production of testosterone in men (and other tissues as described in Chapter 15). The hCG/LH receptor, as it has become known, binds regular hCG and partially binds hyperglycosylated hCG [1]. Although the receptor is not activated by the hCG free α-subunit or free β-subunit, there is clear evidence showing that the α-subunit has a role in receptor binding and that the β-subunit has a function in receptor specificity [2,3].

The human hCG/LH receptor is a protein composed of 675 amino acids [4,5]. The rat receptor is similar (90% homology) and is composed of 674 amino acids [6]. The hCG/LH receptor is coded for by a gene 70,000 bases in length on human chromosome 2p21, with 11 exons and 10 introns [7–9], as a single transcript; exons 1–10 and a portion of exon 11 encode the extracellular domain [10]. When thyroid-stimulating hormone, follicle-stimulating hormone, and LH receptors are compared, it is noted that, within the primary sequence, there is conservation of nine cysteines that appear in their extracellular domains and nine that appear in their transmembrane/cytoplasmic domains. This suggests a common folding pattern through the disulfide pairing that exists among all these glycoprotein hormone receptors [11].

LH/hCG receptor cloning studies indicate that this cell surface protein belongs to a large family of guanine nucleotide-binding protein (G-protein) coupled membrane receptors [6,12]. It is an asymmetrical glycoprotein consisting of one polypeptide chain and six candidate N-linked glycoprotein sites. The large, 340-amino-acid, leucine-rich extracellular hydrophilic N-terminus is distinct among this class of receptor and is responsible for high-affinity hCG and LH binding and for binding specificity [10,13,14]. The domain is believed to have evolved in this specific way to ensure high-affinity binding to both hCG and LH [15]. The transmembrane domain has seven α-helical membrane-spanning segments linked by three extracellular loops (exoloops) and three intracellular loops (cytoloops) (Figure 9.1). These transmembrane helix structures play an important role in receptor activity. Transmembrane helices 3, 6, and 7 are critical for receptor activation [16,17]. Together with the

Human Chorionic Gonadotropin (hCG). DOI: http://dx.doi.org/10.1016/B978-0-12-800749-5.00009-2

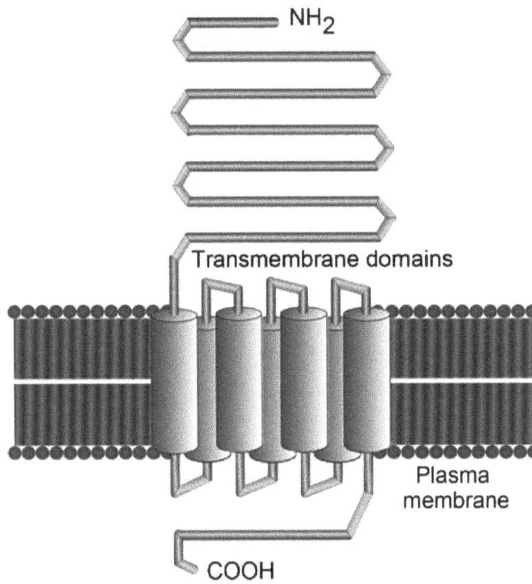

Figure 9.1 Diagrammatic representation of the hCG/LH receptor.

cytoloops, the intracellular hydrophobic C-terminus shares numerous motifs with Ser, Thr, and Tyr residues. This suggests a potential for modulation of receptor function by phosphorylation, serine-threonine protein kinases, and tyrosine kinases [18]. The receptor is palmitoylated on two Cys residues at the C-terminal tail [19,20], and this region also undergoes phosphorylation on 4 Ser residues [7,21]. The role of the phosphorylation is uncertain, although one group has shown that the receptor dissociates from adenylate cyclase after phosphorylation [22].

Only the intact heterodimer of hCG or LH can stimulate the LH/hCG receptor; however, it has been demonstrated that the free subunits will also bind, albeit with much lower affinity [23]. As mentioned, studies have shown that on receptor binding, the α-subunit of hCG is internalized and is associated with stimulation of adenylate cyclase activity. This implies that the binding model is not simple and that hormone/receptor binding involves a topographical determinant extending across both ligand subunits [24]. The 40% structural homology between β-subunits of the glycoprotein hormones [23] is thought to relate to the common property of heterodimer formation. Therefore, the heterologous regions are likely to relate to the specificity of hormone receptor binding conferred by the β-subunit [25].

Receptor binding and signaling appear largely concomitant, with both stages being dependent on the same regions of both the hormone and the receptor molecules. Receptor stimulation also appears to involve multiple steps whereby signal transduction occurs as a result of low-affinity site interaction after the initial high-affinity binding of the ligand. Amino acid Asp 397 of the receptor has been implicated in signal transduction, but not necessarily binding; therefore, it could be involved in low-affinity coupling [26]. Asp 383 is found within the second transmembrane domain and has also been implicated in ligand binding and cAMP second-messenger responses [27].

The hormonal glycosylations were initially thought to be essential for signal transduction [28]; however, although it is agreed that the sugar moieties affect ligand binding, more recent studies have shown that deglycosylated hCG can still stimulate the LH/hCG receptor [29]. Similarly, hCG-H has been shown to bind the receptor as well, albeit with diminished activity [1]. Both of these observations indicate that optimal receptor stimulation is dependent on very specific glycosylation of hCG.

On stimulation by hCG (or LH), activation of the heterotrimeric (α-, β-, and γ-subunits) Gs-protein [18] results in stimulation of the membrane-bound adenylate cyclase by the α-subunit. The sixth transmembrane segment and/or cytoloop-3 are believed to be responsible for Gs-protein interaction [30,31]. Activation of adenylate cyclase catalyzes the conversion of ATP to cAMP, thus elevating intracellular levels. After upregulation of cAMP, activation of phosphokinase A ensues, resulting in phosphorylation and activation of the cAMP-responsive element (CRE) binding proteins and subsequent CREs [5].

Although cAMP is considered the major second messenger involved in the LH/hCG-R cascade, there is additional stimulation of phospholipase C: elevated inositol phosphates, diacylglycerol, and intracellular calcium [32]. In both cases, the culmination of these events leads to enhanced steroid production by catalysis of the conversion of cholesterol to pregnenolone by 20 α-hydroxylase side-chain cleavage enzyme, which then leads to progesterone production by 3β-hydroxysteroid dehydrogenase-isomerase via 17α-hydroxyprogesterone [33].

It is thought that the activation of protein kinase activates mitogen protein kinase pathways and a Janus-kinase signaling pathway [16]. Clearly, all hormonal function involves DNA transcription or generation of mRNA. Promotion of progesterone production in corpus luteal cells involves the synthesis of cholesterol side-chain cleavage enzymes. Fetal tissue growth involves protein synthesis. A parallel mechanism promotes the synthesis of an LHRBP. In turn, this activates exonuclease and endonuclease and leads to the destruction of receptor mRNA (Figure 9.2).

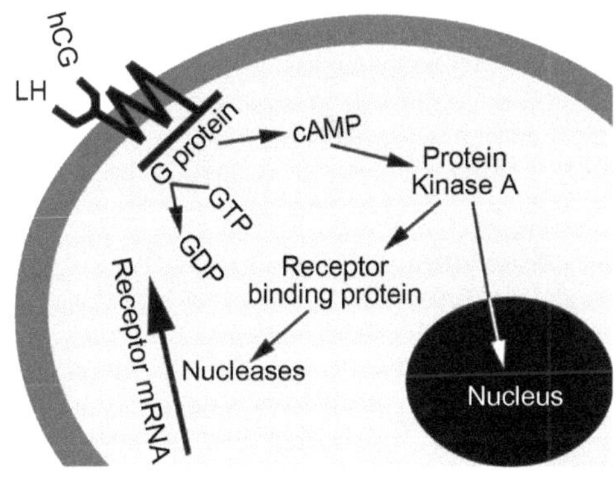

Figure 9.2 Activation of hCG/LH receptor, G-protein and cAMP, protein kinase expression, and production of LH/hCG receptor binding protein (LHRBP). Synthesis of LHRBP activates exonucleases and endonucleases that destroy receptor mRNA, limiting expression and downregulating the receptor.

This mechanism limits receptor expression, effectively downregulating the receptor [17]. Transmembrane helix structures play an important role in receptor activity. Transmembrane helices 3, 6, and 7 are critical to receptor activation [18,19].

References

[1] Cole LA, Kardana A, Andrade-Gordon P, Gawinowicz MA, Morris JC, Bergert ER, et al. The heterogeneity of hCG: III. The occurrence, biological and immunological activities of nicked hCG. Endocrinology 1991;129:1559–67.
[2] Rao CV. Differential properties of human chorionic gonadotropin and human luteinizing hormone binding to plasma membranes of bovine corpora lutea. Acta Endocrinol 1979;90:696–710.
[3] Dufau ML, Kusuda S. Purification and characterization of ovarian LH/hCG and prolactin receptors. J Recept Res 1987;7:167–93.
[4] Jia XC, Oikawa M, Bo M, Tanaka T, Ny T, Boime I, et al. Expression of human luteinizing hormone (LH) receptor: interaction with LH and chorionic gonadotropin from human, but not equine, rat, and ovine species. Mol Endocrinol 1991;5:759–68.
[5] Segaloff DL, Ascoli M. The lutropin/choriogonadotropin receptor—4 years later. Endocr Rev 1993;14:324–47.
[6] McFarland KC, Sprengel R, Phillips HS, Kohler M, Resemblit N, Nikolics K, et al. Lutropin-choriogonadotropin receptor: an unusual member of the G protein-coupled receptor family. Science 1989;245:494–9.
[7] Ascoli M, Fanelli F, Segaloff DL. The lutropin/choriogonadotropin receptor, a 2002 perspective. Endocr Rev 2002;23:141–74.
[8] Fanelli F, Puett D. Structural aspects of luteinizing hormone receptor. Endocrine 2002;18:285–93.
[9] Fanelli F, Themman APN, Puett D. Lutropin receptor function: insights from natural, engineered, and computer-simulated mutations. IUBMB Life 2001;51:149–55.
[10] Puett D, Li Y, Agelova K, DeMars G, Meehan TP, Ganelli F, et al. Structure–function relationships of the luteinizing hormone receptor. Ann NY Acad Sci 2005;1061:41–54.
[11] Dias JA. Recent progress in structure–function and molecular analyses of the pituitary/placental glycoprotein hormone receptors. Biophys Acta 1982;1135:278–94.
[12] Loosfelt H, Salesse R, Jallal B, Garnier J, Milgrom E. Cloning and sequencing of porcine LH-hCG receptor cDNA: variants lacking transmembrane domain. Science 1989;245:525–8.
[13] Fredriksson R, Lagerstrom MC, Lundin LG, Schioth HB. The G-protein-coupled receptors in the human genome form five main families. Phylogenetic analysis, paralogon groups, and fingerprints. Mol Pharmacol 2003;63:1256–72.
[14] Puett D, Li Y, DeMars G, Angelova K, Fanelli F. A functional transmembrane complex: the luteinizing hormone receptor with bound ligand and G-protein. Mol Cell Endocrinol 2007;2:126–36.
[15] Combarnous Y. Molecular basis of the specificity of binding of glycoprotein hormones to their receptors. Endocr Rev 1992;13:671–91.
[16] Angelova K, Fanelli F, Puett D. A model for constitutive lutropin receptor activation based on molecular simulation and engineered mutations in transmembrane helices 6 and 7. J Biol Chem 2002;277:32202–13.

[17] Angelova K, Narayan P, Simon JP, Puett D. Functional role of transmembrane helix 7 in the activation of the heptahelical lutropin receptor. Mol Endocrinol 2000;14:459–71.
[18] Davis JS. Mechanisms of hormone action: luteinizing hormone receptors and second-messenger pathways. Curr Opin Obstet Gynaecol 1994;6:254–61.
[19] Zhu H, Wang H, Ascoli M. The lutropin/choriogonadotropin receptor is palmitoylated at intracellular cysteine residues. Mol Endocrinol 1995;9:141–50.
[20] Kewate N, Menon KMJ. Palmitoylation of luteinizing hormone/human chorionic gonadotropin receptors in transfected cells. Abolition of palmitoylation by mutation of Cys621 and Cys-622 residues in the cytoplasmic tail increases ligand-induced internalization of the receptor. J Biol Chem 1994;269:30651–8.
[21] Hipkin RW, Wang Z, Ascoli M. Human chorionic gonadotropin- and phorbol ester stimulated phosphorylation of the luteinizing hormone/CG receptor maps to serines 635, 639, 649 and 652 in the C-terminal cytoplasmic tail. Mol Endocrinol 1995;9:151–8.
[22] Wang Z, Liu X, Ascoli M. Phosphorylation of the lutropin/choriogonadotropin receptor facilitates uncoupling of the receptor from adenylyl cyclase and endocytosis of the bound hormone. Mol Endocrinol 1997;11:183–92.
[23] Pierce JG, Parsons TF. Glycoprotein hormones: structure and function. Ann Rev Biochem 1981;50:465–95.
[24] Ryan RJ, Keutmann HT, Charlesworth MC, McCormick DJ, Milius RP, Calvo FO, et al. Structure–function relationships of gonadotropins. Recent Prog Horm Res 1987;43:383–417.
[25] Sairam MR, Manjunath P. Hormonal antagonistic properties of chemically deglycosylated hCG. J Biol Chem 1983;258:445–9.
[26] Ji I, Ji TH. Receptor activation is distinct from hormone binding in intact lutropinchoriogonadotropin receptors and Asp397 is important for receptor activation. J Biol Chem 1993;268:20851–4.
[27] Ji I, Ji TH. Asp383 in the second transmembrane domain of the lutropin receptor is important for high-affinity hormone binding and cAMP production. J Biol Chem 1991;266:14953–7.
[28] Keene JL, Matzuk MM, Boime I. Expression of recombinant human chorionic gonadotropin in chinese hamster ovary glycosylation mutants. Mol Endocrinol 1989;3:2011–17.
[29] van Loenen HJ, van Gelderen-Boele S, Flinterman JF, Merz WE. Rommerts FFG. The relative importance of the oligosaccharide units in hCG for LH/CG receptor activation in rat leydig cells and mouse leydig cells. J Endocrinol 1995;147:367–75.
[30] Lefkowitz RJ, Cotecchia S, Samama P, Costa T. Constitutive activity of receptors coupled to guanine nucleotide regulatory proteins. Trends Pharmacol Sci 1993;14:303–7.
[31] Abell AN, McCormick DJ, Segaloff DL. Certain activating mutations within helix 6 of the human luteinizing hormone receptor might be explained by alterations that allow transmembrane regions to activate Gs. Mol Endocrinol 1998;12:1857–69.
[32] Iles RK, Chard T. Molecular insights into structure and function of human chorionic gonadotropin. J Mol Endocrinol 1993;10:217–34.
[33] Marton I, Oakey RE. 3 Beta-hydroxysteroid dehydrogenase-isomerase activity in placentae from pregnancies complicated by steroid sulphatase deficiency. J Steroid Biochem 1980;13:475–9.

TGFβ-II autocrine receptor

10

Laurence A. Cole
USA hCG Reference Service, Angel Fire, NM, USA

The TGFβ-II receptor is currently a matter of controversy. Although all parties agree that hyperglycosylated hCG, hyperglycosylated hCG free β-subunit, and free β-subunit bind and act on a TGFβ-II receptor [1–20], Butler and colleagues assert that this action involves antagonism [1], whereas Berndt and colleagues insist that it acts as an agonist [2]. This controversial matter may simply be resolved by what cell types were examined. Butler and colleagues examined bladder cancer culture cells [1], whereas Berndt and colleagues used LHCGR mouse aortic ring cells. I am told that this may be a matter of invasive and non-invasive cells, with the molecules antagonizing invasive cancer cells and invasive cytotrophoblast cells.

Considering the invasive model, hyperglycosylated hCG, hyperglycosylated hCG free β-subunit, or free β-subunit have all been shown to block apoptosis in multiple cancer models [4–7] and in cytotrophoblast cell models [8]. Agonism of TGFβ receptor has been shown to promote apoptosis, so that antagonism would inhibit apoptosis. Multiple authors show that these promoters enhance cell growth [4–7,9–11]. Other authors show that agonism of the TGFβ receptor blocks collagenase and metalloproteinases, whereas antagonism in cancer leads to production of these invasive enzymes [12–15].

A model of hyperglycosylated hCG, hyperglycosylated hCG free β-subunit, or hCG free β-subunit antagonizing cytotrophoblast cells in blastocyst implantation or cancer malignancies is shown in Figure 10.1. As illustrated, TGFβ-II receptors act through SMAD and cAMP intermediates, which permit nuclear penetration. SMADs have been demonstrated in the response of TGFβ-II receptor to hyperglycosylated hCG [2]. Angiogenesis has also been demonstrated in response to hyperglycosylated hCG [2]. That hyperglycosylated hCG and hyperglycosylated hCG free β-subunit are interchangeable cancer promoters with similar potency has also been demonstrated [3].

Lustabader et al., Lapthorn et al., and Wu et al. [17–20] have demonstrated a common cystine knot structure in hCG (on hCG α-subunit and β-subunit) linking the structure of hCG and TGFβ. That the hCG amino acid sequence with different glycosylation (hyperglycosylation) can bind the LH/hCG hormone receptor and the TGFβ-II autocrine receptor is rather unusual, giving the molecule two distinctly different functions (hCG and hyperglycosylated hCG) and two distinct sets of actions.

Human Chorionic Gonadotropin (hCG). DOI: http://dx.doi.org/10.1016/B978-0-12-800749-5.00010-9

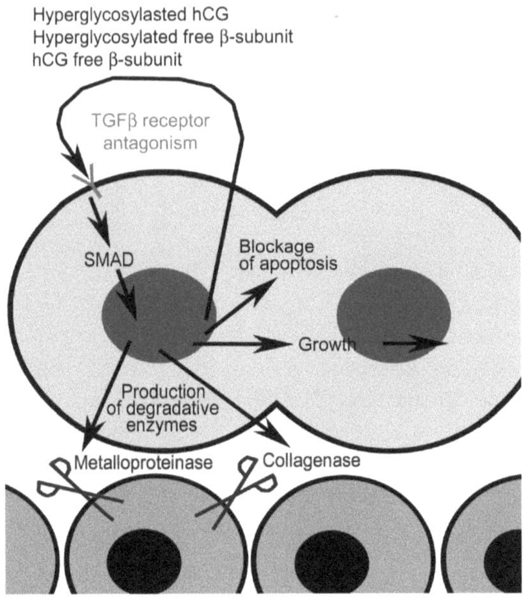

Figure 10.1 Action of hyperglycosylated hCG, hyperglycosylated hCG free β-subunit, and hCG free β-subunit on TGFβ-II receptor.

References

[1] Butler SA, Ikram MS, Mathieu S, Iles RK. The increase in bladder carcinoma cell population induced by the free beta subunit of hCG is a result of an anti-apoptosis effect and not cell proliferation. Br J Cancer 2000;82:1553–6.

[2] Berndt S, Blacher S, Munuat C, Detilleux J, Evain-Brion D, Noel A, et al. Hyperglycosylated human chorionic gonadotropin stimulates angiogenesis through TGF-ß receptor activation. FASEB J 2013; Abstract 12-213686

[3] Cole LA, Butler SA. Hyperglycosylated hCG, hCGß and hyperglycosylated hCGß: interchangeable cancer promoters. Mol Cell Endocrinol 2012;349:232–8.

[4] Carter WB, Sekharem M, Coppola D. Purified hCG induces apoptosis in breast cancer. Breast Cancer Res Treat 2006;100:S243–4.

[5] Li D, Wen X, Ghali L, Al-Shalabi FM, Docherty SM, Purkis P, et al. hCG beta expression by cervical squamous carcinoma—*in vivo* histological association with tumour invasion and apoptosis. Histopathology 2008;53:147–55.

[6] Iles RK. Ectopic hCGß expression by epithelial cancer: malignant behavior metastasis and inhibition of tumor cell apoptosis. Mol Cell Endocrinol 2007;260:264–70.

[7] Butler SA, Iles RK. Ectopic human chorionic gonadotrophin ß secretion by epithelial tumors and human chorionic gonadotrophin ß-induced apoptosis in Karposi's sarcoma is there a connection? Clin Cancer Res 2003;9:4666–73.

[8] Hamade AL, Nakabayashi K, Sato A, Kiyoshi K, Takamatsu Y, Laoag-Fernandez JB, et al. Transfection of antisense chorionic gonadotropin ß gene into choriocarcinoma cells suppresses the cell proliferation and induces apoptosis. J Clin Endocrinol Metab 2005;90:4873–9.

[9] Jankowska A, Gunderson SI, Andrusiewicz N, Burczynska B, Szczerba A, Jarmolowski A, et al. Reduction of human chorionic gonadotropin beta subunit expression by modified U1 snRNA caused apoptosis in cervical cancer cells. Mol Cancer 2008;7:26–9.
[10] Kocdor H, Kocdor MA, Russo J, Snider KE, Vanegas JE, Russo IH, et al. Human chorionic gonadotropin (hCG) prevents the transformed phenotypes induced by 17 beta-estradiol in human breast epithelial cells. Cell Biol Int 2009;33:1135–43.
[11] Bellet D, Lazar V, Bleche I, Paradis V, Giovangrandi Y, Paterliru P. Malignant transformation of nontrophoblastic cells in association with the expression of chorionic gonadotropin ß genes normally transcribed in trophoblastic cells. Cancer Res 1997;57:516–23.
[12] Murphy G, Reynolds JJ, Whitham SE, Docherty AJ, Angel P, Heath JK. Transforming growth factor beta modulates the expression of collagenase and metalloproteinase inhibitor. Eur Mol Biol Org J 1987;6:1899–904.
[13] Karmakar S, Das C. Regulation of trophoblast invasion by IL-1β and TGF- β1. Am J Reprod Immun 2002;48:210–19.
[14] Shooner C, Caron PC, Fréchette-Frigon G, Leblanc V, Déry M-C, Asselin E. TGF-beta expression during rat pregnancy and activity on decidual cell survival. Reprod Biol Endocrinol 2005;3:20.
[15] Wu D, Luo S, Wang Y, Zhuang L, Chen Y, Peng C. Smads in human trophoblast cells: expression, regulation and role in TGF-beta-induced transcriptional activity. Mol Cell Endocrinol 2001;175:111–21.
[16] Schiller M, Dennler S, Anderegg U, Kokot A, Simon JC, Luger TA, et al. Increased cAMP levels modulate transforming growth factor-beta/Smad-induced expression of extracellular matrix components and other key fibroblast effector functions. J Biol Chem 2010;285:409–21.
[17] Wu H, Lustbader JW, Liu Y, Canfield RE, Hendrickson WA. Structure of human chorionic gonadotropin at 2.6 Å resolution from MAD analysis of the selenomethionyl protein. Structure 1994;2:545–58.
[18] Lapthorn AJ, Harris DC, Littlejohn A, Lustbader JW, Canfield RE, Machin KJ. Crystal structure of hCG. Nature 1994;369:455–61.
[19] Lustbader JW, Birken S, Pileggi NF, Kolks MAG, Pollak S, Cuff ME, et al. Crystallization and characterization of human chorionic gonadotropin in chemically deglycosylated and enzymatically desialylated states. Biochemistry 1989;28:9239–43.
[20] Lustbader JW, Wang C, Zhang X, Birken S, Wu H, Brown JM, et al. Human chorionic gonadotropin: progress in determining its tertiary structure. Glycoprotein hormones. New York, NY: Springer-Verlag; 1994. p. 81–102.

Part C

Continuous Dissociation and Degradation

Degradation products of hCG, hyperglycosylated hCG, and free β-subunit

11

Laurence A. Cole
USA hCG Reference Service, Angel Fire, NM, USA

Cytotrophoblast and syncytiotrophoblast cells make excess amounts of hCG and hyperglycosylated hCG. The amounts of hCG and hyperglycosylated hCG produced promote progesterone production, promote invasion as part of implantation, and promote uterine spiral artery angiogenesis and hemochorial placentation, uterine growth, trophoblast differentiation, myometrial muscle relaxation, fetal growth, and immune and macrophage suppression [1]. hCG has an extremely long circulating half-life of 36 h [2]. Humans have intricate mechanisms constantly working to degrade and remove long circulating hCG from the body's circulation. These mechanisms are particularly evident during the third trimester of pregnancy, when hCG no longer plays a significant role in placental function.

Five competing pathways work constantly in humans to clear both hCG and hyperglycosylated hCG. Each one reduces hCG's extensive long circulating half-life and extensively reduces its biological activity [1]. It should be realized that hCG is cleared from the circulation mostly through the liver. The liver clears 78% of hCG-related molecules. The kidneys only clear the remaining 22% of total hCG through the urine [2]. In the first clearance pathway, hCG is directly cleared by the liver or kidney (Figure 11.1).

In the second clearance pathway, hCG is an acidic glycoprotein (pI 3.5) and has an overall circulating half-life of 36 h [2–4]. In contrast, the neutral glycoprotein hormone LH (pI 8.0) only has a circulating half-time of 0.4 h [5]. When hCG is desialylated, the acidic sugar is removed. This reduces the clearance rate of hCG more than 300-fold, from 36 h to a matter of approximately more than 10 min [3]. hCG is normally synthesized in pregnancy with significant sialic acid or acid sugars content, 1.63 ± 0.39 pmol per pmol N-linked or O-linked oligosaccharide on the hCG α-subunit or β-subunit, or 13 sialic acid per molecule [6]. Over time, sialic acid residues are lost from hCG and cleaved by macrophage neuraminidases; these generate an asialo hCG or hCG with reduced sialic acid and exposed galactose sugar (Figure 11.1). This leads to liver clearance by the galactose receptor [7–9]. With advancement of pregnancy, hCG is generally less sialylated (less acidic) and, as a result, clears more rapidly in the latter trimesters of pregnancy.

In the third pathway, hCG (particularly hyperglycosylated hCG) undergoes continuous dissociation (Figure 11.1). The dissociated free β-subunit and free α-subunit

Human Chorionic Gonadotropin (hCG). DOI: http://dx.doi.org/10.1016/B978-0-12-800749-5.00011-0

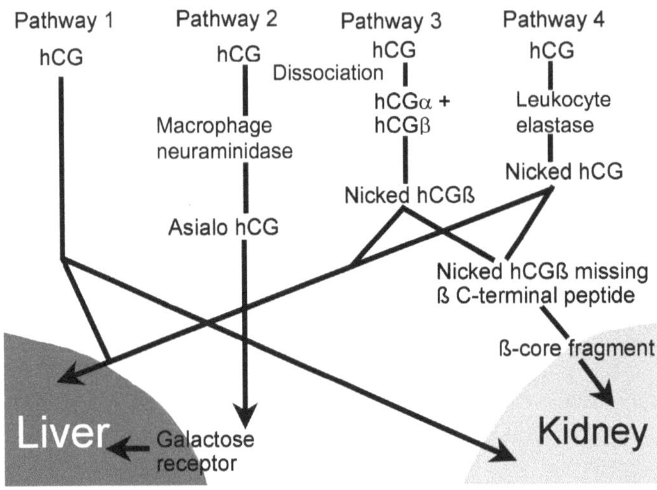

Figure 11.1 The five pathways leading the degradation, dissociation, and clearance of hCG.

Table 11.1 **The clearance of hCG and its degradation products in circulation [3–13]**

1. Time of dissociation of dimer in serum in the laboratory at 37°C	
Third I.S. hCG	Dissociation half-time 240 h
Hyperglycosylated hCG	Dissociation half-time 120 h
Nicked hCG	Dissociation half-time 44 h
Nicked hyperglycosylated hCG	Dissociation half-time 22 h
2. Effect of leukocyte elastase in human blood on nicking at 37°C	
Third I.S. hCG, 5 nmol, 1 h incubation	30% nicked in β44–48 region
CR129 hCG β-subunit, 5 nmol, 1 h incubation	100% nicked in β44–48 region
3. Time for injected molecules in humans to leave the circulation	
Third I.S. hCG	Slow phase half-time 36 h
Third I.S. β-subunit	Slow phase half-time 3.9 h
Third I.S. α-subunit	Slow phase half-time 1.3 h
Asialo Third I.S. hCG	Slow phase half-time 0.096 h
β-Core fragment	Slow phase half-time 0.38 h

components are biologically inactive at the LH/hCG receptor and are cleared from the circulation many fold more rapidly than hCG [10,11] (Table 11.1). Dissociated free β-subunit may alternatively be cleaved by leukocyte elastase causing nicking; this will be cleared even more rapidly. Further cleavage will lose the β-subunit

C-terminal peptide making free β-subunit missing the C-terminal peptide; this will also be cleared more rapidly. This is part of the degradation pathway to β-core fragment and to renal clearance (Figure 11.1).

In the fourth pathway, hCG is continuously nicked/cleaved by leukocyte elastase on secretion at β47–48. The nicking deactivates hCG and opens the structure of hCG, leading to five-fold more rapid dissociation into the free β-subunit and the free α-subunit [12,13] (Figure 11.1) (Table 11.1). Nicking is the first cleavage step in the synthesis of the β-core fragment (the end product of hCG degradation found only in urine). Nicked hCG may be cleared from the circulation through the liver or kidney. Alternatively, nicked hCG may be further cleaved by leukocyte elastase losing the β-subunit C-terminal peptide. This is part of the degradation pathway to β-core fragment and to renal clearance (Figure 11.1).

A few published facts regarding hCG, hyperglycosylated hCG, nicked hCG, the hCG α-subunit, and β-subunit clearance are presented. These are facts that modulate hCG and hyperglycosylated hCG dissociation and degradation (Table 11.1).

11.1 Pure hCG preparations

As described, the hCG degradation and clearance processes are always at work. All natural urinary hCG preparations and standards contain a heterogeneous mixture of intact, dissociated, nicked, and cleaved molecules. The only exception in hCG commercial preparations is recombinant hCG synthesized in Chinese hamster ovary cells. In one publication [14], 13 purified hCG and hyperglycosylated hCG preparations were examined: 11 of the 13 preparations were nicked/cleaved between β44–45 or β47–48; 2 of 13 preparations were missing the β-subunit C-terminus (cleaved at β92–93); and 7 of 13 preparations were cleaved at the N-terminal section of α-subunit. In a later study [6] that investigated the structure of individual hCG and hyperglycosylated hCG preparations as well as the CR series preparation (hCG standards for World Health Organization), 12 of 14 individual hCG preparations and all the CR series standards were nicked to some extent, 3 of 14 were missing the β-subunit C-terminal peptide, and 13 of 14 (which included all CR series standards) were cleaved at the N-terminus of the α-subunit [6]. As a result of these two studies, it is now acceptable to conclude that the amino acid sequences of all hCG and hCG-H preparations (except recombinant hCG) contain molecules that are cleaved to some extent. Thus, hCG preparations are always a mixture of intact hCG, nicked hCG, hCG missing the C-terminal peptide, and α-subunit cleaved N-terminus hCG.

Similarly, all hCG standards sold are invariably nicked and do not contain pure hCG (1st RR, 3rd I.S., 4th I.S., CR127, CR129). The only pure hCG preparation sold that is free of nicked hCG is Serono Ovidrel or Chinese hamster ovary cell recombinant hCG.

11.2 Nicking and enzyme cleavage

Figure 11.2 shows common proteolytic enzyme cleavage site on hCG subunits. As illustrated in Figure 11.3, trophoblastic villi contain monocytes and placental macrophages (Hofbauer cells) and are associated with classic macrophage structures [6,11]. Both of these are white blood cell types that produce a leukocyte elastase that cleaves proteins at Leu, Gly, and Val residues. Leukocyte elastase acts on newly synthesized cytotrophoblast hyperglycosylated hCG and syncytiotrophoblast hCG molecules [12,13,15].

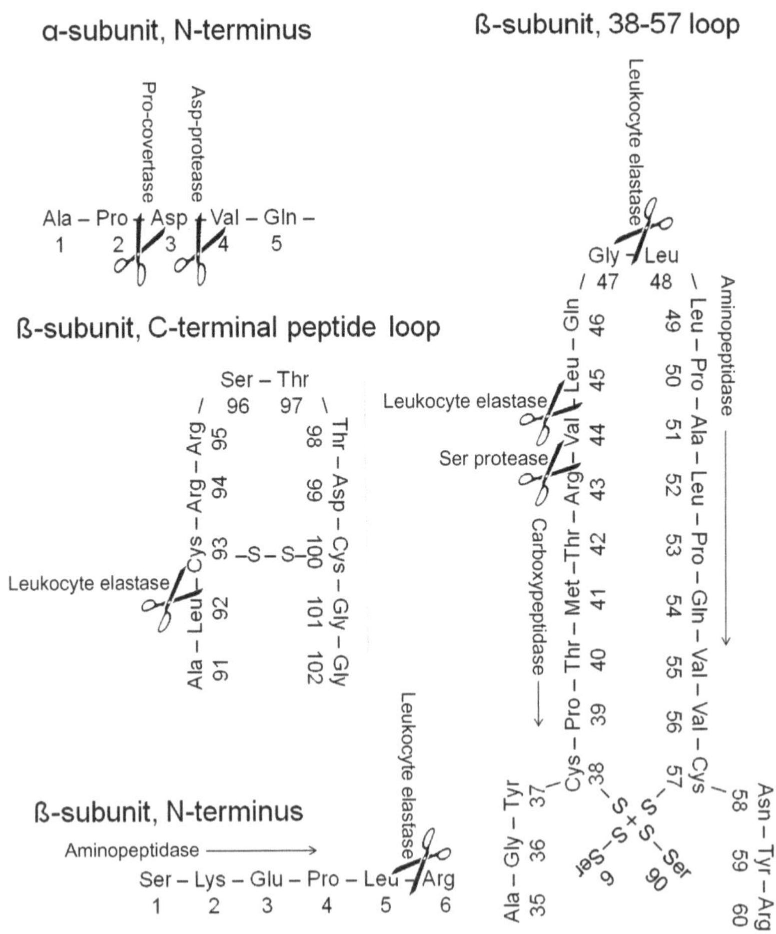

Figure 11.2 Cleavage sites in the degradation of hCG, hyperglycosylated hCG, and hCG free β-subunit.

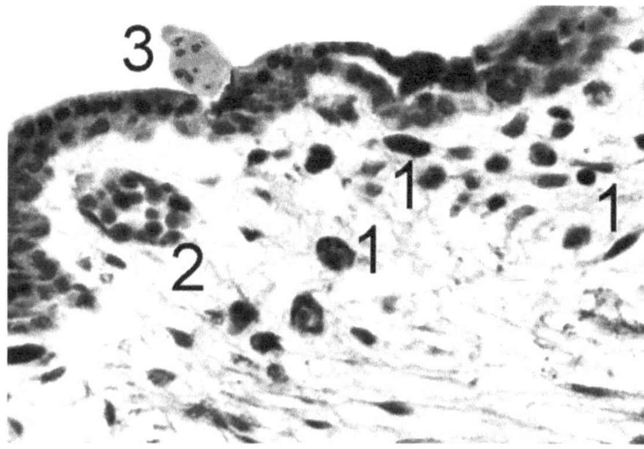

Figure 11.3 Villous placenta stained with antibody B204 (binds β-subunit, nicked β-subunit, β-subunit missing C-terminus, and β-core fragment). The villi contain monocytes stained by B204 [1] as well as placental macrophages (Hofbauer cell complexes) [2].

Figure 11.2 shows cleavage sites on the α-subunits and β-subunits of hCG. These include the α-subunit N-terminal cleavage sites at α2,3 and α3,4, the β-subunit β38–57 loop cleavage site, the β-subunit β5–6 N-terminal cleavage site, and the β-subunit β92–93 C-terminal cleavage site. Many of these sites are cleaved by leukocyte elastase. It has been shown [12,13] that incubating hCG or β-subunit with serum leads to nicking at β44–45 and β47–48. Longer incubation periods lead to cleavage of the C-terminus at β92–93, which also appears to be due to leukocyte elastase. These actions were repeated using purified human leukocyte elastase and yielded similar results [12,15].

β-core fragment is the terminal urinary degradation product of hCG. It comprises β-subunit residues 6–40 that are linked by disulfides to residues 55–92. The β-core fragment appears to be generated by cleavage of the C-terminus at β92–93 and nicking at β44–45 or β47–48 by leukocyte elastase. Carboxypeptidases and aminopeptidases then remove β40–47 and β48–55, taking cleavage from the nicking site to two amino acids away from a disulfide bridge on either side (Figure 11.2). How the amino terminus of β-subunit is cleaved has not been established. It could be the result of an aminopeptidase action or of further leukocyte elastase cleavage at β5–6.

In terms of hCG clearance, the cleavage process is exponential. Once hCG is nicked by leukocyte elastase, it dissociates five-times more rapidly and clears from the circulation seven-times more rapidly (Table 11.1) (Figure 11.2). It should be noted that if the nicked and dissociated molecules are cleared through the kidney, then they will become β-core fragments as an end product. Nicked β-subunit and nicked β-subunit missing the C-terminus are both commonly detected in serum [16], but β-core fragment is only detected in urine [17]. It is thought that the aminopeptidase and carboxypeptidase stages of β-core fragment synthesis are confined strictly to the kidneys. It has been shown that administration of pure Chinese hamster ovary recombinant hCG to humans leads to 13% of the administered amount being degraded to urine β-core fragment. It is assumed that circulating leukocyte elastase and dissociation start the degradation process [17].

11.3 Dissociation

Dissociation is the central event in the degradation and clearance processes of hCG. The dissociation half-life of intact hCG is extremely slow, 240 h in serum at 37°C (Table 11.1). Hyperglycosylated hCG dissociates approximately twice as fast as regular hCG, but nicked hCG and nicked hyperglycosylated hCG dissociate five-times faster (Table 11.1), leading to rapid clearance of nicked hCG β-subunits and α-subunits. Each event in the clearance process appears to be exponential. If hCG is nicked, for example, then it dissociates five-times faster and is cleared from the circulation seven-times faster (Table 11.1). If hCG or hyperglycosylated hCG is dissociated, then the free β-subunit is subsequently nicked much more rapidly (Table 11.1) (Figure 11.1). It appears that nicking and dissociation work together, each speeding the clearance processes of the other's products.

11.4 Liver clearance

Most hCG is cleared from the circulation through the liver (78%), but some is cleared by the kidneys (22%) [2]. hCG is an acidic glycoprotein (pI 3.5). When it is desialylated (removing all the acidic sugar), it speeds clearance to more than 300-times the base value [4]. hCG can be synthesized either without the complete component of sialic acid residues or with exposed galactose residues. Over time, sialic acid residues are lost and are cleaved by macrophage neuraminidases. This leads to rapid liver clearance and utilization of the liver galactose receptor [7–9].

11.5 Degradation with pregnancy advancement

A general parallelism is observed between increasing concentration of degradation products during advancing pregnancy (Figures 11.4). During 4–40 weeks of pregnancy, a continuous increase in free α-subunit and β-core fragment in urine is noted. This increase is apparently due to the dissociation and degradation of hCG. A similar increase is observed in nicked hCG and the free α-subunit in serum [12]. During the first 2 months of pregnancy, 9% of hCG is nicked in serum, 13% is

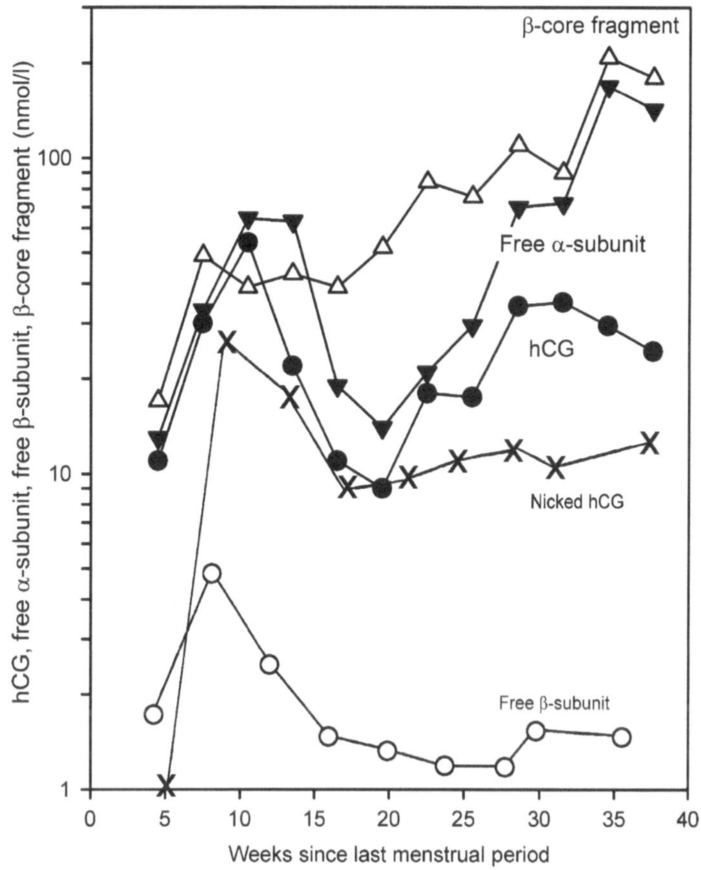

Figure 11.4 Urine concentrations of hCG, free α-subunit, free β-subunit, nicked hCG, and β-core fragment during the course of pregnancy [11].

nicked in the next 2 months, 17% is nicked in the next 2 months, and 21% is nicked in the final 3 months of pregnancy [12].

In urine, β-core fragment concentration is higher than that of hCG (nmol/l) throughout most of pregnancy. In the third trimester, the median β-core fragment concentration is approximately 10-fold greater than the median hCG concentration. As pregnancy advances, increased proportions of hCG are degraded through nicking, dissociation, cleavage of β-subunit C-terminus, and the processes described in this chapter (Figure 11.4). All the clearance pathways are summarized in Figure 11.1. All lead to the rapid clearance of hCG from the circulation. As shown in Table 11.1, following parturition of pregnancy, hCG disappears from the circulation more rapidly than free β-subunits and α-subunits [18]. This is a direct result of the degradation protocol, when free β-subunits and α-subunits are being continuously generated. In the weeks following parturition, the degradation protocol encompasses nicking, dissociation, cleavage of β-subunit C-terminal peptide, and aminopeptidase and carboxypeptidase action, as indicated in Figures 11.1, 11.2, and 11.4.

It is estimated that following first-trimester termination of pregnancy, it takes an average of 30 days for hCG to clear the body [19]; following an ectopic pregnancy, it can take an average of 24 days to clear; and following a term pregnancy, it takes an average of 28 days to clear [18]. The more acidic hCG-H can take as long as 65 days to clear the body. Longer clearance times have been observed with hydatidiform mole, choriocarcinoma, and testicular cancer.

11.6 Degradation in cancer

Two molecules are produced in cancer, hCG free β-subunit and hyperglycosylated hCG free β-subunit. Lesser amounts of hyperglycosylated hCG are produced (see Chapters 20–22). Evidence suggests an alternative degradation pathway in these cases. The first indication is that urine β-core fragment is the optimal marker for cancers and the end product of hCG free β-subunit degradation. Although hCG free β-subunit is detected in serum in 30% of cases, β-core fragment in urine is detected in 55% of cases [20]. Certain cleavage processes are unique to hCG produced in cancers. For example, both nicking at Arg residue β43–44 and cleavage of the C-terminus, part of the β-core fragment digestion process, are processes unique to cancer cases [6]. Figure 11.5 shows the proposed cleavage pathways in cancer cases.

Ten of 11 automated hCG tests used in modern laboratories incorporate an antibody to the β-subunit C-terminal peptide to specifically bind the hormone. Tests that use these antibodies should be avoided when monitoring cancer cases because the C-terminal peptide may be absent [12] (see Chapter 20). The only automated suitable hCG test is the Siemens Immulite Total hCG test. As shown recently in a USA hCG Reference Service study, all other tests give negative or unduly low or incorrect hCG values in cancer, gestational trophoblastic disease, spontaneous abortion, and other cases (Table 11.2).

hCG Free β-subunit	Hyperglycosylated hCG free β-subunit
↓	↓
Nicked free β-subunit	Nicked hyperglycosylated hCG free β-subunit
↓	↓
Nicked free β-subunit missing C-terminal peptide	Nicked hyperglycosylated hCG free β-subunit missing C-terminal peptide
↓	↓
β-core fragment	β-core fragment 2

Figure 11.5 Cleavage of hCG in cancer cases.

Table 11.2 **Detection of serum total hCG by the immulite and 10 other automated assays**

Diagnosis	Total hCG, immulite assay (mIU/ml)	Total hCG, other assay (mIU/ml)	Other assay, % immulite result	Other assay used
Normal term pregnancy	441,624	292,136	66%	Siemens Dimension
Normal term pregnancy	9358	16,621	56%	Siemens Dimension
Normal term pregnancy	7370	5743	78%	Siemens Dimension
Normal term pregnancy	275	263	96%	Siemens Dimension
Spontaneous abortion	6	5	83%	Siemens Dimension
Spontaneous abortion	11	11	100%	Siemens Dimension
Spontaneous abortion	46	38	83%	Siemens Dimension
Spontaneous abortion	2	1	50%	Siemens Dimension
Spontaneous abortion	2.4	<0.5	<21%	Beckman Access 2
Spontaneous abortion	140	66	47%	Ortho Vitros Eci
Spontaneous abortion	444	103	23%	Siemens Centaur
Spontaneous abortion	148	38	26%	Siemens Centaur
Spontaneous abortion	140	66	47%	Siemens Centaur
Spontaneous abortion	16	8	50%	Siemens Centaur
Spontaneous abortion	156	103	66%	Siemens Centaur
Spontaneous abortion	137	93	68%	Tosoh A1A
hCG doping	80,400	78,000	97%	Siemens Centaur
hCG doping	106	75	71%	Abbott AxSym
Menopausal hCG	10.0	9.0	90%	Siemens Centaur

(*Continued*)

Table 11.2 (Continued)

Diagnosis	Total hCG, immulite assay (mIU/ml)	Total hCG, other assay (mIU/ml)	Other assay, % immulite result	Other assay used
Cancer	2315	1875	81%	Abbott Architect
Cancer	717	651	91%	Siemens Dimension
Cancer	3.0	2.0	67%	Siemens Dimension
Cancer	204	101	50%	Siemens Centaur
Cancer	107	88	83%	Siemens Centaur
Familial hCG syndrome	216	24	11%	Beckman Dxi
Familial hCG syndrome	201	29	14%	Siemens Centaur
Familial hCG syndrome	168	50	30%	Siemens Centaur
Familial hCG syndrome	130	39	30%	Siemens Centaur
Familial hCG syndrome	43	12	28%	Siemens Centaur
Familial hCG syndrome	17	<2.0	<12%	Siemens Centaur
Familial hCG syndrome	2.8	<2.0	<71%	Siemens Centaur
Molar pregnancy	47	38	81%	Siemens Dimension
Molar pregnancy	235	48	20%	Abbott AxSym
Molar pregnancy	118	70	59%	Abbott AxSym
Molar pregnancy	118	66	56%	Beckman Access 2
Molar pregnancy	9.2	7.1	77%	Roche Elecsys hCG + ß
Molar pregnancy	5.9	4.6	78%	Siemens ACS180
Average	**Median 130**	**Median 68**	**Mean 60%**	

References

[1] Cole LA. Biological functions of hCG and hCG-related molecules. Reprod Biol Endocrinol 2010;8:102–16.
[2] Nisula BC, Blithe DL, Akar A, Lefort G, Wehmann RE. Metabolic fate of human chorionic gonadotropin. J Steroid Biochem 1989;33:733–7.
[3] Strickland TW, Puett D. The kinetic and equilibrium parameters of subunit association and dissociation. J Biol Chem 1982;257:2954–60.
[4] Rosa C, Amr S, Birken S, Wehmann R, Nisula B. Effects of desialylation on human chorionic gonadotropin on its metabolic clearance rate in humans. J Clin Endocrinol Metab 1984;59:1215–19.
[5] Veldhuis JD, Fraioli F, Rogol AD, Dufau ML. Metabolic clearance of biologically active luteinizing hormone in man. J Clin Invest 1986;77:1122–8.
[6] Elliott MM, Kardana A, Lustbader JW, Cole LA. Carbohydrate and peptide structure of the α- and β-subunits of human chorionic gonadotropin from normal and aberrant pregnancy and choriocarcinoma. Endocrine 1997;7:15–32.
[7] Lefort GP, Stolk JM, Nisula BC. Evidence that desialylation and uptake by hepatic receptors for galactose-terminated glycoproteins are immaterial to the metabolism of human choriogonadotropin in the rat. Endocrinology 1984;115:1551–7.
[8] Van Hall EV, Vaitukaitis JL, Ross GT, Hickman JW, Ashwell G. Effects of progressive desialylation on the rate of disappearance of immunoreactive hCG from plasma in rats. Endocrinology 1971;89:11–15.
[9] Wide L, Hobson B. Some qualitative differences of hCG in serum from early and late pregnancies and trophoblastic diseases. Acta Endocrinol 1987;116:465–72.
[10] Wehmann RE, Nisula BC. Metabolic clearance rates of the subunits of human chorionic gonadotropin. J Clin Endocrinol Metab 1979;48:753–9.
[11] Wemann RE, Nisula BC. Metabolic and renal clearance rates of purified human chorionic gonadotropin. J Clin Invest 1981;68:184–94.
[12] Cole LA, Karadana A, Park S-Y, Braunstein GD. The deactivation of hCG by nicking and dissociation. J Clin Endocrinol Metab 1993;76:704–15.
[13] Kardana A, Cole LA. Human chorionic gonadotropin β-subunit nicking enzymes in pregnancy and cancer patient serum. J Clin Endocrinol Metab 1994;79:761–7.
[14] Kardana A, Elliott ME, Gawinowicz MA, Birken S, Cole LA. The heterogeneity of hCG: I. Characterization of peptide variations in 13 individual preparations of hCG. Endocrinology 1991;129:1541–50.
[15] Cole LA, Kardana A, Andrade-Gordon P, Gawinowicz MA, Morris JC, Bergert ER, et al. The heterogeneity of hCG: III. The occurrence, biological and immunological activities of nicked hCG. Endocrinology 1991;129:1559–67.
[16] Cole LA, Shahabi S, Butler S, Mitchell H, Newlands ES, Behrman HR, et al. Utility of commonly used commercial hCG immunoassays in the diagnosis and management of trophoblastic diseases. Clin Chem 2001;47:308–915.
[17] Norman RJ, Buchholtz MW, Somogyl AA, Amato F. hCG-β core fragment is a metabolite of hCG: evidence from the infusion of recombinant hCG. J Endocrinol 2000;164:299–305.
[18] Korhonen J, Alfhan H, Ylostalo P, Veldhuis J, Stenmam U-H. Disappearance of human chorionic gonadotropin and its α- and β-subunits after term pregnancy. Clin Chem 1997;43:2155–63.
[19] Aral K, Gunrkan Zorlu C, Gokman O. Plasma human chorionic gonadotropin levels after induced abortion. Adv Contracept 1996;12:11–14.
[20] Cole LA. hCG variants, the growth factors which drive human malignancies. Am J Cancer Res 2011;2:22–35.

Part D

Biological Function: Pregnancy

Pregnancy-1, functions during preimplantation and during blastocyst implantation

12

Laurence A. Cole
USA hCG Reference Service, Angel Fire, NM, USA

12.1 Preimplantation

Several articles show that signaling occurs between the unimplanted blastocyst and the decidua tissue [1–4]. Four separate reports show that the blastocyst preimplantation secretes human chorionic gonadotropin (hCG) into the uterine space, which somehow reaches hCG/LH hormone receptors on the endometrial surface. This prepares the endometrium for impending implantation [1–4]. These nonvascular communications by hCG are a critical part of successful pregnancy. Recent studies show the importance of a receptive endometrium and of hCG preimplantation signaling [5–7]. hCG signaling seemingly causes immunotolerance and angiogenesis at the endometrial maternal – fetal interface. hCG has been shown to increase the number of uterine natural killer cells that play a key role in the establishment of pregnancy [5–7].

The lack of proper control of implantation creates a major obstacle to *in vitro* fertilization and other techniques in obtaining successful pregnancies. Implantation is a complex event involving positioning of the 60–100 cell blastocysts and adhesion of the blastocyst to the decidua. The final event is invasion, which is described later in this chapter. Although implantation could occur in any human tissue, the endometrial decidua is the only tissue in which embryo implantation cannot occur except during a restricted period called the "implantation window." During this window, the endometrium is highly receptive to the embryo. Maternal – embryo crosstalk caused by hCG is a likely event favoring the implantation window [5,6].

Ongoing research suggests that hCG may act as a growth factor and differentiation factor during pregnancy [7]. Research used an intrauterine microdialysis system to administer low concentrations of hCG to the endometrium of women in the luteal phase of the menstrual cycle. hCG administration provoked profound effects on paracrine parameters of differentiation (IGFBP-1, prolactin) and implantation (LIF, M-CSF). VEGF, a cytokine important for neoangiogenesis, was significantly stimulated by hCG, suggesting a role for hCG in the control of endometrial vascularization and placentation. The investigation of endometrial parameters of tissue remodeling revealed a significant increase of MMP-9 ($P < 0.05$) but not of TIMP-1 after hCG infusion.

Human Chorionic Gonadotropin (hCG). DOI: http://dx.doi.org/10.1016/B978-0-12-800749-5.00012-2

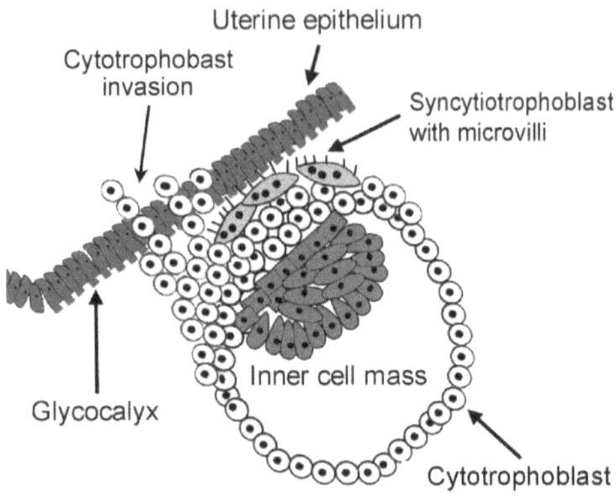

Figure 12.1 Blastocyst implantation at 3 weeks of gestation [4–8].

Figure 12.1 shows blastocyst implantation at 3 weeks of gestation [4–8]. As shown, syncytiotrophoblast microvilli link onto a glycocalx on the uterine epithelium. The concept that the blastocyst secretes hCG prior to implantation is a bit odd. Yet, the blastocyst contains a few differentiated syncytiotrophoblast cells, so secretion of the hormone hCG is feasible. hCG secretion may wait and occur on attachment and on linkage of blastocyst syncytiotrophoblast microvilli to the glycocalyx (Figure 12.1). At this point it is easy to consider hCG communicating between the blastocyst and receptors on the uterine epithelium. It is difficult to consider effective communication prior to attachment, at which point the hCG has no effective means of being transmitted through the uterine space.

12.2 Hyperglycosylated hCG function

Recent research both demonstrates [9,10] and confirms [11,12] that hyperglycosylated hCG promotes both the growth and invasion by cytotrophoblast cells. Other recently confirmed studies [13 – 15] show that hyperglycosylated hCG achieves this objective by being an autocrine and binding a TGFβ type II receptor. As demonstrated [13–15], these TGFβ functions are restricted to hyperglycosylated hCG. Regular hCG has no TGFβ or cytotrophoblast implantation actions.

As shown in Table 12.1, hyperglycosylated hCG, but not hCG, significantly promotes cytotrophoblast growth. This was demonstrated in JEG-3 choriocarcinoma cytotrophoblast cells and in primary cytotrophoblast cultures [9–12]. As shown in Table 12.2, examining Matrigel extracellular matrix invasion chambers and hyperglycosylated hCG, but not hCG, significantly promotes invasion.

With the understanding that hyperglycosylated hCG is an autocrine and functions through binding at TGFβ receptor [13–15], it is thought that hyperglycosylated hCG functions through blocking cytotrophoblast cell apoptosis and directly promoting cell growth [11,12], and it promotes invasion by driving the production of metalloproteinases and collagenases [16].

Table 12.1 Action of hCG and hyperglycosylated hCG on cytotrophoblast cell growth

Cell line and additives	Cell count after 24 h culture
Term pregnancy primary cytotrophoblast cells	
No supplement	5427 ± 834
Hyperglycosylated hCG added, 19.8 ng (10×)	7633 ± 777[a] (141%)
With regular hCG added, 19.8 ng (10×)	4227 ± 769 (78%)
JEG-3 choriocarcinoma cytotrophoblast cells	
No supplement	7114 ± 553
Hyperglycosylated hCG added, 81 ng (10×)	10,315 ± 1477[a] (145%)
With regular hCG added, 81 ng (10×)	5620 ± 657 (79%)

Concentrations of hCG and hyperglycosylated hCG used to promote invasion were 10 times that normally produced by the cells, 19.8 ng total for term pregnancy cytotrophoblast and 81 ng total for JEG-3 choriocarcinoma cells.
[a]A significant difference observed compared to no supplement, $P < 0.03$.

Table 12.2 Action of hCG and hyperglycosylated hCG on invasion by cytotrophoblast cells

	Mean penetration ± standard deviation
(A) Term pregnancy primary cytotrophoblast cells	
Control cultures	40 ± 10%
Hyperglycosylated hCG, 19.8 ng (10×)	76 ± 13%[a] (190%)
Regular hCG, 19.8 ng/ml (10×)	32 ± 9.1% (85%)
(B) JEG-3 choriocarcinoma cytotrophoblast cells	
Control cultures	46 ± 9.8%
Hyperglycosylated hCG, 81 ng (10×)	84 ± 6.2%[a] (183%)
Regular hCG, 81 ng (10×)	39 ± 4.5% (85%)

Isolated cytotrophoblast cells were prepared from fresh placenta. Isolated primary cytotrophoblast cells and 70% confluent JEG-3 choriocarcinoma cells, subcultured, were separately cultured 24 h (5000 cells) on Matrigel basement membranes and control inserts in quadruplicate. Concentrations of hCG and hyperglycosylated hCG used to promote invasion were 10 times that normally produced by the cells, 19.8 ng total for term pregnancy cytotrophoblast and 81 ng total for JEG-3 choriocarcinoma cells. The underside of Matrigel basement membranes, containing penetrated or invaded cells, was stained and counted. Cell penetration was compared with that of control inserts. The percentage penetration or invasion was calculated using the formula described by the manufacturer.

12.3 Blastocyst implantation

What is the evidence that hyperglycosylated hCG controls human blastocyst implantation? As proven [9–12], hyperglycosylated hCG drives invasion. Other data examine pregnancy failure and implantation. First, inappropriate blastocyst implantation is a proven cause of pregnancy failure [17,18]. Second, pregnancy failure is directly linked with significant deficiency in cytotrophoblast hyperglycosylated hCG at implantation [19,20]. It is therefore inferred that hyperglycosylated hCG drives implantation, and its deficiency, or inappropriate implantation signal, causes pregnancy failures.

As described previously [19,20], 133 women who achieved pregnancy were monitored on the first day of implantation of pregnancy (first day of hCG production, >1 mIU/ml [21]). The proportion of hyperglycosylated hCG (hyperglycosylated hCG ÷ total hCG) was determined. Of these, 70 had normal term pregnancies (53%), 35 had spontaneous abortions (SAB) (26%), and 28 had biochemical pregnancies (21%) (Figure 12.2). Of those with normal term pregnancies, 70% or 100% produced more than 40% hyperglycosylated hCG. Of those 35 with SAB outcomes, only 6 of 35 produced more than 40% hyperglycosylated hCG (mean hyperglycosylated hCG). Of the 28 with biochemical pregnancies, only 8 of 28 produced more than 40% hyperglycosylated hCG.

Hyperglycosylated hCG clearly marks SAB or failing pregnancy. Quest Diagnostics offers the only absolute test for term pregnancy outcome.

One final indication confirms the relationship between hyperglycosylated hCG production and pregnancy implantation. Published data show that at the time of blastocyst implantation, or with the first hCG-related molecule detection, the mean term pregnancy case produces 88 ± 17% hyperglycosylated hCG [20]. At 4 weeks of gestation 51% is hyperglycosylated hCG, at 5 weeks of gestation 43% is hyperglycosylated hCG, at 6 weeks of gestation 36% is hyperglycosylated hCG, and at 7 weeks of gestation 16% of the hCG is hyperglycosylated hCG [22]. Clearly, hyperglycosylated hCG production is focused on the third week of gestation or time of implantation.

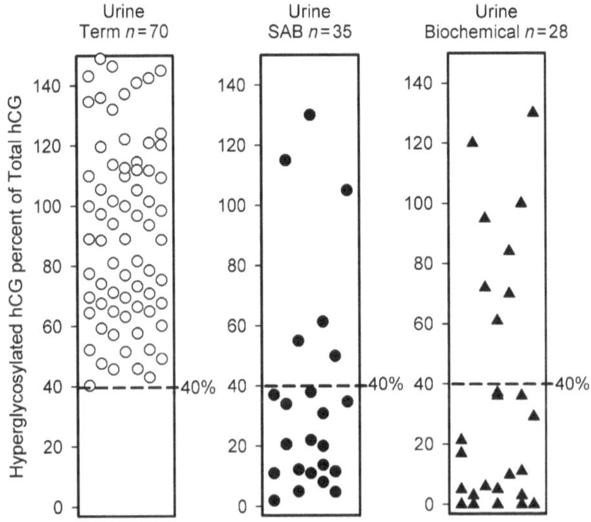

Figure 12.2 Determination of proportion of hyperglycosylated hCG (hyperglycosylated hCG ÷ total hCG) in 133 pregnancies with term, SAB, and biochemical pregnancy outcomes.

12.4 Hyperglycosylated hCG the human time bomb

Trophoblast cells make hCG, produced by syncytiotrophoblast cells, and hyperglycosylated hCG, made by cytotrophoblast cells. Just one cell type, root cytotrophoblast cells, evolved and started expressing the genes for hCG. However, root cytotrophoblast cells continuously fuse to make 3–50 fused cell syncytiotrophoblast cells. As such, both hCG and hyperglycosylated hCG are made by trophoblast cells.

As discussed in Chapter 17, hCG and hyperglycosylated hCG have an essential role in human evolution from primates. First, early simian primates evolve the molecules hCG and hyperglycosylated hCG from a deletion mutation in the LHß gene [23,24]. The expression of hCG and hyperglycosylated hCG caused the appearance of hemochorial placentation, a more efficient placenta mechanism in these primates. Later evolved advanced simian primates produced a more acidic hCG with longer circulating half-life, an hCG+. More efficient hemochorial placentation+ occurs. Then, evolved hominids produced even more acidic hCG with an even longer circulating half-life, a super-hCG. Even more efficient super-hemochorial occurred. This evolutionary process of hCG improving continued. Then, evolved humans produced more acidic hCG with an even longer circulating half-life, a super-ultra-hCG. Even more efficient super-ultra-hemochorial occurred.

It was through this process of hemochorial placentation, hemochorial placentation+, super-hemochorial placentation, and super-ultra-hemochorial placentation that the brains in primates, hominids, and humans were able to grow bigger. A byproduct of this development was the generation of a super-ultra-hyperglycosylated hCG, or a super-ultra-growth factor and invasion stimulus. This super-ultra problem maker resides in our genome. It is our human time bomb. As is now proven, human cancers use this super-ultra-growth factor and invasion factor to drive most malignancies [13,14,25,26]. This molecule designed as a driver of human implantation of pregnancy drives human cancers.

Cancer cells during transformation start to express hCG α-subunit and β-subunit genes. Cancer cells lack the enzymes needed to complete disulfide bridges in hCG. The end result is that β-subunit cannot combine with α-subunit in cancers. Cancer cells secrete a hyperglycosylated hCG free β-subunit [14]. This cancer molecule, like hyperglycosylated hCG, binds an autocrine TGFβ receptor [13] and promotes growth and invasion of cancer cells [13,14,25,26]. Hyperglycosylated hCG and its free β-subunit are human time bombs. Yes, they did well in creating humans and the human brain, but in humans they are just a time bomb.

References

[1] Ohlsson R, Larsson E, Nilsson O, Wahlstrom T, Sundstrom P. Blastocyst implantation precedes induction of insulin-like growth factor II gene expression in human trophoblasts. Development 1989;106:555–9.

[2] d'Hauterivea SP, Berndtb BS, Tsampalasa M, Charlet-Renarda C, Duboisb M, Bourgainc C, et al. Dialogue between blastocyst hCG and endometrial hCG/LH receptor: which role in implantation? Gyncol Obstet Invest 2007;64:156–60.
[3] Joshi NJ, Nandedkar TD. Effects of intrauterine instillation of antiserum to hCG during early pregnancy in mice. Acta Endocrinol 1984;107:268–74.
[4] Srisuparp S, Strakova Z, Fazleabas AT. The role of chorionic gonadotropin (CG) in blastocyst implantation. Arch Med Res 2001;32:627–34.
[5] Tsampalasa M, Grideleta V, Berndt S, Foidart J-M, Geenena V, d'Hauterive SP. Human chorionic gonadotropin: a hormone with immunological and angiogenic properties. J Repod Immunol 2010.
[6] d'Hauterive SP. Implantation: the first maternal-embryo crosstalk. <http://hdl.handle.net/2268/28418>.
[7] Licht P, Russu V, Wildt L. On the role of human chorionic gonadotropin (hCG) in the embryo-endometrial microenvironment: implications for differentiation and implantation. Semin Reprod Med 2001;19(1):37–47.
[8] Enders AC. The implantation chamber, blastocyst and blastocyst imprint of the rat; a scanning electron microscope study. Anat Rec 1975;182:137–49.
[9] Cole LA, Dai D, Butler SA, Leslie KK, Kohorn EI. Gestational trophoblastic diseases: 1. Pathophysiology of hyperglycosylated hCG-regulated neoplasia. Gynecol Oncol 2006;102:144–9.
[10] Cole LA, Khanlian SA, Riley JM, Butler SA. Hyperglycosylated hCG (hCG-H) in gestational implantation, and in choriocarcinoma and testicular germ cell malignancy tumorigenesis. J Reprod Med 2006;51:919–29.
[11] Guibourdenche J, Handschuh K, Tsatsaris V, Gerbaud P, Leguy MC, Muller F, et al. Hyperglycosylated hCG is a marker of early human trophoblast invasion. J Clin Endocrinol Metab 2010;95:E240–4.
[12] Handschuh K, Guibourdenche J, Tsatsari V, Guesnon M, Laurendeau I, Evain-Brion D, et al. Human chorionic gonadotropin produced by the invasive trophoblast but not the villous trophoblast promotes cell invasion and is down-regulated by peroxisome proliferator-activated receptor-a. Endocrinology 2007;148:5011–19.
[13] Butler SA, Ikram MS, Mathieu S, Iles RK. The increase in bladder carcinoma cell population induced by the free beta subunit of hCG is a result of an anti-apoptosis effect and not cell proliferation. Brit J Cancer 2000;82:1553–6.
[14] Cole LA, Butler SA. Hyperglycosylated hCG, hCGß and hyperglycosylated hCGß: interchangeable cancer promoters. Molec Cell Endocrinol 2012;349:232–8.
[15] Berndt S, Blacher S, Munuat C, Detilleux J, Evain-Brion D, Noel A, et al. Hyperglycosylated human chorionic gonadotropin stimulates angiogenesis through TGF-ß receptor activation. FASEB J 2013; Abstract 12-213686.
[16] Murphy G, Reynolds JJ, Whitham SE, Docherty AJ, Angel P, Heath JK. Transforming growth factor beta modulates the expression of collagenase and metalloproteinase inhibitor. Euro Molec Biol Org J 1987;6:1899–904.
[17] Norwitz ER, Schust DJ, Fisher SJ. Implantation and the survival of early pregnancy. New Engl J Med 2001;345:1400–8.
[18] Semprini AE, Simon G. Not so efficient reproduction. Lancet 2000;356:257–8.
[19] Sasaki Y, Ladner DG, Cole LA. Hyperglycosylated hCG the source of pregnancy failures. Fertil Steril 2008;89:1781–6.
[20] Cole LA. Hyperglycosylated hCG and pregnancy failures. J Reprod Immunol 2012;93:119–22.

[21] Wilcox AJ, Baird DD, Weinberg CR. Time of implantation of the conceptus and loss of pregnancy. New Engl J. Med 1999;340:1796–9.
[22] Cole LA. hCG, the wonder of today's science. Reprod Biol Endocrinol 2012;10:1–18.
[23] Fiddes JC, Goodman HM. The cDNA for the β-subunit of hCG suggests evolution of a gene by readthrough into the 3′ untranslated region. Nature 1980;286:684–7.
[24] Maston GA, Ruvolo M. Chorionic gonadotropin has a recent origin within primates and an evlutionary history of selection. Mol Biol Evol 2002;19:320–35.
[25] Carter WB, Sekharem M, Coppola D. Human chorionic gonadotropin induces apoptosis in breast cancer. Breast Cancer Res Treat 2006;100:S243–4.
[26] Iles RK. hCGβ expression by cervical squamous carcinoma: *in vivo* histological association with tumour invasion and apoptosis. Histopathology 2008;53:147–55.

Pregnancy-2, maintenance of gestation

13

Laurence A. Cole
USA hCG Reference Service, Angel Fire, NM, USA

hCG comprises an α-subunit and a β-subunit. The α-subunit is common to hCG, the autocrine/paracrine hyperglycosylated hCG and the hormone pituitary hCG, and the hormones luteinizing hormone (LH), follicle-stimulating hormone (FSH), and thyroid-stimulating hormone (TSH). The hCG β-subunit, although structurally similar to the β-subunit of LH, differentiates hCG, hyperglycosylated hCG, and pituitary hCG from other molecules. hCG and LH both bind and function through a common hCG/LH receptor. The biggest difference is that LH, pI 8.0, has a circulating half-life of just 25 min [1], whereas hCG, pI 3.5, has a circulating half-life of 36 h [2], or 80-fold longer than that of LH. In many respects hCG is a super LH produced in pregnancy, with 80-times the biological activity of LH. Although LH, FSH, and TSH are made by the anterior lobe of the pituitary, hCG is produced by differentiated placental syncytiotrophoblast cells.

The original biological activity of hCG was revealed in the 1960s [3–6]. hCG takes over from LH in promotion of progesterone production by ovarian corpus luteal cells in pregnant women [7–25]. As we now know, hCG only promotes progesterone production for 3–4 weeks after pregnancy implantation. This function exists for less than 10% of the length of pregnancy. As shown in Table 13.1, hCG reaches a peak at 10 weeks of pregnancy, or almost 1 month after progesterone promotion is complete, and then continues to be produced through the length of pregnancy. Clearly, corpus luteal progesterone production is not the major purpose of hCG. As illustrated in Figure 13.1, hCG has been found in recent years to have numerous functions in the placenta, uterus, and fetus during pregnancy.

The research groups of Rao et al., Zygmunt et al., and Noel et al. have each shown and confirmed that hCG functions to promote angiogenesis or vasculogenesis in the uterine vasculature during pregnancy (Figure 13.1). This insures optimal blood supply to the invading placenta and hemochorial placentation structures and optimal nutrition to the fetus [24–30]. hCG/LH receptor gene is expressed by uterine arteries and veins, and hCG/LH receptor presence has been demonstrated. This is probably a major function of hCG during pregnancy, insuring optimal blood supply from the mother to the hemochorial placentation to be passed on to the fetus. hCG also has an important function at the trophoblast tissue level, promoting the fusion of cytotrophoblast cells and their differentiation to syncytiotrophoblast cells [31,32].

Four research teams have shown and confirmed that hCG promotes an anti-macrophage inhibitory factor or a macrophage migration inhibitory factor, a cytokine that controls the immune response during pregnancy that reduces macrophage

Human Chorionic Gonadotropin (hCG). DOI: http://dx.doi.org/10.1016/B978-0-12-800749-5.00013-4

Table 13.1 Concentration of total hCG and hyperglycosylated hCG (hCG-H) in 496 serum samples from 310 women with term pregnancies measured using the Siemens Immulite 1000 total hCG assay

Gestation age (weeks since start of menstrual period)	*N*	Median total hCG (ng/ml)	Range total hCG (ng/ml) (variation)	Median hCG-H (ng/ml)	Range hCG-H (ng/ml) (variation)	hCG-H %
3 weeks	$n = 42$	0.26	0.04–5.5	0.20	0.01–6.45 (645×)	87%
4 weeks	$n = 42$	3.4	0.21–173 (824×)	2.5	0.18–160 (888×)	51%
5 weeks	$n = 67$	65	1.86–1308 (704×)	8.6	0.96–698 (731×)	43%
6 weeks	$n = 29$	252	3.80–855 (225×)	86	0.76–629 (827×)	36%
7 weeks	$n = 30$	3278	203–7766 (38×)	359	27–931 (34×)	16%
8 weeks	$n = 33$	4331	1064–10,057 (9.4×)	386	67–1050 (15.6×)	7.0%
9 weeks	$n = 24$	5832	1031–11,586 (11.2×)	430	102–1158 (11.3×)	5.1%
10 weeks	$n = 20$	10,352	1952–19,958 (10.2×)	521	188–1855 (9.9×)	4.3%
11–13 weeks	$n = 41$	5953	1440–15,318 (10.6×)	137	24–330 (13.7×)	2.3%
14–17 weeks	$n = 57$	2934	311–4757 (15.2×)	26	6.7–129 (19.3×)	1.3%
18–26 weeks	$n = 62$	1931	210–6223 (30.3×)	15.8	5.3–95 (17.9×)	0.65%
27–40 weeks	$n = 49$	1911	184–8530 (46.4×)	2.95	0.3–12.2 (40.6×)	0.14%

Source: Data from 50 pregnancies that failed due to miscarriage were excluded from this table. Hyperglycosylated hCG is indicated hCG-H, hCG-H % is the proportion of hCG-H of total hCG.

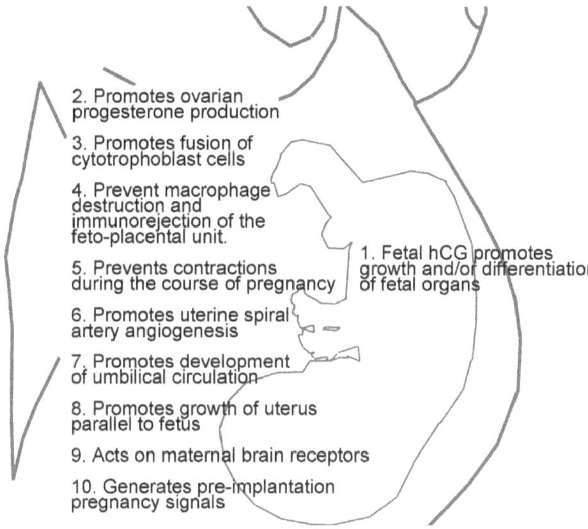

Figure 13.1 Actions of the hormone hCG during the course of pregnancy.

phagocytosis at the placenta–uterus interface, preventing destruction of foreign fetoplacental tissue [33–35]. Three other groups have shown that hCG may suppress any immune action against the invading foreign tissue [36–38]. hCG appears to be one of the factors acting to prevent rejection of the fetoplacental tissue. Most observations suggest that hCG has an inhibitory or suppressive function on macrophage activity. Wan et al. [35] demonstrated that hCG can directly enhance innate immunity by promoting macrophage function.

Various groups have found hCG/LH receptor in the myometrium of the uterus (Figure 13.1). It has been indicated by two groups that uterine growth in line with fetal growth may be stimulated by hCG [39,40]. Four groups have shown that hCG relaxes myometrial contractions during pregnancy. hCG acts on a BK-Ca calcium-activated channel to relax to myometrium during pregnancy [39,41–43]. hCG levels drop during the final weeks of pregnancy. It has been suggested that this drop may be the cause of increased contractions in the weeks before parturition.

Researchers are now finding hCG/LH receptors in fetal organs. Goldsmith et al. [44] have found hCG/LH receptors in the fetal kidney and liver. Rao et al. [45–49] have located hCG/LH receptors in the lung, liver, kidneys, spleen, and small and large intestines. Interestingly, this hCG/LH receptor is present in the fetal organs but absent in the adult organs. It is concluded that hCG may promote organ growth and differentiation in the fetus. The human fetus produces its own independent source of hCG from the fetal kidneys and liver [44,50]. The concentrations in fetal circulation, however, are much lower than maternal concentrations, suggesting that placental hCG secretion is directed toward the maternal circulation, and it is prevented from entering into fetal circulation [50]. Although hCG/LH receptors have been demonstrated in fetal organs, no function has been directly demonstrated, only indicated. As such, all findings regarding the fetus have to be considered only as indications at this time. Unfortunately, most animals do not make a form of hCG, making the role of hCG in the human fetus difficult to prove.

Other new research shows that hCG influences the eicosanoid metabolism in fetal membranes [46–49]. Seemingly, this action contributes to the progressive weakening of fetal membranes during the progression of labor, effecting fetal membranes that extend into the cervical canal and contributing to eicosanoid softening of the membranes, which is a prerequisite for a successful delivery.

hCG has been shown to function in umbilical cord growth and development [51,52]. This may be part of hemochorial placentation (see Chapter 14). Other new data show the pre-pregnancy implantation function of hCG. Studies by Rao et al. [53–55] and Gawronska et al. [55] show the presence of an hCG/LH receptor (shown by presence of mRNA and demonstration of receptor action) in human sperm and in the fallopian tubes. The function of the hCG/LH receptor in sperm is uncertain. It possibly has some relationship to fertility. The hCG/LH receptor in the fallopian tubes may be that which is normally acted on by LH, which relaxes the fallopian tube for fertilization to take place.

It has long been thought that hCG may have a role in implantation of pregnancy [56–59]. Articles suggest an autocrine or paracrine function of hCG in implantation of pregnancy. hCG of implantation is seemingly produced by cytotropblast cells. However, hCG is an endocrine. We know from recent research that a variant of hCG, hyperglycosylated hCG, rather than hCG itself, is produced by cytotrophoblast cells [8,9]. Hyperglycosylated hCG is an autocrine or paracrine and has been shown to directly promote implantation of pregnancy (see Chapter 14) [8,10,11]. This is seemingly what was considered the hCG implantation function. A recent study by Fluhr et al. [57] suggests a direct role of hCG in cytotrophoblast cell metalloproteinase production; this could be true and needs careful investigation.

Finally, the hCG/LH receptor has been found in the brains of adult women. CNS receptors are present in several areas of the brain such as the hippocampus, hypothalamus, and brain stem [60,61]. The finding of an hCG receptor in these parts of the brain may explain the hyperemesis, or nausea and vomiting, that occurs during normal pregnancy.

hCG has a wide range of actions through the hCG/LH receptor to control pregnancy (Figure 13.1). They seemingly start their action early, with blastocysts signaling the endometrium of forthcoming implantation. Hyperglycosylated hCG then promotes implantation and growth of cytotrophoblast cells. hCG promotes the differentiation of cytotrophoblast cells to syncytiotrophoblast cells and so the villous structures which are mixture of the two cell types are formed. During pregnancy, hCG promotes the uterine vasculature and the umbilical circulation to maximally provide blood to the hemochorial placentation structure and to the fetus. hCG has critical functions in suppressing macrophage attack on the maternally foreign fetoplacental unit, and in suppressing contractions during the term of the pregnancy. hCG also seemingly acts on the fetus to promote growth and differentiation of fetal organs during pregnancy. During this time hCG acts on the maternal brain, causing hyperemesis gravidarum. Considering everything together, hCG and hyperglycosylated hCG are seemingly a master hormone and an autocrine that control pregnancy.

References

[1] Schalch DS, Parlow AF, Boon RC, Reichlin S. Measurement of human luteinizing hormone in plasma by radioimmunoassay. J Clin Invest 1968;47:665–78.

[2] Faiman C, Ryan RJ, Zwirek SJ, Rubin ME. Serum FSH and HCG during human pregnancy and puerperium. J Clin Endocrinol Metab 1968;28:1323–9.

[3] Aschner B. Ueber die function der hypophyse. Pflug Arch Gest Physiol 1912;146:1–147.

[4] Fellner OO. Experimentelle untersuchungen uber die wirkung von gewebsextrakten aus der plazenta und den weiblichen sexualorganen auf das genital. Arch Gynakol 1913;100:641.

[5] Hirose T. Experimentalle histologische studie zur genese corpus luteum. Mitt Med Fakultd Univ ZU 1919;23:63–70.

[6] Aschheim S, Zondek B. Das Hormon des hypophysenvorderlappens: testobjekt zum Nachweis des hormons. Klin Wochenschr 1927;6:248–52.

[7] Hoshina M, Boime I, Mochizuki M. Cytological localization of hPL, hCG, and mRNA in chorionic tissue using *in situ* hybridization. Acta Obstet Gynaecol Japonica 1984;36:397–404.

[8] Handschuh K, Guibourdenche J, Tsatsaris V, Guesnon M, Laurendeau I, Evain-Brion D, et al. Human chorionic gonadotropin produced by the invasive trophoblast but not the villous trophoblast promotes cell invasion and is down-regulated by peroxisome proliferator-activated receptor-a. Endocrinology 2007;148:5011–19.

[9] Kovalevskaya G, Genbacev O, Fisher SJ, Cacere E, O'Connor JF. Trophoblast origin of hCG isoforms: cytotrophoblasts are the primary source of choriocarcinoma-like hCG. Mol Cell Endocrinol 2002;194:147–55.

[10] Sasaki Y, Ladner DG, Cole LA. Hyperglycosylated hCG the source of pregnancy failures. Fertil Steril 2008;89:1781–6.

[11] Cole LA, Dai D, Butler SA, Leslie KK, Kohorn EI. Gestational trophoblastic diseases: 1. Pathophysiology of hyperglycosylated hCG-regulated neoplasia. Gynecol Oncol 2006;102:144–9.

[12] Cole LA. Structures of free α-subunit and free β-subunit. In: Cole LA, editor. Human chorionic gonadotropin (hCG). Oxford: Elsevier; 2010.

[13] Butler SA, Ikram MS, Mathieu S, Iles RK. The increase in bladder carcinoma cell population induced by the free beta subunit of hCG is a result of an anti-apoptosis effect and not cell proliferation. Brit J Cancer 2000;82:1553–6.

[14] Iles RK. Ectopic hCGβ expression by epithelial cancer: malignant behavior metastasis and inhibition of tumor cell apoptosis. Mol Cell Endocrinol 2007;260:264–70.

[15] Cole LA, Ladner DG. Background hCG in non-pregnant individuals: need for more sensitive point-of-care and over-the-counter pregnancy tests. Clin Biochem 2009;42:168–75.

[16] Cole LA. Background hCG. In: Cole LA, editor. Human chorionic gonadotropin (hCG). Oxford: Elsevier; 2010.

[17] Rao CV, Griffin LP, Carman Jr. FR. Prostaglandin F2 alpha binding sites in human corpora lutea. J Clin Endocrinol Metab 1977;44:1032–7.

[18] Strott CA, Yoshimi T, Ross GT, Lipsett MB. Ovarian physiology: relationship between plasma LH and steroidogenesis by the follicle and corpus luteum; effect of HCG. J Clin Endocrinol Metab 1969;29:1157–67.

[19] Cedard L, Varangot J, Yannotti S. The metabolism of estrogens in human placentas artificially maintained in survival by perfusion *in vitro*. C R Hebd Seances Acad Sci 1962;254:1870–1.

[20] Azuma K, Calderon I, Besanko M, MacLachlan V, Healy DL. Is the luteo-placental shift a myth? Analysis of low progesterone levels in successful art pregnancies. J Clin Endocrinol Metab 1993;77:195–8.

[21] Mayerhofer A, Fritz S, Grunert R, Sanders SL, Duffy DM, Ojeda SR, et al. D1-Receptor, DARPP-32, and PP-1 in the primate corpus luteum and luteinized granulosa cells: evidence for phosphorylation of DARPP-32 by dopamine and human chorionic gonadotropin. J Clin Endocrinol Metab 2000;85:4750–7.

[22] Pierce JG, Parsons TF. Glycoprotein hormones: structure and function. Ann Rev Biochem 1981;50:65–95.

[23] Rao CV. Differential properties of human chorionic gonadotropin and human luteinizing hormone binding to plasma membranes of bovine corpora luteal. Acta Endocrinol 1979;90:696–710.

[24] Berndt S, Blacher S, d'Hauterive PS, Thiry M, Tsampalas M, Cruz A, et al. Chorionic gonadotropin stimulation of angiogenesis and pericyte recruitment. J Clin Endocrinol Metab 2009;94:4567–74.

[25] Toth P, Li X, Rao CV, Lincoln SR, Sanfillipino JS, Spinnato JA, et al. Expression of functional human chorionic gonadotropin/human luteinizing hormone receptor gene in human uterine arteries. J Clin Endocrinol Metab 1994;79:307–15.

[26] Lei ZM, Reshef E, Rao CV. The expression of human chorionic gonadotropin/luteinizing hormone receptors in human endometrial and myometrial blood vessels. J Clin Endocrinol Metab 1992;75:651–9.

[27] Zygmunt M, Herr F, Keller-Schoenwetter S, Kunzi-Rapp K, Munstedt K, Rao CV, et al. Characterization of human chorionic gonadotropin as a novel angiogenic factor. J Clin Endocrinol Metab 2002;87:290–5296.

[28] Herr F, Baal N, Reisinger K, Lorenz A, McKinnon T, Preissner KT, et al. hCG in the regulation of placental angiogenesis. Results of an *in vitro* study. Placenta 2007;28 (Suppl. A):S85–93.

[29] Zygmunt M, Herr F, Munstedt K, Lang U, Liang OD. Angiogenesis and vasculogenesis in pregnancy. Eur J Obstet Gynecol Reprod Biol 2003;110(Suppl. 1):S10–18.

[30] Toth P, Lukacs H, Gimes G, Sebestyen A, Pasztor N, Paulin F, et al. Clinical importance of vascular hCG/LH receptors—a review. Reprod Biol 2001;1:5–11.

[31] Shi QJ, Lei ZM, Rao CV, Lin J. Novel role of human chorionic gonadotropin in differentiation of human cytotrophoblasts. Endocrinology 1993;132:387–95.

[32] Cronier L, Bastide B, Herve JC, Deleze J, Malassine A. Gap junctional communication during human trophoblast differentiation: influence of human chorionic gonadotropin. Endocrinology 1994;135:402–8.

[33] Akoum A, Metz CN, Morin M. Marked increase in macrophage migration inhibitory factor synthesis and secretion in human endometrial cells in response to human chorionic gonadotropin hormone. J Clin Endocrinol Metab 2005;90:2904–10.

[34] Matsuura T, Sugimura M, Iwaki T, Ohashi R, Kanayama N, Nishihira J. Anti-macrophage inhibitory factor antibody inhibits PMSG-hCG-induced follicular growth and ovulation in mice. J Assist Reprod Genet 2002;19:591–5.

[35] Wan H, Marjan A, Cheung VW, Leenen PJM, Khan NA, Benner R, et al. Chorionic gonadotropin can enhance innate immunity by stimulating macrophage function. J Leukocyte Biol 2007;82:926–33.

[36] Kamada M, Ino H, Naka O, Irahara M, Daitoh T, Mori K, et al. Immunosuppressive 30-kDa protein in urine of pregnant women and patients with trophoblastic diseases. Eur J Obstet Gynecol Reprod Biol 1993;50:219–25.

[37] Noonan FP, Halliday WJ, Morton H, Clunie GJA. Early pregnancy factor is immunosuppressive. Nature 1879;278:649–51.
[38] Majumdar S, Bapna BC, Mapa MK, Gupta AN, Devi PK, Subrahmanyam D. Pregnancy specific proteins: suppression of *in vitro* blastogenic response to mitogen by these proteins. Int J Fertil 1982;27:66–9.
[39] Reshef E, Lei ZM, Rao CV, Pridham DD, Chegini N, Luborsky JL. The presence of gonadotropin receptors in nonpregnant human uterus, human placenta, fetal membranes, and decidua. J Clin Endocrinol Metab 1990;70:421–30.
[40] Zuo J, Lei ZM, Rao CV. Human myometrial chorionic gonadotropin/luteinizing hormone receptors in preterm and term deliveries. J Clin Endocrinol Metab 1994;79:907–11.
[41] Eta E, Ambrus G, Rao V. Direct regulation of human myometrial contractions by human chorionic gonadotropin. J Clin Endocrinol Metab 1994;79:1582–6.
[42] Doheny HC, Houlihan DD, Ravikumar N, Smith TJ, Morrison JJ. Human chorionic gonadotrophin relaxation of human pregnant myometrium and activation of the BKCa channel. J Clin Endocrinol Metab 2003;88:4310–15.
[43] Edelstam G, Karlsson C, Westgren M, Löwbeer C, Swahn ML. Human chorionic gonadotropin (hCG) during third trimester pregnancy. Scand J Clin Lab Invest 2007;67:519–25.
[44] Goldsmith PC, McGregor WG, Raymoure WJ, Kuhn RW, Jaffe RB. Cellular localization of chorionic gonadotropin in human fetal kidney and liver. J Clin Endocrinol Metab 1983;57:54–61.
[45] Abdallah MA, Lei ZM, Li X, Greenwold N, Nakajima ST, Jauniaux E, et al. Human fetal nongonadal tissues contain human chorionic gonadotropin/luteinizing hormone receptors. J Clin Endocrinol Metab 2004;89:952–6.
[46] Rao CV. Chapter 15: Paradigm shift on the targets of hCG actions. In: Cole LA, editor. Human chorionic gonadotropin (hCG). Oxford: Elsevier; 2010.
[47] Rao CV. Nongonadal actions of LH and hCG in reproductive biology and medicine. Semin Reprod Med 2001;19:1–119.
[48] Rao CV. An overview of the past, present and future of nongonadal hCG/LH actions in reproductive biology and medicine. Semin Reprod Endocrinol 2001;19:7–17.
[49] Rao CV, Lei ZM. The past, present and future of nongonadal hCG/LH actions in reproductive biology and medicine. Mol Cell Endocrinol 2007;269:2–8.
[50] McGregor WG, Raymoure WJ, Kuhn RW, Jaffe RB. Fetal tissues can synthesize a placental hormone. Evidence for chorionic gonadotropin β-subunit synthesis by human fetal kidney. J Clin Invest 1981;68:306–9.
[51] Rao CV, Li X, Toth P, Lei ZM. Expression of epidermal growth factor, transforming growth factor-alpha and their common receptor genes in human umbilical cords. J Clin Endocrinol Metab 1995;80:1012–20.
[52] Rao CV, Li X, Toth P, Lei ZM, Cook VD. Novel expression of functional human chorionic gonadotropin/luteinizing hormone receptor in human umbilical cords. J Clin Endocrinol Metab 1993;77:1706–14.
[53] Lei ZM, Toth P, Rao CV, Pridham D. Novel coexpression of human chorionic gonadotropin (hCG)/human luteinizing hormone receptors and their ligand hCG in human fallopian tubes. J Clin Endocrinol Metab 1993;77:863–72.
[54] Rao CV. Physiological and pathological relevance of human uterine hCG/LH receptors. J Soc Gynecol Invest 2006;13:77–8.
[55] Gawronska B, Paukku T, Huhtaniemi I, Wasowicz G, Ziecik AJ. Oestrogen-dependent expression of hCG/LH receptors in pig fallopian tube and their role in relaxation of the oviduct. J Reprod Fertil 1999;115:293–301.

[56] Licht P, Fluhr H, Neuwinger J, Wallwiener D, Wildt L. Is human chorionic gonadotropin directly involved in the regulation of human implantation? Mol Cell Endocrinol 2007;269:85–92.
[57] Fluhr H, Bischof-Islami D, Krenzer S, Licht P, Bischof P, Zygmunt M. Human chorionic gonadotropin stimulates matrix metalloproteinases-2 and -9 in cytotrophoblastic cells and decreases tissue inhibitor of metalloproteinases-1, -2, and -3 in decidualized endometrial stromal cells. Fertil Steril 2008;90:1390–5.
[58] Reis FM, Cobellis L, Luisi S, Driul L, Florio P, Faletti A, et al. Paracrine/autocrine control of female reproduction. Gynecol Endocrinol 2000;14:464–75.
[59] Ticconi C, Zicari A, Belmonte A, Realacci M, Rao ChV, Piccione E. Pregnancy-promoting actions of HCG in human myometrium and fetal membranes. Placenta 2009;28:S137–43.
[60] Lei ZM, Rao CV, Kornyei J, Licht P, Hiatt ES. Novel expression of human chorionic gonadotropin/luteinizing hormone receptor gene in brain. Endocrinology 1994;132:262–70.
[61] Rao CV. Immunocytochemical localization of gonadotropin and gonadal steroid receptors in human pineal glands. J Clin Endocrinol Metab 1997;82:2756–7.

Pregnancy-3, creation and continuation of hemochorial placentation

14

Laurence A. Cole
USA hCG Reference Service, Angel Fire, NM, USA

There is one very solid piece of data showing, beyond question, that chorionic gonadotropin (CG) and/or hyperglycosylated CG controls and initiates hemochorial placentation, that is, the simultaneous appearance of cytotrophoblast hyperglycosylated CG and syncytiotrophoblast CG (prior species, i.e., prosimian primates, made no form of CG) and hemochorial placentation (prior species, i.e., prosimian primates, relied on epitheliochorial placentation) in early simian primates [1,2]. No other hormone, autocrine, or molecular initiation mechanism appears with the evolution of early simian primates [1,2].

In support of this assumption, CG and hyperglycosylated CG function in humans in the synthesis of every component of hemochorial placentation, growth of trophoblast cells [3–5], differentiation of trophoblast cells [6] or growth of villous trophoblast [7], growth of maternal uterine spiral artery supplying blood [8–11], and growth of umbilical vessels distributing blood to fetus [11–14]. Considering everything together, CG and hyperglycosylated CG clearly appear to promote and maintain hemochorial placentation [7].

Looking carefully, step-by-step, at the process of human hemochorial placentation, the CG and hyperglycosylated CG relationship is observed. As proven, hyperglycosylated CG promotes growth of root placental cells: cytotrophoblast cells [3–5]. hCG promotes the fusion of cytotrophoblast cells to multinuclear syncytiotrophoblast cells [6]. These two processes lead to synthesis of cytotrophoblast cells and syncytiotrophoblast cells or villous tissue.

Figure 14.1 illustrates the engulfment of the blastocyst by the decidua at 3.5 weeks of gestation. It is cytotrophoblast hyperglycosylated CG that drives this initial invasion into the decidua by the blastocyst. The blastocyst takes on notable features at this time, including a cytotrophoblast column (Figure 14.1) [15]. This produces hyperglycosylated CG, which in turn inhibits cytotrophoblast apoptosis, promotes growth, and seemingly promotes continual invasion through transforming growth factor (TGFβ type II) receptors.

The early syncytiotrophoblast cells make CG, which enters the circulation and maintains progesterone production by the corpus luteum. CG also promotes growth of uterine spiral arteries, as well as the fusion of cytotrophoblast cells to

Human Chorionic Gonadotropin (hCG). DOI: http://dx.doi.org/10.1016/B978-0-12-800749-5.00014-6

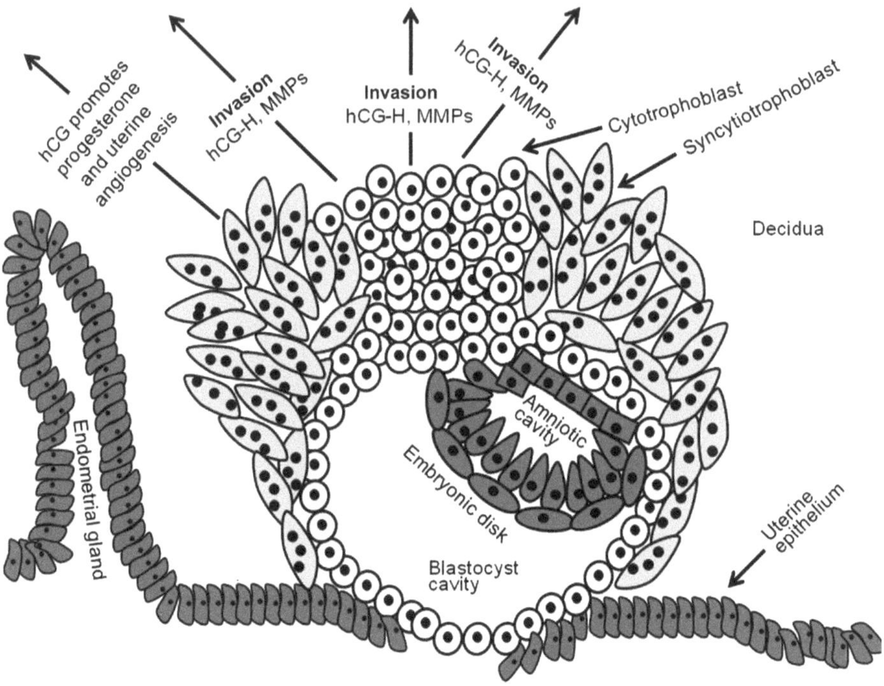

Figure 14.1 Complete implantation of blastocyst. Approximately 3–4 days after attachment, or at 3.5 weeks of gestation, the blastocyst becomes wholly implanted into the decidua. The blastocyst forms notable features, including a cytotrophoblast extension or extravillous cytotrophoblast. This produces hCG-H, which inhibits cytotrophoblast apoptosis, promoting growth invasion through the decidua. The syncytiotrophoblast cells immediately start making hCG, which maintains progesterone production by the corpus luteum and promotes growth of myometrial and decidual spiral arteries. The inner cell mass, or amnion, forms an amniotic cavity. Cytotrophoblast cells are shown in white, syncytiotrophoblast cells are in light gray, and uterine epithelium and the inner cell mass (amnion) are in dark gray.

syncytiotrophoblast cells, and inhibits immune rejection of the foreign cells invading the uterus. Figure 14.2 illustrates the amnion and chorion at 4 weeks of gestation. Amnion differentiation occurs with formation of a clear yolk sac. Multiple cytotrophoblast extensions or columns form. These are the roots of villous trophoblast structures that form in the following weeks.

Figure 14.3 illustrates the amnion and chorion at 5 weeks of gestation. The cytotrophoblast columns (promoted by hyperglycosylated CG) combine via differentiation to syncytiotrophoblast cells (promoted by CG) and lead to the initiation of villus formation. At 6 weeks of gestation, complete villus structures are observed (Figure 14.4). Villous structures are primarily composed of cytotrophoblast cells overlaid by syncytiotrophoblasts.

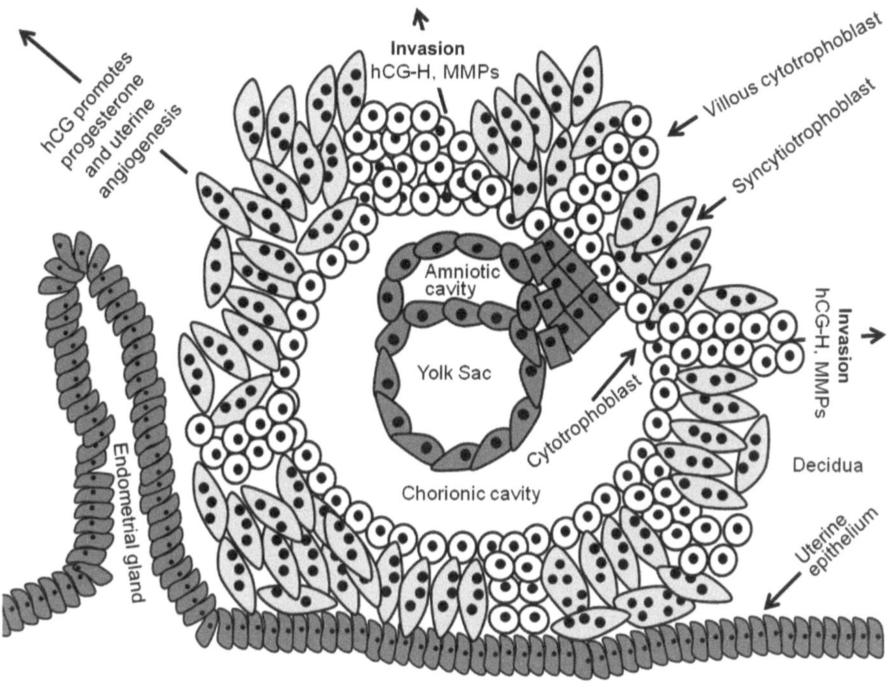

Figure 14.2 At 4 weeks of gestation, amnion differentiation occurs with formation of a clear yolk sac. Multiple cytotrophoblast extensions become the roots of villous structures. hCG-H-driven invasion continues through the decidua and into the myometrium. hCG promotes corpus luteal progesterone production and myometrial and decidual spiral artery angiogenesis. Cytotrophoblast cells are shown in white, syncytiotrophoblast cells are in light gray, and uterine epithelium and amnion cells are in dark gray.

Two types of villi are formed within each structure: floating villi, which are the active villi for maternal/fetal nutrient exchange, and the anchor villi for invasion and anchoring of villi (Figure 14.4). The anchor villi are terminated by a mass of extravillous cytotrophoblast cells. Hyperglycosylated CG produced by the extravillous cytotrophoblast cells drives invasion of the villous structures to as deep as one-third the thickness of the myometrium. CG produced by syncytiotrophoblast cells continuously promotes angiogenesis in decidual and myometrial vasculature, so that vasculature can meet the deeply invading placental villi.

By 6 weeks of gestation, CG-promoted corpus luteal progesterone production comes to an end. At this time, the placental syncytiotrophoblast cells take over progesterone production. CG continues to immune-suppress the invading trophoblast cells, promote uterine growth, and relax contraction of the myometrial muscle cells. As illustrated in Figure 14.4, fetal circulation begins at this stage, feeding into a primitive umbilical cord. The villus structure is now soaking in maternal blood; the

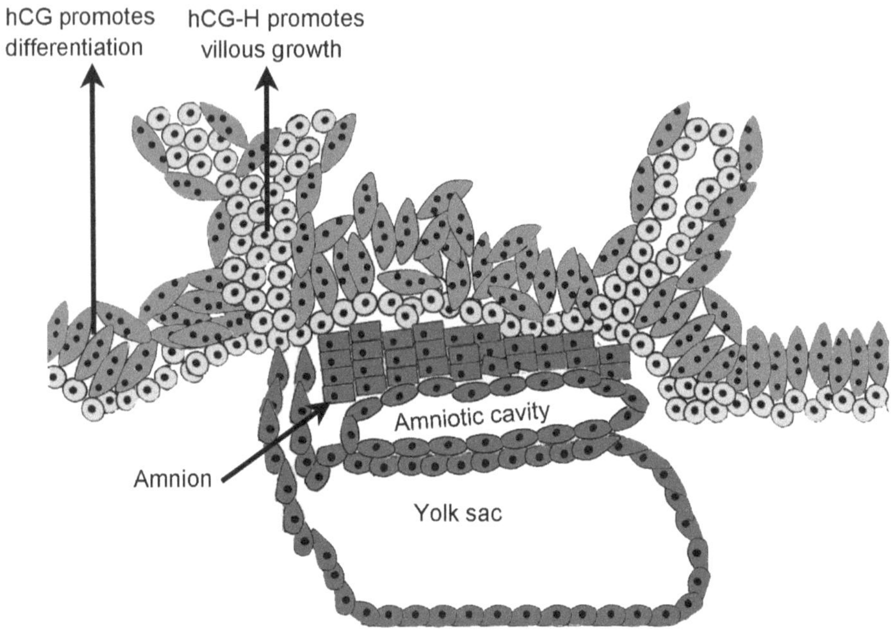

Figure 14.3 Trophoblast cells and amnion cells at 5 weeks of gestation. Cytotrophoblast columns grow because of hCG-H promotion; with hCG differentiation, they form primitive extensions or villus structures. Amnion grows and the yolk sac expands.

floating villus is transferring oxygen, glucose, and other nutrients from the maternal blood into the circulation of the developing fetus.

By 8–10 weeks of pregnancy, the villus structures have attached to the decidua parientalis and hemochorial placentation is now fully active (Figures 14.4 and 14.5). Villous structures have maximally invaded the myometrium and floating villi are now optimally filtering surrounding maternal blood into the fetal circulation. Invasion by the extravillous cytotrophoblast cells or hyperglycosylated CG continues. Figure 14.5 shows hemochorial placentation in place and active for the remainder of the pregnancy. Recent research shows that hyperglycosylated CG is critical through the length of the second trimester. Deficiency leads to preeclampsia, indicating that villous growth promotion of hemochorial placentation by hyperglycosylated CG through the second trimester of pregnancy is critical [16].

Hemochorial placentation as controlled and promoted by CG and hyperglycosylated CG absorbs maternal nutrients and passes them into the fetal circulation efficiently.

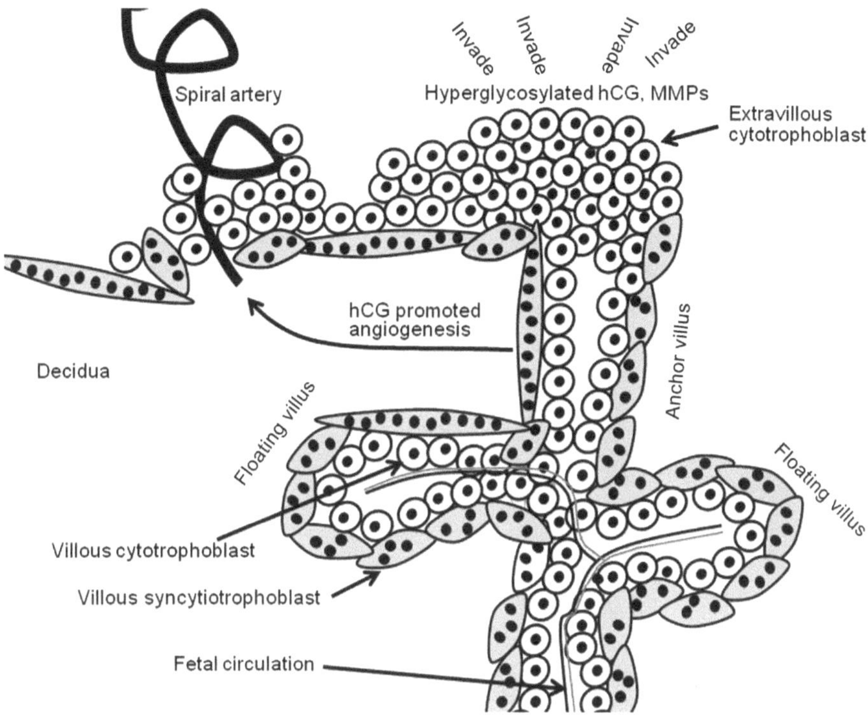

Figure 14.4 A fully formed villus structure at 6 weeks of gestation. This structure is formed by the growing cytotrophoblast column (promoted by hCG-H) and differentiated by hCG. Two types of villi are formed: floating villi and an anchor villus. The anchor villus is terminated by a mass of extravillous cytotrophoblast cells. hCG-H (produced by the extravillous cytotrophoblast cells) drives invasion to as deep as one-third the thickness of the myometrium. hCG (produced by syncytiotrophoblast cells) continuously promotes angiogenesis in decidual and spiral arteries. At this point, syncytiotrophoblast cells now produce progesterone themselves and take over hCG production from corpus luteal cells. As illustrated, active fetal circulation begins; the syncytiotrophoblast cells of floating villi extract nutrition from surrounding maternal blood and filter it into the fetal circulation. Cytotrophoblast cells are shown in white and syncytiotrophoblast cells are in light gray.

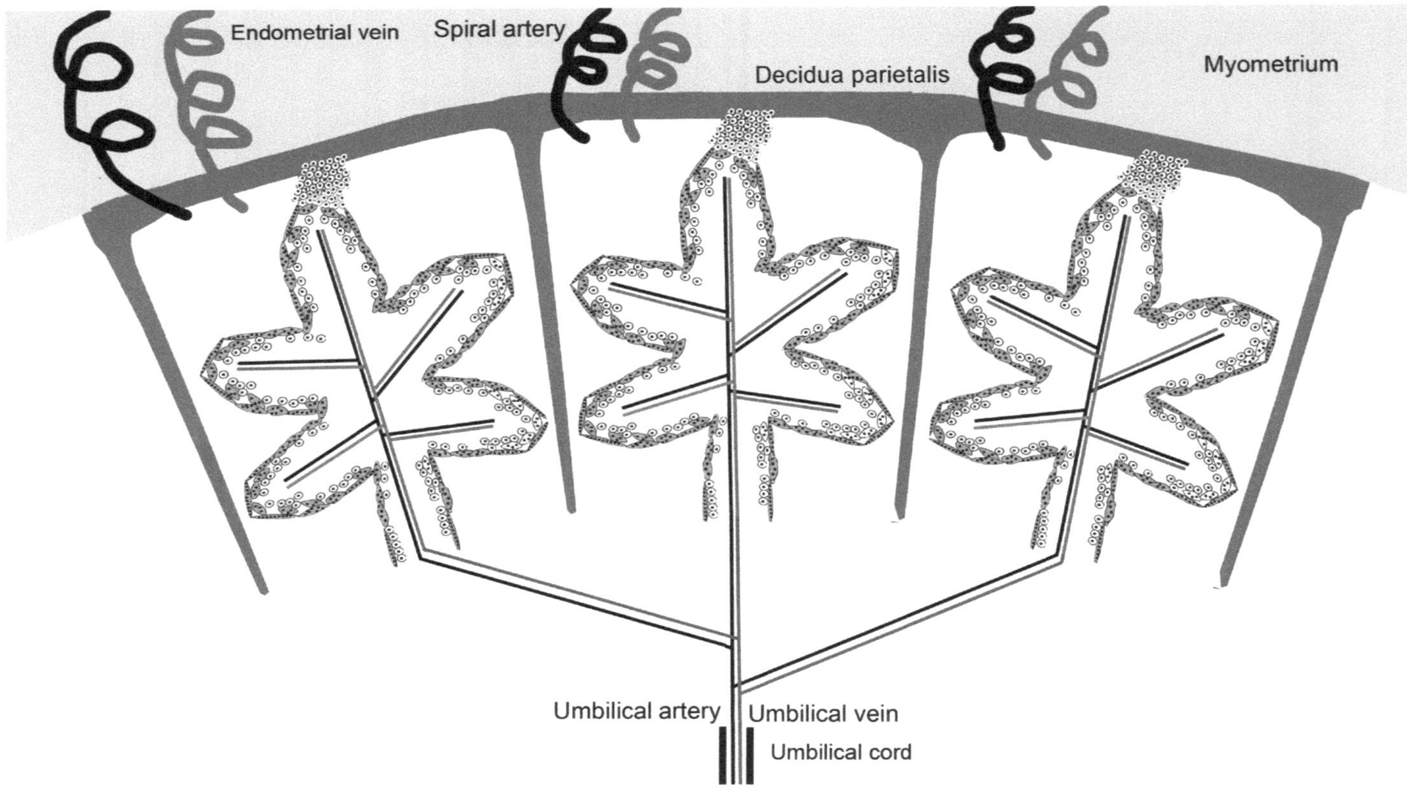

Figure 14.5 Active hemochorial placentation at 8–10 weeks of gestation. This figure shows the complete fetal nutrition system, with villous structures implanted into the decidua parientalis within the myometrium.

References

[1] Fiddes JC, Goodman HM. The cDNA for the ß-subunit of human chorionic gonadotropin suggests evolution of a gene by readthrough into the 3′-untranslated region. Nature 1980;286:684–7.

[2] Maston GA, Ruvolo M. Chorionic gonadotropin has a recent origin within primates and an evolutionary history of selection. Mol Biol Evol 2002;19:320–34.

[3] Cole LA, Dai D, Leslie KK, Butler SA, Kohorn EI. Gestational trophoblastic diseases: 1. Pathophysiology of hyperglycosylated hCG-regulated neoplasia. Gynecol Oncol 2006;102:144–9.

[4] Handshuh K, Guibourdenche J, Tsatsaris V, Guesnon M, Laurendeau I, Evain-Brion D, et al. Human chorionic gonadotropin produced by the invasive trophoblast but not the villous trophoblast promotes cell invasion and is down-regulated by peroxisome proliferator-activated receptor-a. Endocrinology 2007;148:5011–19.

[5] Cole LA, Khanlian SA, Riley JM, Butler SA. Hyperglycosylated hCG (hCG-H) in gestational implantation, and in choriocarcinoma and testicular germ cell malignancy tumorigenesis. J Reprod Med 2006;51:919–29.

[6] Shi QJ, Lei ZM, Rao CV, Lin J. Novel role of human chorionic gonadotropin in differentiation of human cytotrophoblasts. Endocrinology 1993;132:1387–95.

[7] Cole LA. hCG and hyperglycosylated hCG, promoters of villous placenta and hemochorial placentation. Placenta: functions, development and disease. Nova Publishers; 2013.

[8] Lei ZM, Reshef E, Rao CV. The expression of human chorionic gonadotropin/luteinizing hormone receptors in human endometrial and myometrial blood vessels. J Clin Endocrinol Metab 1992;75:651–9.

[9] Herr F, Baal N, Reisinger K, Lorenz A, McKinnon T, Preissner KT, et al. HCG in the regulation of placental angiogenesis. Results of an *in vitro* study. Placenta 2007;28 (Suppl. A):S85–93.

[10] Zygmunt M, Herr F, Munstedt K, Lang U, Liang OD. Angiogenesis and vasculogenesis in pregnancy. Eur J Obstet Gynecol Reprod Biol 2003;110:S10–18.

[11] Toth P, Lukacs H, Gimes G, Sebestyen A, Pasztor N, Paulin F, et al. Clinical importance of vascular LH/hCG receptors: a review. Reprod Biol 2001;1:5–11.

[12] Rao CV, Li X, Toth P, Lei ZM. Expression of epidermal growth factor, transforming growth factor-alpha and their common receptor genes in human umbilical cords. J Clin Endocrinol Metab 1995;80:1012–20.

[13] Rao CV, Li X, Toth P, Lei ZM, Cook VD. Novel expression of functional human chorionic gonadotropin/luteinizing hormone receptor in human umbilical cords. J Clin Endocrinol Metab 1993;77:1706–14.

[14] Wasowicz G, Derecka K, Stepien A, Pelliniemi L, Doboszynska T, Gawronska B, et al. Evidence for the presence of luteinizing hormone–chorionic gonadotrophin receptors in the pig umbilical cord. J Reprod Fertil 1999;117:1–9.

[15] Norwitz ER, Schust DJ, Fisher SJ. Implantation and the survival of early pregnancy. N Engl J Med 2001;345:1400–8.

[16] Brennan MC, Wolfe MD, Murray-Krezan CM, Cole LA, Rayburn WF. First trimester hyperglycosylated human chorionic gonadotropin and development of hypertension. Prenat Diagn 2013.

Paradigm shift on the targets of hCG actions

15

C.V. Rao
Herbert Wertheim College of Medicine, Florida International University, Miami, FL, USA

hCG is a heterodimeric glycoprotein hormone that belongs to the cystine-knot growth factor family [1,2]. In addition, it has properties of cytokines and chemokines [3]. Its α-subunit is encoded by a single gene and its β-subunit is encoded by a gene cluster [1,3]. Although human placental syncytiotrophoblasts produce large quantities, various normal and cancerous tissues can secrete small amounts of hCG that are usually below the detection limits of conventional hCG assays [3]. hCG binds to a receptor that is shared with LH [1]. For a long time, this so-called hCG/LH receptor was thought to be present only in gonads [1,3]. However, the data published from around the world in the past 20 years from various laboratories have demonstrated that many nongonadal reproductive and nonreproductive organs and cells also contain them [4–6]. The activation of these receptors results in functional consequences that vary with the tissue, cell type, and physiological state [4–6]. The receptors are encoded by a single copy gene [7,8]. The receptor binds hCG with somewhat higher affinity than LH and only dimeric hormones bind [9]. Isolated hCG subunits have occasionally been reported to bind, but this binding can usually be attributed either to contaminating dimer hCG or to homodimerization of β-subunits, which mimics the conformation of the native hormone [1,3]. There is no convincing evidence to support hCG binding to any other type of receptor.

Some previous studies, conducted before the paradigm shift, indicated that hCG receptors might be present in nongonadal tissues [6]. These were thought to be where hCG binds to human placenta and regulates its endocrine functions and thought to be why there is lower uterine uptake after radioiodinated hCG is injected into superovulated rodents [6]. These findings suggested the presence of the receptor, but the possibility was not further investigated because of overwhelmingly accepted dogma at the time.

The paradigm shift was further fueled by the new technologies that emerged in the 1980s and 1990s, which allowed receptor detection at DNA, mRNA, and protein levels [4–6]. Table 15.1 shows a list of hCG receptor – positive female and male nongonadal reproductive and nonreproductive organs and cells [4–6]. These receptors were detected in various laboratories using multiple techniques that minimized the possibility of artifacts [4–6]. The receptor levels are lower in nongonadal tissues than in gonads [4–6]. However, the total number of receptors, for example, in uterine tissue, far exceeds the number in the gonads [4–6]. Although it has not been determined for every tissue, the nongonadal receptor affinity appears to be

Human Chorionic Gonadotropin (hCG). DOI: http://dx.doi.org/10.1016/B978-0-12-800749-5.00015-8

Table 15.1 hCG-receptor-positive nongonadal reproductive and nonreproductive tissues and cells

Fallopian tubes	Neural retina
Uterus	Breast
Cervix	Skin
Oocyte/early embryo/blastocyst	Bone
Placenta	Urinary bladder
Fetal membranes	Adrenals-zona reticularis
Decidua	Blood vessels in target tissues
Umbilical cord	Suppressor and regulatory T cells/monocytes/ macrophages/dendritic cells
Brain	Prostate
Pineal gland	Epididymis
Spinal cord	Seminal vesicles
Cavernours sinus–carotid rete vascular complex	Sperm

similar to gonadal tissue [4–6]. Receptor detection by traditional ligand binding assays had been difficult because of the lower receptor numbers in nongonadal tissues. Many of the hCG receptor – positive nongonadal tissues are of human origin [4–6]. However, as exemplified by the uterus, the nongonadal receptors appear to be well-conserved across species [4–6].

hCG receptors have also been identified in lower organisms [10]. It is likely that the molecular nature of these receptors is different from those present in vertebrates, and they could have evolved to serve different functions in vertebrates. Thus, hCG receptors are probably ancient molecules that were active during the evolutionary process.

Lungs, liver, kidneys, spleen, and small and large intestines, which are hCG receptor – negative in adults, are hCG receptor – positive in the human fetus [11]. This finding suggests that hCG might play growth-promoting and differentiation-promoting roles in fetal organs; however, once these actions are completed, hCG receptors might disappear.

hCG is one of the earliest signals that a developing embryo can sense, and the levels of hCG rapidly increase when cells of the inner cell mass differentiate into tissues of all three germ layers [3]. Because of its unique properties [12], it should not be surprising that hCG might have a role in the differentiation of cells of the inner cell mass, along with the other known factors. Trophoectoderm, which develops into placenta, is likely to contain hCG receptors, suggesting that the hCG regulation of trophoblast functions might go back to the earliest stages of placental formation [3].

The hCG actions in the mother vary with the tissue and stage of pregnancy [4–6]. hCG receptors are present in gametes and could be involved in their maturation and fertilization [4–6]. The receptors in the early embryo might promote its growth and development [4–6]. The receptors in blastocysts might promote implantation, independent of estradiol and progesterone [4–6]. The process of implantation involves bidirectional communication between the implanting blastocyst and the receptive endometrium, and hCG derived from both these sources might play a role in this communication [4–6,13–17]. The human fetus might produce its own hCG from the kidneys and liver [18,19]. The levels in fetal circulation, however, are much lower than in the mother, suggesting that placental hCG secretion is almost always directed toward the maternal circulation and is prevented from entering into fetal circulation [3,20]. There could be a good reason for this. For instance, if maternal hCG floods fetal circulation, especially at certain critical times, it could possibly interfere with growth and differentiation of fetal tissues.

The hCG actions in the reproductive tract are dependent on region, reproductive phase, and cell type. For example, hCG might regulate the secretion and motility of Fallopian tubes that is conducive to gamete maturation, fertilization, early embryonic growth and development, and timely transport of the embryo into the uterus for implantation [4–6]. hCG in the uterus promotes blastocyst implantation by increasing the blastocyst's ability to attach and invade the endometrium and superficial myometrium. This increases blood supply to the uterus through a combination of new vessel formation and vasodilation of the existing vessels, inhibits myometrial activity, and prevents rejection of the newly implanted blastocysts by the maternal immune system [4–6,13–15,21–25].

The functions in the fetoplacental unit, which consists of placenta, fetal membranes, decidua, and umbilical cord, contribute to pregnancy maintenance, fetal well-being, and the successful initiation and progression of labor at the end of pregnancy [3–6]. In placenta, hCG regulates its own synthesis, promotes the differentiation of cytotrophoblasts into syncytiotrophoblasts, and promotes the invasion of extravillous trophoblasts. It also regulates the secretion of various cytokines and upregulates indoleamine 2,3-dioxygenase, which catalyzes the breakdown of tryptophan, a potent activator of T cells, thereby preventing T-cell attack on the fetal tissues [3–6,24,26–30].

All the changes that are now attributed to hCG are also induced by other regulatory agents, which raises the question of which comes first. This question has been answered for the differentiation of cytotrophoblasts. Although many other regulatory agents promote the differentiation, they are ineffective when the hCG actions are blocked, indicating that hCG actions are central to the formation of syncytiotrophoblasts from cytotrophoblasts [3].

hCG influences the eicosanoid metabolism in fetal membranes [4–6]. It is possible that this action contributes to the progressive weakening of fetal membranes during the progression of labor. Moreover, fetal membranes that extend into the cervical canal during active labor might contribute eicosanoids for cervical softening, which is a prerequisite for successful delivery.

The presence of hCG receptors in umbilical cord cells and blood vessels indicates that hCG might play a role in maintaining plasticity of the cord and regulating blood flow [4–6]. If it does perform such regulation, then hCG could be an important determinant of nutrient delivery to the fetus and of removal of metabolic waste from the fetus. Several cells of the immune system, such as suppressor and regulatory T cells, dendritic cells, monocytes, and macrophages, contain hCG receptors [4–6,22,26,27,29]. The functions that hCG regulates in these cells include influencing cytokine secretion, chemoattraction, and upregulating indoleamine dioxygenase [4,24,26–30]. These actions broadly fall into categories of establishing pregnancy, inducing apoptosis, engulfing the dead cells by activated macrophages along the path of trophoblast invasion in the uterus, and protecting the fetus from immunorejection.

hCG receptors are present in several areas of the central nervous system (CNS), such as the hippocampus, hypothalamus, brainstem, and anterior pituitary [4–6]. Consistent with this broad receptor distribution, hCG might have diverse CNS functions. For example, nausea and vomiting during pregnancy could be hCG-driven, because the brainstem, which contains the centers for nausea and vomiting reflexes, has no blood–brain barrier, allowing free in-and-out diffusion of hCG. Although this finding formalized the positive relationship between hCG levels and nausea and vomiting, several previous reports suggested this possibility. For example, the frequency and intensity of nausea and vomiting are higher in twin pregnancies and gestational trophoblastic diseases than in singleton pregnancies, and are higher during the first trimester than during the second and third trimesters in singleton pregnancies. These findings, along with the facts that not every pregnant woman has these episodes and that these episodes can extend into the second and third trimesters, could mean that the threshold levels that trigger these episodes might vary from pregnant woman to pregnant woman.

In preparation for lactation after the pregnancy, breast tissue grows and differentiates, and these changes can be induced in nonpregnant rodent models by administering hCG [4–6]. These actions do not seem to be mediated by increasing ovarian estradiol and/or progesterone secretions alone. Moreover, breast epithelial cells contain hCG receptors, and their activation results in numerous changes, including the nonreversible differentiation of proliferative to secretory epithelial cells [4–6]. The secretory epithelial cells are relatively more resistant to carcinogenic transformation. As a result, women who complete a full-term pregnancy before 20 years of age have a significantly decreased risk of breast cancer in later life. This pregnancy-induced protection gradually decreases with increasing maternal age and disappears by approximately 36 years of age. There are also many other hCG-driven changes, including adrenal steroid secretion and urinary incontinence during pregnancy [4–6,31].

Table 15.2 summarizes the potential therapeutic possibilities of the newly discovered hCG actions. There is a clear rationale for these possibilities, and preclinical data exist for some of them [4–6]. hCG is inexpensive, nontoxic, and has a few or no physiologic side effects. Moreover, if hCG does not work in a given patient, for whatever reason, one can fall back on conventional treatments, which are generally expensive, toxic, have side effects, and are not well-tolerated by all patients. Even if these conventional treatments must be used, perhaps hCG can be used in a combination therapy to reduce the dose of conventional therapeutic agents to decrease their expense and toxicity and to increase tolerability.

Table 15.2 Therapeutic potential from the newly discovered hCG actions

Increasing pregnancy rates in assisted reproductive technologies
Decreasing miscarriages
Preventing preterm births
Breast cancer treatment and prevention
Treatment of gynecologic infections and HIV/AIDS
Treatment of certain autoimmune diseases, including rheumatoid arthritis
Treatment of spinal cord injuries
Treatment of urinary incontinence

If hCG is proven useful in novel therapeutic applications, then there is a possibility of developing orally active hCG analogs—analogs that survive in the circulation much longer than hCG itself—and nanoparticle technology that might deliver hCG for weeks at a time after a single injection.

15.1 Summary and perspectives

hCG should no longer be considered just a hormone with a limited role of maintaining corpus luteum secretion of progesterone during the first weeks of pregnancy. It is a multifaceted molecule with properties of cystine-knot growth factors and cytokines. It plays roles starting from very early pregnancy to the end of pregnancy, most likely because hCG is present all through pregnancy. Logically, then, there should be no pregnancy in the absence of hCG. It is unethical to manipulate hCG levels during pregnancy. Simply by default, pregnancy complications under the conditions of low or high hCG levels have been ascribed to other hormones and regulatory agents. It is possible, however, that these hormones and regulatory agents might have a modifying role in the actions of hCG. Obviously, much further research is needed to investigate these possibilities.

References

[1] Pierce JG, Parsons TF. Glycoprotein hormones: structure and function. Annu Rev Biochem 1981;50:465–95.
[2] Lapthorn AJ, Harris DC, Littlejohn A, Lustbader JW, Canfield RE, Machin KJ, et al. Crystal structure of human chorionic gonadotropin. Nature 1994;369:455–61.
[3] Lei ZM, Rao CV. Endocrinology of the trophoblast tissue. In: Becker KL, editor. Principles and practice of endocrinology and metabolism. Philadelphia, PA: Lippincott Williams & Wilkins; 2001. p. 1096–102.
[4] Rao CV. Nongonadal actions of LH and hCG in reproductive biology and medicine. Semin Reprod Med 2001;19:1–119.

[5] Rao CV. An overview of the past, present and future of nongonadal LH/hCG actions in reproductive biology and medicine. Semin Reprod Endocrinol 2001;19:7–17.

[6] Rao CV, Lei ZM. The past, present and future of nongonadal LH/hCG actions in reproductive biology and medicine. Mol Cell Endocrinol 2007;269(2):8.

[7] McFarland KC, Sprengel R, Phillips HS, Kohler M, Rosemblit N, Nikolics K, et al. Lutropin-choriogonadotropin receptor: an unusual member of the G-protein-coupled receptor family. Science 1989;245:494–9.

[8] Loosfelt H, Misrahi M, Atger M, Salesse R, Vu Hai-Luu Thi MT, Jolivet A, et al. Cloning and sequencing of porcine LH/hCG receptor cDNA: variants lacking transmembrane domain. Science 1989;245:525–8.

[9] Rao CV. Differential properties of human chorionic gonadotropin and human luteinizing hormone binding to plasma membranes of bovine corpora lutea. Acta Endocrinol 1979;90:696–710.

[10] Castellanos Sanchez VO, Gomez-Conde E, Rocha-Gracia RD, Pimentel A, Aluja AS, Hernandez-Jauregui P, et al. Chorionic gonadotropin hormone receptors on Taenia solium and Taenia crassiceps cysticerci in culture. J Parasitol 2009;5:1.

[11] Abdallah MA, Lei ZM, Li X, Greenwold N, Nakajima ST, Jauniaux E, et al. Human fetal nongonadal tissues contain human chorionic gonadotropin/luteinizing hormone receptors. J Clin Endocrinol Metab 2004;89:952–6.

[12] Baal N, Reisinger K, Jahr H, Bohle RM, Liang O, Munstedt K, et al. Expression of transcriptional factor oct-4 and other embryonic genes in CD133 positive cells from human umbilical cord blood. Throm Haemost 2004;92:767–75.

[13] d'Hauterive SP, Berndt S, Tsampalas M, Charlet-Renard C, Dubois M, Bourgain C, et al. Dialogues between blastocyst hCG and endometrial LH/hCG receptor: which role in implantation? Gynecol Obstet Invest 2007;64:156–60.

[14] Fluhr H, Bischof-Islami D, Krenzer S, Licht P, Bischof P, Zygmunt M. Human chorionic gonadotropin stimulates matrix metalloproteinases-2 and -9 in cytotrophoblastic cells and decreases tissue inhibitor of metalloproteinases-1, -2, and -3 in decidualized endometrial stromal cells. Fertil Steril 2008;90:1390–5.

[15] Fluhr H, Carli S, Deperschmidt M, Wallwiener D, Zygmunt M, Licht P. Differential effects of human chorionic gonadotropin and decidualization on insulin-like growth factors-I and -II in human endometrial stromal cells. Fertil Steril 2008;90:1384–9.

[16] Zimmerman G, Baier D, Majer J, Alexander H. Expression of beta hCG and alpha CG mRNA and hCG hormone in human decidual tissue in patients during tubal pregnancy. Mol Hum Reprod 2003;9:81–9.

[17] Zimmermann G, Ackermann W, Alexander H. Epithelial chorionic gonadotropin is expressed and produced in human secretory endometrium during normal menstrual cycle. Biol Reprod 2009;80:1053–65.

[18] McGregor WG, Raymoure WJ, Kuhn RW, Jaffe RB. Fetal tissues can synthesize a placental hormone: evidence for chorionic gonadotropin β-subunit synthesis by human fetal kidney. J Clin Invest 1981;68:306–9.

[19] Goldsmith PC, McGregor WG, Raymoure WJ, Kuhn RW, Jaffe RB. Cellular localization of chorionic gonadotropin in human fetal kidney and liver. J Clin Endocrinol Metab 1983;57:654–61.

[20] Katabuchi H, Ohba T. Human chorionic villous macrophages as a fetal biological shield from maternal chorionic gonadotropin. Dev Growth Differ 2008;50:299–306.

[21] Zygmunt M, Herr F, Keller-Schoenwetter S, Kunzi-Rapp K, Munsteadt K, Rao CV, et al. Characterization of human chorionic gonadotropin as a novel angiogenic factor. J Clin Endocrinol Metab 2002;87:5290–6.

[22] Filicori M, Fazleabas AT, Huhtaniemi IT, Licht P, Rao ChV, Tesarik K, et al. Novel concepts of human chorionic gonadotropin reproductive system interactions and potential in the management of infertility. Fertil Steril 2005;84:275–84.

[23] Berndt S, d'Hauterive SP, Blacher S, Pequeux C, Lorquet S, Munaut C, et al. Angiogenic activity of human chorionic gonadotropin through LH receptor activation on endothelial and epithelial cells of the endometrium. FASEB J 2006;20:2189–98.

[24] Lei ZM, Yang M, Li X, Takikawa O, Rao CV. Upregulation of placental indoleamine 2,3-dioxygenase by human chorionic gonadotropin. Biol Reprod 2007;76:639–44.

[25] Berndt S, Blacher S, d'Hauterive SP, Thiry M, Tsampalas M, Cruz A, et al. Chorionic gonadotropin stimulation of angiogenesis and pericyte recruitment. J Clin Endocrinol Metab 2009;94:4567–74.

[26] Wan H, Versnel MA, Leijten LME, van Helden-Meeuwsen CG, Fekkes D, Leenen PJM, et al. Chorionic gonadotropin induces dendritic cells to express a tolerogenic phenotype. J Leukoc Biol 2008;83:894–901.

[27] Wan H, Versnel MA, Cheung WY, Leenen PJM, Khan NA, Benner R. Chorionic gonadotropin can enhance innate immunity by stimulating macrophage function. J Leukoc Biol 2007;82:1–9.

[28] Kayisli UA, Selam B, Guzeloglu-Kayisli O, Demir R, Arici A. Human chorionic gonadotropin contributes to maternal immunotolerance and endometrial apoptosis by regulating Fas-Fas ligand system. J Immunol 2003;171:2305–13.

[29] Schumacher A, Brachwitz N, Sohr S, Engeland K, Langwisch S, Dolaptchieva M, et al. Human chorionic gonadotropin attracts regulatory T cells into the fetal–maternal interface during early human pregnancy. J Immunol 2009;182:5488–97.

[30] Mightmo JL, Perez AP, Sanchez-Margalet V, Duenas JL, Calvo JC, Varone CL. Up-regulation of placental leptin by human chorionic gonadotropin. Endocrinol 2009;150:304–13.

[31] Rizk DEE, Osman NA, Shafiullah MM, Nagelkerke NJD, Fahim MA. Effect of human chorionic gonadotropin on *in vitro* contractions of stimulated detrusor muscle strips of female rats. J Obstet Gynaecol Res 2009;35:835–41.

Part E

Biological Function: Normal Pituitary

Pituitary sulfated hCG

16

Laurence A. Cole
USA hCG Reference Service, Angel Fire, NM, USA

Numerous older publications from the 1960s, 1970s, and 1980s suggested bacteria, crabs, and other unlikely sources to explain detection of hCG outside of pregnancy [1–8]. We now have a better understanding of the various sources of hCG when a person is not pregnant. This is explained in Chapter 28, Positive hCG Tests: Causes Other than Pregnancy, as quiescent gestational trophoblastic disease, as false-positive hCG assays, and as sulfated pituitary hCG. From this chapter we understand that men and women can have positive hCG tests because of circulating heterophilic or interfering antibodies. These cause false-positive hCG results and explain some of the strange findings in non-pregnant individuals [9–11]. We also now know that men and women with cancer normally produce the total hCG assay immunoreactive free β-subunit of hCG [12–14]. Here we examine the evidence that shows that the pituitary gland normally produces sulfated hCG during the menstrual cycle and in menopause [15–19]. This is a further cause of positive hCG results in non-pregnant individuals.

It has now been 33 years since hCG production was first demonstrated to come from the pituitary gland [20]. Since then, more than 40 publications have confirmed pituitary production and described how low levels of hCG (<3 mIU/ml) accompany luteinizing hormone (LH) during the menstrual cycle [15–29]. In amenorrhea or menopause, and partially in perimenopause, the ovary fails to provide estrogen to feed back to the hypothalamus to inhibit and control gonadotropin-releasing hormone (GnRH) production (Figure 16.1). This leads to GnRH pulses constantly stimulating the pituitary gonadotrope cells, to increased production of LH and follicle-stimulating hormone, and to the increased incidental production of human chorionic gonadotropin (<1.0 to 33.6 mIU/ml) (Figure 16.1).

It is thought that the increased production of hCG may be incidental because the single LH β-subunit gene of chromosome 19 is buried in the middle of eight parallel hCG β-subunit genes. Excess common α-subunit made in the pituitary combines with the hCG β-subunit. It is possible that the increased levels of hCG (extremely low concentrations) may result from the incidental potent action of GnRH on that group of eight hCGβ genes and one LHβ gene in menopausal women.

As such, elevated levels of hCG (>1 mIU/ml) may be evident in all women with amenorrhea or menopause (age older than 48), in women after bilateral salpingooophorectomy (BSO), and in women during oligomenorrhea in perimenopause [15–18,21,24–29]. In medical practice, a positive hCG test result prior to menopause suggests pregnancy or gestational trophoblastic disease [23,24,27,28]. Positive hCG test results in perimenopausal and menopausal women, however, suggest elevated pituitary hCG (<1.0 to 33.6 mIU/ml).

Human Chorionic Gonadotropin (hCG). DOI: http://dx.doi.org/10.1016/B978-0-12-800749-5.00016-X

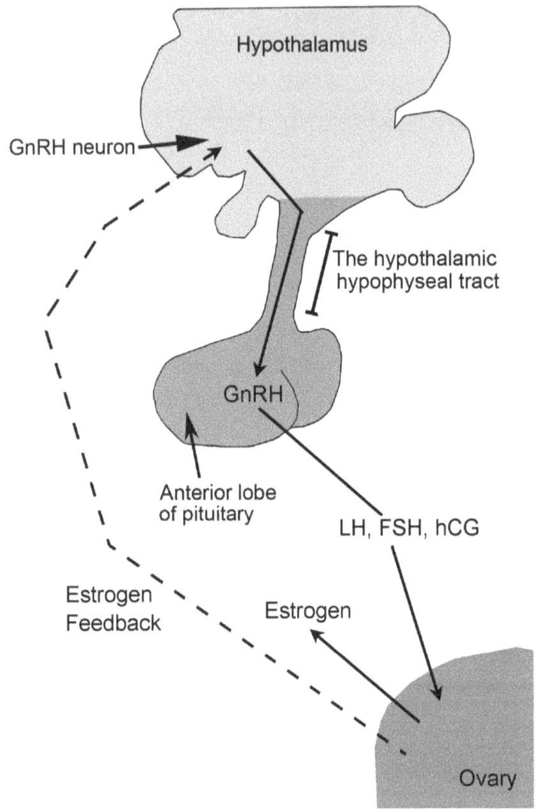

Figure 16.1 Endocrine feedback pathway linking the ovary, the hypothalamus, and anterior pituitary (at base of brain) in women.

Today, testing for pregnancy is routinely performed before surgical procedures, before imaging procedures, before administration of certain medications that might harm a fetus, and with admission to emergency departments. When hCG is detected, all planned procedures, whether critical or non-critical, are halted pending investigation of pregnancy and the source of hCG, seemingly regardless of patient age or menopausal status. After excluding regular and ectopic pregnancies, most non-pregnant positive hCG cases, regardless of age or menopausal status, are referred to gynecologic oncologists. Numerous older obstetrics and gynecology textbooks state that the only source of hCG, if pregnancy and ectopic pregnancy have been eliminated, is choriocarcinoma or cancer. Many physicians still adhere to these guidelines, which generally leads to needless hysterectomies or unnecessary chemotherapy in menopausal patients naturally producing pituitary hCG. Studies show that, extremely rarely, cases of positive hCG outside of pregnancy involve a malignancy or choriocarcinoma [30–34]. In most cases it is just pituitary hCG.

Sixty million women are now in perimenopause or menopause in the United States. Considering the frequency of hCG testing before surgery, at the emergency department, before administration of certain drugs, and before X-rays, many of these 60 million have to experience the stress and alarm of a non-pregnancy positive hCG test. Here we carefully consider all aspects of pituitary hCG, a key form of hCG present in menopausal women. We examine its structure, its biological function, and its occurrence.

16.1 Structure of pituitary sulfated hCG

Pituitary hCG has an amino acid structure identical to placental hCG [25]. It is unique, however, in having a variable portion of sulfated oligosaccharides [25]. The sulfate groups are attached to *N*-acetylgalactosamine residues, which replace galactose and sialic acid residues in N-linked and O-linked oligosaccharides (Figure 16.2). As published, the β-subunit of pituitary hCG has 2.7 mole/mole oligosaccharides terminating in sulfate groups and 4.6 mole/mole oligosaccharides terminating in sialic acid groups, or 37% sulfation [25]. The α-subunit of pituitary hCG has 0.8 mole/mole oligosaccharides terminating in sulfate and 1.7 mole/mole oligosaccharides terminating in sialic acid groups, or 32% sulfation.

Sulfated hCG is less acidic than sialylated or normal pregnancy hCG. As such, it has a lower circulating half-life (20 h versus 36 h) and 50% of the biological activity of normal pregnancy hCG [25].

16.2 Biological function of pituitary sulfated hCG

The USA hCG Reference Service recently examined more than 8300 urine samples from women with normal menstrual periods [18]. Low concentrations (>1 mIU/ml) of hCG were detectable at the time of the mid-cycle LH peak in 232 of 277 menstrual cycles [18]. The mean hCG level at the time of LH peak was 1.54 ± 0.90 mIU/ml and the range was <1.0–9.2 mIU/ml. Considering the mean and the distribution of the range of hCG values, it is logical to assume that the 45 of 277 menstrual cycles in which hCG was not detected produced lower hCG concentrations (>1 mIU/ml)—too low to be detected by the assay used (the Siemens Immulite 1000 assay, sensitivity ≥1.0). It was inferred that these low concentrations of hCG (<1.0–9.2 mIU/ml) supplement LH during all human menstrual cycles and are part of normal physiology.

As shown by Odell and Griffin [35,36], using an ultra-sensitive sandwich assay for hCG, pituitary hCG is produced at very low levels (0.01 mIU/ml) in healthy men, with a wide range from 0.03 to 1.7 mIU/ml [35,36]. Pituitary hCG was detected in women in pulses of the luteal and follicular phases of the menstrual cycle that paralleled LH levels [35,36]. Injections of GnRH were shown to directly promote circulating pituitary hCG levels in men and women, just as it similarly

Sulfated hCG

N-linked oligosaccharides

NeuAc α2,3 Gal ß1,4 GlcNAc ß1,2 Man α1,
6
Man ß1,4 GlcNAc ß1,4 GlcNAc-N-Asn
3
NeuAc α2,3 Gal ß1,4 GlcNAc ß1,2 Man α1,

SO_4-GalNAc ß1,4 GlcNAc ß1,2 Man α1,
6
Man ß1,4 GlcNAc ß1,4 GlcNAc-N-Asn
3
NeuAc α2,3 Gal ß1,4 GlcNAc ß1,2 Man α1,

SO_4-GalNAc ß1,4 GlcNAc ß1,2 Man α1,
6
Man ß1,4 GlcNAc ß1,4 GlcNAc-N-Asn
3
SO_4-GalNAc ß1,4 GlcNAc ß1,2 Man α1,

O-linked oligosaccharides

NeuAc α2,3 Gal ß1,3 GalNAc-O-Ser

SO_4-GalNAc ß1,4 GalNAc-O-Ser

NeuAc α2,6
SO_4-GalNAc ß1,4 GalNAc-O-Ser

NeuAc α2,6
NeuAc α2,3 GalNAc ß1,4 GalNAc-O-Ser

Figure 16.2 The N-linked and O-linked sugar structures on pituitary sulfated hCG.

promoted LH levels [35,36]. It is inferred that pituitary hCG supplements pituitary LH in healthy men and women [18,25,36].

It is yet unknown whether there is a specific function for pituitary hCG, independent of LH, during the menstrual cycle. Pituitary hCG could have functions separate from those of LH. But even if pituitary hCG has no specific function, there is a natural explanation for its production. There is a single LH β-subunit gene buried among the eight back-to-back hCG β-subunit genes on human chromosome 19 [37]. hCG and LH share a single common α-subunit. It is possible that a small amount of hCGβ is promoted by GnRH along with specific LH β-subunit stimulation in pituitary gonadotrope cells during normal menstrual cycle physiology in women and in men.

LH has multiple functions during the LH peak and ovulation period. We know that both LH and hCG act on the LH/hCG receptor to promote progesterone production by corpus luteum cells [26,29,38]. We assume that LH and hCG act on the same receptor on follicles. As established, with the appearance of an LH/hCG receptor on granulosa cells of the Graafian follicle or primary follicle, LH first promotes follicular growth [39,40], and then it stimulates diploid cell meiosis [41,42]. LH causes the follicle to form a stigma or protrusion [43], and then it promotes collagenase production to degrade and penetrate the stigma [43,44]. The penetration of the stigma causes bursting of the follicle (ovulation) to occur. LH then acts to differentiate the burst or ovulated follicle into a corpus luteum [45,46]. It is not clear whether hCG just incidentally assists LH in each of these five steps or if it has specific functions of its own in one or more of these steps.

Consider that the mean hCG at the time of the LH peak is 1.54 ± 0.90 mIU/ml [18]. Consider that the mean LH, at the time of peak, is 70–90 mIU/ml or 50-fold greater. Then, consider that the mean biological activity of pituitary hCG is approximately 50-times that of hCG. The 50-times greater biological activity is balanced by the 50-fold lesser concentration. As such, pituitary hCG and LH probably contribute as equal partners to ovulation at the time of LH peak. Similarly, the mass of hCG in an individual woman's or man's pituitary gland is 0.5–1.1 μg/gland. This is approximately 25- to 50-fold less than the mass of LH in a woman's pituitary gland [29]. Publications show that pituitary hCG has approximately half the biological activity in promoting progesterone production of placental hCG [25]. In conclusion, the pituitary hCG seems to parallel LH in an approximately 50 bioactivity:50 bioactivity relationship during the course of the menstrual cycle in women and in men. The saying that LH promotes ovulation may not be correct.

16.3 Occurrence of pituitary sulfated hCG

As shown by the USA hCG Reference Service with more than 8300 urine samples, women produce 1.54 ± 0.90 mIU/ml (mean $\pm$ standard deviation) of pituitary hCG during the LH peak of the normal menstrual cycle [18]. Odell and Griffin showed, in contrast, that a miniscule level of 0.01 mIU/ml (average) pituitary hCG is detectable in all men [35,36]. Similar miniscule levels of pituitary hCG are detected during the luteal and follicular phase of the menstrual cycle [35,36]. Extremely low levels of hCG can be detected in healthy men and in women during the luteal and follicular phases of menstrual cycle. These tiny concentrations are close to or below the detection limits of most laboratory hCG/pregnancy assays. Rarely do these tiny levels cause alarm regarding possible pregnancy or halt scheduled surgeries or necessary procedures.

During menopause (age older than 42) and as early as perimenopause (age older than 38), the ovary partially or completely fails, thus reducing or halting production of progestagens and estrogens. Similarly, if a woman receives a bilateral oophorectomy, then production of progestagens and estrogens will halt and she will be in a state of menopause. Estrogens modulate GnRH pulses by the hypothalamus. In the absence or reduced presence of estrogen, GnRH pulses go wild and become as frequent as possible. This leads to highly increased serum LH and follicle-stimulating hormone (FSH) concentrations. Pituitary hCG is increased significantly as well. In menopausal women, the mean pituitary hCG level in serum is 9.66 ± 6.52 mIU/ml and the range is 1.3–33.6 mIU/ml [19].

As defined, perimenopausal women are still fertile but have increased FSH levels (>30 mIU/ml) [17] and oligomenorrhea (extremely variable menstrual periods) [15]. Menopausal woman have amenorrhea or no menstrual cycle. Oophorectomy and primary amenorrhea induce the onset of menopause at an early age. The USA hCG Reference Service's experience with 103 cases is presented in Table 16.1. As shown, the mean serum concentration of pituitary hCG was 5.5 ± 3.2 mIU in perimenopausal women (range 1.4–12 mIU/ml), 11.0 ± 6.3 mIU/ml in menopausal women (range 2.1–32.9 mIU/ml), and 9.8 ± 9.5 mIU/ml in women with a history of oophorectomy or primary amenorrhea (range 1.8–33.6 mIU/ml). A significant difference was observed between perimenopause and menopause cases ($P = 0.000064$), and between perimenopause and oophorectomy plus primary amenorrhea cases ($P = 0.029$).

The USA hCG Reference Service consults on numerous pituitary hCG cases each year (Table 16.1). These cases seem to mostly derive from routine hCG testing performed on women before surgery, X-ray procedures, and emergency department visits. It appears to the USA hCG Reference Service that these services wrongly concluded from a positive hCG test that these menopausal women were pregnant. Women are routinely tested for pregnancy by ultrasound, are tested further for ectopic pregnancy by ultrasound and other methods, are administered methotrexate to eliminate ectopic pregnancy, and, finally, are referred to an oncologist or gynecologic oncologist for cancer investigation. All too often these women undergo an unnecessary hysterectomy or chemotherapy when hCG levels do not disappear [15,16].

More than 40 publications have described the normality of pituitary hCG in menopausal women, but it appears that physicians either have not read these publications or they cannot believe that a positive hCG test can mean anything but pregnancy. Time and again, after excluding cancer, choriocarcinoma, and other possible explanations (see Chapter 28), we conclude that a patient has pituitary hCG (Table 16.1). These are the kinds of cases that get referred to the USA hCG Reference Service. We know of four cases in which menopausal women were waiting years for a renal transplant. Finally, a matching kidney became available and, when they entered the hospital for the transplantation, they were administered the hCG test. It was positive. Even though, based on age, it was not physically possible for them to be pregnant their transplant procedures were canceled and the kidneys were given to other patients. All because of the positive hCG test.

The USA hCG Reference Service has no specific assay for sulfated pituitary hCG because there is no known specific antibody that only detects sulfated pituitary hCG. We simply exclude other possibilities like gestational trophoblastic disease, familial hCG syndrome, quiesced gestational trophoblastic disease, and cancer. Hyperglycosylated hCG (hCG-H) is not produced by the pituitary (Table 16.1).

Table 16.1 The USA hCG Reference Service experience with 103 woman demonstrated to be producing pituitary hCG

Case #, Age	hCG (mIU/ml)	hCG-H (mIU/ml)	History
A. Perimenopausal woman in oligomenhorea			
Case 455, Age 38	2.0	<0.05	
Case 339, Age 39	2.0	<0.05	
Case 194, Age 39	5.8	<0.05	
Case 199, Age 39	2.8	<0.05	
Case 563, Age 40	6.0	<0.05	
Case 150, Age 40	7.6	<0.05	
Case 561, Age 40	10.1	<0.05	
Case 36, Age 40	12.0	<0.05	
Case 216, Age 40	4.4	<0.05	
Case 427, Age 40	11.6	<0.05	
Case 422, Age 42	8.3	<0.05	
Case 442, Age 42	7.2	<0.05	
Case 377, Age 43	8.4	<0.05	
Case 421, Age 43	3.8	<0.05	
Case 317, Age 44	3.2	<0.05	
Case 305, Age 44	4.2	<0.05	
Case 410, Age 44	9.2	<0.05	
Case 257, Age 44	2.2	<0.05	
Case 173, Age 46	3.0	<0.05	
Case 3, Age 46	3.3	<0.05	
Case 313, Age 46	3.7	<0.05	
Case 506, Age 46	5.0	<0.05	
Case 534, Age 47	1.4	<0.05	
23 cases, Mean	5.5 ± 3.2 Range 1.4–12		
B. Menopausal women experiencing amenorrhea			
Case 331, Age 42	20.0	<0.05	
Case 538, Age 42	12.9	<0.05	
Case 397, Age 43	16.1	<0.05	
Case 489, Age 43	12.7	<0.05	
Case 499, Age 44	6.4	<0.05	
Case 17, Age 45	7.2	<0.05	
Case 432, Age 47	10.6	<0.05	
Case 187, Age 48	12.2	<0.05	
Case 350, Age 48	4.1	<0.05	
Case 450, Age 48	7.7	<0.05	
Case 328, Age 48	6.8	<0.05	
Case 300, Age 49	2.1	<0.05	
Case 406, Age 49	19.2	<0.05	
Case 533, Age 49	7.7	<0.05	
Case 540, Age 49	9.9	<0.05	

(*Continued*)

Table 16.1 **(Continued)**

Case #, Age	hCG (mIU/ml)	hCG-H (mIU/ml)	History
Case 22, Age 50	16.2	<0.05	
Case 412, Age 50	7.5	<0.05	
Case 425, Age 50	12.5	<0.05	
Case 306, Age 50	7.7	<0.05	
Case 376, Age 51	9.8	<0.05	
Case 169, Age 51	7.4	<0.05	
Case 191, Age 51	13.7	<0.05	
Case 201, Age 51	4.3	<0.05	
Case 79, Age 51	13.7	<0.05	
Case 206, Age 51	12.3	<0.05	
Case 411, Age 51	8.1	<0.05	
Case 413, Age 51	4.1	<0.05	
Case 434, Age 51	14.6	<0.05	
Case 564, Age 51	7.3	<0.05	
Case 21, Age 52	4.7	<0.05	
Case 32, Age 52	11.0	<0.05	
Case 151, Age 52	9.0	<0.05	
Case 193, Age 52	10.2	<0.05	
Case 143, Age 52	10.6	<0.05	
Case 344, Age 52	14.5	<0.05	
Case 469, Age 52	7.5	<0.05	
Case 164, Age 53	27.2	<0.05	
Case 183, Age 53	5.9	<0.05	
Case 192, Age 53	10.6	<0.05	
Case 20, Age 53	6.0	<0.05	
Case 90, Age 53	10.4	<0.05	
Case 329, Age 53	7.0	<0.05	
Case 478, Age 53	9.3	<0.05	
Case 49, Age 54	6.1	<0.05	
Case 336, Age 54	6.7	<0.05	
Case 517, Age 54	39.0	<0.05	
Case 467, Age 54	6.0	<0.05	
Case 476, Age 54	12.1	<0.05	
Case 127, Age 55	5.3	<0.05	
Case 320, Age 55	8.0	<0.05	
Case 565, Age 55	8.2	<0.05	
Case 445, Age 55	9.2	<0.05	
Case 465, Age 55	4.6	<0.05	
Case 269, Age 55	9.6	<0.05	
Case 357, Age 55	14.2	<0.05	
Case 498, Age 55	11.5	<0.05	
Case 115, Age 56	7.3	<0.05	
Case 188, Age 57	7.9	<0.05	
Case 34, Age 57	14.9	<0.05	
Case 162, Age 59	15.6	<0.05	

(Continued)

Table 16.1 **(Continued)**

Case #, Age	hCG (mIU/ml)	hCG-H (mIU/ml)	History
Case 429, Age 59	16.1	<0.05	
Case 34, Age 60	32.9	<0.05	
Case 514, Age 61	28.5	<0.05	
Case 517, Age 62	15.2	<0.05	
Case 543, Age 66	17.1	<0.05	
Case 548, Age 68	2.5	<0.05	
Case 182, Age 69	4.5	<0.05	
Case 18, Age 70	31.5	<0.05	
Case 428, Age 54	16	<0.05	
69 cases, Mean	11.4 ± 7.1 Range 2.1–39.0		
C. Bilateral oophorectomy and primary amennhorea			
Case 444, Age 34	3.2	<0.05	Oophorectomy
Case 349, Age 35	11.5	<0.05	Oophorectomy
Case 519, Age 35	2.3	<0.05	Oophorectomy
Case 454, Age 36	3.9	<0.05	Oophorectomy
Case 460, Age 37	6.9	<0.05	Oophorectomy
Case 322, Age 40	1.8	<0.05	Oophorectomy
Case 468, Age 29	6.1	<0.05	Amenorrhea
Case 415, Age 42	19.4	<0.05	Amenorrhea
Case 319, Age 28	12.7	<0.05	Oophorectomy
Case 319, Age 28	33.6	<0.05	Oophorectomy
Case 471, Age 34	6.4	<0.05	Oophorectomy
11 cases. Mean	9.8 ± 9.5 Range 1.8–33.6		

There are, however, two other tests that are invaluable. The first test is the FSH test. As shown by Gronowski et al. [17], increased serum FSH of >30 mIU/ml illustrates perimenopause and menopause, or a lack of sufficient steroid feedback to the hypothalamus. An FSH level of >30 mIU/ml means that there is likely to be elevated pituitary hCG present. In the experience of the USA hCG Reference Service, an FSH test (>30 mIU/ml) correctly predicated pituitary hCG in 29 of 30 cases tested. The second test involves treating patients with a high-estrogen oral contraceptive pill for 3 weeks, which suppresses pituitary hCG and proves whether the hCG is, in fact, from a pituitary origin [15,16,24,28].

These two methods make it easy to prove pituitary hCG is the source of circulating hCG. We emphasize again, however, that production of pituitary hCG in menopause is normal and natural, and it does not mean pregnancy or cancer. Hospitals need to start using the FSH test along with the hCG tests when evaluating women before surgery. If the FSH is >30 mIU/ml, then they need to consider that a positive hCG result is attributable to pituitary hCG. All pregnancies will still be identified by the hCG test.

References

[1] Acevedo HF, Slifkin M, Pouchet GR, Pardo M. Immunochemical localization of a choriogonadotropin-like protein in bacteria isolated from cancer patients. Cancer 1978;41:1217–29.

[2] Cohen H, Strampp A. Bacterial synthesis of substances similar to human chorionic gonadotropin. Proc Soc Exp Biol Med 1976;152:408–10.

[3] Dominigue G, Johson E. Isolation of subcellular fractions containing immunogenic endobacterial common antigen. Z Immunitatsforsh Exp Klinical Immunol 1974;148:23–8.

[4] Backus BT, Affronti LF. Tumor-associated bacteria capable of producing human choriogonadotropin-like substance. Infect Immun 1981;32:1211–15.

[5] Maruo T, Segal SJ, Koide SS. Studies on the apparent human chorionic gonadotropin-like factor in crab *Ovalipes ocellatus*. Endocrinology 1979;104:932–9.

[6] Yoshimot Y, Wolfsen AR, Odell WD. Human chorionic gonadotropin-like substance in nonendocrine tissues of normal subjects. Science 1977;197:1977–81.

[7] Braunstein GD, Kamdar V, Rasort J, Swaminathan N, Wade ME. Widespread distribution of chorionic gonadotropin-like substance in normal human tissues. Endocrinology 1978;102(Suppl):96.

[8] Chen H-C, Hodgen GC, Matsuura S, Lin LJ, Gross E, Reichert LE, et al. Evidence for a gonadotropin from nonpregnant subjects that has physical immunological and biological similarities to human chorionic gonadotropin. Proc Natl Acad Sci USA 1976;73:2885–9.

[9] Knight AK, Bingemann T, Cole L, Cunningham-Rundles C. Frequent false positive beta human chorionic gonadotropin in immunoglobulin A deficiency. Clin Exp Immunol 2005;141:333–7.

[10] Rotmensch S, Cole LA. False diagnosis and needless therapy of presumed malignant disease in women with false-positive human chorionic gonadotropin concentrations. Lancet 2000;355:712–15.

[11] Hussa RO, Rinke ML, Schweitzer PG. Discordant human chorionic gonadotropin results: causes and solutions. Obstet Gynecol 1985;65:211–19.

[12] Regelson W. Have we found the "definitive cancer biomarker"? The diagnostic and therapeutic implications of human chorionic gonadotropin-beta statement as a key to malignancy. Cancer 1995;76:1299–301.

[13] Butler SA, Iles RK. Ectopic human chorionic gonadotropin β secretion by epithelial tumors and human chorionic gonadotropin β-induced apoptosis in Karposi's sarcoma. Is there a connection? Clin Cancer Res 2003;9:4666–73.

[14] Iles RK. Human chorionic gonadotropin and its fragments as markers of prognosis in bladder cancer. Tum Mark Upd 1995;7:161–6.

[15] Cole LA, Khanlian SA, Muller CY. Detection of hCG peri- or post-menopause an unnecessary source of alarm. Am J Obstet Gynecol 2008;198:275–9.

[16] Cole LA, Sasaki Y, Muller CY. Normal production of hCG in menopause: a medical management dilemma. N Engl J Med 2007;356:1184–6.

[17] Gronowski AM, Fantz CR, Parvin CA, Sokoll LJ, Wiley CL, Wener MH, et al. Use of serum FSH to identify perimenopausal women with pituitary hCG. Clin Chem 2008;54:652–6.

[18] Cole LA, Ladner DG. Background hCG in non-pregnant individuals: need for more sensitive point-of-care and over-the-counter pregnancy tests. Clin Biochem 2009;42:168–75.

[19] Cole LA, Laidler L, Muller C. USA hCG Reference Service, 10 year report. Clin Biochem 2010;43:1013–22.
[20] Matsuura S, Ohashi M, Chen HC, Shownkeen RC, Hartree AS, Reichert Jr LE, et al. Physicochemical and immunological characterization of an hCG-like substance from human pituitary glands. Nature 1980;286:740–1.
[21] Huang SC, Chen HC, Chen RJ, Hsieh CY, Wei PY, Ouyang PC. The secretion of human chorionic gonadotropin-like substance in women employing contraceptive measures. J Clin Endocrinol Metab 1984;58:646–53.
[22] Wahlstrom T, Stenman UH, Lundqvist C, Tanner P, Schroder J, Seppala M. The use of monoclonal antibodies against human chorionic gonadotropin for immunoperoxidase.
[23] Seki K, Matsui H, Sekiya S. Advances in the clinical laboratory detection of gestational trophoblastic disease. Clin Chim Acta 2004;349:1–13.
[24] Cole LA, Khanlian SA. Inappropriate management of women with persistent low hcg results. J Reprod Med 2004;49:423–32.
[25] Birken S, Maydelman Y, Gawinowicz MA, Pound A, Liu Y, Hartree AS. Isolation and characterization of human pituitary chorionic gonadotropin. Endocrinology 1996;137:1402–11.
[26] Hoermann R, Spoettl G, Moncayo R, Mann K. Evidence for the presence of human chorionic gonadotropin (hCG) and free beta-subunit of hCG in the human pituitary. J Clin Endocrinol Metab 1990;71:179–86.
[27] Snyder JA, Haymond S, Parvin CA, Gronowski AM, Grenache DG. Diagnostic considerations in the measurement of human chorionic gonadotropin in aging women. Clin Chem 2005;51:1830–5.
[28] Cole LA, Khanlian SA, Giddings A, Butler SA, Muller CY, Hammond C, et al. Gestational trophoblastic diseases. 4. Presentation with persistent low positive human chorionic gonadotropin. Gynecol Oncol 2006;102:165–72.
[29] Hartree AS, Shownkeen RC, Stevens VC, Matsuura S, Ohashi M, Chen H-C. Studies of human chorionic gonadotropin-like substance of human pituitary glands and its significance. J Clin Endocrinol Metab 1983;96:115–26.
[30] Louhimo J, Kokkola A, Alfthan H, Stenman UH, Haglund C. Preoperative hCG β and CA 72-4 are prognostic factors in gastric cancer. Int J Cancer 2004;111:929–33.
[31] Carpelan-Holmstrom M, Louhimo J, Stenman UH, Alfthan H, Jarvinen H, Haglund C. Estimating the probability of cancer with several tumor markers in patients with colorectal disease. Oncology 2004;66:296–302.
[32] Louhimo J, Alfthan H, Stenman UH, Haglund C. Serum hCG beta and CA 72-4 are stronger prognostic factors than CEA, CA 19-9 and CA 242 in pancreatic cancer. Oncology 2004;66:126–31.
[33] Hotakainen K, Haglund C, Paju A, Nordling S, Alfthan H, Rintala E, et al. Chorionic gonadotropin beta-subunit and core fragment in bladder cancer: mRNA and protein expression in urine, serum and tissue. Eur Urol 2002;41:677–85.
[34] Hotakainen K, Ljungberg B, Paju A, Rasmuson T, Alfthan H, Stenman UH. The free beta-subunit of human chorionic gonadotropin as a prognostic factor in renal cell carcinoma. Br J Cancer 2002;86:185–9.
[35] Odell WD, Griffin J. Pulsatile secretion of human chorionic gonadotropin in normal adults. N Engl J Med 1987;317:1688–91.
[36] Odell WD, Griffin J. Pulsatile secretion of chorionic gonadotropin during the normal menstrual cycle. J Clin Endocrinol Metab 1989;69:528–32.
[37] Policastro PF, Daniels-McQueen S, Carle G, Boime I. A map of the hCG beta-LH beta gene cluster. J Biol Chem 1986;13:5907–16.

[38] Lei ZM, Reshef E, Rao CV. The expression of human chorionic gonadotropin/luteinizing hormone receptors in human endometrial and myometrial blood vessels. J Clin Endocrinol Metab 1992;75:651–9.
[39] Gougeon A. Dynamics of follicular growth in the human: a model from preliminary results. Hum Reprod 1986;1:81–7.
[40] Bogovich K, Richards JS, Reichert Jr LE. Obligatory role of luteinizing hormone (LH) in the initiation of preovulatory follicular growth in the pregnant rat: specific effects of human chorionic gonadotropin and follicle-stimulating of hormone on LH receptors and steroidogenesis in theca, granulosa, and luteal cells. Endocrinology 1981;109:860–7.
[41] Motola S, Popliker M, Tsafriri A. Are steroids obligatory mediators of LH/hCG triggered resumption of meiosis in mammals? Biol Reprod 2007;77:167–8.
[42] Hegele-Hartung C, Grütznera M, Lessla M, Grøndahl C, Ottesen JL, Brännström M. Activation of meiotic maturation in rat oocytes after treatment with follicular fluid meiosis-activating sterol *in vitro* and *ex vivo*. Biol Reprod 2001;64:418–24.
[43] Robkera RL, Russella DL, Yoshiokab S, Sharmaa SC, Lydona JP, O'Malleya BW, et al. Ovulation: a multi-gene, multi-step process. Steroids 2000;65:559–70.
[44] Butler TA, Zhu C, Mueller RA, Fuller GC, Lemaire WJ, Woessner Jr JF. Inhibition of ovulation in the perfused rat ovary by the synthetic collagenase inhibitor SC 44463. Biol Reprod 1991;44:1183–8.
[45] Acosta JT, Miyamoto A. Vascular control of ovarian function: ovulation, corpus luteum formation and regression. Anim Reprod Sci 2004;82–83:127–40.
[46] Nalbandov AV, Bahr JM. Ovulation, corpus luteum formation, and steroidogenesis. Basic Life Sci 1974;4:399–407.

Part F

Biological Function: Evolution

Evolution

17

Laurence A. Cole
USA hCG Reference Service, Angel Fire, NM, USA

The glycoprotein hormone family of molecules all evolved from TGFβ over hundreds of millions of years. In many respects, the stories of humans and evolution of hCG and hyperglycosylated hCG are all interrelated. Unfortunately, these stories are also one and the same with the horror stories of the evolution of pregnancy failures and the evolution of human cancers. This is because the evolution of hCG and hyperglycosylated hCG led to the evolution of humans, and led to human pregnancy failures and human malignancies. All are all driven by variants of hCG and hyperglycosylated hCG.

The evolution of chorionic gonadotropin (CG) is a biological marvel in which two separate molecules evolved from a single deletion mutation in early primates. Dozens of incredible changes in amino acid sequences occurred, and amazing differences in glycosylation happened during the evolution of CG and hyperglycosylated CG in early primates, advanced primates, and humans. These changes took CG and hyperglycosylated CG from nonacidic molecules with minimal activity in early primates to very acidic molecules with a long circulating half-life and maximal activity in humans. In so doing, CG and hyperglycosylated CG evolved, making human hCG and hyperglycosylated hCG ultra-potent, or the growth factors at the root of human evolution and at the root of human cancers.

17.1 TGFβ and evolution of hCG

How did TGFβ-binding hyperglycosylated hCG evolve? Why does it bind the TGFβ receptor? As shown by Lapthorn et al. [1], the three-dimensional crystal structure of hCG α- and β-subunits includes distinct cystine knot three-dimensional structures common to or virtually identical to that on TGFβ. In fact, the entire family of glycoprotein hormones, hCG, luteinizing hormone (LH), follicle-stimulating hormone (FSH), and thyroid-stimulating hormone (TSH) all have this cystine knot and all evolved from TGFβ [1–3]. TGFβ is a monomer and the glycoprotein hormones are all dimers. Both the α-subunit and the β-subunit of glycoprotein hormones separately evolved from TGFβ, so both contain cystine knot structures [1–3]. The hCG β-subunit three-dimensional cystine knot structure is shown in Figure 17.1. It comprises four intercepting specific-shape peptides linked or tied together by three disulfide bridges (Figure 17.1).

The evolution of hCG from TGFβ and the evolution of the other glycoprotein hormones is an evolutionary event that occurred over approximately 500 million years

Human Chorionic Gonadotropin (hCG). DOI: http://dx.doi.org/10.1016/B978-0-12-800749-5.00017-1

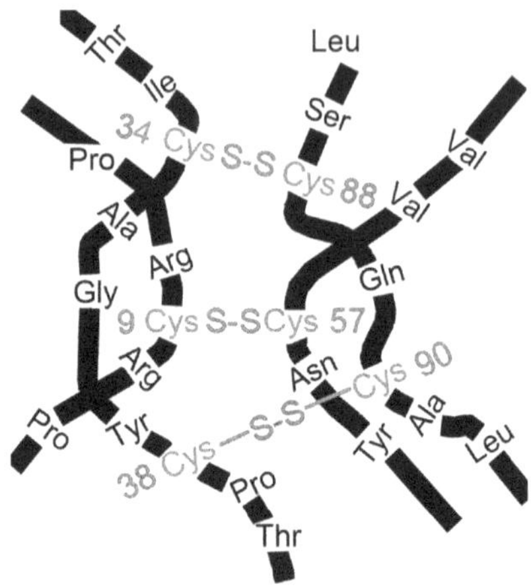

Figure 17.1 The complex four-peptide cystine knot structure on the hCG β-subunit.

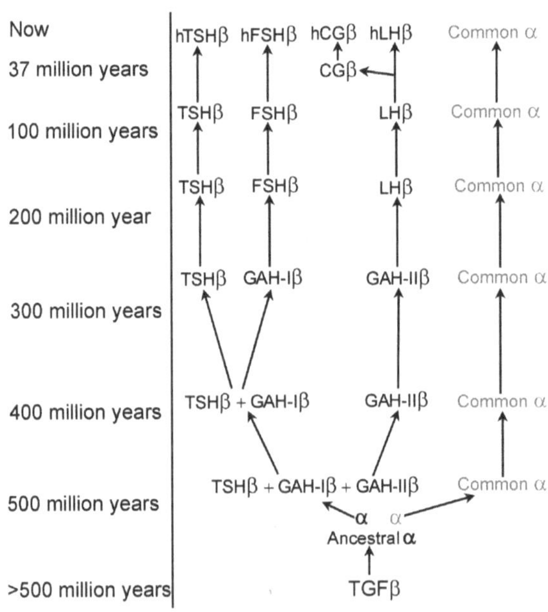

Figure 17.2 The evolution of TGFβ to glycoprotein hormones. GAH-I and GAH-II are gonadotropin ancestral hormones I and II.

(Figure 17.2). The first variant of TGFβ to evolve was ancestral α-subunit. This molecule evolved into the common α-subunit of glycoprotein hormones, and into the root β-subunit of glycoprotein hormones. Each of these molecules then mutated and evolved as separate molecules. First, the common α-subunit and GAH-Iβ, GAH-IIβ, and TSHβ evolved approximately 480 million years ago (GAH is gonadotropin ancestral hormone). Approximately 200 million years ago, GAH-Iβ and the common

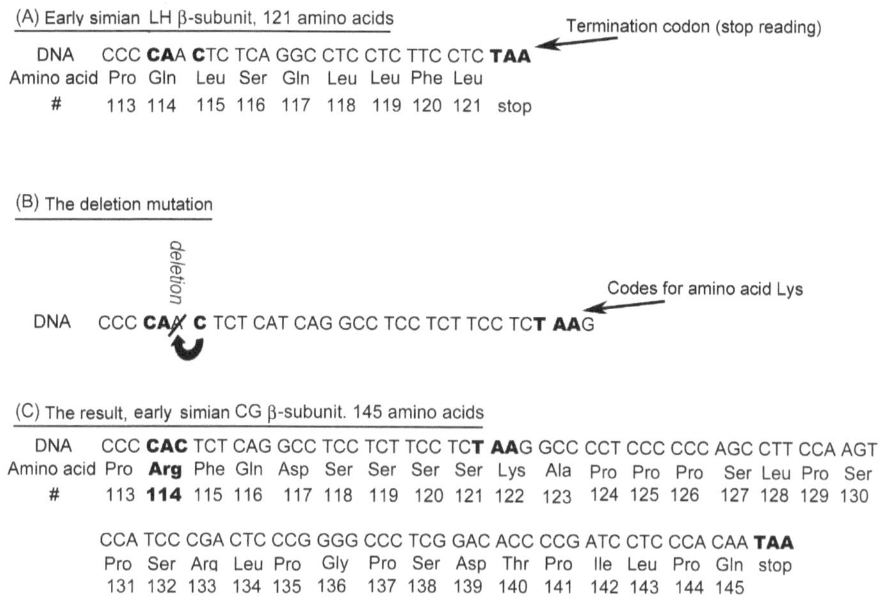

Figure 17.3 CG evolved from LH through a deletion mutation in the β-subunit. Amino acid sequence of early simian LH and CG. The sequence of a deletion mutation is shown: (A) LH β-subunit before a deletion mutation, (B) the deletion mutation, and (C) CG formed by the deletion mutation.

α-subunit became the hormone FSH, and GAH-IIβ and the common α-subunit became the hormone LH (Figure 17.2). hCG is the most recent member of the glycoprotein hormone family evolving in early simian primates approximately 37 million years ago.

The strange thing is that the TGFβ cystine knot remained on the α-subunit and β-subunit of the four glycoprotein hormones, LH, TSH, FSH, and hCG, with extreme stability and little alteration or few mutations. It was always hidden within the middle of the three-dimensional structure of the molecules and was never expressed. It was only after 500 million years, with the super-glycosylation of hCG in cytotrophoblast cells (hyperglycosylated hCG), that these TGFβ elements were finally expressed and the molecule bound to or acted on the TGFβ receptor [4,5].

17.2 Evolution of hCG and hyperglycosylated hCG

The story of CG evolution was first described by Fiddes and Goodman in 1979 [6,7]. It was detailed, refined, and extended by Maston and Ruvolo [8], Hallast et al. [9], and Policastro et al. [10]. Before CG molecules were produced by primates, no CG or hyperglycosylated CG form was made by earlier mammalian species. Even the most primitive primates, prosimian primates (e.g., lemurs) did not produce any form of CG or hemochorial placentation promoter. CG forms came about with the evolution of early simian primates (e.g., cebus monkey or platyrrhine). A deletion mutation occurred in the gene coding for the β-subunit of LH (Figure 17.3). With

Table 17.1 **The evolution of hemochorial placentation and humans [6–19]**

Species	Implantation characteristics	Depth of invasion	Sugar structures, acidity or pI	Brain mass (% body weight)
(1) Primates				
Humans	Hemochorial placentation	1/3rd myometrium	CG, 8 oligosaccharides, pI 3.5	2.4%
Advanced simian primate	Hemochorial placentation	1/10th myometrium	CG, 6 oligosaccharides, pI 4.9	0.74%
Early simian primate	Hemochorial placentation	Through 2 mm decidua thickness	CG, 5 oligosaccharides, pI 6.3	0.17%
Prosimian primate	Epitheliochorial placentation	No invasion	No CG forms produced	0.07%
(2) Mammals				
Cetacea (whale)	Epitheliochorial placentation	No invasion	No CG forms produced	0.08%
Equus (horse)	Epitheliochorial placentation	No invasion	No CG forms produced	0.11%
Bos taurus (cow)	Epitheliochorial placentation	No invasion	No CG forms produced	0.09%

the evolution of early simian primates, hemochorial placentation also evolved as a consequence of the evolution of CG and hyperglycosylated hCG (see Chapter 14). Although all previous mammals and prosimian primates utilized inefficient epitheliochorial placentation to provide nutrition to the fetus, with the evolution of CG a more efficient hemochorial placentation developed (Table 17.1).

As illustrated in Figure 17.3, this deletion mutation in the codon corresponding to LH β-subunit amino acid residue 114 caused a frame shift in the LH β-subunit gene. The termination codon TAA (immediately after codon for amino acid 121) was lost. Read-through then occurred until an additional TAA or termination codon was reached. This next termination codon was immediately after the codon for amino acid 145 (Figure 17.3), 72 nucleotides and 24 amino acids downstream [6]. This led to the synthesis of a 145-amino-acid CG β-subunit compared with the 121-amino-acid β-subunit originally on LH. This 145-amino-acid β-subunit combines with the common α-subunit to LH, CG, FSH, and TSH. The C-terminal extension significantly extended the circulating half-life and the biological activity of early simian CG (half-life ~2.4 h) compared with LH (half-life 0.4 h).

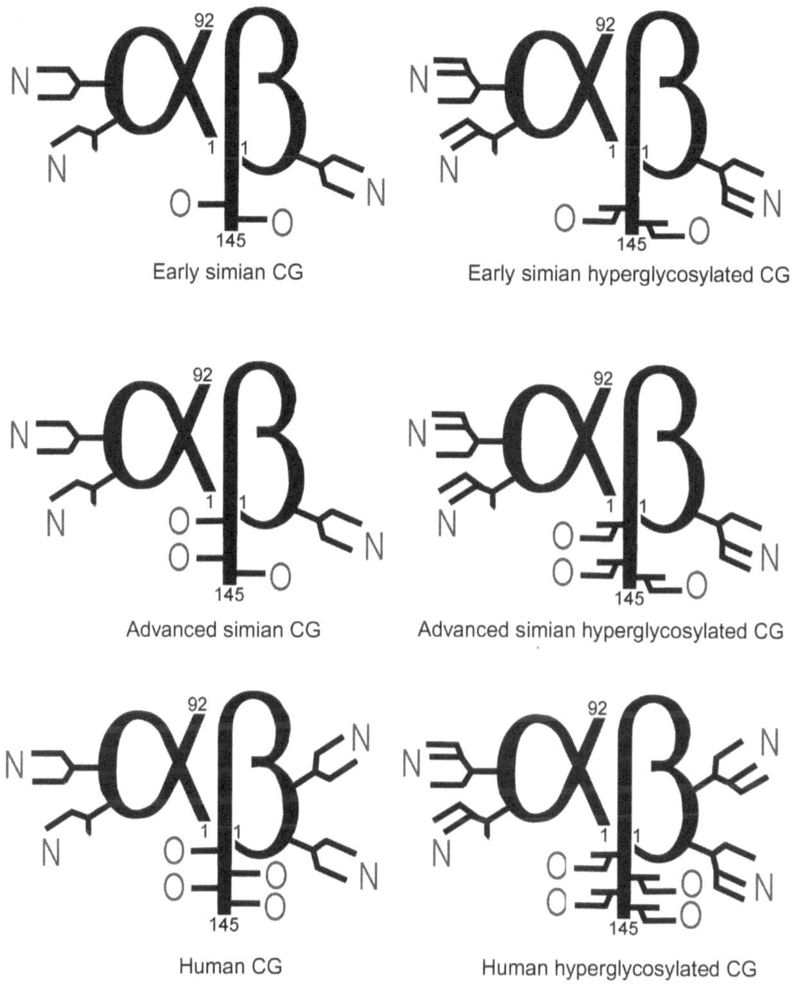

Figure 17.4 Structure of early simian, advanced simian, and human CG and hyperglycosylated CG.

With the evolution of advanced simian primates and humans, the CG β-subunit gene became duplicated. Humans have as many as eight back-to-back copies of the hCG β-subunit gene surrounding the single LH β-subunit gene on chromosome 19 [10]. In early simian primates, the CG β-subunit gained two O-linked oligosaccharides on the new C-terminal peptide sequence, in addition to a single N-linked oligosaccharide inherent to its LH β-subunit origin. Although the LH β-subunit was produced in pituitary gonadotrope tissue, CG was produced by placental or trophoblast tissues. In pituitary tissue, oligosaccharides were sulfated; in placental tissue, oligosaccharides were terminated with very acidic sialic acid sugar residues (Figure 17.4).

In advanced simian primates, further mutations occurred in the C-terminal extension of CG, or additional Ser residues were added. CG β-subunit gained a third acidic serine or O-linked oligosaccharide (Figure 17.4). With the evolution of humans, further mutations occurred in the C-terminal extension of CG, additional Ser residues were added, and the CG β-subunit gained a fourth acidic O-linked oligosaccharide. With further human evolution mutations (additional Asn residue), the CG β-subunit also gained a second N-linked oligosaccharide (Figure 17.4) [8]. With this advancing addition of O- and N-linked oligosaccharides, CG became more and more acidic, increasing its circulating half-life and taking its potency to the extreme [6–11] (Table 17.5).

Starting with early simian primates, the entire evolutionary story is centered on the generation of human CG and human hyperglycosylated CG—extremely potent forms of LH and a TGFβ antagonist produced by the placenta. Root placenta cells are cytotrophoblast cells that fuse to form multinucleated syncytiotrophoblast cells. Because of the presence of specific *N*-acetylglucosaminyltransferases (IV and VI), cytotrophoblast cells glycosylate molecules differently than syncytiotrophoblast cells [20]. As such, the cytotrophoblast cells of simian primates made a hyperglycosylated CG [21] (Figure 17.4). Hyperglycosylated CG has larger, double-sized oligosaccharide structures that are produced in cytotrophoblast cells. Because hyperglycosylated CG is an autocrine and binds a TGFβ receptor on cytotrophoblast cells [5,22], it has functions completely different from those of regular CG. Hyperglycosylated CG promotes pregnancy-like implantation invasion (see Chapter 14). Regular CG functions like LH, promoting corpus luteal progesterone production for 3 weeks in very early pregnancy. Regular CG also has critical functions in differentiating trophoblast cells to form villi, in promoting uterine spiral artery growth to provide fetal nutrition, and in uterine development throughout the course of pregnancy [21,23]. The two molecules, CG and hyperglycosylated CG, seemingly work together in pregnancy to promote invasion at implantation, uterine growth, and villi formation, and provide the nutrition needed from maternal spiral arteries (see Chapter 14). Humans evolved CG and hyperglycosylated CG with a long circulating half-life and high biological potency, which developed into the ultra-efficient nutritional transfer system we have today.

17.3 hCG, hyperglycosylation, and the evolution of humans

For many years, evolution specialists wondered how the human brain evolved [14–19]. How did placentation become efficient enough to provide the nutrients needed for the human brain to evolve? It seemed to logically defy all reason [13–19,24–28]. The answer is relatively simple, CG and hyperglycosylated CG together drive hemochorial placentation (see Chapter 14). Hyperglycosylated hCG through antagonizing an autocrine TGFβ receptor drove placental cytotrophoblast cells to invade (as in implantation) and to grow. With fusion and differentiation prompted by hCG [28], villous structures formed.

Under stimulation by hyperglycosylated hCG, villous structures continued to grow. Under the influence of hCG, a spiral artery blood supply in the uterus was extended to the villous structure and umbilical circulation was developed. Together, hCG and hyperglycosylated hCG drive hemochorial placentation.

The CG and hyperglycosylated hCG produced by early simian primates, from which CG and hyperglycosylated CG evolved, only had five oligosaccharides (Figure 17.4, Table 17.1). This was not a very acidic CG (pI 6.3) and it seemingly had a short circulating half-life (~2 h). With the evolution of advanced simian primates (e.g., chimpanzee, orangutan), the CG molecule contained additional oligosaccharide attachment points (Figure 17.2, Table 17.1). This CG and hyperglycosylated CG contained six oligosaccharides (pI of 4.9). This CG had a circulating half-life of approximately 6 h. Finally, with the evolution of humans emerged a CG and a hyperglycosylated CG with multiple additional oligosaccharide attachment points (eight oligosaccharides total) (Figure 17.4, Table 17.1). This CG and hyperglycosylated CG were extremely acidic, with a pI of 3.5. This CG was super-potent and had a circulating half-life of 36 h (extreme biological activity).

Summarizing Table 17.1, early simian primates formed a CG with minimal biological activity, and these only implanted in the uterus to the thickness of the decidua (to a depth of 2 mm). This depth did not permit establishment of an efficient form of hemochorial placentation involving a tank of maternal blood (Chapter 14). In advanced simian primates, the blastocyst implanted to a depth of 10% of the myometrial thickness. This permitted a more efficient form of hemochorial placentation. In humans, the blastocyst implants to a depth of one-third of the uterine thickness, and hyperglycosylated hCG promotes continual hemochorial placentation growth. The result was that hemochorial placentation was efficient in early simian primates, super-efficient in advanced simian primates, and ultra-super-efficient in humans. This permitted development, super-development, and ultra-super-development of the brain. This is how the large human brain came about.

It is through CG and hyperglycosylated CG becoming ultra-super-growth factors that extreme biological activity allowed the development of the large human brain (Table 17.1). It is thought that the human brain uses 60% of nutritional resources [13–19], leaving some organs, such as the human digestive system, deficient in nutrition.

17.4 Hominids and the evolution of the brain

Research indicates that the development of the human brain occurred in three clear steps, the evolution of lower simian primates, the evolution of advanced simian primates, and the evolution of humans. With this evolution, the lower simian CG had five oligosaccharides, the advanced simian CG had six oligosaccharides, and the human CG had eight oligosaccharides.

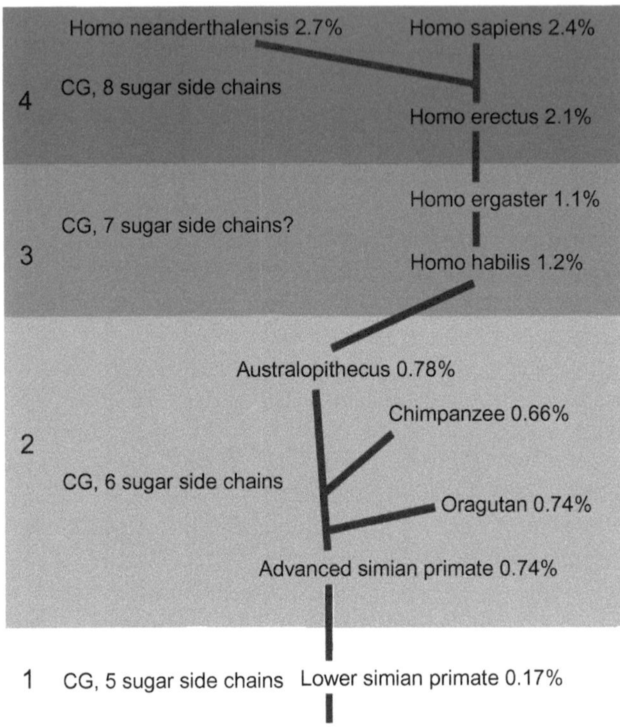

Figure 17.5 Evolution of brain size in primates, hominids, and humans. Lower simian.

Studies of brain size in primates and hominids suggest four steps in brain size advancement (Figure 17.5): early simian primates have a brain size that is 0.17% of body weight; advanced simian primates have a brain size that is 0.66–0.78% of body weight (Figure 17.5); hominids have a brain size that is 1.1–1.2% of body weight; and advanced hominids and humans have a brain size that is 2.1–2.7% of body weight. This four-step advancement includes an in-between step that involves hominids. Oligosaccharide size in our model jumps from six oligosaccharides (advanced primates) to eight oligosaccharides (humans), leaving space for an in-between step comprising hominids with seven oligosaccharides. Unfortunately, we cannot get blood from hominids, homo ergaster, or homo habilis, or determine pI or circulating half-life of CG because the species are extinct. However, it does appear that brain evolution or CG evolution occurred in four distinct steps and not three.

17.5 hCG evolution and the development of pregnancy disorders

hCG and hyperglycosylated hCG are the molecules that permitted human evolution [29,30]. Hyperglycosylated hCG is also, unfortunately, the molecule whose deficiency is the principal cause of failures such as miscarriages in pregnancy.

Deficiencies of hyperglycosylated hCG in pregnancy are also the cause of hypertension in pregnancy, the biggest killer of mothers in pregnancy.

One-hundred and thirty-seven woman were monitored with daily assessments of total hCG and hyperglycosylated hCG until achieving pregnancy [31,32]. Measurements were assessed until the day of pregnancy implantation. Among the 70 women with term outcome pregnancies (deliveries), hyperglycosylated hCG on the day of implantation (3 weeks gestation) was >40% of total hCG in 70 of 70 cases (Figure 17.6). In the 35 women with miscarried pregnancies, only 6 of 35 women had hyperglycosylated hCG >40%. In the 28 women with biochemical pregnancies, only eight had hyperglycosylated hCG concentration >40% of total hCG. Among the four women with ectopic pregnancies, none had hyperglycosylated hCG concentrations >40%. It was inferred that hyperglycosylated hCG was an absolute marker of term outcome pregnancies [31,33]. Unduly low hyperglycosylated hCG marked all failed pregnancies, whether they were miscarriages, biochemical pregnancies, or ectopic pregnancies. The remaining positive >40% of pregnancies were probably genetic abnormalities.

As established, most miscarried or spontaneous abortion (SAB) pregnancies are the direct result of inappropriate implantation [32,34]. It is inferred that most miscarried pregnancies may be a result of a deficient supply of hyperglycosylated hCG, the invasion signal at pregnancy implantation (Chapter 14).

Research indicates that hypertensive pregnancy, preeclampsia, or pregnancy-induced hypertension (PIH) are the result of deficient hyperglycosylated hCG [35,36]; as demonstrated, deficient hyperglycosylated hCG is associated with imminent preeclampsia and PIH. As such, preeclampsia and PIH can be predicted by unduly low hyperglycosylated hCG at 10–20 weeks of gestation [36,37]. Recent studies specifically show that unduly low hyperglycosylated hCG in the second trimester (10–20 weeks gestation) of pregnancy predicts preeclampsia/PIH [37]. It is inferred that low production of hyperglycosylated hCG in the second trimester of pregnancy leads to poor hemochorial placentation growth, nutritional deficiency, and preeclampsia/PIH.

17.6 hCG and cancer

Conflicting studies indicate opposite roles for hCG, free β-subunit, hyperglycosylated hCG, and its free β-subunit in cancer. As reported by six separate groups, administration of hCG and pregnancy significantly reduces a woman's risk for development of breast cancer [38–43]. In contrast, as reported by seven independent groups, hyperglycosylated hCG, hyperglycosylated hCG free β-subunit, and free β-subunit, which are autocrines acting on a TGFβ receptor [4,5,44,45], drive most malignancies [4,44–50], promoting growth, blocking apoptosis, and promoting invasion (Figure 17.7). What do hCG-related molecules do? Do they block cancer or drive cancer?

It is possible that both sets of highly confirmed research are correct. hCG may act on the LH/hCG hormone receptor to do one thing, and hyperglycosylated hCG, hyperglycosylated hCG free β-subunit, and free β-subunit may be autocrines that

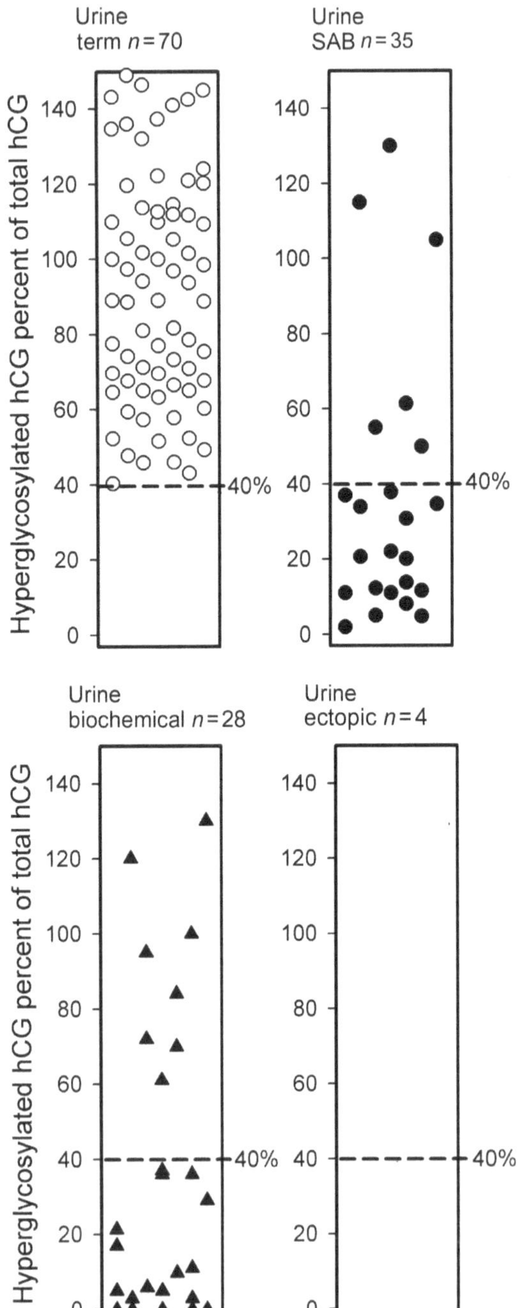

Figure 17.6 Measurement of hyperglycosylated hCG on the day of implantation in 137 pregnancies with normal term, miscarriage (SAB), biochemical pregnancy, and ectopic pregnancy outcomes. Proportion of hyperglycosylated hCG is concentration divided by total hCG concentration.

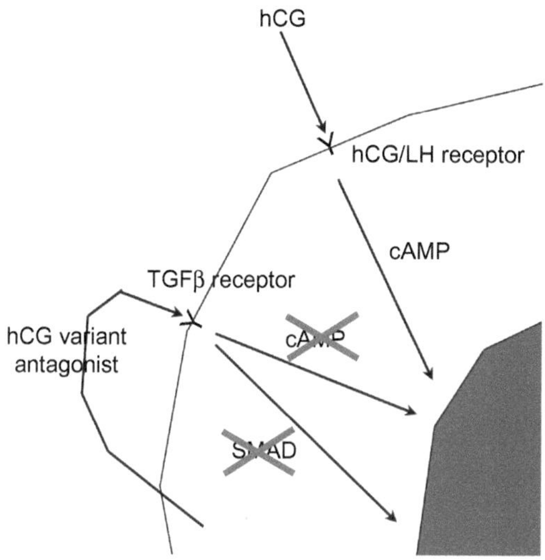

Figure 17.7 hCG and hCG variants acting on breast cancer cells, hCG/LH, and TGFβ receptors in a divergent manner (hCG partially opposing hCG variant antagonistic action).

antagonize the TGFβ receptor to promote the opposite. As illustrated in Figure 17.4, hCG may act on the LH/hCG receptor of breast cancer cells. One of a few cancers that may express such receptors, promoting intracellular cyclic AMP (cAMP), hyperglycosylated hCG, hyperglycosylated hCG free β-subunit, and free β-subunit, may be an autocrine that antagonizes a TGFβ receptor on these cells [4,5,44,45], reducing intracellular cAMP. The two signals may oppose each other, increasing and decreasing intracellular cAMP. By this process, antagonism of the TGFβ receptor may promote malignancy, whereas hCG promoting intracellular cAMP may inhibit growth, thus blocking apoptosis and invasion actions.

Consider hCG variants, hyperglycosylated hCG, hyperglycosylated hCG free β-subunit, and free β-subunit as cancer promoters. Confirmed research shows that these three variants are autocrines that act on and antagonize a TGFβ type 2 receptor [4,5,44]. The antagonism of receptor blocks cAMP and SMAD production, effectively blocking TGFβ-enhanced apoptosis, activating TGFβ-blocked cell growth, blocking TGFβ-enhanced differentiation, and activating TGFβ-blocked production of invasive enzymes, collagenases, and metalloproteinases [51,52]. Nude mouse cancer xenograft studies show that blocking hCG variants, hyperglycosylated hCG, hyperglycosylated hCG free β-subunit, and hCG free β-subunit leads to a complete halt in cancer cell growth (oncostasis) [45–47]. This shows that hCG variants are the prime or only signal for growth and invasion of malignancy or cancer cells.

Studies show that hCG variants may control cancer malignancy in 15 widely variable histologies of cancer [4,44–50]. It is inferred that hCG variants may make many, most, or all cancers malignant. It appears that expression of hCG genes may be a key part of cancer carcinogenesis or transformation. Expression of hCG variants or activation of the hCG variant pregnancy implantation invasion pathway may be a key part of a cell becoming cancerous.

17.7 All tied together

It is intriguing that the same course of evolution used to generate humans causes human miscarriages of pregnancy, preeclampsia/PIH in pregnancy, and drives human malignancies. This appears to be the molecule of life (the ability to have a baby) and death (miscarriages and cancer). It appears that having the extreme growth promoter and invasion promoter hyperglycosylated hCG in the human genome can be a problem. The single molecule that promoted the evolution of humans can also be evil.

References

[1] Lapthorn AJ, Harris DC, Littlejohn A, Lustbader JW, Canfield RE, Machin KJ. Crystal structure of hCG. Nature 1994;369:455–61.
[2] Sun PD, Davies DR. The cystine-knot growth-factor superfamily. Ann Rev Biophys Biomol Struct 1995;24:269–91.
[3] Chen J, Deng M, He Q, Meng E, Jiang L, Liao Z, et al. Molecular diversity and evolution of cystine knot toxins of the tarantula *Chilobrachys jingzhao*. Cell Mol Life Sci 2008;65:2431–44.
[4] Butler SA, Ikram MS, Mathieu S, Iles RK. The increase in bladder carcinoma cell population induced by the free beta subunit of hCG is a result of an anti-apoptosis effect and not cell proliferation. Brit J Cancer 2000;82:1553–6.
[5] Berndt S, Blacher S, Munuat C, Detilleux J, Evain-Brion D, Noel A, et al. Hyperglycosylated human chorionic gonadotropin stimulates angiogenesis through TGF-β receptor activation. FASEB J 2013; Abstract 12-213686.
[6] Fiddes JC, Goodman HM. The cDNA of the β-subunit of human chorionic gonadotropin suggests evolution of a gene by readthrough into the 3′ untranslated region. Nature 1979;281:351–6.
[7] Talmadge K, Vamvakopoulos NC, Fiddes JC. Evolution of the genes for the β-subunits of human chorionic gonadotropin and luteinizing hormone. Nature 1984;307:37–40.
[8] Maston GA, Ruvolo M. Chorionic gonadotropin has a recent origin within primates and an evolutionary history of selection. Mol Biol Evol 2002;19:320–35.
[9] Hallast P, Rull K, Laan M. The evolution and genomic landscapes of CGB1 and CGB2 genes. Mol Cell Endocrinol 2007;260:2–11.
[10] Policastro P, Ovitt CE, Hoshina M, Fukuoka H, Boothby MR, Boime I. The β-subunit of human chorionic gonadotropin is encoded by multiple genes. J Biol Chem 1983;258:11492–9.
[11] Bambra CS. Purification and properties of baboon chorionic gonadotropin. J Reprod Fertil 1987;19:421–30.
[12] Nisula BC, Wehmann RE. Distribution, metabolism, and excretion of human chorionic gonadotropin and its subunits in man. In: Segal SJ, editor. Chorionic gonadotropin. New York, NY: Plenum Press; 1980. p. 199–212.
[13] Robillard PY, Hulset TC, Deka CG. Preeclampsia and human reproduction. An essay of a long term reflection. J Reprod Immunol 2003;59:93–100.
[14] Martin RD. Scaling of the mammalian barrier: the maternal energy hypothesis. News Physiol Sci 1996;4:149–54.
[15] Cunnane SC, Herbige LS, Crawford MA. The importance of energy and nutrient supply in human brain evolution. Nutr Health 1993;9:19–35.

[16] Martin RD. Human reproduction: a comparative background for medical hypotheses. J Reprod Immunol 2003;59:111–35.
[17] Martin RD. Relative brain size and basal metabolic rate in terrestrial vertebrates. Nature 1981;293:57–60.
[18] Pijnenborg R. The placental bed. Hypertens Pregn 1996;15:7–23.
[19] Jordaan HVF. Newborn brain: body weight ratios. Am J Phys Anthropol 2005;44:2279–84.
[20] Endo T, Nishimura R, Mochizuki M, Kochibe N, Kobata A. Altered glycosylation is induced in both α- and β-subunits of human chorionic gonadotropin produced by choriocarcinoma. J Biochem 1988;103:1035–8.
[21] Zygmunt M, Herr F, Keller-Schoenwetter S, Kunzi-Rapp K, Munstedt K, Rao CV, et al. Characterization of human chorionic gonadotropin as a novel angiogenic factor. J Clin Endocrinol Metab 2002;87:5290–6.
[22] Gibbons A. Solving the brain's energy crisis. Science 1998;280:1345–7.
[23] Cole LA. Hyperglycosylated hCG. Placenta 2007;28:977–86.
[24] Kliman HJ. Uteroplacental blood flow. The story of decidualization, menstruation, and trophoblast invasion. Am J Pathol 2000;157:1759–68.
[25] Jauniaux E, Poston L, Burton GJ. Placental-related diseases of pregnancy: involvement of oxidative stress and implications in human evolution. Hum Reprod Update 2006;12:747–55.
[26] Burton GJ. Early placental development. Placenta 2006;27:A2.
[27] Luckett WP. Comparative development and evolution of the placenta in primates. Contrib Primatol 1974;3:142–234.
[28] Shi QJ, Lei ZM, Rao CV, Lin J. Novel role of human chorionic gonadotropin in differentiation of human cytotrophoblasts. Endocrinology 1993;132:1387–95.
[29] Cole LA, Khanlian SA, Kohorn EI. Evolution of the human brain, chorionic gonadotropin and hemochorial implantation of the placenta. J Reprod Med 2008; 53:549–57.
[30] Cole LA. hCG and hyperglycosylated hCG in the establishment and evolution of hemochorial placentation. J Reprod Immunol 2009;82:112–18.
[31] Cole LA. Hyperglycosylated hCG and pregnancy failures. J Reprod Immunol 2012; 93:119–22.
[32] Norwitz ER, Schust DJ, Fisher SJ. Implantation and the survival of early pregnancy. N Engl J Med 2001;345:1400–8.
[33] Sasaki Y, Ladner DG, Khanlian SA, Cole LA. Hyperglycosylated hCG and the source of pregnancy failures. Fertil Steril 2008;89:1871–86.
[34] Semprini AE, Simoni G. Not so efficient reproduction. Lancet 2000;356:257–8.
[35] Bahado-Singh RO, Oz AU, Kingston JM, Shahabi S, Hsu CD, Cole LA. The role of hyperglycosylated hCG in trophoblast invasion and the prediction of subsequent preeclampsia. Prenat Diagn 2002;22:478–81.
[36] Burton GJ. Placental oxidative stress: from miscarriage to preeclampsia. J Soc Gynecol Invest 2004;11:342–52.
[37] Brennan MC, Wolfe MD, Murray-Krezan CM, Cole LA, Rayburn WF. First trimester hyperglycosylated human chorionic gonadotropin and development of hypertension. Prenat Diagn 2013.
[38] Russo IH, Russo J. Use of human chorionic gonadotropin in the prevention of breast cancer. Women's Health 2008;4:1–5.
[39] Lopez D, Sekharam M, Coppola D, Carter WB. Purified human chorionic gonadotropin induces apoptosis in breast cancer. Mol Cancer Ther 2008;7:2837–44.

[40] Kocdor H, Kocdor MA, Russo J, Snider KE, Vanegas JE, Russo IH, et al. Human chorionic gonadotropin (hCG) prevents the transformed phenotypes induced by 17 beta-estradiol in human breast epithelial cells. Cell Biol Int 2009;33:1135–43.
[41] Tuma R. Mimicking pregnancy to reduce breast cancer risk. J Natl Cancer Inst 2010;102:517–18.
[42] Toniolo P, Grankvist K, Wulff M, Chen T, Johansson R, Schock H, et al. Human chorionic gonadotropin in pregnancy and maternal risk of breast cancer. Cancer Res 2010;70:6779–86.
[43] Boukaidi SA, Cooley A, Hardy A, Matthews L, Zelivianski S, Jeruss JS. Impact of infertility regimens on breast cancer cells: follicle-stimulating hormone and luteinizing hormone lack a direct effect on breast cell proliferation *in vitro*. Fertil Steril 2012;97:440–4.
[44] Cole LA, Butler SA. Hyperglycosylated hCG hCGβ and hyperglycosylated hCGβ: interchangeable cancer promoters. Mol Cell Endocrinol 2012;349:232–8.
[45] Cole LA, Iles RK, Butler SA. Hyperglycosylated hCG and its β-subunit: pervasive promoters of invasion and growth in cancer. Mol Cell Endocrinol 2013.
[46] Cole LA, Dai D, Leslie KK, Butler SA, Kohorn EI. Gestational trophoblastic diseases. 1. Pathophysiology of hyperglycosylated hCG-regulated neoplasia. Gynecol Oncol 2006;102:144–9.
[47] Carter WB, Sekharem M, Coppola D. Purified hCG induces apoptosis in breast cancer. Breast Cancer Res Treat 2006;100:S243–4.
[48] Iles RK. Ectopic hCGβ expression by epithelial cancer: malignant behavior metastasis and inhibition of tumor cell apoptosis. Mol Cell Endocrinol 2007;260:264–70.
[49] Li D, Wen X, Ghali L, Al-Shalabi FM, Docherty SM, Purkis P, et al. hCG beta expression by cervical squamous carcinoma-*in vivo* histological association with tumour invasion and apoptosis. Histopathology 2008;53:147–55.
[50] Jankowska A, Andrusiewicz M, Grabowski J, Nowak-Markwitz E, Warchol JB. Coexpression of human chorionic gonadotropin beta subunit and its receptor in nontrophoblastic gynecological cancer. Int J Gynecol Cancer 2008;18:1102–7.
[51] Wu D, Luo S, Wang Y, Zhuang L, Chen Y, Peng C. Smads in human trophoblast cells: expression regulation and role in TGF-beta-induced transcriptional activity. Mol Cell Endocrinol 2001;175:111–21.
[52] Massague J. TGF-beta signal transduction. Ann Rev Biochem 1998;67:753–91.
[53] Crawford CRJ, Tegear GW, Niall HD. The nucleotide sequence of baboon chorionic gonadotropin β-subunit genes have diverged from the human. Gene 1986;46:161–9.
[54] Baldwin DM, Highsmith RF, Ramey JW, Krummen LA. An *in vitro* study of LH release, synthesis and heterogeneity in pituitaries from proestrous and short-term ovariectomized rats. Biol Reprod 1986;34:304–15.
[55] Gam L-H, Latiff A. SDS-PAGE electrophoretic property of human chorionic gonadotropin (hCG) and its β-subunit. Int J Biol Sci 2005;1:103–9.
[56] Zygmunt M, Herr F, Munstedt K, Lang U, Liang OD. Angiogenesis and vasculogenesis in pregnancy. Eur J Obstet Gynecol Reprod Biol 2003;110:S10–18.
[57] Murphy G, Reynolds JJ, Whitham SE, Docherty AJ, Angel P, Heath JK. Transforming growth factor beta modulates the expression of collagenase and metalloproteinase inhibitor. EMBO J 1987;6:1899–904.
[58] Kovalevskaya G, Genbacev O, Fisher SJ, Cacere E, O'Connor JF. Trophoblast origin of hCG isoforms: cytotrophoblasts are the primary source of choriocarcinoma-like hCG. Mol Cell Endocrinol 2002;194:147–55.
[59] Massague J. TGFβ in cancer. Cell 2008;134:215–30.

Part G

Biological Function: Gestational Trophoblastic Disease

Biological functions of hyperglycosylated hCG

18

Laurence A. Cole
USA hCG Reference Service, Angel Fire, NM, USA

18.1 Discovery of hyperglycosylated hCG

The discovery started out in 1985–1987, when I proposed to Raymond Ruddon, PhD, my postdoctoral adviser at the time, that lectin studies indicated that an O-linked oligosaccharide was attached to the Jar cell line free α-subunit. He asked me to prove it. This is when I first established long, thin, fine Sephadex G10, G15, and G25 columns for separating the O-linked oligosaccharides released from pure hCG preparations. O-linked oligosaccharides were released by β-elimination. At first, I radioactively labeled the released oligosaccharides. With time, I developed other methods for detecting them.

Having confirmed the presence of a single O-linked oligosaccharide on choriocarcinoma hCG free α-subunit [1], I went on to examine the structures of purified hCG subunits produced in pregnancy and choriocarcinoma. Choriocarcinoma molecules had a unique double size hexasacharide O-linked structure on their β-subunit [2,3], which comprised all or most of the O-linked oligosaccharides. This very much marked choriocarcinoma molecules.

At this point that I remember being invited to present my research in the Biochemistry Department at Tokyo University. The chairman at that time was Akira Kobata, PhD (see Chapter 7). I also remember presenting this discovery and, at question time, Dr. J. Amano arose [4]: "I have found almost identical result for the structures of pregnancy and choriocarcinoma O-linked oligosaccharides." True enough, her confirmatory findings were almost identical [4]. I thank Dr. Amano and Dr. Kobata in Japan for their contributions to this discovery [4].

It was 10 years before I formally confirmed this discovery with six pregnancy urine hCG preparations, four hydatidiform mole hCG preparations, and six choriocarcinoma hCG preparations [5]. The discovery of the hexasaccharide structure was confirmed, as were the N-linked oligosaccharide structures (Chapters 4 and 5). They have both been confirmed more recently using mass spectrometry methods [6].

I remember initially inferring that the hCG with alternative O-oligosaccharides and N-oligosaccharides was a molecule separate from regular pregnancy hCG. I named the choriocarcinoma molecule "invasive trophoblast antigen." I published a few articles on this variant of hCG using this name [7,8]. I recall receiving a strong letter from WHO informing me that the name had to include the title hCG because

Human Chorionic Gonadotropin (hCG). DOI: http://dx.doi.org/10.1016/B978-0-12-800749-5.00018-3

the amino acid structure, as we had shown [5], was the same as hCG. I renamed the choriocarcinoma molecule "hyperglycosylated hCG."

In 2013, we now know that hyperglycosylated hCG is a molecule completely independent of hCG, binding a separate TGFβ type II receptor to hCG [9–11] and having completely independent functions [12–14]. We still hear confusion today about the molecule being called hCG. Would the WHO, today, still have problems with that old name "invasive trophoblast antigen?"

18.2 Hyperglycosylated hCG, independent molecule to hCG

Data presented in Figures 18.1 and 18.2 are previously published data [12]. Similar data have been published by Guibourdenche et al. [13] and Handschuh et al. [14].

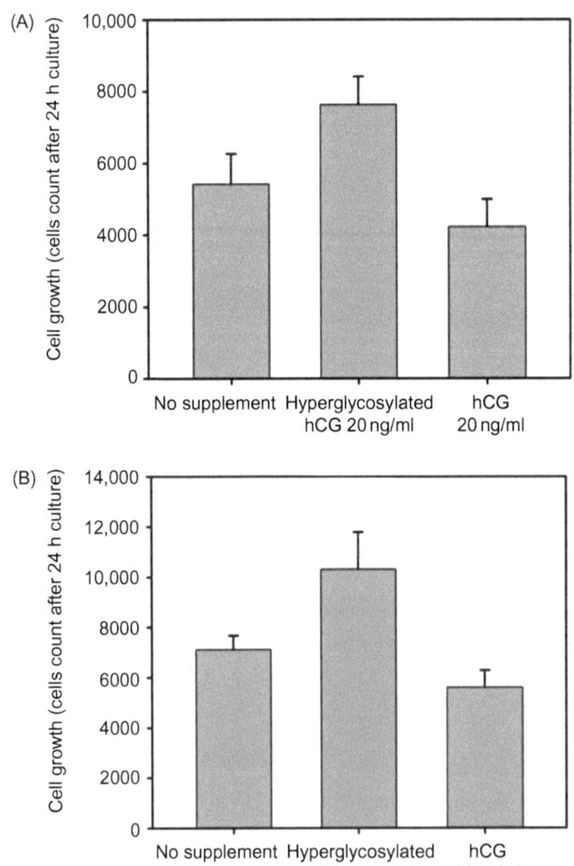

Figure 18.1 Effect of hCG and hyperglycosylated hCG on growth (cell count) of (A) term pregnancy cytotrophoblast primary cultures and (B) JEG-3 choriocarcinoma culture cells.

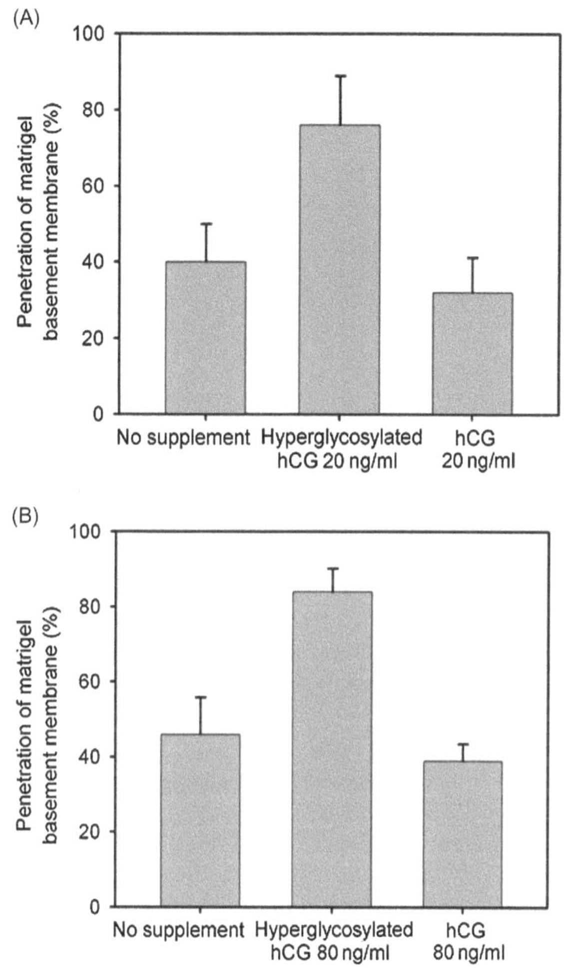

Figure 18.2 Effect of hCG and hyperglycosylated hCG on the invasion (penetration of basement membrane) Matrigel by (A) term pregnancy cytotrophoblast primary cultures and (B) JEG-3 choriocarcinoma culture cells.

As shown, hyperglycosylated hCG significantly enhances primary pregnancy cytotrophoblast cell growth and Jar choriocarcinoma cultured cell growth (Figure 18.1), and the regular hormone hCG significantly inhibits normal cytotrophoblast and choriocarcinoma cell growth. Similarly, hyperglycosylated hCG significantly promoted primary pregnancy cytotrophoblast cell invasion of a basement membrane and Jar choriocarcinoma cultured cell invasion of a basement membrane (Figure 18.2). The regular hormone hCG significantly inhibits normal cytotrophoblast and choriocarcinoma cell invasion of a basement membrane. It is inferred that hCG and hyperglycosylated hCG bind separate receptors and have separate functions.

As published, hyperglycosylated hCG, but not total hCG (hyperglycosylated hCG plus hCG), significantly drives pregnancy implantation [15,16]. Furthermore, a significant difference in hyperglycosylated hCG, but not total hCG, is found between preeclampsia and pregnancy-induced hypertension cases [17]. This confirms the independent functions of hyperglycosylated hCG and regular hCG. As demonstrated, hyperglycosylated hCG, but not regular hCG, can antagonize a TGFβ receptor, confirming its independence [9,10].

In conclusion, hyperglycosylated hCG is not simply a variant of hCG that shares a common amino acid sequence [5,6], but rather a completely independent molecule with independent functions binding a separate receptor [9–11].

18.3 Hyperglycosylated hCG binds a TGFβ receptor

As first shown by Butler et al. in 2000 [9], hCG free β-subunit promotes bladder cancers. As shown, hCG free β-subunit acts as an autocrine in bladder cancer cells by binding and antagonizing a TGFβ type II receptor [9]. As shown by Cole and Butler in 2012 [10], hCG free β-subunits are interchangeable with hyperglycosylated hCG free β-subunits and with hyperglycosylated hCG. All three molecules can equally promote bladder cancer, normally promoted by hCG free β-subunit [9], and choriocarcinoma, normally promoted by hyperglycosylated hCG [12]. Although all three promoters are produced by separate cancer cells, it was concluded, based on their interchangeability, that all three must act on the TGFβ type II receptor. As confirmed by Berndt and colleagues in 2013 [11], hyperglycosylated hCG functions as an autocrine through a TGFβ type 2 receptor.

This is a unique situation with regular hCG being a hormone acting solely on an hCG/LH receptor. Hyperglycosylated hCG is a carbohydrate variant of hCG with identical amino acid sequence to hCG [5,6]. Hyperglycosylated hCG acts on a separate receptor, the TGFβ type II receptor. These are two completely independent molecules only varying in N-linked and O-linked oligosaccharide side chain structures [1–6].

References

[1] Cole LA. Distribution of O-linked sugar units on hCG and its free alpha-subunit. Mol Cell Endocrinol 1987;50:45–57.
[2] Cole LA. O-Glycosylation of proteins in the normal and neoplastic trophoblast. Troph Res 1987;2:139–48.
[3] Cole LA. The O-linked oligosaccharides are strikingly different on pregnancy and choriocarcinoma hCG. J Clin Endocrinol Metab 1987;65:811–13.
[4] Amano J, Nishimura R, Mochizuki M, Kobata A. Comparative study of the mucin-type sugar chains of human chorionic gonadotropin present in the urine of patients with trophoblastic diseases and healthy pregnant women. J Biol Chem 1988;263:1157–65.

[5] Elliott MM, Kardana A, Lustbader JW, Cole LA. Carbohydrate and Peptide structure of the α- and β-subunits of human chorionic gonadotropin from normal and aberrant pregnancy and choriocarcinoma. Endocrine 1997;7:15–32.
[6] Valmu L, Alfthan H, Hotakainen K, Birken S, Stenman UH. Site-specific glycan analysis of human chorionic gonadotropin beta-subunit from malignancies and pregnancy by liquid chromatography-electrospray mass spectrometry. Glycobiology 2006;16:1207–18.
[7] Cole LA, Shahabi S, Oz UA, Bahado-Singh RO, Mahoney MJ. Hyperglycosylated hCG (invasive trophoblast antigen) immunoassay: a new basis for gestational down syndrome screening. Clin Chem 1999;45:2109–19.
[8] Cole LA, Khanlian SA, Sutton JM, Davies S, Stephens N. Hyperglycosylated hCG (invasive trophoblast antigen, ITA) a key antigen for early pregnancy detection. Clin Biochem 2003;36:647–55.
[9] Butler SA, Ikram MS, Mathieu S, Iles RK. The increase in bladder carcinoma cell population induced by the free beta subunit of hCG is a result of an anti-apoptosis effect and not cell proliferation. Br J Cancer 2000;82:1553–6.
[10] Cole LA, Butler SA. Hyperglycosylated hCG hCGß and hyperglycosylated hCGß: interchangeable cancer promoters. Mol Cell Endocrinol 2012;349:232–8.
[11] Berndt S, Blacher S, Munuat C, Detilleux J, Evain-Brion D, Noel A, et al. Hyperglycosylated human chorionic gonadotropin stimulates angiogenesis through TGF-ß receptor activation. FASEB J 2013; Abstract 12-213686.
[12] Cole LA, Dai D, Leslie KK, Butler SA, Kohorn EI. Gestational trophoblastic diseases: 1. Pathophysiology of hyperglycosylated hCG-regulated neoplasia. Gynecol Oncol 2006;102:144–9.
[13] Guibourdenche J, Handschuh K, Tsatsaris V, Gerbaud P, Leguy MC, Muller F, et al. Hyperglycosylated hCG is a marker of early human trophoblast invasion. J Clin Endocrinol Metab 2010;95:E240–4.
[14] Handschuh K, Guibourdenche J, Tsatsari V, Guesnon M, Laurendeau I, Evain-Brion D, et al. Human chorionic gonadotropin produced by the invasive trophoblast but not the villous trophoblast promotes cell invasion and is down-regulated by peroxisome proliferator-activated receptor-A. Endocrinology 2007;148:5011–19.
[15] Sasaki Y, Ladner DG, Cole LA. Hyperglycosylated hCG the source of pregnancy failures. Fertil Steril 2008;89:1781–6.
[16] Cole LA. Hyperglycosylated hCG and pregnancy failures. J Reprod Immunol 2012;93:119–22.
[17] Bahado-Singh RO, Oz AU, Kingston JM, Shahabi S, Hsu CD, Cole LA. The role of hyperglycosylated hCG in trophoblast invasion and the prediction of subsequent pre-eclampsia. Prenat Diagn 2002;22:478–81.

Gestational trophoblastic diseases 19

Laurence A. Cole
USA hCG Reference Service, Angel Fire, NM, USA

19.1 Gestational trophoblastic diseases

Gestational trophoblast diseases are an oddity of human biology. A hydatidiform mole is what happens, for instance, when two male gametes and their chromosomal haploid components form a pregnancy. The zygote formed does not differentiate appropriately and leads to a pregnancy comprising all or mostly placenta tissue. Hydatidiform mole can be invasive, attacking the uterus and associated organs. Normal placenta tissue and hydatidiform mole tissue can become a kind of malignancy. This is known as gestational trophoblastic neoplasm (GTN). The most common malignancy is choriocarcinoma, a type of over-invading pregnancy implantation. Less common neoplasms include placental site trophoblastic tumor (PSTT) and epithelial trophoblastic tumor.

19.2 Complete hydatidiform mole

A complete hydatidiform is a meiosis complication of the menstrual cycle and pregnancy. Normally in every menstrual cycle, one competing follicle becomes the dominant follicle or Graafian follicle. In the process of ovulation, the hormones luteinizing hormone (LH) and sulfated hCG promote the first stage of meiosis, generation of a haploid set of chromosomes, to combine with the haploid set coming from the fertilizing sperm.

Rarely, meiosis goes very wrong. This is seemingly because of an inappropriate LH or sulfated hCG peak. In complete hydatidiform, mole meiosis or generation of a haploid set of chromosomes fails. The result is that an empty egg, or an egg without a haploid set of chromosomes to donate, is ovulated. This egg can be fertilized by one sperm; the sperm then finds no oocyte haploid set to pair with. In this case, the sperm haploid set duplicates itself, making two identical haploid sets, then fertilizes itself, creating two analogous sperm origin haploid sets. Alternatively, two sperm can provide haploid sets if cortical granules fail to block sperm penetration of the egg (see "partial mole") [1–3]. The end result is a fertilized embryo of diandrogenous origin [1–3]. This forms a zygote (a fertilized single cell being) and grows and develops to become a morula with 8 – 16 identical cells. At the morula stage, the first differentiation must occur in fetal and placental tissue [4]. This is seemingly signaled by a gene on the female X chromosome. This is absent in a

Human Chorionic Gonadotropin (hCG). DOI: http://dx.doi.org/10.1016/B978-0-12-800749-5.00019-5

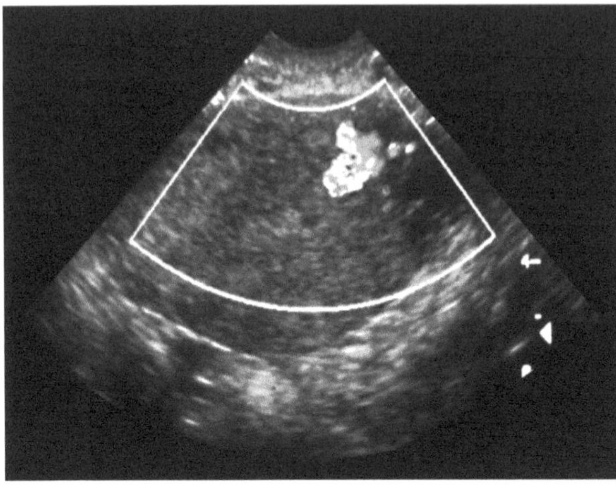

Figure 19.1 Ultrasound of a complete hydatidiform mole.

hydatidiform mole zygote. Without differentiation, by default it becomes 100% placental tissue. This is a complete hydatidiform mole. A complete hydatidiform mole is identified in pregnancy by ultrasound. Complete moles comprise villous trophoblast tissue as in normal pregnancies.

A classical snowflake picture is seen by ultrasound with no fetus (see Figure 19.1). The complete hydatidiform mole placental mass must then be terminated by dilation and curettage. One complication of complete hydatidiform moles is invasive or persistent moles. This happens in complete hydatidiform mole; cytotrophoblast cells become invasive, resembling the invasive process of implantation of pregnancy, driven by hyperglycosylated hCG and the TGFβ receptor (see Chapter 12). Invasive or persistent moles occur in approximately 20–30% of cases of complete hydatidiform mole. Complete hydatidiform moles occur in the United States at an incidence of 1 in 900 pregnancies [5].

19.3 Partial hydatidiform mole

Partial hydatidiform mole is another odd complication of pregnancy. It is a complication that most commonly causes a spontaneous abortion or miscarriage of pregnancy. Every woman's eggs have a shell called zona pellucida (see Figure 19.2), and multiple small granules inside the shell are called cortical granules in the perivitteline space (see Figure 19.2).

Normally, when one sperm undergoes the acrosome reaction and can penetrate the zona pellucida egg shell or penetrate or fertilize the egg, on infiltration it causes the cortical granules to release enzymes into the perivitteline space (see Figure 19.2). These enzymes cross-link or tie together the glycoproteins that comprise the zonal pellucida egg shell, such that it becomes completely unpenetratable by any

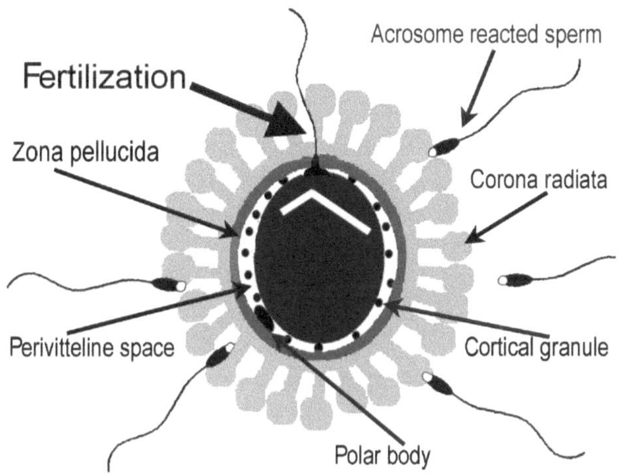

Figure 19.2 Human fertilization.

further sperm. This way, only the haploid set of chromosomes in one sperm can combine with the haploid set of chromosomes in an egg to make a new diploid embryo with no interference.

Approximately once in 141 fertilizations, the cortical granules fail to release the enzymes on penetration by sperm of the zona pellucida. As such, without those enzymes, the egg can be penetrated by more than one sperm. The result is that the haploid set from more than one sperm may enter into the fertilization equation, and one ends up with a faulty triploid or polypoid embryo. This is a partial hydatidiform mole. It has two haploid or half sets of chromosomes from male sperm and one haploid set from the female egg. A zygote is formed and divides. Somehow, when growth and differentiation start, the process goes crazy and one ends up with a mess of shuffled, or disconnected, fetal components and placental components. This is a partial mole. If a partial mole is not miscarried, then it will show up on ultrasound as a partial mole with no clear fetus. Partial moles comprise differentiated villous trophoblast tissue as in normal pregnancies.

At this time, termination by dilation and curettage is recommended. However, in some countries, physicians just wait for a partial mole to miscarry. Rarely does a partial hydatidiform mole lead to a persistent or invasive hydatidiform mole or become GTN. It is clearly a cause of failing pregnancy or miscarriage. The incidence of partial hydatidiform mole is 1 in 330 pregnancies [5].

19.4 Persistent or invasive hydatidiform mole

Complete hydatidiform moles commonly persist or become invasive (1 in 3 cases). Rarely do partial hydatidiform moles become invasive (1 in 200 cases) [5]. Partial and complete hydatidiform moles comprise villous trophoblast tissue as in normal pregnancies. This is a mixture of differentiated syncytiotrophoblast cells and root cytotropoblast cells. The root cytotrophoblast cells are similar to root cytotrophoblast cells in decidua pre-implantation (see Chapter 12).

Like the normal pregnancy cytotrophoblast cells, the hydatidiform mole cells produce hyperglycosylated hCG, which acts through a TGFβ receptor to drive growth and invasion. This may drive growth and invasion or malignancy of this villous hydatidiform mole tissue. The villous cytotrophoblast cells differentiate, forming invasive villous tissue. This is persistent or invasive hydatidiform mole. This invasive tissue can invade the thickness of the uterus, with metastasis occurring in the pulmonary, vaginal, and pelvic region. Chemotherapy is recommended in treating these cases. The villous tissue structure of this malignant disease distinguishes it from the nonvillous nature of other neoplasic gestational trophoblastic diseases.

19.5 Gestational trophoblastic neoplasm

An invasive nonvillous cytotrophoblast cell lesion is called GTN. If surgery and histology studies are performed and malignancy is identified, then it is called choriocarcinoma, or cancer of the placenta. Cells can fuse and differentiate and form noninvasive syncytiotrophoblast cells. GTN or choriocarcinoma occurs after parturition of a normal term pregnancy (1 in 30,000 pregnancies) and after complete hydatidiform moles (1 in 50); considering the 1 in 900 incidence of complete hydatidiform mole, GTN/choriocarcinoma occurs in 1 in 45,000 pregnancies [6].

When GTN/choriocarcinoma occurs after normal term pregnancy, tissue starts growing and invading at parturition, which is detected as a widely spread cancer, commonly as brain and lung metastases, approximately 2 months after delivery. Similarly, when GTN/choriocarcinoma occurs after complete hydatidiform mole, it is commonly detected as a widely spread malignancy.

It is questioned whether GTN/choriocarcinoma is technically a cancer (see Chapter 22). Normally, a cancer by definition involves carcinogenesis or transformation. This seemingly does not occur in GTN/choriocarcinoma. Cytotrophoblast cells naturally produce hyperglycosylated hCG, which promotes their growth and invasion. Other cells need to be transformed to produce invasive hyperglycosylated hCG. It appears that GTN/choriocarcinoma may result from pregnancies and hydatidiform moles that produce an excess of hyperglycosylated hCG (see Chapter 22).

GTN/choriocarcinoma should be treated with chemotherapy. Treatment should involve methotrexate and actinomycin D as single-agent therapies or EMA-CO (etoposide, methotrexate, actinomycin D in sequence with cyclophosphamide and onvocin [vincristine]) or EMA-EP (etoposide, methotrexate, actinomycin D in sequence with etoposide and cisplatinin) combination chemotherapies.

19.6 Placental site trophoblastic disease

PSTT is a separate malignancy of placental trophoblastic cells. This disease seemingly involves cancer transformation of placental trophoblastic cells. In that respect, it is very reminiscent of a nontrophoblastic cancer. Like a nontrophoblastic cancer, it produces primarily small quantities of the hCG free β-subunit or the hyperglycosylated free β-subunit. It does not produce high concentrations of total hCG and hyperglycosylated hCG like choriocarcinoma and GTN.

PSTT is best diagnosed by histology, by an expert placental pathologist (such as Robert J. Kurman, MD, of Johns Hopkins Medical Center), or by a specialist at the Brewer Trophoblast Disease Center (Chicago, IL), New England (Boston, MA) or Charing Cross (London, UK) Trophoblast Disease Specialty Center. The only other method to specifically diagnose PSTT is to demonstrate that 30–100% of the total hCG immunoreactivity in blood is due to the hCG free β-subunit. Unfortunately, laboratories only perform the hCG free β-subunit tests in the United States for Down syndrome screening. Under a patent agreement promoted by Johnson and Johnson, they are not allowed to use this test for cancer measurements. The only centers that can measure the hCG free β-subunit are the USA hCG Reference Service and laboratories in Canada.

PSTT develops from trophoblastic cells in pregnancy or hydatidiform mole. Such a malignancy may occur months or even years after pregnancy or hydatidiform mole. This is a very slow-growing transformed malignancy and it is usually very curable. Standard treatment is a hysterectomy to remove PSTT mass.

19.7 Epithelioid trophoblastic disease

Epithelioid malignancy of placental cells is a cancer separate to invasive mole (tumor comprising villous tissue), choriocarcinoma (tumor comprising mostly cytotrophoblast cells), and placenta site trophoblastic tumor (tumor comprising transformed trophoblastic cells). Epithelioid malignancy appears as a hemorrhagic solid and cystic mass. Microscopically it appears to involve mononucleated intermediate trophoblast cells surrounded by extensive necrosis [5,7].

Epithelioid malignancy is best diagnosed by histology, by an expert placental pathologist (such as Robert J. Kurman, MD, of Johns Hopkins Medical Center), or by a specialist at the Brewer (Chicago, IL), New England (Boston, MA), or Charing Cross (London, UK) Trophoblast Disease Specialty Center. Epitheliod malignancy of placental cells is an extremely rare disease. Epitheliod malignancy like PSTT is a very slow-growing malignancy and is usually very curable. The best treatment currently is surgery, such as hysterectomy [5,7].

19.8 hCG and gestational trophoblastic disease

A total hCG assay will detect all forms of gestational trophoblastic disease. In an untreated complete hydatidiform mole or partial hydatidiform mole case, the common serum total hCG ranges from 75,000 to 600,000 mIU/ml. In GTN/choriocarcinoma, before treatment the total hCG ranges, depending on the size of the tumor and its advancement, from 50,000 to 2,000,000 mIU/ml. In PSTT and epithelioid malignancy cases, much lower hCG β-subunit titers are detected, ranging from 3.0 to 2000 mIU/ml. In every gestational trophoblastic disease case, the syncytiotrophoblast cells secrete regular hCG, and the cytotrophoblast cells secrete hyperglycosylated hCG. In PSTT and epithelioid malignancies, primarily hCG free β-subunit or hyperglycosylated hCG free β-subunit may be produced.

19.9 Hyperglycosylated hCG and gestational trophoblastic diseases

Hyperglycosylated hCG, produced by cytotrophoblast cells in complete and partial hydatidiform mole, persistent or invasive hydatidiform mole, GTN, or choriocarcinoma cases, is the invasion and growth signal. Hyperglycosylated hCG is an autocrine that will feed-back and bind a THGβ receptor, driving both growth and invasion (promotes production of collagenases and metalloproteinases) [8]. In GTN/choriocarcinoma cases, the higher the proportion of hyperglycosylated hCG produced (as a percentage of total hCG), the more invasive and vicious the malignancy [9,10]. A new case of GTN/choriocarcinoma with brain and lung metastasis may present with 100% of the total hCG being hyperglycosylated hCG [9,10]. By comparison, in a slow-growing minimally aggressive case of GTN/choriocarcinoma, hyperglycosylated hCG may only represent 50% of the total hCG. In general, the proportion of hyperglycosylated hCG in GTN/choriocarcinoma cases represents the aggressiveness of the disease [9,10].

References

[1] Van Blerkom J, Henry G. Dispermic fertilization of human oocytes. J Electron Microsc Tech 1991;17:437–49.

[2] Ohama K, Kajii T, Okamoto Y, Fukuda Y, Imaizumi K, Tzekahara M, et al. Dispermic origin of XY hydatidiform moles. Nature 1981;251:551–2.

[3] Pattillo RA, Sasaki S, Katayama KP, Roesler M, Mattingly RF. Genesis of 46,XY hydatidiform mole. Am J Obstet Gynecol 1981;141:104–5.

[4] Red-Horse K, Zhou Y, Genbacev O, Prakobphol A, Foulk R, McMaster M, et al. Trophoblast differentiation during embryo implantation and formation of the maternal-fetal interface. J Clin Invest 2004;114:744–54.

[5] Kenny L, Seckl MJ. Treatments for gestational trophoblastic disease. Exp Rev Obstet Gynecol 2010;5:215–25.

[6] Hancock BW, Berkowitz RS, Cole LA. Gestational trophoblastic disease, eds. 2nd ed. 450 pages, <http://www.isstd.org/isstd/book.html>; 2007.

[7] Shih IM, Kurman RJ. Epithelioid trophoblastic tumor: a neoplasm distinct from choriocarcinoma and placental site trophoblastic tumor simulating carcinoma. Am J Surg Pathol 1998;22:1393–403.

[8] Cole LA, Dai D, Leslie KK, Butler SA, Kohorn EI. Gestational trophoblastic diseases: 1. Pathophysiology of hyperglycosylated hCG-regulated neoplasia. Gynecol Oncol 2006; 102:144–9.

[9] Cole LA. Hyperglycosylated hCG, the most valuable test in the management of gestational trophoblastic neoplasms. Troph Dis Upd 2010;10:5–22.

[10] Cole LA. hCG the centerpiece of life and death. Intl J Endocrinol Metab 2011; 9:225–42.

Part H

Biological Function: Cancer

Cancer-1, hCG variants as tumor markers

20

Laurence A. Cole
USA hCG Reference Service, Angel Fire, NM, USA

Research for this book's author, Laurence Cole, PhD, occurred during 1977–1981 with his PhD on the structure of the hCG variant produced by DoT and CasKi cervical endometroid cervical cancer cells. As previously demonstrated, both these cell lines secreted an hCG immunoreactive molecule into culture medium. As found by Laurence Cole's PhD work, they both produced a hyperglycosylated variant of the hCG free β-subunit (hyperglycosylated hCGβ) [1–4].

In the years that followed, Laurence Cole, PhD showed that hCGβ was seemingly a general tumor marker for all malignancies (Table 20.1) present in the serum and urine in all trophoblastic malignancies (testicular germ cell and ovarian germ cell malignancies adopt trophoblastic histology). Total hCG and hCG free β-subunit are detected in serum in 30% of other malignancies, and β-core fragment, the urinary degradation product of hCGβ, is a urine tumor marker detectable in the urine of 55% of other malignancy cases (Table 20.1) [5–9].

It was noted at this time that 100% of trophoblastic and germ cell malignancies produced are detected in serum hyperglycosylated hCG or its free β-subunit. Other malignancies produced are detected in serum hyperglycosylated hCG free β-subunit or hCG free β-subunit. Laurence Cole then patented this discovery; total hCG, hCG free β-subunit, and β-core fragment are general tumor markers for all malignancies but not for any specific malignancy. Between 1991 and 1999, Ciba-Corning in the United States and Wako and Toagosei Chemicals in Japan marketed these markers as urine tumor markers using Cole's patent.

Many centers then showed that most cancer patients and most cancer cell lines produced a variant of hCG free β-subunit [4,10–14]. It was inferred at that time that many cancers made hCG free β-subunit. Around this time, Acevedo and colleagues [15] examined the plasma membranes of a large number of cancer tissue membranes by flow cytometry methods and concluded that all cancers (100%) made a variant of hCG free β-subunit. As determined [56], the plasma membranes of all cancers stain for hCGβ, but not all cancers secrete hCG free β-subunit.

It is now known that cancer cells produce a mixture of hyperglycosylated hCG free β-subunits and hCG free β-subunits [16]. It is also known that these molecules function as autocrines antagonizing a TGFβ type-II receptor on cancer cells [10,16,17]. Data showing production and nonproduction of hCG free β-subunit (Table 20.1), together with data indicating that 100% cancers have hCG free β-subunits in membranes, indicate that cancer cells may be "simple autocrines" and "complex

Human Chorionic Gonadotropin (hCG). DOI: http://dx.doi.org/10.1016/B978-0-12-800749-5.00020-1

Table 20.1 Use of serum hCGβ (hCGβ plus hyperglycosylated hCG free β-subunit) and urine hCG β-core fragment as tumor markers for detection of malignancies

Malignancy	Serum total hCG		Urine hCG β-core fragment	
	Cases	**Sensitivity (>3 fmol/ml)**	**Cases**	**Sensitivity (>3 fmol/ml)**
A. Trophoblastic cancers				
Choriocarcinoma	20	100%	20	100%
Testicular germ cell cancer	17	100%	17	100%
Ovarian germ cell cancer	30	100%	30	100%
Total	67	100%	67	100%
B. Other cancers				
Ovarian adenocarcinoma	150	38%	207	70%
Cervical cancer	60	37%	410	48%
Endometrial cancer	55	33%	53	47%
Vulvar	50	38%	8	50%
Bladder cancer	170	35%	102	48%
Uterine cancer			9	67%
Lung cancer	143	18%	90	28%
Colorectal cancer	136	17%		
Pancreatic cancer			29	55%
Total	764	226 Positive (30%)	908	495 Positive (55%) Range

Averages are determined by combining total positive cases from multiple reports [5–9].

autocrines" [18,19]. Simple autocrines occur when cells produce low concentrations of molecules; they are secreted and act directly on the plasma membrane receptor. Those producing higher concentrations may use them as "complex autocrines," which may loop by following the circulation through the body before acting back on the cell's plasma membrane receptor [18,19].

It appears that trophoblast cancers function with hyperglycosylated hCG through a "complex autocrine" pathway, with 100% of cancer detected in serum. It appears that other cancers use hCG free β-subunits through "simple autocrine" or "complex autocrine" pathways, explaining the 55% detection in urine and 100% detection in the membranes of cancers.

In 2013, 10 of 11 automated total hCG tests, the Abbott Axsym, the Abbott Architect, the Beckman Access 2, the Beckman Dxi 800, the Ortho Vitros EciQ, the Roche Elecsys hCG + β, the Siemens Centaur, the Siemens ACS180, the

Siemens Dimension, and the Tosoh A1A all use the hCG β-subunit C-terminal peptide capture antibody [20,21]. This type of assay does not bind hCG free β-subunit missing the β-subunit C-terminal peptide, which is common in cancer cases [20,21]. The exception is the Siemens Immulite total hCG assay, which detects hCG free β-subunits and molecules missing the C-terminal peptide equally [20,21]. For monitoring cancer cases, the Siemens Immulite total hCG assay should always be used.

References

[1] Cole LA, Hussa RO, Rao CV. Discordant synthesis and secretion of human chorionic gonadotropin and subunits by cervical cancer cells. Cancer Res 1981;41:1615–19.
[2] Cole LA, Birken S, Sutphen S, Hussa RO, Pattillo RA. Absence of the COOH-terminal peptide on ectopic human chorionic gonadotropin β-subunit (HCG-β). Endocrinology 1982;110:2198–200.
[3] Cole LA, Hussa RO. Use of glycosidase digested human chorionic gonadotropin β-subunit to explain the partial binding of glycoprotein hormones to Con A. Endocrinology 1981;109:2276–9.
[4] Cole LA, Hartle RJ, Laferla JJ, Ruddon RW. Detection of the free β subunit of human chorionic gonadotropin in cultures of normal and malignant trophoblast cells, pregnancy sera, and sera of patients with choriocarcinoma. Endocrinology 1983;113:1176–8.
[5] Nam JH, Cole LA, Chambers JT, Schwartz PE. Urinary gonadotropin fragment a new tumor marker: I. Assay development and cancer-specificity. Gynecol Oncol 1990;36:383–90.
[6] Schwartz PE, Chambers JT, Taylor KJ, Cole LA, Makuch R. Urinary gonadotropin fragments. Anticancer Res 1993;13:1722–5.
[7] Cole LA. β-core fragment (β-core UGP or UGF). Tumor Marker Upd 1994;6:69–75.
[8] Cole LA, Nam J-H, Park S-Y, Koh MW, Tanaka A. Urinary beta core fragment: 7 years later. J Tumor Marker Oncol 1994;9:53–60.
[9] Cole LA, Tanaka A, Kim GS, Park S-Y, Koh MW, Schwartz PE, et al. Beta core fragment (β-core/UGF/UGP), a tumor marker: seven year report. Gynecol Oncol 1996;60:264–70.
[10] Butler SA, Ikram MS, Mathieu S, Iles RK. The increase in bladder carcinoma cell population induced by the free beta subunit of hCG is a result of an anti-apoptosis effect and not cell proliferation. Br J Cancer 2000;82:1553–6.
[11] Ruddon RW, Hanson CA, Bryan AH, Putterman GJ, White EL. Synthesis and secretion of human chorionic gonadotropin subunits by cultured human malignant cells. J Biol Chem 1980;255:1000–7.
[12] Carter PG, Iles RK, Neven P, Ind TE, Shepherd JH, Chard T. Measurement of urinary beta core fragment of human chorionic gonadotrophin in women with vulvovaginal malignancy and its prognostic significance. Br J Cancer 1995;71:350–3.
[13] Iles RK. Ectopic hCGβ expression by epithelial cancer: malignant behavior metastasis and inhibition of tumor cell apoptosis. Mol Cell Endocrinol 2007;264–70.
[14] Cosgrove DE, Campain JA, Cox GS. Chorionic gonadotropin synthesis by human tumor cell lines: examination of subunit accumulation steady-state levels of mRNA and gene structure. Biochem Biophys Acta 1989;1007:44–54.

[15] Acevedo HF, Krichevsky A, Campbell-Acevedo EA, Galyon JC, Buffo MJ, Hartsock RJ. Flow cytometry method for the analysis of membrane-associated human chorionic gonadotropin, its subunits, and fragments on human cancer cells. Cancer 1992;69:1818–28.

[16] Cole LA, Butler SA. Hyperglycosylated hCG hCGβ and hyperglycosylated hCGβ: interchangeable cancer promoters. Mol Cell Endocrinol 2012;349:232–8.

[17] Berndt S, Blacher S, Munuat C, Detilleux J, Evain-Brion D, Noel A, et al. Hyperglycosylated human chorionic gonadotropin stimulates angiogenesis through TGF-β receptor activation. FASEB J 2013; Abstract 12-213686.

[18] Fairbank M, St-Pierre P, Nabi IR. The complex biology of autocrine motility factor/phosphoglucose isomerase (AMF/PGI) and its receptor, the gp78/AMFR E3 ubiquitin ligase. Mol Biosyst 2009;5:793–801.

[19] Offermann MK, Faller DV. Autocrine induction of major histocompatibility complex class I antigen expression results from induction of beta interferon in oncogene-transformed BALB/c-3T3 cells. Mol Cell Biol 1989;9:1969–77.

[20] Cole LA, Shahabi S, Butler SA, Mitchell H, Newlands ES, Behrman HR, et al. Utility of commercially used human chorionic gonadotropin immunoassays in the diagnosis and management of trophoblastic diseases. Clin Chem 2001;47:308–15.

[21] Cole LA, Sutton JM, Higgins TN, Cembrowski GS. Between-method variation in hCG test results. Clin Chem 2004;50:874–82.

Cancer-2, hCG variants drive malignancies

21

Laurence A. Cole
USA hCG Reference Service, Angel Fire, NM, USA

It is well-established that all cancers make a free β-subunit or a hyperglycosylated hCG free β-subunit [1–15]. As shown, there are two separate types of β-subunit genes on chromosome 19 among the eight head-to-head similar hCG β-subunit genes: type I and type II [15–19]. Type I genes include β6 and β7 and express a glycoprotein β-subunit with Ala at β117 [15–19]. Type II genes include β1, β5, β8, and β9 and express a glycoprotein β-subunit with Asp at β117 [15–19]. As demonstrated, trophoblastic cells (i.e., in pregnancy) express type II genes [15,20]. Nontrophoblastic malignancies or other cancers express type I genes or both type I and type II genes [15,20].

Regular hCG acts on an LH/hCG hormone receptor. As demonstrated, the free β-subunit or hyperglycosylated hCG either does not bind this receptor or is biologically inactive [21]. With this knowledge, cancer cell production of hCG free β-subunit and hyperglycosylated hCG free β-subunit has generally been laughed at and classed as an epiphenomenon [22,23].

In 1995 and 1996, Lunardi-Iskandar et al. [24] and Gillott et al. [25] found that the hCG free β-subunit made by cancer cells is actually functional and not an epiphenomenon. As shown, hCG free β-subunit promotes cancer cell growth and invasion and blocks apoptosis in cancer cells [24,25]. In 2000, Butler (editor of this book), whose PhD was focused on hCG free β-subunit as a tumor promoter, and colleagues demonstrated the mechanism whereby hCG free β-subunit functions in cancer cells [26]. As demonstrated, hCG free β-subunit acts as an autocrine, binding and antagonizing the TGFβ type II receptor on cancer cells [26]. This promotes cancer cell growth and invasion and blocks apoptosis. Intriguingly, these functions exactly fit with proven roles of a TGFβ type II receptor. Normally, TGFβ enhances cell apoptosis and blocks cell growth, and it blocks cell production of collagenases and metalloproteinases, the degradative enzymes of cell invasion [27–30]. The concept that hCG free β-subunit antagonizes this TGFβ receptor exactly matches the concept that this molecule blocks apoptosis, promotes cancer cell growth, and promotes cancer cellular invasion. This finding explains everything, including how hCG free β-subunit had a separate function from the regular hormone hCG. Recent studies demonstrate that hyperglycosylated hCG, hyperglycosylated hCG free β-subunit, and free β-subunit function equally as TGFβ-II antagonists, whereas the hormone hCG has no activity at this receptor [31].

Human Chorionic Gonadotropin (hCG). DOI: http://dx.doi.org/10.1016/B978-0-12-800749-5.00021-3

These findings have all been repeated, confirmed, double-confirmed, and triple-confirmed [8,31–39]. The extent to which these studies have been confirmed is listed in Table 21.1. Considering this large accumulation of findings together (Table 21.1), it is appropriate to say that most or all cancers are driven in malignancy by hCG variants. Hyperglycosylated hCG drive choriocarcinoma and germ cell malignancies, and hCG free β-subunit and hyperglycosylated hCG free β-subunit or hCG free β-subunit drive all other malignancies. As shown in nude mouse xenograft models [32,39], blocking the hCG β-subunit brings cancer to oncostasis (a complete stand-still). It is inferred that cancer cells have no alternative antagonist to take the place of hyperglycosylated hCG or free β-subunit at the TGFβ-II receptor when hyperglycosylated hCG or free β-subunit are blocked, and that hyperglycosylated hCG, hyperglycosylated hCG free β-subunit, or free β-subunit are the principal driving mechanisms in human cancers.

In pregnancy, hyperglycosylated hCG acts on the cytotrophoblast cell TGFβ-II receptor to drive invasion and implantation (see Chapter 12). This is the advanced or ultra-efficient hyperglycosylated hCG mechanism developed from the evolution that permitted development of the human brain (see Chapter 17). Carcinogenesis involves expression of the hCG β-subunit gene and, with it, its TGFβ-II mechanism and the pregnancy implantation mechanism, which is the only invasive mechanism that is part of normal human biology. In essence, cancers going through carcinogenesis steal the human pregnancy implantation mechanism as a means to make them invasive or malignant. This may be the principal part of the carcinogenesis process.

The clearest clinical confirmations of these studies have been made by the ongoing clinical trials with a synthetic hCG β-subunit-related vaccine. Starting in 2000, three companies, Celldex, CG Therapeutics, and MCI BioPharma, started clinical trials using this synthetic hCG β-subunit vaccine with advanced cancer cases [40–45]. Results have been very exciting, showing that the presence of an intact immune system was critical and that a single injection of the vaccine can considerably extend the lives of patients with advanced cancer. For instance, when one injection of hCG vaccine was administered to a group of patients with advanced terminal colorectal cancer (stage 4+), who would normally live a median of 24 weeks, they lived a median of 45 weeks [41]. The hCG β-subunit vaccine seemingly bound and blocked hyperglycosylated hCG/hyperglycosylated hCG free β-subunit/hCG free β-subunit, which are the critical cancer promoters.

Antibody therapy, unlike vaccine treatment, using a human anti-hCG free β-subunit or anti-hyperglycosylated hCG is not subject to the need for an intact immune system and is not limited to one therapy treatment (like this vaccine is). Multiple antibody treatment, as shown in nude mouse studies, is considered to completely suppress these cancers. These vaccine data support multiple injections of specific monoclonal antibodies as an effective treatment or a possible cure for cancer.

In one study of bladder cancer [46], hCG free β-subunit was shown to specifically promote cancer cell growth and to block apoptosis. Treatment of mice with the hCG β-subunit vaccine reversed this process, promoting apoptosis and blocking tumor growth. This research confirmed the vaccine – tumor relationship and that

Table 21.1 Demonstration (as shown in publication) that hCG variants promotes cancer cell grow and invasion or block apoptosis, table list cell lines tested and their cancer histology

AIDS-KS Kaposis's sarcoma	Lunardi-Iskandar et al. [24]
SCaBER bladder epithelial carcinoma	Gillott et al. [25]
T24 bladder epithelial carcinoma	
RT112 bladder epithelial carcinoma	
5637 bladder epithelial carcinoma	
SCaBER bladder epithelial carcinoma	Butler et al. [26]
T24 bladder epithelial carcinoma	
RT112 bladder epithelial carcinoma	
5637 bladder epithelial carcinoma	
J82 bladder epithelial carcinoma	
SCaBER bladder epithelial carcinoma	Iles [8]
T24 bladder epithelial carcinoma	
RT112 bladder epithelial carcinoma	
5637 bladder epithelial carcinoma	
J82 bladder epithelial carcinoma	
SKBR3 nude mice xenograft, breast adenocarcinoma	Carter et al. [39]
JEG-III nude mouse xenograft, choriocarcinoma	Cole et al. [32]
JEG-III, choriocarcinoma	
71 squamous cervical cancers, hCGβ correlated with apoptosis	Li et al. [37]
12 cervical adenocarcinomas, hCGβ correlated with apoptosis	
7 adenosquamous cervical cancer, hCGβ correlated with apoptosis	
15 epithelial cervical cancers. Complementary hCGβ mRNA raised apoptosis	Jankowska et al. [36]
JAR choriocarcinoma	Cole and Butler [31]
JEG-III choriocarcinoma	
SCaBER bladder epithelial carcinoma	
T24 bladder epithelial carcinoma	
KLE endometrial adenocarcinoma	
Hec1A endometrial epithelial carcinoma	
JAR choriocarcinoma	Cole and Butler [34]
JEG-III choriocarcinoma	
SCaBER bladder epithelial carcinoma	
T24 bladder epithelial carcinoma	
KLE endometrial adenocarcinoma	
Hec1A endometrial epithelial carcinoma	
NTERA testicular embryonal germ cell	
OVCA-3 ovarian cystadenocarcinoma	
Caski cervical endometriod carcinoma	
HeLa cervical adenocarcinoma	
HOG-1 cervical squamous carcinoma	
IGROV-1 ovarian adenocarcinoma	
Ishikawa endometrial adenocarcinoma	
SiHA cervical squamous carcinoma	
C33 cervical epithelial carcinoma	

the vaccine works through the hCG free β-subunit − TGFβ-II pathway [46]. Considering the hyperglycosylated hCG, hyperglycosylated hCG free β-subunit, hCG free β-subunit, cancer studies, and the human vaccine clinical trials together, it appears that these molecules mark the major pathway driving most cancers in humans. It appears that with the generation of a human antibody to these hCG variants, we may have a new treatment for advanced cancer in humans or a possible cure for many cancers in humans.

References

[1] Cole LA, Birken S, Sutphen S, Hussa RO, Pattillo RA. Absence of the COOH-terminal peptide on ectopic human chorionic gonadotropin β-subunit (HCG-β). Endocrinology 1982;110:2198−200.

[2] Schwartz PE, Chambers JT, Taylor KJ, Cole LA, Makuch R. Urinary gonadotropin fragments. Anticancer Res 1993;13:1722−5.

[3] Cole LA. β-core fragment (β-core UGP or UGF). Tumor Marker Upd 1994;6:69−75.

[4] Cole LA, Nam J-H, Park S-Y, Koh MW, Tanaka A. Urinary beta core fragment: 7 years later. J Tumor Marker Oncol 1994;9:53−60.

[5] Cole LA, Tanaka A, Kim GS, Park S-Y, Koh MW, Schwartz PE, et al. Beta core fragment (β-core/UGF/UGP), a tumor marker: seven year report. Gynecol Oncol 1996; 60:264−70.

[6] Ruddon RW, Hanson CA, Bryan AH, Putterman GJ, White EL. Synthesis and secretion of human chorionic gonadotropin subunits by cultured human malignant cells. J Biol Chem 1980;255:1000−7.

[7] Carter PG, Iles RK, Neven P, Ind TE, Shepherd JH, Chard T. Measurement of urinary beta core fragment of human chorionic gonadotrophin in women with vulvovaginal malignancy and its prognostic significance. Br J Cancer 1995;71:350−3.

[8] Iles RK. Ectopic hCGβ expression by epithelial cancer: malignant behavior metastasis and inhibition of tumor cell apoptosis. Mol Cell Endocrinol 2007;264−70.

[9] Cosgrove DE, Campain JA, Cox GS. Chorionic gonadotropin synthesis by human tumor cell lines: examination of subunit accumulation steady-state levels of mRNA and gene structure. Biochem Biophys Acta 1989;1007:44−54.

[10] Acevedo HF, Krichevsky A, Campbell-Acevedo EA, Galyon JC, Buffo MJ, Hartsock RJ. Flow cytometry method for the analysis of membrane-associated human chorionic gonadotropin, its subunits, and fragments on human cancer cells. Cancer 1992;69: 1818−28.

[11] Nagelberg SB, Cole LA, Rosen SW. A novel form of ectopic human chorionic gonadotrophin β-subunit in the serum of a woman with epidermoid cancer. J Endocrinol 1985;107:403−8.

[12] Rosen SW, Calvert I, Weintraub BD, Tseng JS, Rabson AS. Stimulation of N6O2′-dibutyryl cyclic adenosine 3′:5′-monophosphate of ectopic production of the free beta subunit of cho rionic gonadotropin by a human brain tumor cell line. Cancer Res 1980;40:4325−8.

[13] Ozturk M, Bellet D, Isselbacher KJ, Wands J. Ectopic beta-human chorionic gonadotropin production by a human hepatoma cell line (FOCUS): isolation and immunochemical characterization. Endocrinology 1987;120:559−66.

[14] Hussa RO, Fein HG, Tattillo RA, Nagelberg SB, Rosen SW, Weintraub BD, et al. A distinctive form of human chorionic gonadotropin β-subunit-like material produced by cervical cancer cells. Cancer Res 1986;46:1948–54.

[15] Bellet D, Lazar V, Bleche I, Paradis V, Giovangrandi Y, Paterliru P. Malignant transformation of nontrophoblastic cells in association with the expression of chorionic gonadotropin β genes normally transcribed in trophoblastic cells. Cancer Res 1997; 57:516–23.

[16] Butler SA. The molecular genetic of hCG [chapter 4] In: Cole LA, Butler SA, editors. Human chorionic gonadotropin (hCG). Burlington, MA: Elsevier; 2010. p. 37–49.

[17] Jameson JL, Hollenberg AN. Regulation of human chorionic gonadotropin gene expression. Endocr Rev 1993;14:203–21.

[18] Talamadge K, Boorstein WR, Vamvakopoulos NC, Gething MJ, Fiddes JC. Only three og the seven human chorionic gonadotropin β-subunit genes can be expressed in the placenta. Nucleic Acid Res 1984;12:8415–36.

[19] Bo M, Boime I. Identification of the transcriptionally active genes of the chorionic gonadotropin β gene cluster *in vitro*. J Biol Chem 1992;267:3179–84.

[20] Lazar V, Diez SG, Laurent A, Gevangrandi J, Radvanyi F, Chopin D, et al. Cancer Res 1995;55:3735–8.

[21] Pierce JG, Parsons TF. Glycoprotein hormones: structures and function. Ann Rev Biochem 1981;50:465–95.

[22] Braunstein GD. hCG expression in trophoblastic and nontrophoblastic tumours. In: Fisman WH, editor. Oncodevelopmenta markers: biologic diagnostic and monitoring aspects. New York, NY: Academic Press; 1983. p. 35.

[23] Iles RK, Chard T. Immunochemical analysis of the human chorionic gonadotropin-like material secreted by "normal" and neoplatic uotheial cells. J Mol Endocrinol 1989; 2:107–12.

[24] Lunardi-Iskandar Y, Bryant H, Zeman RA, Lam VH, Samaniego F, Besnier JM, et al. Tumorigenesis and metastasis of Kaposi's sarcoma cell line in immunodeficient mice blocked by a human pregnancy hormone. Nature 1995;375:64–8.

[25] Gillott DJ, Iles RJ, Chart T. The effect of βhCG on the *in vitro* growth of bladder cancer cells. Br J Cancer 1996;73:323–8.

[26] Butler SA, Ikram MS, Mathieu S, Iles RK. The increase in bladder carcinoma cell population induced by the free beta subunit of hCG is a result of an anti-apoptosis effect and not cell proliferation. Br J Cancer 2000;82:1553–6.

[27] Murphy G, Reynolds JJ, Whitham SE, Docherty AJ, Angel P, Heath JK. Transforming growth factor beta modulates the expression of collagenase and metalloproteinase inhibitor. Eur Mol Biol Org J 1987;6:1899–904.

[28] Wu D, Luo S, Wang Y, Zhuang L, Chen Y, Peng C. Smads in human trophoblast cells: expression regulation and role in TGF-beta-induced transcriptional activity. Mol Cell Endocrinol 2001;175:111–21.

[29] Massague J. TGFβ in cancer. Cell 2008;134:215–30.

[30] Massague J. TGF-beta signal transduction. Ann Rev Biochem 1998;67:753–91.

[31] Cole LA, Butler SA. Hyperglycosylated hCG hCGβ and hyperglycosylated hCGβ: interchangeable cancer promoters. Mol Cell Endocrinol 2011;349:232–8.

[32] Cole LA, Dai D, Leslie KK, Butler SA, Kohorn EI. Gestational trophoblastic diseases: 1. Pathophysiology of hyperglycosylated hCG-regulated neoplasia. Gynecol Oncol 2006; 102:144–9.

[33] Hamade AL, Nakabayashi K, Sato A, Kiyoshi K, Takamatsu Y, Laoag-Fernandez JB, et al. Transfection of antisense chorionic gonadotropin β gene into choriocarcinoma

cells suppresses the cell proliferation and induces apoptosis. J Clin Endocrinol Metab 2005;90:4873–9.

[34] Cole LA, Butler SA. Hyperglycosylated hCG and its β-subunit: pervasive promoters of invasion and growth in cancer. Mol Cell Endocrinol 2013;2013.

[35] Butler SA, Iles RK. Ectopic human chorionic gonadotrophin β secretion by epithelial tumors and human chorionic gonadotrophin β-induced apoptosis in Kaposi's sarcoma: is there a connection? Clin Cancer Res 2003;9:4666–73.

[36] Jankowska A, Gunderson SI, Andrusiewicz N, Burczynska B, Szczerba A, Jarmolowski A, et al. Reduction of human chorionic gonadotropin beta subunit expression by modified U1 snRNA caused apoptosis in cervical cancer cells. Mol Cancer 2008;7:26–9.

[37] Li D, Wen X, Ghali L, Al-Shalabi FM, Docherty SM, Purkis P, et al. hCG beta expression by cervical squamous carcinoma—*in vivo* histological association with tumour invasion and apoptosis. Histopathology 2008;53:147–55.

[38] Berndt S, Blacher S, Munuat C, Detilleux J, Evain-Brion D, Noel A, et al. Hyperglycosylated human chorionic gonadotropin stimulates angiogenesis through TGF-β receptor activation. FASEB J 2013; Abstract 12-213686.

[39] Carter WB, Sekharem M, Coppola D. Purified hCG induces apoptosis in breast cancer. Breast Cancer Res Treat 2006;100:S243–4.

[40] Triozzi PL, Stevens VC. Human chorionic gonadotropin as a target for cancer vaccines. Oncol Rep 1999;6:7–17.

[41] Moulton HM, Yoshihara PH, Mason DH, Iversen P, Triozzi PL. Active specific immunotherapy with β-human chorionic gonadotropin peptide vaccine in patients with metastatic colorectal cancer: antibody response is associated with improved survival. Clin Cancer Res 2002;8:2044–51.

[42] Ivers PL, Mourich DV, Moulton HM. Monoclonal antibodies to two epitopes of β-human chorionic gonadotropin for the treatment of cancer. Cur Opin Mol Ther 2003; 5:156–60.

[43] Delves PJ, Iles RK, Roitt I,M, Lund T. Designing a new generation of anti-hCG vaccines for cancer therapy. Mol Cell Endocrinol 2007;260:276–81.

[44] He L-Z, Ramakrishna V, Connolly JE, Wang X-T, Smith P, Jones CL, et al. A novel human cancer vaccine elicits cellular responses to the tumor-associated antigen human chorionic gonadotropin β. Clin Cancer Res 2004;10:1920–7.

[45] Morse MA, Chapman R, Powderly J, Blackwell K, Keler T, Green J, et al. Phase I study utilizing a novel antigen-presenting cell-targeted vaccine with toll-like receptor stimulation to induce immunity to self-antigens in cancer patients. Clin Cancer Res 2011;17:4844–53.

[46] Butler SA, Staite EM, Iles RK. Reduction in bladder cancer cell growth in response to hCGβ CTP37 vaccinated mouse serum. Oncol Res 2003;14:93–100.

Cancer-3, tying everything together

22

Laurence A. Cole
USA hCG Reference Service, Angel Fire, NM, USA

As editor of this book, thinking about hCG and cancer, about my own research contribution, I realize it has been 35 years of digging up facts, testing facts, confirming facts, and then tying everything together. To me, by far, the best confirmation that hCG variants have a key role in human cancers (Chapter 21) is how well all the research of choriocarcinoma and other cancers ties together with the newer cancer data or makes sense. In some respects the findings of how hCG variant controls human cancer is a keynote discovery that, hopefully, someday will lead to new treatments for cancers or possible cures for cancers in humans. I try my best in this chapter to illustrate everything from my own perspective or how everything ties together.

For me, everything started with my PhD at Medical College of Wisconsin, working with Robert O. Hussa (see Chapter 1). As found in my PhD (1977–1981), DoT and CaSki cervical cancer cells were producing a hyperglycosylated form of the free β-subunit of hCG [1–4]. At that time, hCG free β-subunit production by cancers was considered an epiphenomenom, something made by cancer cells with no possible biological activity.

After obtaining my PhD, I spent 4 years at University of Michigan as a postdoctoral fellow. My career continued at Yale University with multiple NIH grants. I was determined to continue the hCG and cancer investigation and examined serum hCG free β-subunit and urine β-core fragment as tumor markers. As found after 7 years of research (1984–1991) [5–9], 100% of trophoblastic malignancies, choriocarcinoma, and ovarian and testicular germ cell malignancies (germ cell malignancies take on trophoblastic histology) produce a variant of hCG detectable in blood or urine. Structure studies showed this variant of hCG produced by trophoblastic cancers was hyperglycosylated hCG. Only 30% of other cancers (epithelial ovarian, cervical, endometrial vulvar, bladder, uterine, lung, colorectal, and pancreatic) were detectable in blood ($n = 764$). In these cases, hCG free β-subunit and hyperglycosylated hCG free β-subunit were being produced [5–9]. In urine, the breakdown product of hCG free β-subunit was detected: β-core fragment [5–9]. We detected 55% of cancer cases by testing β-core fragment in urine from 904 patients (see Table 20.1).

It appeared to us at that time that approximately half of all cancers tested were producing an hCG variant, with 30% detection is serum and 55% detection of the breakdown product in urine. Around this time, Acevedo and colleagues [10] showed through flow cytometry methods that hCG free β-subunit immunoreactivity was

Human Chorionic Gonadotropin (hCG). DOI: http://dx.doi.org/10.1016/B978-0-12-800749-5.00022-5

present in 100% of cancer plasma membranes. At first, I did not believe this and thought that somehow he was exaggerating. But when tying everything together, this somehow made sense.

We now know hyperglycosylated hCG, hCG free β-subunit, and hyperglycosylated hCG free β-subunit are autocrines binding a TGFβ-II receptor [11]. There are two kinds of autocrines, "simple autocrines" and "complex autocrines" [12,13]. Everything may be explained by these two kinds of autocrines. Simple autocrines occur when cells produce low concentrations of an autocrine and molecules are secreted and act directly on the plasma membrane receptor. Cells producing higher concentrations or "complex autocrines" producing an autocrine may circulate through the body in the bloodstream before acting back on the cell's plasma membrane receptor [12,13]. In the case of "simple autocrine," no molecule would be detected in the blood or urine, whereas in the case of "complex autocrines" molecules would be detected in the blood and urine.

It appears that trophoblast cancers function with hyperglycosylated hCG through a "complex autocrine" pathway, with 100% of cancers detected as hyperglycosylated hCG in serum. It appears that other cancers use hCG free β-subunit through "simple autocrine" or "complex autocrine" pathways, explaining the 55% detection in urine and 100% detection in the membranes of cancers. Thus, tying together our 7 years of tumor marker studies with the cell membrane studies of Acevedo et al. [10], we can explain how all cancers seemingly produce an hCG variant.

In the years that followed, our research specialized in hyperglycosylated hCG, the molecule produced by choriocarcinoma cells and testicular and ovarian germ cell malignancies, and the biological function of hyperglycosylated hCG. Our research started with examining the structure of regular hCG and hyperglycosylated hCG. Preliminary studies showed that hyperglycosylated hCG was produced in choriocarcinoma. As shown by us in 1987, choriocarcinoma produced a molecule with double-size O-linked oligosaccharides [14–16]. I had somebody working for me, Wang Yixun, MD, from a hospital that specialized in choriocarcinoma. I went to Liaoning Tumor Institute in Shenyang, Manchuria, China, a one-of-a-kind cancer hospital with an unheard of 40–50 patients in-house with choriocarcinoma.

I flew to Beijing and then proceeded to the train station where I boarded a steam train traveling 1000 miles north to Shenyang, Manchuria, China. When the train arrived in Shenyang, I was greeted by 100 physicians all smartly lined up on the platform to greet me. It was there that I collected 35 l of urine from seven patients with advanced choriocarcinoma. I brought with me a tangential flow dialysis machine. I was able to use this to concentrate the urine. The final concentrates were frozen on dry ice and transported back to my laboratory at Yale University. I am told that this would not be possible today, because China now prohibits the export of human fluids.

Back at my laboratory the hCG in these seven urine samples, along with four concentrates from hydatidiform mole and nine concentrates from normal pregnancy, were purified by a combination of acetone precipitation, ethanol precipitation, gel filtration on Sephacryl S100, DEAE ion exchange, and gel filtration on Sephacryl S100. The purified hCG preparations are those described at P1–P9 (pregnancy),

M1–M4 (hydatidiform mole), and C1–C7 (choriocarcinoma) in our structure publication [17]. As proven, choriocarcinoma hCG is hyperglycosylated hCG [17]. I still keep stocks of these fully characterized choriocarcinoma preparations and use them as standards today.

As shown, all choriocarcinoma, hydatidiform mole, and pregnancy hCG preparations contained an identical amino acid sequence [17]. The pregnancy hCG preparation contained 12.5–18.5% hexasaccharide O-linked oligosaccharide on the β-subunit, whereas the choriocarcinoma hCG preparations contained 48–100% hexasaccharide. The pregnancy preparation contained 5.0–17% triantennary oligosaccharide on the α-subunit and β-subunit. In contrast, the choriocarcinoma preparations contained 26–40% triantennary structure. It was concluded that hyperglycosylated hCG, as produced in choriocarcinoma, contains >20% hexasaccharide and >20% triantennary N-linked oligosaccharides. It was very clear from these structural studies that choriocarcinoma hCG, or hyperglycosylated hCG, had an absolute structural difference from pregnancy hCG.

Soon after returning, I talked with Stephen Birken, PhD, at Columbia University about making a monoclonal antibody to hyperglycosylated hCG. They used my purified preparation C5 to make a monoclonal antibody, B152. B152 is the only antibody ever made that totally distinguishes regular hCG from hyperglycosylated hCG [18]. An assay was established using antibody B152 to specifically detect hyperglycosylated hCG [19]. It was using this assay that O'Connor and colleagues first showed that hyperglycosylated hCG was present in early pregnancy fluids [20]. That is, in extremely early pregnancy, at the time of pregnancy implantation (3–5 weeks of gestation), solely or mostly hyperglycosylated hCG is produced [20].

This finding made me very confused. Why would hyperglycosylated hCG, the choriocarcinoma from hCG, be produced in very early pregnancy? My laboratory then used the B152 assay to examine very early pregnancy. We showed, like O'Connor et al. [20], that hyperglycosylated hCG predominated at 3–5 weeks of gestation [21]. This confused me further. I kept asking, why is it made in pregnancy? Choriocarcinoma is an invasive condition. What is invasive in early pregnancy? Several publications suggest that a variant of hCG present in standard preparations, such as hyperglycosylated hCG, has a role in blastocyst implantation [22–26]. Yet, pure pregnancy hCG has no implantation role. This is when I started investigating the role of hyperglycosylated hCG, or what some people called choriocarcinoma hCG, in pregnancy implantation.

We first looked into whether hyperglycosylated hCG could be a growth and invasion signal in choriocarcinoma cells and normal cytotrophoblast cells as present in the blastocyst at implantation of pregnancy. As illustrated in Tables 12.1 and 12.2, hyperglycosylated hCG seemingly controls growth and implantation in both normal pregnancy cytotrophoblast cells and choriocarcinoma cytotrophoblast cells. Hyperglycosylated hCG, but not regular hCG, significantly enhances choriocarcinoma and pregnancy cytotrophoblast cell growth (Table 12.1) and invasion, as shown in Matrigel basement membrane chambers (Table 12.2) [27,28]. It was concluded that hyperglycosylated hCG, but not hCG, promoted growth and invasion as in implantation and choriocarcinoma. It was also concluded that hyperglycosylated hCG is a molecule independent of regular hCG with separate biological functions, as now confirmed by the discovery that hyperglycosylated hCG binds the TGFβ type II receptor [11,29].

This concept that hyperglycosylated hCG or choriocarcinoma hCG is independent of regular hCG was supported by the findings of Hamade et al. [30]. As shown, the molecule produced in choriocarcinoma cases blocked apoptosis. Blockage of apoptosis is not a property of regular hCG, suggesting that the choriocarcinoma or hyperglycosylated molecules are different from regular hCG. This very much supported our findings regarding growth promotion and invasion promotion [27,28].

More evidence was needed, other than its role as a promoter of growth and invasion, to show that hyperglycosylated hCG controls blastocyst implantation. It is well-known that poor implantation is the principal cause of biochemical pregnancies and miscarriages or pregnancy failures, which occur in the first and second trimesters of pregnancy [31–33]. We examined the relationship between miscarriages, biochemical pregnancies, and concentration of hyperglycosylated hCG. It was found that 70 of 70 term outcome pregnancies produced >40% hyperglycosylated hCG (of total hCG) on the day of implantation (first day of production >1 mIU/ml total hCG) [34,35]. However, only 6 of 35 miscarriages and 8 of 28 biochemical pregnancies produced >40% hyperglycosylated hCG [34,35]. It was inferred that most miscarriages and biochemical pregnancies are attributable to insufficient production of hyperglycosylated hCG causing poor implantation. Blastocyst implantation was clearly driven by hyperglycosylated hCG.

An amazing similarity existed between the normal pregnancy process of blastocyst implantation and the supposed cancer process of choriocarcinoma. Both seemingly used analogous mechanisms of growth and invasion. Choriocarcinoma was first discovered in Germany by Hans Chiari in 1877 [36], then in 1889 by Max Saenger [37], and in 1898 by Felix Marchand [38]. Hans Chiari, Max Saenger, Felix Marchant, or any physician or scientist since have never demonstrated that choriocarcinoma is a cancer or malignancy per se by showing a transformation or any major change between pregnancy and choriocarcinoma cytotrophoblast cells [39]. It has always been assumed that choriocarcinoma is a cancer because its cells are rapidly growing or are part of a highly invasive disease. Nobody has demonstrated, however, that choriocarcinoma cytotrophoblast cells differ in any way from pregnancy cytotrophoblast cells, which are naturally extremely invasive cells. Furthermore, nobody has demonstrated by biochemistry or histology methods that choriocarcinoma cells appear or function differently than normal pregnancy extravillous cytotrophoblast cells.

Biologically, choriocarcinoma cells appear in every way analogous to normal pregnancy cytotrophoblast cells. Not surprisingly, choriocarcinoma cytotrophoblast cells seem to naturally use the inherent hyperglycosylated hCG and invasion and growth pathway of implantation cytotrophoblast cells. This is a property of all cytotrophoblast cells. It is inferred based on the cell similarity that no carcinogenesis or transformation occurs in choriocarcinoma, and that this disease called cancer is not really a cancer or does not involve cancer transformation.

It is well-established that delivery at pregnancy parturition leads to a tearing of the large ball of extravillous cytotrophoblast cells, the cells that attach villous tissue to the uterus at hemochorial placentation (Figure 22.1). It is thought that the remaining hunk of implanted cytotrophoblast tissue normally dies through apoptosis after delivery. In choriocarcinoma cases, excessive hyperglycosylated hCG may be present in serum. This has been proven to block apoptosis [30] and seemingly promotes the growth and invasion of this tissue as in implantation and choriocarcinoma. Although it has been concluded that choriocarcinoma is not a transformed tissue or cancer per se, it is the only noncancerous invasive or malignant disease. In that respect, it can be considered the root malignancy.

Our studies of choriocarcinoma indicate that it is not a cancer per se, but probably the root malignancy. It is seemingly driven by hyperglycosylated hCG using a mechanism separate from that of pregnancy hCG. Nude mouse xenograft studies

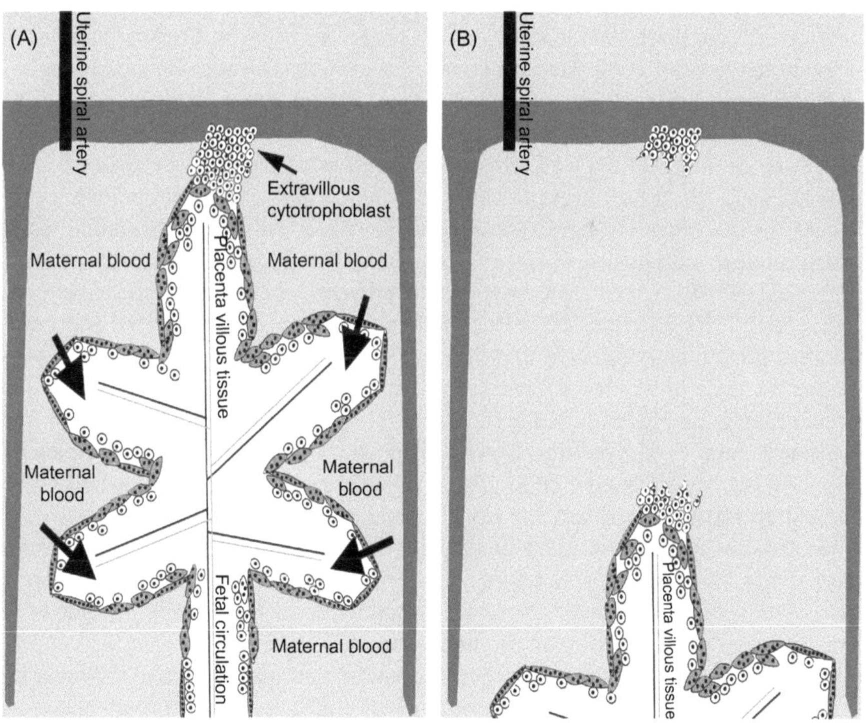

Figure 22.1 Hemochorial placentation in the second and third trimesters of pregnancy (1–4). Multinucleated syncytiotrophoblast cells are shown in gray and mononuclear cytotrophoblast cells are illustrated in white. Panel A shows active hemochorial placentation in a single chamber. Each placenta contains four to seven chambers of hemochorial placentation. As illustrated, the nutrients in maternal blood cross the syncytiotrophoblast cells of floating villi. Implantation of extravillous cytotrophoblast cell in decidua parietalis, the compartment casing, is illustrated. Panel B shows tearing of the extravillous cytotrophoblast cells at parturition and the remaining extravillous cytotrophoblast ball of cells after delivery.

were performed with JEG-III choriocarcinoma cells. Cells grew rapidly under mouse skin from metastasis. As demonstrated, antibody to hyperglycosylated hCG B152 completely blocked cancer cell growth and invasion or created a state of oncogenesis [27]. Control nonspecific IgG had no action. It was inferred from this and the studies shown in Tables 12.1 and 12.2 that hyperglycosylated hCG must drive or power this disease.

As shown by the USA hCG Reference Service, choriocarcinoma can be classified as aggressive disease, minimally aggressive disease, or quiescent disease solely based on the proportion of hyperglycosylated hCG produce [40,41]. An analogy was made between aggressive and minimally aggressive choriocarcinoma and between term and miscarrying pregnancy in implantation. Hyperglycosylated hCG appears to be the solitary signal behind pregnancy implantation and the solitary signal behind choriocarcinoma. Choriocarcinoma should be simply renamed "hyperglycosylated hCG syndrome."

In 1999, I hired Stephen Butler, PhD, as a postdoctoral fellow. He remained working with me until 2004. Stephen was engaged to Paulo in England and I could not keep him beyond 2004. Stephen coedited this book with me. Stephen brought to my laboratory many new ideas and thoughts. It is from these ideas we advanced to the next stage and tied everything together. Stephen kept telling me about his studies with bladder cancer: how hCG free β-subunit drove bladder cancer in an analogous manner to how hyperglycosylated hCG drove choriocarcinoma (see Chapter 21 for details); how others confirmed that hCG free β-subunit drives growth of other malignancies [42–47]; how hCG free β-subunit functions by antagonizing a TGFβ-II receptor; and how this antagonism blocked apoptosis, like hyperglycosylated hCG function in choriocarcinoma blocked apoptosis [30], and how it promoted growth and promoted invasion.

After a couple of months, I started repeating many of Stephen's studies and found myself running my JEG-III and Jar choriocarcinoma and cytotrophoblast primary cell experiment parallel to experiments with Stephen's bladder and cervical carcinoma cells. I found myself testing hCG free β-subunit and hyperglycosylated hCG free β-subunit in experiments parallel to my hCG and hyperglycosylated hCG studies.

Month after month, my mind plodded over Stephen's results and my findings. One night at 4:00 am, my mind was full of all these facts. I got out of bed rapidly because suddenly everything somehow fit together, it all made sense. I rushed over to my office and reread many of Stephen's papers.

Hyperglycosylated hCG is the driving signal in choriocarcinoma. As shown, it is also the driving signal in testicular germ cell malignancies [28]. As shown by Beebe et al. [48] and Ruddon et al. [49], placental trophoblast cells and pituitary gonadotrope cells make hCG dimer. Efficient dimer formation in these cells is facilitated by enzymes that rapidly complete the last two critical disulfide bridges β93-100 and β26-110 on hCG β-subunit, critical to dimer formation [48,49]. Other cancer cells, in contrast, do not have these enzymes and therefore secrete free β-subunit in place of dimer. Thus, choriocarcinoma and testicular and ovarian germ cell malignancies secrete hyperglycosylated hCG, whereas other cancers secrete hCG free β-subunit or hyperglycosylated hCG free β-subunit.

In our studies, we repeated experiments examining the action of hCG and hyperglycosylated hCG on JEG-III and Jar choriocarcinoma cell growth, We also examined hCG free β-subunit and hyperglycosylated hCG free β-subunit, and multiple bladder cancer cell lines and cervical cancer cell lines. Hyperglycosylated hCG, hCG free β-subunit, and hyperglycosylated hCG free β-subunit are interchangeable, all acting in a similar manner on all trophoblastic and all other cell lines [11]. As shown by Stephen Butler, PhD, hCG free β-subunit functions in bladder cancer cells through antagonizing a TGFβ-II receptor [29]. Seemingly all cancers examined interchangeably use this process [29]. This is when everything started fitting together.

Hyperglycosylated hCG is the driving signal behind the root malignancy choriocarcinoma. This malignancy uses the cytotrophoblast cell (principal cell of choriocarcinoma) implantation process or the TGFβ-II antagonism pathway. Seemingly multiple other malignancies steal this process to drive malignancies, just as hyperglycosylated hCG drives choriocarcinoma. The malignancies can be driven by hyperglycosylated hCG, hyperglycosylated hCG free β-subunit, or hCG free β-subunit. These other cancer findings were proven or confirmed by different groups using bladder epithelial carcinoma, Kaposi's sarcoma, cervical squamous carcinoma, cervical adenocarcinoma, cervical adenosquamous carcinoma, endometrial adenocarcinoma, endometrial epithelial carcinoma, and breast adenocarcinoma [42–47] (see Table 21.1).

I have repeatedly asked, does this implantation mechanism drive all cancers? As listed in Table 21.1, we tested 15 cancer cell lines of widely varying histology (we tested, for instance, Caski cervical endometrioid carcinoma, HeLa cervical adenocarcinoma, HOG-1 cervical squamous carcinoma, SiHA cervical squamous carcinoma, and C33 cervical epithelial carcinoma cell); 15 of 15 were driven to grow by these hCG-related cancer promoters [50].

It is amazing how the choriocarcinoma and other cancer facts blend together. As shown, choriocarcinoma was driven to grow and invade by hyperglycosylated hCG, and apoptosis was blocked [30] through a mystery mechanism other than the hormone LH/hCG receptor mechanism used by hCG. As found by Stephen Butler, PhD, a TGFβ-II autocrine mechanism is active in cancer cells and works interchangeably with hyperglycosylated hCG, free β-subunit, and hCG free β-subunit [11,29]. This mystery mechanism appeared to be that active in choriocarcinoma. Amazingly, this mechanism antagonizes an established pathway or fits exactly with pathways in choriocarcinoma, with TGFβ promoting apoptosis, inhibiting cell growth, and blocking invasive enzymes [51–54].

As described in this book in Chapter 10 on the TGFβ Autocrine Receptor, we discovered that the α-subunit blocks the TGFβ cystine knot sequences, or sequence analogous between hCGβ and TGFβ [50]. Hyperglycosylated hCG unfolds this structure, exposing the cysteine knot region; the region is naturally exposed by free β-subunit and hyperglycosylated hCG free β-subunit missing an α-subunit [55]. As shown, placental trophoblast and pituitary gonadotrope have specialized enzymes that complete hCG β-subunit disulfide bond formation, as is critical for α-subunit – β-subunit combination to form a dimer [48,49]. Cancer cells lack these enzymes and therefore produce an hCG free β-subunit or a hyperglycosylated hCG free β-subunit. These studies all explain (why cancer cells produce a free β-subunit and how hyperglycosylated hCG, free β-subunit, and hyperglycosylated hCG free β-subunit share a common TGFβ-II receptor binding) the close relationship between hyperglycosylated hCG action and choriocarcinoma, and between hCG free β-subunit action and other cancers.

As I mentioned, everything now ties together into a simple story. Choriocarcinoma in the base malignancy is seemingly not a cancer because no transformation occurs. In this cancer, hyperglycosylated hCG is a complete autocrine detected in 100% of cases and antagonizes a TGFβ-II receptor. Other cancers, either most cancers or all cancers, only produce a free β-subunit of hCG or hyperglycosylated hCG because of the lack of combination enzymes. This is either a simple autocrine or complex autocrine explaining the 55% detection rate. This free β-subunit acts in a way similar to hyperglycosylated hCG as an autocrine to antagonize a TGFβ-II receptor and to block apoptosis, promoting cell growth and cell invasion [42–47]. It appears that carcinogenesis or cancer transformation requires activation of the hCG β-subunit gene and stealing the pregnancy implantation TGF-II pathway.

This hCG variant pathway controlling human malignancies has been very much confirmed by ongoing vaccine clinical trials showing how a generation of human hCG β-subunit antibodies can treat advanced malignancies [56–60]. As proven, these vaccines work by blocking the described hCG variant TGFβ pathways [61]. Figure 22.2 illustrates the proposed pathways whereby choriocarcinoma and other malignancies function.

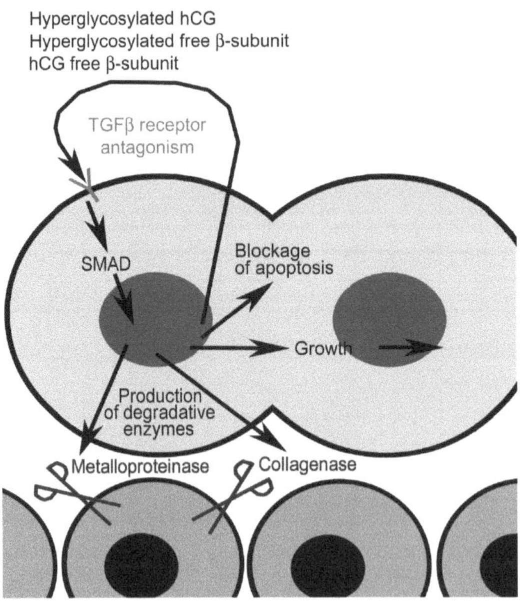

Figure 22.2 Function of choriocarcinoma and most other cancers using hyperglycosylated hCG, hCG free β-subunit, and hyperglycosylated hCG free β-subunit signals.

As found in nude mice xenograft models, choriocarcinoma cells grow and invade rapidly [27]. Antibody to hyperglycosylated hCG B152 completely suppresses malignancy or induces oncostasis. We have shown using antibody B152 that it suppresses eight of eight other cancers and trophoblastic cancers tested [50]. Research suggests that a humanized B152 or antibody to hyperglycosylated hCG and its β-subunit may be valuable as a treatment for all or most cancers or as a possible cure for cancer. Research by Carter et al. [47] shows that breast mammary adenocarcinoma cells grow rapidly in nude mice. Growth was blocked, invoking oncostasis by an opposing hCG β-subunit RNA sequence [47]. This further confirms that human antibodies may have a future in the treatment or core of trophoblastic and other malignancies.

A company named Omnimmune has licensed all the patents regarding hyperglycosylated hCG, hCG free β-subunit, and hyperglycosylated hCG free β-subunit and cancer, and it is hoping to raise money to generate human antibodies. Future research should focus on continued nude mouse cancer studies and on developing human antibodies and clinical trials.

To fully understand the hCG variant cancer and cancer story, it is suggested that you carefully read Chapters 20 and 21 of this book.

References

[1] Cole LA, Hussa RO, Rao CV. Discordant synthesis and secretion of human chorionic gonadotropin and subunits by cervical cancer cells. Cancer Res 1981;41:1615–19.

[2] Cole LA, Birken S, Sutphen S, Hussa RO, Pattillo RA. Absence of the COOH-terminal peptide on ectopic human chorionic gonadotropin β-subunit (HCG-β). Endocrinology 1982;110:2198–200.

[3] Cole LA, Hussa RO. Use of glycosidase digested human chorionic gonadotropin β-subunit to explain the partial binding of glycoprotein hormones to Con A. Endocrinology 1981;109:2276–9.

[4] Cole LA, Hartle RJ, Laferla JJ, Ruddon RW. Detection of the free β subunit of human chorionic gonadotropin in cultures of normal and malignant trophoblast cells, pregnancy sera, and sera of patients with choriocarcinoma. Endocrinology 1983;113:1176–8.

[5] Nam JH, Cole LA, Chambers JT, Schwartz PE. Urinary gonadotropin fragment a new tumor marker: I Assay development and cancer-specificity. Gynecol Oncol 1990;36:383–90.

[6] Schwartz PE, Chambers JT, Taylor KJ, Cole LA, Makuch R. Urinary gonadotropin fragments. Anticancer Res 1993;13:1722–5.

[7] Cole LA. β-core fragment (β-core UGP or UGF). Tumor Marker Upd 1994;6:69–75.

[8] Cole LA, Nam J-H, Park S-Y, Koh MW, Tanaka A. Urinary beta core fragment: 7 years later. J Tumor Marker Oncol 1994;9:53–60.

[9] Cole LA, Tanaka A, Kim GS, Park S-Y, Koh MW, Schwartz PE, et al. Beta core fragment (β-core/UGF/UGP), a tumor marker: seven year report. Gynecol Oncol 1996;60:264–70.

[10] Acevedo HF, Krichevsky A, Campbell-Acevedo EA, Galyon JC, Buffo MJ, Hartsock RJ. Flow cytometry method for the analysis of membrane-associated human chorionic gonadotropin, its subunits, and fragments on human cancer cells. Cancer 1992;69:1818–28.

[11] Cole LA, Butler SA. Hyperglycosylated hCG hCGβ and hyperglycosylated hCGβ: interchangeable cancer promoters. Mol Cell Endocrinol 2011;349:232–8.

[12] Fairbank M, St-Pierre P, Nabi IR. The complex biology of autocrine motility factor/phosphoglucose isomerase (AMF/PGI) and its receptor, the gp78/AMFR E3 ubiquitin ligase. Mol BioSyst 2009;5:793–801.

[13] Offermann MK, Faller DV. Autocrine induction of major histocompatibility complex class I antigen expression results from induction of beta interferon in oncogene-transformed BALB/c-3T3 cells. Mol Cell Biol 1989;9:1969–77.

[14] Cole LA, Birken S, Perini F. The structures of the serine-linked sugar chains on human chorionic gonadotropin. Biochem Biophys Res Comm 1985;126:333–9.

[15] Cole LA. O-Glycosylation of proteins in the normal and neoplastic trophoblast. Troph Res 1987;2:139–48.

[16] Cole LA. The O-linked oligosaccharides are strikingly different on pregnancy and choriocarcinoma hCG. J Clin Endocrinol Metab 1987;65:811–13.

[17] Elliott MM, Kardana A, Lustbader JW, Cole LA. Carbohydrate and Peptide structure of the α- and β-subunits of human chorionic gonadotropin from normal and aberrant pregnancy and choriocarcinoma. Endocrine 1997;7:15–32.

[18] Birken S, Krichevsky A, O'Connor J, Schlatterer J, Cole LA, Kardana A, et al. Development and characterization of antibodies to a nicked and hyperglycosylated form of hCG from a choriocarcinoma patient. Endocrine J 1999;10:137–44.

[19] Cole L, Omrani A, Cermik D, Bahado-Singh RO, Mahoney RO. Hyperglycosylated hCG, a potential alternative to hCG in down syndrome screening. Prenat Diagn 1998;8:926–33.

[20] O'Connor JF, Ellish N, Kakuma T, Schlatterer J, Kovalevskaya G. Differential urinary gonadotrophin profiles in early pregnancy and early pregnancy loss. Prenat Diagn 1998;18:1232–40.

[21] Cole LA, Khanlian SA, Sutton JM, Davies S, Stephens N. Hyperglycosylated hCG (invasive trophoblast antigen, ITA) a key antigen for early pregnancy detection. Clin Biochem 2003;36:647–55.

[22] Ohlsson R, Larsson E, Nilsson O, Wahlstrom T, Sundstrom P. Blastocyst implantation precedes induction of insulin-like growth factor II gene expression in human trophoblasts. Development 1989;106:555–9.

[23] d'Hauterivea SP, Berndt BS, Tsampalasa M, Charlet-Renarda C, Dubois M, Bourgain C, et al. Dialogue between blastocyst hCG and endometrial hCG/LH receptor: which role in implantation? Gyncol Obstet Invest 2007;64:156–60.

[24] Rao CV. Paradigm shift on the targets of hCG action. In: Cole LA, editor. Human chorionic gonadotropin (hCG). Burlington, MA: Elsevier; 2010. p. 131–5.

[25] Nepomnaschy PA, Weinberg CR, Wilcox AJ, Baird DD. Urinary hCG patterns during the week following implantation. Hum Reprod 2008;23:271–7.

[26] Mansour R, Tawab N, Kamal O, El-Faissal Y, Serour A, Aboulghar M, et al. Intrauterine injection of human chorionic gonadotropin before embryo transfer significantly improves the implantation and pregnancy rates in *in vitro* fertilization/intracytoplasmic sperm injection: a prospective randomized study. Fertil Steril 2011;96:1370–4.

[27] Cole LA, Dai D, Leslie KK, Butler SA, Kohorn EI. Gestational trophoblastic diseases: 1. Pathophysiology of hyperglycosylated hCG-regulated neoplasia. Gynecol Oncol 2006;102:144–9.
[28] Cole LA, Khanlian SA, Riley JM, Butler SA. Hyperglycosylated hCG (hCG-H) in gestational implantation and in choriocarcinoma and testicular germ cell malignancy tumorigenesis. J Reprod Med 2006;51:919–29.
[29] Butler SA, Ikram MS, Mathieu S, Iles RK. The increase in bladder carcinoma cell population induced by the free beta subunit of hCG is a result of an anti-apoptosis effect and not cell proliferation. Br J Cancer 2000;82:1553–6.
[30] Hamade AL, Nakabayashi K, Sato A, Kiyoshi K, Takamatsu Y, Laoag-Fernandez JB, et al. Transfection of antisense chorionic gonadotropin β gene into choriocarcinoma cells suppresses the cell proliferation and induces apoptosis. J Clin Endocrinol Metab 2005;90:4873–9.
[31] Norwitz ER, Schust DJ, Fisher SJ. Implantation and the survival of early pregnancy. New Engl J Med 2001;345:1400–8.
[32] Semprini AE, Simon G. Not so efficient reproduction. Lancet 2000;356:257–8.
[33] Wilcox AJ, Weinberg CR, O'Connor JF, Baird DD, Schlatterer JP, Canfield RE, et al. Incidence of early loss of pregnancy. N Engl J Med 1988;319:189–94.
[34] Sasaki Y, Ladner DG, Cole LA. Hyperglycosylated hCG the source of pregnancy failures. Fertil Steril 2008;89:1781–6.
[35] Cole LA. Hyperglycosylated hCG and pregnancy failures. J Reprod Immunol 2012;93:119–22.
[36] Chiari H. Uber drei Falle von primarem kacino in findus und corpus des uterus. Med Jahrb 1877;7:364–7.
[37] Saenger M. Deciduoma malignum. Zbl Gyak 1889;167:537.
[38] Marchand FJ. Uber die sogenannten "decidualen" geshwulskeim im anshluss an normale geburt, abort, blasenmole und extrauterineschwanggerahaft. Monatsschr Geburtshilfe Gynakol 1895;1:419–38.
[39] Ober WB, Fass RO. The early history of choriocarcinoma. J Hist Med Allied Sci 1961;16:49–73.
[40] Cole LA, Laidler L, Muller C. USA hCG reference service, 10 year report. Clin Biochem 2010;43:1013–22.
[41] Cole LA. Minimally-aggressive gestational trophoblastic neoplasms. Gynecol Oncol 2012;126:145–50.
[42] Lunardi-Iskandar Y, Bryant H, Zeman RA, Lam VH, Samaniego F, Besnier JM, et al. Tumorigenesis and metastasis of Kaposi's sarcoma cell line in immunodeficient mice blocked by a human pregnancy hormone. Nature 1995;375:64–8.
[43] Gillott DJ, Iles RJ, Chart T. The effect of βhCG on the *in vitro* growth of bladder cancer cells. Br J Cancer 1996;73:323–8.
[44] Butler SA, Iles RK. Ectopic human chorionic gonadotrophin β secretion by epithelial tumors and human chorionic gonadotrophin β-induced apoptosis in Kaposi's sarcoma is there a connection? Clin Cancer Res 2003;9:4666–73.
[45] Jankowska A, Gunderson SI, Andrusiewicz N, Burczynska B, Szczerba A, Jarmolowski A, et al. Reduction of human chorionic gonadotropin beta subunit expression by modified U1 snRNA caused apoptosis in cervical cancer cells. Mol Cancer 2008;7:26–9.
[46] Li D, Wen X, Ghali L, Al-Shalabi FM, Docherty SM, Purkis P, et al. hCG beta expression by cervical squamous carcinoma—*in vivo* histological association with tumour invasion and apoptosis. Histopathology 2008;53:147–55.

[47] Carter PG, Iles RK, Neven P, Ind TE, Shepherd JH, Chard T. Measurement of urinary beta core fragment of human chorionic gonadotrophin in women with vulvovaginal malignancy and its prognostic significance. Br J Cancer 1995;71:350–3.

[48] Beebe JS, Huth JR, Ruddon RW. Combination of the chorionic gonadotropin free beta-subunit with alpha. Endocrinology 1990;126:384–91.

[49] Ruddon RW, Krzesicki RF, Norton SE, Beebe JS, Peters BP, Perini F. Detection of a glycosylated, incompletely folded form of chorionic gonadotropin beta subunit that is a precursor of hormone assembly in trophoblastic cells. J Biol Chem 1987;262:12533–40.

[50] Cole LA, Butler SA. Hyperglycosylated hCG and its β-subunit: pervasive promoters of invasion and growth in cancer. Mol Cell Endocrinol 2013.

[51] Murphy G, Reynolds JJ, Whitham SE, Docherty AJ, Angel P, Heath JK. Transforming growth factor beta modulates the expression of collagenase and metalloproteinase inhibitor. Eur Mol Biol Organ J 1987;6:1899–904.

[52] Wu D, Luo S, Wang Y, Zhuang L, Chen Y, Peng C. Smads in human trophoblast cells: expression regulation and role in TGF-beta-induced transcriptional activity. Mol Cell Endocrinol 2001;175:111–21.

[53] Massague J. TGFβ in cancer. Cell 2008;134:215–30.

[54] Massague J. TGF-beta signal transduction. Ann Rev Biochem 1998;67:753–91.Triozzi PL, Stevens VC. Human chorionic gonadotropin as a target for cancer vaccines. Oncol Rep 1999;6:7–17.

[55] Cole LA. hCG structure: a logical perspective. Asian Pacific J Reprod 2012;1:287–92.

[56] Moulton HM, Yoshihara PH, Mason DH, Iversen P, Triozzi PL. Active specific immunotherapy with β-human chorionic gonadotropin peptide vaccine in patients with metastatic colorectal cancer: antibody response is associated with improved survival. Clin Cancer Res 2002;8:2044–51.

[57] Ivers PL, Mourich DV, Moulton HM. Monoclonal antibodies to two epitopes of β-human chorionic gonadotropin for the treatment of cancer. Curr Opin Mol Ther 2003;5:156–60.

[58] Delves PJ, Iles RK, Roitt IM, Lund T. Designing a new generation of anti-hCG vaccines for cancer therapy. Mol Cell Endocrinol 2007;260:276–81.

[59] He L-Z, Ramakrishna V, Connolly JE, Wang X-T, Smith P, Jones CL, et al. A novel human cancer vaccine elicits cellular responses to the tumor-associated antigen human chorionic gonadotropin β. Clin Cancer Res 2004;10:1920–7.

[60] Morse MA, Chapman R, Powderly J, Blackwell K, Keler T, Green J, et al. Phase I study utilizing a novel antigen-presenting cell-targeted vaccine with toll-like receptor stimulation to induce immunity to self-antigens in cancer patients. Clin Cancer Res 2011;17:4844–53.

[61] Butler SA, Staite EM, Iles RK. Reduction in bladder cancer cell growth in response to hCGβ CTP37 vaccinated mouse serum. Oncol Res 2003;14:93–100.

Expression and Biological Function of the Free β-Subunit in Cancer: Expression and Treatment Target in Cancer

23

Stephen A. Butler and Ray K. Iles
MAP Diagnostics, BioPark, Broadwater Road, Welwyn Garden City, Hertfordshire, UK

Reports of ectopic hCG molecules expressed *in vivo* by nongestational tumors were noted as early as 1904 [1]. A rare case of a bladder tumor was described (called *chorioepithelioma*) that contained syncytiotrophoblast elements, had widely metastasized, and expressed biologically active gonadotropin. The tumor occurred in a postmenopausal woman whose ovaries were atrophic, despite a hyperplastic endometrium. Therefore, 23 years before the discovery of hCG by Aschheim and Zondeck [2], it was correctly concluded that these changes were due to a gonadotrophic hormone produced by the tumor. Ectopic production of biologically active hCG produced by nongerm-cell tumors was next reported in 1946 [3].

It is not uncommon for ectopic hCGb production to be explained by de-differentiation (trophoblastic differentiation), whereby it is assumed that the tissue has reverted to pluripotence, taking on the characteristics of the syncytiotrophoblast and thus expressing hCG. In almost all cases, however, the sole criterion for the trophoblastic differentiation claim is the detection of hCG; this detection is often a result of a misinterpreted hCGβ-positive assay. Because common epithelial tumors will express hCGβ [4], most of these claims are the result of false dogma. It is, in fact, quite distinct, because germ-cell tumors will express both hCGα and hCGβ. Resulting in the production of the gonadotropic holo-hormone hCG, ectopic expression by common epithelial tumors consists almost exclusively of the free β-subunit. Only rarely is the holo-hormone found in advanced-stage carcinomatosis [5], and it has been only sporadically noted in liver and lung cancers [6]. Thus, such de-differentiation is a much rarer event than has been claimed and has more to do with confusion over assay specificity and hCG/hCGβ terminology than de-differentiation or carcinomatosis.

Although the holo-hormone hCG is produced by placental and germ-cell tumors, the free β-subunit (hCGβ) is produced by epithelial tumors and is (more often than not) independent of glycoprotein hormone alpha gene expression [7,8]. Ectopic production of free hCGβ by bladder carcinoma is well-described, and the majority of our work has been concentrated in this field [9]. However, expression of hCGβ is not exclusive to bladder carcinoma, and it has also been shown in cervical and endometrial carcinoma, as well as many other nongerm-cell tumors of the breast, colon, lung, ovary, oral/facial tissue, prostate, pancreas, vulva/vagina, kidney, and neuroendocrine tissue (Table 23.1).

Human Chorionic Gonadotropin (hCG). DOI: http://dx.doi.org/10.1016/B978-0-12-800749-5.00023-7

Table 23.1 **Summary of the frequencies of hCGβ protein expression and *CGB* gene expression by non-germ cell epithelial cancers reported over the past 25 years**

Publication year	Tissue of origin	hCGβ protein expression			*CGB* gene expression					Prognostic
		Serum hCGβ	Urine hCGβ/βcf	Immunohisto-chemistry hCGβ	All *CGB*	Type II/*CGB*3,5,8	*CGB* 3,5,7,8	Type I/CG117 index	*CGB1/CGB2*	
[10] 1989	Bladder	76% ($n = 21$)	44% ($n = 25$)	–	–	–	–	–	–	–
[5] 1990	Bladder	–	93% ($n = 28$)	–	–	–	–	–	–	–
[11] 1993	Bladder	–	–	–	45% ($n = 75$)	–	–	–	–	Yes
[12] 1995	Bladder	–	–	–		61% ($n = 34$)	–		–	–
[13] 1995	Bladder	–	60% ($n = 63$)	–	–	–	–	–	–	–
[14] 1996	Bladder	76% ($n = 33$)	–	–	–	–	–	–	–	–
[15] 1996	Bladder	–	73% ($n = 237$)	–	–	–	–	–	–	–
[16] 1996	Bladder	30% ($n = 76$)	–	–	–	–	–	–	–	Yes
[17] 1996	Bladder	–	35% ($n = 52$)	–	–	–	–	–	–	Yes
[18] 1997	Bladder	–	–	–	61% ($n = 34$)	–	–	21% ($n = 34$)	–	–
[19] 1998	Bladder	–	–	35% ($n = 104$)	–	–	–	–	–	–
[20] 2002	Bladder	29% ($n = 66$)	8% ($n = 72$)	–	50% ($n = 84$)	–	–	–	–	–
[21] 2010	Bladder	–	–	–	38% ($n = 34$)	–	–	–	–	–
[22] 1992	Breast	–	–	56% ($n = 27$)	–	–	–	–	–	–
[23] 1996	Breast	–	–	–	80% ($n = 32$)	–	–	–	–	–
[18] 1997	Breast	–	–	–	–	46% ($n = 30$)		14% ($n = 30$)		
[24] 1997	Breast	–	19% ($n = 72$)	–	–	–	–	–	–	–
[25] 1998	Breast	–	–	–	–	–	–	48.5% ($n = 99$)	–	Yes
[26] 2005	Breast	–	–	13% ($n = 334$)	–	–	–	–	–	–
[27] 1994	Cervical	–	67% ($n = 61$)	–	–	–	–	–	–	Yes
[28] 1995	Cervical	35% ($n = 40$)	–	–	–	–	–	–	–	–
[24] 1997	Cervical	–	29% ($n = 49$)	–	–	–	–	–	–	–
[29] 1998	Cervical	–	33% ($n = 46$)	26% ($n = 46$)	–	–	–	–	–	Yes
[30] 2008	Cervical	–	–	78% ($n = 89$)	–	–	–	–	–	–
[31] 1995	Colorectal	–	–	54% ($n = 377$)	–	–	–	–	–	Yes
[32] 1996	Colorectal	–	–	37% ($n = 123$)	–	–	–	–	–	Yes
[33] 2000	Colorectal	17% ($n = 232$)	–	22% ($n = 232$)	–	–	–	–	–	Yes

[34] 2002	Colorectal	16% ($n = 204$)	–	–	–	–	–	–	–	Yes
[28] 1995	Endometrial	30% ($n = 39$)	–	–	–	–	–	–	–	–
[24] 1997	Endometrial	–	32% ($n = 25$)	–	–	–	–	–	–	–
[35] 2002	Kidney	23% ($n = 177$)	–	–	–	–	–	–	–	Yes
[36] 2003	Kidney	27% ($n = 256$)	–	15% ($n = 229$)	40% ($n = 20$)	–	–	–	–	Yes
[37] 2004	Kidney	–	–	0% ($n = 55$)	–	–	–	–	–	No
[38] 2006	Kidney	–	–	–	40% ($n = 104$)	–	15% ($n = 42$)	95% ($n = 42$)	–	Yes
[39] 1995	Lung	–	–	93% ($n = 44$)	–	–	–	–	–	–
[40] 1997	Lung	–	–	64% ($n = 14$)	–	–	–	–	–	–
[41] 1998	Lung	22% ($n = 2$)	–	–	–	–	–	–	–	–
[42] 1999	Lung	14% ($n = 85$)	–	28% ($n = 85$)	–	–	–	–	–	Yes
[43] 2000	Lung	–	–	9% ($n = 90$)	–	–	–	–	–	–
[44] 2008	Neuro-endocrine	12% ($n = 360$)	–	–	–	–	–	–	–	Yes
[45] 1999	Oral/facial	20% ($n = 59$)	–	–	–	–	–	–	–	Yes
[46] 1999	Oral/facial	–	–	64% ($n = 45$)	–	–	–	–	–	Yes
[28] 1995	Ovarian	41% ($n = 27$)	–	–	–	–	–	–	–	–
[47] 1997	Ovarian	36% ($n = 73$)	–	–	–	–	–	–	–	Yes
[48] 2008	Ovarian	33% ($n = 173$)	–	–	–	–	–	–	–	Yes
[49] 2011	Ovarian	–	–	36% ($n = 539$)	–	–	–	–	–	Yes
[50] 2013	Ovarian	–	–	–	100% ($n = 32$)	100% ($n = 32$)	–	–	41% ($n = 32$)	–
[51] 1998	Pancreas	42% ($n = 36$)	–	–	–	–	–	–	–	Yes
[52] 2001	Pancreas	–	–	56% ($n = 107$)	–	–	–	–	–	Yes
[53] 2004	Pancreas	50% ($n = 160$)	–	–	–	–	–	–	–	Yes
[54] 1996	Prostate	–	–	15% ($n = 80$)	–	–	–	–	–	Yes
[18] 1997	Prostate	–	–	–	–	20% ($n = 10$)	–	20% ($n = 10$)	–	–
[55] 2006	Prostate	4% ($n = 104$)	10% ($n = 104$)	–	–	–	–	–	–	No
[18] 1997	Thyroid	–	–	–	–	25% ($n = 12$)	–	3% ($n = 12$)	–	–
[56] 1995	Vulval/vaginal	38% ($n = 50$)	–	–	–	–	–	–	–	Yes
[24] 1997	Vulval/vaginal	–	17% ($n = 12$)	–	–	–	–	–	–	–
[57] 1997	Vulval/vaginal	10% ($n = 118$)	–	–	–	–	–	–	–	Yes

A total of 57 studies revealed expression of elevated levels of hCGβ (as free hCGβor hCGβ-core fragment or CGB genes) in serum, urine, immunohistochemistry or by PCR. Of the 28 studies that investigated prognosis, 93% indicated that hCGβ expression was a good marker for poor prognosis. (–) indicates that this specific element was not studied in this case.

23.1 hCGβ gene expression in cancer

In 1987, Hussa proposed several hypotheses for nontrophoblastic production of hCG. These included what he referred to as gene depression, blocked ontogeny, incomplete suppression of the hCG genome, natural DNA recombination, and vestigial autocrine–isocrine secretion [58]. As discussed in Chapter 4, the control of hCG gene expression has not been completely elucidated; it is still unclear which *CGB* genes are truly active in terms of the protein they ultimately translate and the effects that the protein produced might have, whether eutopically or ectopically expressed. Considering some of the hypotheses emerging in the field and alluded to or detailed in this book, it is tempting to suggest that the proteins arising from our *CG* paralogs may have evolved new functions; however, the true hierarchy of expression within the *CGB* gene cluster has yet to be established, as do the mechanisms that control them. What we do know is that the *LH-CGB* gene cluster is not amplified or rearranged in any way, certainly not in bladder cancer [59]. These findings strongly indicate that the regulation of transcription and/or translation is likely to hold the key to ectopic hCGβ production.

After complete cloning of the *LH-CGB* gene cluster, two groups of hCGβ producers (termed type I and type II) were described [18]. Type I refers to the gene product of *CGB7*, which appears to lead to the synthesis of an hCGβ protein with an alanine residue at position 117 (see Chapter 3); this gene appears to be largely inactive in pregnancy [60,61]. Type II refers to the gene products of *CGB*, *CGB5*, and *CGB8;* these encode an aspartic acid residue at position 117. The difference in which *CGB* type was expressed was explained through malignant transformation, suggesting that normal urothelium expresses *CGB7* and bladder carcinoma expresses genes *CGB*, *CGB5*, and *CGB8* [18], and that a "CGB117 test" (involving a simple equation—CGB7 over CGB7 + *CGB*, *CGB5*, *CGB8*) could provide prognostic information (as shown in Table 23.1). More recently, other studies of breast, lung, and renal carcinoma all indicate that *CGB7* is expressed at a comparatively low level [12,62,63].

Interestingly, CGB1 and CGB2 are often overlooked in these molecular studies because they had been largely regarded (incorrectly) as nonfunctional pseudogenes [64]. It is, of course, dangerous to start comparing choriocarcinoma and placental data with that derived from epithelial cancer studies, because their respective origins are so distinct; however, new data now appear to implicate these CGB gene products in the invasion process of trophoblast tissues [61]. Although this has been shown in first-trimester pregnancies only, it is tempting to start drawing parallels between placental implantation and tumor invasion. Most recent studies indicated that *CGB1* and *CGB2* genes were transcriptionally active in ovarian cancer [50] and that transcripts of these genes could only be detected in epithelial cells that secreted hCGβ protein [65]. It is interesting to note that the product of *CGB2* shares a common amino acid substitution with CGB7, an alanine at position 117 (see Figure 3.7). Suggesting that any studies using the CGB117 test also detected the product of CGB2 and that the type I and type II categorization of CGB genes may now need revising. This putative protein is shown in Figure 23.1.

CGB1 protein	1	MSTFPVLAEDIPLRERHVKEAAAVAAAEHGRDMGIQGAASATVPPHQCHP	50
CGB2 protein	1	MSTSPVLAEDIPLRERHVKGAAAVAAAEHGRDMGIQGAASATVPPHQCHP	50
CGB1 protein	51	GCGEGGLPRVHHRQHHHLCRLLPHHDPRAAGGPAGPASGGVQLPRCALRV	100
CGB2 protein	51	GCGEGGLPRVHHRQHHHLCRLLPHHDPRAAGGPAGPASGGVQLPRCALRV	100
CGB1 protein	101	HPAPWLPARREPRGLLRRGSQLSMCTLPPQHH	132
CGB2 protein	101	HPAPWLPARREPRGLLRRGSQLSMCTLPPQHH	132

Figure 23.1 Hypothetical hCGβ protein−proposed sequence of a hypothetical protein as products of splicing variants discussed in Chapter 3. NCBI was used to deduct amino acid sequence. Pairwise Sequence Alignment (EMBOSS Water) used for alignment was in silico. These have now been detected in epithelial cancer cell lines as described previously [65].

23.2 hCGβ expression in epithelial cancer

Well-documented evidence for ectopic production of hCG/hCGβ by bladder tumors was not published until 1973 [66]. Subsequently, more case reports were reviewed in 1978 [67] and again in 1983 [68], whereby ectopic hCG expression became a generally accepted phenomenon in cancer. Two years later, one article stated that the production of an hCG-like material by bladder carcinoma was "probably not rare" [69]; however, it was not until *in vitro* studies in 1987 [70] and 1989 [10] that the free β-subunit of hCG was identified as a potential bladder tumor marker; at this time, the hCG-like product was determined to be hCGβ and not hCG [7]. Bladder cancer is not unique in expressing hCGβ, it has merely been our focus. Over the past three decades, there have been many reports highlighting the ectopic production of hCGβ from nongerm-cell or nonplacental origins [68,70−80]; this has culminated in the latest wave of publications linking hCGβ expression to poor prognosis in 93% of studies (see Table 23.1).

Since 1989, 57 significant studies (omitting case reports) of the detection of hCGβ in various epithelial cancers have been published (see Table 23.1). Each study investigated the expression of hCGβ by immunohistochemistry, serum analysis, or urinalysis (of either hCGβ or hCG β-core fragment) for hCGβ protein or expression of any of *CGB* genes or gene transcription tests (the *CGB*117 test). Positive detection ranges from 4% in prostate cancer to 93% in small cell (SC) lung cancer. In some cases, 100% detection is observed when looking for the expression of "any" *CGB* gene; this is more likely to be due to the sensitivity of modern amplification techniques and not necessarily indicative of disease. There is considerable variation in positivity for hCGβ/*CGB*, not only between studies but also between detection methods used in the same studies of the same sample set. Serum and urine levels rarely match, and there are even more clear differences when looking for the *CGB* gene expression and protein profile.

In bladder cancer, the relative incidence of expression has been disputed in many studies [9]; however, from our own studies, only approximately 35%

(see Table 23.1) of bladder cancer patients have increases in immunoreactive hCGβ material (intact hCG, hCGβ, or hCG β-core fragment). There is good correlation of hCGβ expression by such tumors with grade and stage; metastatic cases often tend to be hCGβ-positive [11]; in those cases, up to 75% of all serum and urine samples have increased levels of immunoreactive hCGβ [81].

With a focus on early detection, however, hCGβ appears to be of little use diagnostically. According to Table 23.1, optimistic assessment could conclude that 93% of SC lung carcinomas [39], 80% of breast tumors [23], or 78% of cervical carcinomas [30] express hCGβ; however, the real picture is probably quite different, because frequencies from the original studies range from 4% to 93% (Table 23.1). More generally, the overall incidence of hCGβ expression in epithelial cancer is likely to be approximately one-third of all cases. We can see that in renal, prostate, vulval/vaginal, and neuroendocrine cases, the frequency is somewhat lower, and in bladder, cervical, and pancreatic carcinoma, the frequency is slightly higher.

With a sensitivity of 50% at best, there is a question of clinical significance that has maintained interest in ectopic hCGβ expression: why would cancer produce a pregnancy hormone subunit? Only since the advent of specific immunoassays [82] has it been possible to identify free β-subunit as the hCG-related antigen and distinguish it from the holo-hormone hCG and free hCGα. Only the intact αβ heterodimer is an active gonadotropic hormone, and for some researchers this strengthens the argument that hCGβ production is not an epiphenomenon.

The prognostic significance of hCGβ expression has not gone unnoticed and, as highlighted in Table 23.1, almost always appears to indicate poor prognosis. The exceptions here were prostate cancer, renal carcinoma, and breast cancer (where hCG is sometimes detected but not hCGβ). Prostate and kidney cancers do not appear to be significantly associated with the expression of hCGβ at all. From our analysis, prostate and kidney cancers are indicated by the lowest expression frequencies of approximately 10% (Table 23.1). This might suggest that a different mechanism is at play, or it could be an artifact associated with the small number of studies with low *n*-values. Breast cancer and hCGβ expression, however, is a complex and well-investigated topic, and one that is often at odds with our own studies. Much of the conflicting evidence could simply come down to hCG/hCGβ confusion, because the molecule here is more likely to be exogenous hCG with a differentiation effect rather than ectopic hCGβ, which has an autocrine tumor-promoting effect. It is with this hypothesis in mind that we recently published a review of the topic [83].

It is not uncommon for authors to report very aggressive tumors in which hCGβ has been detected and to find lower response rates to radiotherapy and chemotherapy in patients in whom hCGβ is elevated. In our own studies, survival analysis indicated a very strong association between early death and hCGβ expression (Figure 23.2), suggesting an autocrine control mechanism that we later explored further [84].

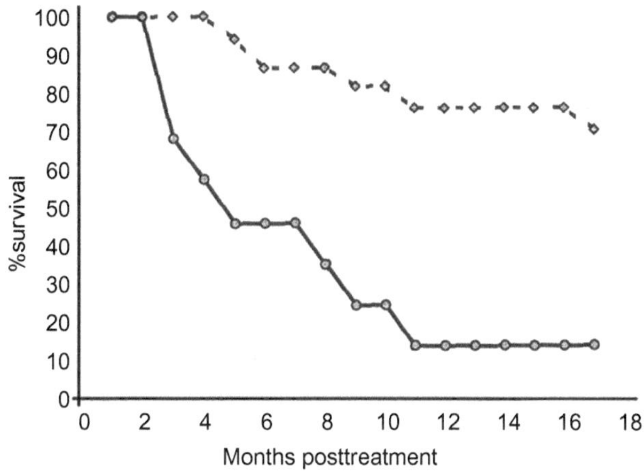

Figure 23.2 Survival curves of hCGβ-positive and hCGβ-negative patients with T2−T4 bladder tumors. The *x*-axis indicates percentage survival of patients with normal (- - - -) and elevated (------) hCGβ as a function of months posttreatment (the *y*-axis). Adapted from Ref. [9]; original data from Ref. [18].

23.3 The biological action of hCGβ on epithelial tumors

Historically, the expression of hCGβ by epithelial cancers has been dismissed as an epiphenomenon, with no oncogenic significance. Thus, hCG (and, by association, hCGβ) expression is often solely regarded as a marker of germ cells, as mentioned. In 1996, we published our first study reporting that bladder cancer cell numbers increased after incubation with hCGβ in a dose-dependent manner [84]. No effect could be seen after treatment with intact hCG, growth-promoting hormone alpha (GPHα), or hCG β-core fragment. Neither effect was observed when hCGβ was coincubated with polyclonal anti-hCGβ antiserum. Furthermore, when this antiserum was added to bladder cancer cell cultures known to produce hCGβ, their growth was significantly inhibited in proportion to the quantity of hCGβ produced from each cell line (i.e., high hCGβ expressers were more adversely affected by the antisera than the low expressers or nonexpressors). This suggested autocrine stimulation [84]. Anti-hCG antibody inhibitory effects on the growth of tumor cells has also been demonstrated *in vivo* in athymic (nude) mice [85] and in cell lines incubated with murine antisera from mice immunized with anti-hCG-CTP vaccines [86]. More recently, antisense oligonucleotides directed to the hCG β-chain have also been shown to have antiproliferative effects on tumor cell lines *in vitro* [87−89]. Recently, this has been shown to be more specific if CGB2 is the target rather than any other genes in the cluster [65] (see Chapter 3).

Further studies showed that the increase in the cancer cell population in response to hCGβ was brought about by inhibition of apoptosis rather than by stimulation of cell replication [90], and that the recombinant hCGβ was at least as potent, if not more potent, than the CR129 preparation used before [84]. Furthermore, given that the LH/hCG receptor is not expressed by these tissues and that only intact hCG can stimulate the LH/hCG receptor (as described in Chapter 15), we have to assume that any activity observed must be mediated by an as-yet-unidentified pathway.

The X-ray crystallographic structure of hCG was revealed [91,92] after we finalized and published our studies of the prognostic significance of hCGβ in bladder cancer [17] and the autocrine function of hCGβ on bladder cancer cells *in vitro* [84]. The most striking feature noted was the cystine knot, which we described in Chapter 10. This conserved structure was identified in a number of cytokines designated as the "cystine-knot growth factor" (CKGF) family. This family includes transforming growth factor β (TGFβ), platelet-derived growth factor B (PDGFB), and nerve growth factor (NGF). Since the identification of the structure of hCG in 1994, the family of growth factors containing a cystine-knot motif has expanded and now includes vascular endothelial growth factor (VEGF) and placental growth factor (PlGF), among others.

Each of the growth factors exists as a functional homodimer, and in 1999 we published our findings suggesting that hCGβ also forms naturally occurring homodimers [93]. We later indicated that the β–β homodimer of hCG was no more bioactive in our cell line model than the monomeric hCGβ [94]. We now suggest that the ability to homodimerize is indicative of structural variation within bioactive hCGβ rather than a requirement for ectopic bioactivity. The latter additions to the CKGF family and the ability of hCGβ to homodimerize further increased the intriguing possibilities that cross-talk may occur between both multiple growth regulatory systems and that ectopic expression of molecules containing a cystine knot may give rise to oncogenic molecules (whose disruption of these regulatory systems favor cancer development).

Homology modeling (using current data regarding the structure of hCG as obtained from the published crystal structure) [91] was used in conjunction with data from the Protein Data Bank (www.pdb.org) and Clustalw2 (http://www.ebi.ac.uk/Tools/clustalw2/index.html). By mapping hCGβ to either TGFβ or VEGF crystal structures, it is possible to generate potentially naturally occurring molecular models of hCGβ–β dimers (Figure 23.3), where each model of hCGβ dimerizing is energetically favorable. Homology model monomers alone have 51% and 69% alignment with TGFβ and VEGF, respectively, without any changes to amino acid sequences. If hCGβ folds in one of these conformations, then we propose that the molecule is likely to interact with the respective CKGF receptor. Exploring these methods further, hCGβ can be demonstrated to be very likely to bind the VEGF receptor kinase insert domain receptor (KDR). This claim is substantiated by VEGF homology-modeled hCGβ with four of the eight KDR receptor-binding hot spots (Ile 47, Ile 53, Ile 87, and Pro 90) in correct alignment for binding and stimulation of the receptor (unpublished data). This is two more than is required by VEGF-C to bind and stimulate the same receptor [95]. Similarly, amino acid regions in the receptor binding loop of TGFβ 3, which are essential for binding TGFβ-RII, have been identified in similar positions in hCGβ (unpublished data). These latest findings add to the hypothesis conceived 15 years ago, that ectopic hCGβ cross-reacted

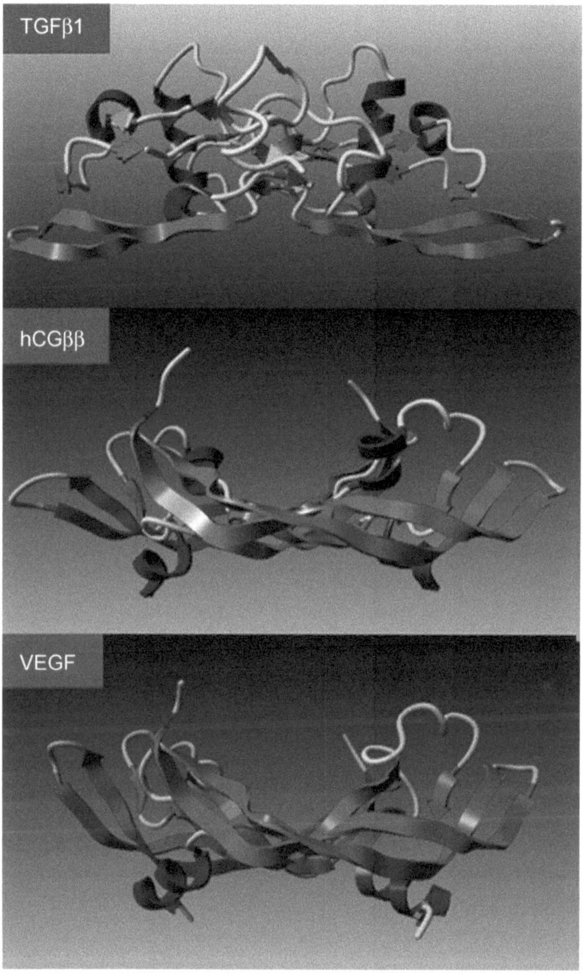

Figure 23.3 CKGF dimers. Images of the TGFβ dimer, hCGββ homodimer, and VEGF dimer after bioinformatics studies of hCGβ folding and receptor interaction. Note the striking resemblances, especially those of hCGβ and VEGF.

in CKGF receptors. The advances and availability of bioinformatics and molecular biology tools mean that teasing-out of the molecular mechanism of the ectopic function of hCGβ is tantalizingly close.

If one examines the roles of VEGF and TGFβ during both placentation and oncogenesis, then our hypotheses extend further than structural similarities alone. Recent studies have suggested roles for VEGF and hCG in placental vasculature [96,97]. hCG has been proposed as a novel angiogenic factor in its own right [98], and VEGF is strongly implicated in the development of blood vessels during placentation as a distinct process involving both VEGF receptor expression and stimulation [99].

VEGF is involved in placental vascularization and neovascularization in oncogenesis, and we have now linked immunohistochemically detected hCGβ expression by cervical cancer with the extent of tumor vascularization [30]; serum hCG levels have recently been linked to neovascularization in nonseminomatous testicular germ-cell tumors [100]. In bladder cancer, TGFβ and its receptors are expressed in high numbers, yet tumors appear to evade apoptosis [101–103]. We now believe hCGβ may antagonize these pathways [90]. It was also noted that the proximity of the TGFβ gene is close to the hCGβ gene cluster on chromosome 19 [104], suggesting possible early evolutionary duplication from TGFβ. It has been shown that there are highly conserved regions between CKGFs with regard to key cystine residues involved in cystine-knot formation [105].

Whatever complex molecular mechanisms are revealed, the most fundamental observation to date is that the immunodepletion of hCGβ in closed *in vitro* cultures increases apoptosis of hCGβ-expressing cancer cells. It is clear that focus should turn to the prognostic significance of hCGβ detection and utilize what we now know to identify patients who are at greater risk for a more rapid decline in health. From this and from the data we have regarding hCG immunodepletion, a patient selection process for studies in which hCGβ-targeted therapeutics are being investigated is a must if we want to see any significant improvements in the potential for new adjuvant therapies.

23.4 hCG cancer vaccines

Investigation into the use of immunotherapy as an alternative approach to cancer treatment is still in its infancy, and any efforts to use vaccinations for the treatment of cancer have met with only minimal success. Different approaches modify either tumor cells with immunomodulatory molecules (such as cytokines) [106] or costimulatory molecules [107], which enhance either their immunogenicity or their ability to immunize with well-defined specific antigens [108]. Many such antigens relate specifically to studies of melanoma [108,109]; however, p53, p21ras, Her2/neu, EGFRVIII, and MUC1 have been described in studies of cancers of the breast, colon, ovaries, and prostate [108,110–113]. The Her2 example is now perhaps the best understood and, in many ways, suggests a model to which hCGβ expression by cancer can be compared, but even here immunogenicity remains a problem [114].

Many of these approaches try to stimulate tumor-specific cytotoxic T-cell-mediated responses. Unfortunately, specifically targeting this arm of the adaptive immune system subjugates the stimulation of B lymphocytes; therefore, the production of good antibody titers are affected [115]. Immunogens alone are often insufficient to stimulate an adequate immunological response. Carriers such as the diptheria toxoid (DT) or keyhole limpet hemocyanin (KLH) must be used along with administrations of toxic adjuvants such as Bacillus Calmette–Guerin (BCG). Cowpea mosaic virus (CPMV) was also used as a successful, and safer, alternative to these [116].

The focus on tumor therapy appears to shift between cell-mediated cytotoxicity and immunodepletive therapy specifically targeting tumor-associated antigens in both the hCG [115] and Her2 arenas [114]. Immunodepletion targets peptides produced by tumors that have been identified to elicit specific tumor effects. Therefore, antibodies to these antigens, in turn, might play a role in preventing metastatic disease. In fact, the greatest logical potential for successful therapy is as an adjuvant treatment of cancer. Purposefully targeting metastasis through their production of cancer-associated bioantigens (i.e., those with a demonstrable biological function), rather than the primary tumor itself, would effectively stave off tumor progression, have a greater effect on overall survival, and at the same time allow current therapies to perhaps be more effective. The concept of immunodepletion exploits this and relies on the stimulation of a B-cell antibody response, which results in the reduction of circulating autocrine/paracrine factors produced and required by the tumor. Therefore, hCG (particularly hCGβ) is an ideal target for this form of cancer immunotherapy because of the control hCGβ appears to have over the proliferation of carcinoma cells *in vitro*.

Interestingly, vaccines targeting hCG were developed as a potential contraceptive, not as a cancer therapy. The first report was written in 1976 [117], and that same research group dominated the field for the remainder of the decade. Not surprisingly, the majority of studies focused on toxicity, immunogenicity, and epitope selection, preparing for phase I trials. Historically, hCGβ-CTP has been the epitope antigen of choice in developing anti-hCG fertility vaccines, because any resulting antibody would fail to recognize LHβ. Although fundamentally important in vaccines for use as infertility treatments, the cross-reaction with LH is not the primary concern in cancer immunotherapy; however, the change in direction from contraception to cancer is a logical and rapid one. Subsequently, the same epitopes selected as targets in fertility control could again be used for specific responses to the hCG (hCGβ) produced by cancer.

There are obvious parallels between the use of anti-hCG vaccines developed as contraception and the potential use in immunodepletion of ectopic hCGβ expression by tumors, particularly when considering the tolerance of such a vaccine in healthy females in whom no serious side effects have been reported [118,119]. In each case, the isolated hCGβ was the target antigen, with the unique carboxy terminal region being the target epitope.

In early studies, CTP was coupled to the powerful adjuvant/carrier diptheria toxin to generate higher titers of antibody [120]. Therefore, it was suggested that it might be possible to further increase the effects by utilizing a more immunogenic receptor − significant hCGβ epitope for vaccine development. It has been suggested that it might be possible to further increase immunogenicity using vaccines displaying peptides involved in hCG/LH receptor binding, such as those within the long loop region at residues 38−57 [81]. It was shown that immunization with this peptide, which contains a conformational epitope, conjugated to DT-generated antibodies that have only low cross-reactivity with LH [121]. When these DT conjugates (including both CTP and hCGβ 38−57) were investigated, however, both higher immunological potency and more marked inhibition of hCG bioactivity were observed [122]. An alternative epitope found within the hCG β-core fragment and distinct from CTP entirely has also achieved some success in fertility trials in India [123].

We have tested numerous hCG antisera, and monoclonal antibodies directed against the hCGβ-CTP or other regions clearly bring about reductions in cell number when introduced to hCGβ-expressing bladder carcinoma cells. In one of our own studies, we investigated the antiserum from mice vaccinated with hCG constructs expressed on CPMV. The expressed constructs were a 37-mer peptide of amino acids 109–145 of the hCGβ C-terminus (TCDDPRFQDSSSSKAPPPSLPSPSRLPGPSDTPILPQ) (long CTP) and a 24-mer peptide comprising amino acids 109–118 and 132–145 from hCGβ C-terminus (TCDDPRFQDSSRLPGPSDTPILPQ) (short CTP) identified as significant epitopes (β8 and β9) within the CTP [124,125]. Pooled sera from mice immunized with long CTP vaccine were shown to markedly reduce the *in vitro* growth of the hCGβ-expressing bladder cancer cell lines used [86]; cell number was reduced by almost 50% in 48 h. In contrast, these sera had no effect on the growth of cells that did not express hCGβ. Interestingly, sera from mice immunized with short CTP vaccine did not significantly inhibit the growth of any cells tested [86]. The reason for this difference is unclear, but the results suggest a difference in affinity for hCGβ between the two antibody populations generated, or they may indicate that certain regions of CTP are involved in the ectopic biological activity of hCGβ.

With this information one might wonder, is CTP the best antigen? CTP plays no known functional role and is assumed to be a peripheral entity. CTP-specific antibodies still react with hCG even when it is bound to the hCG/LH receptor [124]. The epitope is not particularly immunogenic and yields low-affinity antibodies when compared with other hCGβ epitopes [126]. Its selection as a target was, in many ways, only to distinguish hCGβ from LHβ, and there are at least five other unique epitopes on hCGβ [125]—named β1, β6, and β7—in addition to β8 and β9 on CTP. In more recent studies, these epitopes were used to generate high-affinity antibodies using mutant hCGβ with disrupted LHβ common epitopes [115]. One mutant with an Arg Glu substitution at amino acid 68 generated high-affinity antibodies in rabbits [127].

The vaccines are generally effective at generating titers. Fertility vaccine antibodies have been observed to remain at detectable levels in humans for up to 10 months [122]; this would provide ideal postsurgical protection against any metastatic spread. In addition, a method of administering single-dose CTP-DT vaccine in biodegradable microspheres provided measurable hCG antibodies for more than 1 year [128].

Phase I clinical trials using anti-hCGβ vaccines have been performed in patients with advanced-stage epithelial tumors. The initial results have indicated that the diphtheria toxin CTP immunogen is well-tolerated, safe, and immunogenic in humans [129]. The results were also promising in clinical trials of pancreatic and colorectal cancer patients [130], although it was not shown at the time whether the immune response would significantly reduce any tumor development. Most recently, phase II clinical trials of colorectal cancer have been completed using the CTP-DT vaccine [131], with only limited success when a strong immune response

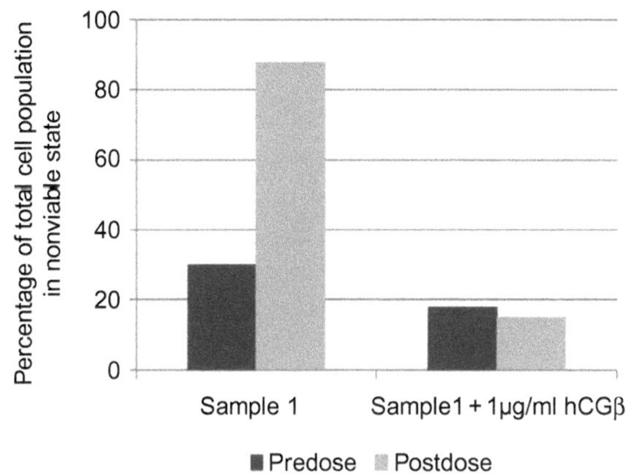

Figure 23.4 Data from flow population analysis (FACS Calibur) indicated all nonviable cell populations (propidium iodide positive) of SCaBER bladder cancer cells after 72-h incubation under standard cell culture conditions with serum from a patient before and after vaccination with CDX-1307 vaccine. Serum was diluted at 1:50 dilution and data shown include data with and without the addition of 1 mg/ml recombinant hCGβ from Sigma Aldrich (as competitor). Original data from Ref. [132].

was associated with improved survival. Most recently, CDX-1307, an antibody-based vaccine fused with hCG-β, was administered to patients with advanced epithelial malignancies. Immune responses were seen in patients with and without levels of serum hCG-β. The antisera from vaccinated trial patients were demonstrated to reduce viability in bladder cancer cell populations *in vitro;* an example of which can be seen in Figure 23.4 [132].

We have proposed a molecular model that accounts for resistance to radiotherapy and associated poor prognosis for hCGβ-expressing cancers. We postulated that removing hCGβ from the circulation should assist in initiating tumor regression *in vivo* through resumption of apoptosis; however, trials conducted to date have yet to show any significant promise for anti-hCG vaccines as an adjuvant treatment for cancer. If multiple cancers are producing these molecules, and if a credible link between their production and tumor growth can be demonstrated, then why have attempts to use vaccines failed to deliver any significant responses in patients? As described in this chapter, only approximately one-third of epithelial cancers express hCGβ. To date, patients have not been preselected based on their hCGβ status when recruited into studies. In the case of the latest trial, vaccine was administered to all colorectal cancer patients and was not restricted to those who tested hCGβ-positive. A vaccine that stimulates immune responses against hCGβ would be expected to benefit only those patients in whom hCGβ is aberrantly expressed, suggesting that the vaccine might have benefited a higher proportion of the trial group if detection of ectopic hCGβ expression had been applied as a fundamental inclusion criterion for entry into the trial. The Her2 paradigm clearly showed that Herceptin® never would have displayed its small but significant advantage in breast cancer patient survival if it had been administered to all breast cancer patients irrespective of Her2 expression status [133].

23.5 Summary

In summary, this chapter reviews and reinforces the collective data supporting the notion that the ectopic secretion of hCGβ by common epithelial tumors is an epiphenomenon. Expression of hCGβ by epithelial cancer is approximately 30%, and in bladder cancer, pancreatic cancer, and colorectal cancer, per se, it may be as high as 50%. The expression is associated with resistance to radiotherapy, rapid development of metastasis, and, thus, poor prognosis. hCGβ influences cell population growth by inhibiting apoptosis in culture—an indirect molecular mechanism consistent with the clinical finding of resistance to radiotherapy. Structural biophysical studies and bioinformatics analyses show that hCGβ can homodimerize; it shares topological and homological features with the cystine-knot family of growth factor proteins. The topological resemblance of TGFβ and VEGF to hCGβ is particularly striking, and hCGβ has been shown to reverse TGFβ-induced apoptosis. We hypothesize that any ectopic hCGβ effect is modulated via the TGFβ-RII receptor and/or the VEGF KDR receptor, bringing about multiple responses associated with the oncogenic process. Whatever the precise mechanism, hCGβ has a growth-modulating function in tumorigenesis that can be removed using specific antibodies. This presents a new adjuvant approach to treatment strategies pertaining to hCGβ-expressing cancers. One such potential approach is immunodepletive therapy in the form of an anti-hCGβ vaccine.

The authors thank Beata B. Burczynska for her assistance with Figure 23.1 and Selam Mahmoud and Jason K. Iles for their assistance with Table 23.1.

References

[1] Djewitzi W. Über einen fall von chorionepithelioma der harnblase. Virchows Arch 1904;178:451–64.
[2] Aschheim S, Zondek B. Das hormon des hypophysenvorderlappens: testobjekt zum nachweis des hormons. Klin Wochenschr 1927;6:248–52.
[3] McFadzean AJS. Feminization associated with carcinoma of the adrenal cortex. Lancet 1946;2:940–3.
[4] Iles RK, Chard T. Human chorionic gonadotropin expression by bladder cancers: biological and clinical potential. J Urol 1991;145:453–8.
[5] Iles RK, Lee CL, Oliver RTD, Chard T. Composition of intact hormone and free subunits in the human chorionic gonadotropin-like material found in serum and urine of patients with carcinoma of the bladder. Clin Endocrinol 1990;33(3):355–64.
[6] Stenman UH, Alfthan H, Hotakainen K. Human chorionic gonadotropin in cancer. Clin Biochem 2004;37:549–61.

[7] Iles RK, Chard T. Immunochemical analysis of human chorionic gonadotropin-like material secreted by "normal" and neoplastic urothelial cells. J Mol Endocrinol 1989;2:107−12.

[8] Iles RK, Purkis PE, Whitehead PC, Oliver RTD, Leigh I, Chard T. Expression of β human chorionic gonadotropin by non-trophoblastic nonendocrine "normal" and malignant epithelial cells. Br J Cancer 1990;61:663−6.

[9] Iles RK, Butler SA. Human urothelial carcinomas—a typical disease of the aged: the clinical utility of human chorionic gonadotropin in patient management and future therapy. Exp Gerontol 1998;33:379−91.

[10] Iles RK, Jenkins BJ, Oliver RT, Blandy JP, Chard T. Beta human chorionic gonadotrophin in serum and urine. A marker for metastatic urothelial cancer. Br J Urol 1989;64 (3):241−4.

[11] Moutzouris G, Yannopoulos D, Barbatis C, Zaharof A, Theodorou C. Is beta-human chorionic gonadotrophin production by transitional cell carcinoma of the bladder a marker of aggressive disease and resistance to radiotherapy? Br J Urol 1993;72 (6):907−9.

[12] Lazar V, Diez SG, Laurent A, Giovangrandi Y, Radvanyi F, Chopin D, et al. Expression of human chorionic gonadotropin beta subunit genes in superficial and invasive bladder carcinomas. Cancer Res 1995;55:3735−8.

[13] Halim A-B, El-Ahmady O, Hamza S, Aboul-Ela M, Oehr P. Urinary β-hCG in benign and malignant urinary tract disease. Dis Markers 1995;12:109−15.

[14] Mora J, Gascon N, Tabernero JM, Rodriguez-Espinosa J, Gonzalez-Sastre F. Different hCG assays to measure ectopic hCG secretion in bladder carcinoma patients. Br J Cancer 1996;74:1081−4.

[15] El-Ahmady O, Halim AB, Mansour O, Salman T, El-Din AG, Walker RP. Urinary gonadotropin peptide (UGP) in Egyptian patients with benign and advanced malignant urological disease. Br J Cancer 1996;73:1486−90.

[16] Pectasides D, Bafaloucos D, Antoniou F, Gogou L, Economides N, Varthalitis J, et al. AFP, β-hCG, PSA, SCC and CA 19-9 for monitoring transitional cell carcinoma of the bladder. Am J Clin Oncol 1996;19:271−7.

[17] Iles RK, Persad R, Trivedi M, Sharma KB, Dickinson A, Smith P, et al. Urinary concentration of human chorionic gonadotropin and its fragments as a prognostic marker in bladder cancer. Br J Urol 1996;77:61−9.

[18] Bellet D, Lazar V, Bieche I, Paradis N, Giovangrandi Y, Paterlini P, et al. Malignant transformation of nontrophoblastic cells is associated with the expression of chorionic gonadotropin β genes normally transcribed in trophoblastic cells. Cancer Res 1997;57:516−23.

[19] Dirnhofer S, Koessler P, Ensinger C, Feichtinger H, Madersbacher S, Berger P. Production of trophoblastic hormones by transitional cell carcinoma of the bladder: association to tumor stage and grade. Hum Pathol 1998;29:377−82.

[20] Hotakainen K, Haglund C, Paju A, Nordling S, Alfthan H, Rintala E, et al. Chorionic gonadotropin β-subunit and core fragment in bladder cancer: mRNA and protein expression in urine, serum and tissue. Eur Urol 2002;103:677−85.

[21] Venyo AK, Herring D, Greenwood H, Maloney DJL. The expression of beta human chorionic gonadotrophin (β-HCG) in human urothelial carcinoma. Pan Afr Med J 2010;7:20.

[22] Agnantis NJ, Patra F, Khaldi L, Filis S. Immunohistochemical expression of subunit beta HCG in breast cancer. Eur J Gynaecol Oncol 1992;13:461−6.

[23] Hoon DS, Sarantou T, Doi F, Chi DD, Kuo C, Conrad AJ, et al. Detection of metastatic breast cancer by beta-hCG polymerase chain reaction. Int J Cancer 1996;69:369−74.

[24] Schwarz-Roeger U, Petzoldt B, Waldschmidt R, Walker RP, Bauknecht T, Kiechle M, et al. UGP-A tumor marker of gynecologic and breast malignancies? Specificity and sensitivity in pretherapeutic patients and the influence of hormonal substitution on the expression of UGP. Anticancer Res 1997;17:3041–5.
[25] Bieche I, Lazar V, Nogues C, Poynard T, Giovangrandi Y, Bellet D, et al. Prognostic value of chorionic gonadotropin beta gene transcripts in human breast carcinoma. Clin Cancer Res 1998;4:671–6.
[26] Chang E, Lee E, Oh SJ, Kim JS, Kang C. The immunoexpressions and prognostic significance of inhibin alpha and beta human chorionic gonadotropins (HCG) in breast carcinomas. Cancer Res Treat 2005;37:241–6.
[27] Carter PG, Iles RK, Neven P, Ind TE, Shepherd JH, Chard T. The prognostic significance of urinary beta core fragment in premenopausal women with carcinoma of the cervix. Gynecol Oncol 1994;55(2):271–6.
[28] Grosssmann M, Hoermann R, Gocze PM, Ott M, Berger P, Mann K. Measurement of human chorionic gonadotropin-related immunoreactivity in serum, ascites and tumor cysts of patients with gynaecologic malignancies. Eur J Clin Invest 1995;25:867–73.
[29] Crawford RA, Iles RK, Carter PG, Caldwell CJ, Shepherd JH, Chard T. The prognostic significance of beta human chorionic gonadotropin and its metabolites in women with cervical carcinoma. J Clin Pathol 1998;51:685–8.
[30] Li D, Wen X, Ghali L, Al-Shalabi FM, Docherty SM, Purkis P, et al. hCG beta expression by cervical squamous carcinoma: *in vivo* histological association with tumor invasion and apoptosis. Histopathology 2008;53:147–55.
[31] Webb A, Scott-Mackie P, Cunningham D, Norman A, Andreyev J, O'Brien M, et al. The prognostic value of CEA, beta HCG, AFP, CA125, CA19-9 and C-erb B-2, beta HCG immunohistochemistry in advanced colorectal cancer. Ann Oncol 1995;6:581–7.
[32] Kido A, Mory M, Adachi Y, Yukaya H, Ishida T, Sugimachi K. Immunohistochemical expression of beta-human chorionic gonadotropin in colorectal carcinoma. Surg Today 1996;26:966–70.
[33] Lundin M, Nording S, Carpelan-Holmstrom M, Louhimo J, Alfthan H, Stenman UH, et al. A comparison of serum and tissue hCGβ as prognostic markers in colorectal cancer. Anticancer Res 2000;20:4949–51.
[34] Louhimo J, Carpelan-Holmstrom M, Alfthan H, Stenman UH, Jarvinen HJ, Haglund C. Serum hCG β, CA 72–4 and CEA are independent prognostic factors in colorectal cancer. Int J Cancer 2002;101:545–8.
[35] Hotakainen K, Ljunberg B, Paju A, Rasmusen T, Althan H, Stenman UH. The free beta subunit of human chorionic gonadotropin as a prognostic factor in renal cell carcinoma. Br J Cancer 2002;86:185–9.
[36] Hotakainen K, Ljungberg B, Haglund C, Nordling S, Paju A, Stenman U-H. Expression of the free β-subunit of human chorionic gonadotrophin in renal cell carcinoma: prognostic study on tissue and serum. Int J Cancer 2003;104:631–5.
[37] Berzal-Cantalejo F, Sabater Marco V, Alonso Hernandez S, Jimenez Pena R, Martorell Cebollada MA. Syncytial giant cell component. Review of 55 renal cell carcinomas. Histol Histopathol 2004;19:113–18.
[38] Hotakainen K, Lintula S, Ljungberg B, Finne P, Paju A, Stenamn U-H, et al. Expression of human chorionic gonadotrophin β-subunit type I genes predicts adverse outcome in renal cell carcinoma. J Mol Diagn 2006;8(5):598–603.
[39] Boucher LD, Yoneda K. The expression of trophoblastic cell markers by lung carcinomas. Hum Pathol 1995;26:1201–6.

[40] Yokotani T, Koizumi T, Taniguchi R, Nakagawa T, Isobe T, Yoshimura M, et al. Expression of alpha and beta genes of human chorionic gonadotropin in lung cancer. Int J Cancer 1997;16(71):539–44.
[41] Slodkwoska J, Szturmowicz M, Rudzinski P, Giedronowicz D, Sakowitcz A, Androsiuk W, et al. Expression of CEA and trophoblastic cell markers by lung carcinoma in association with histological characteristics and serum marker levels. Eur J Cancer Prev 1998;7:51–60.
[42] Szturmowicz M, Wiatr E, Sakowicz A, Slodkowska K, Filipecki S, Rowinska-Zakrzewska ER. The role of human chorionic gonadotropin beta subunit elevation in small-cell lung cancer patients. Tumour Biol 1999;20:99–104.
[43] Dirnhofer S, Freund M, Rogatsch H, Krabichler S, Berger P. Selective expression of trophoblastic hormones by lung carcinoma: neurendocrine tumors exclusively produce human chorionic gonadotropin β-subunit (hCGβ). Hum Pathol 2000;31:966–72.
[44] Shah T, Srirajaskanthan R, Bhogal M, Toubanakis C, Meyer T, Noonan A, et al. α-fetoprotein and human chorionic gonadotropin-β as prognostic markers in neuroendocrine tumor patients. Br J Cancer 2008;99:72–7.
[45] Hedstrom J, Grenman R, Ramsey H, Finne P, Lundin J, Haglund C, et al. Concentration of free hCGbeta subunit in serum as a prognostic marker for squamous-cell carcinoma of the oral cavity and oropharynx. Int J Cancer 1999;84:525–8.
[46] Bhalang K, Kafrawy AH, Miles DA. Immunohistochemical study of the expression of human chorionic gonadotropin-beta in oral squamous cell carcinomas. Cancer 1999;85:757–62.
[47] Ind T, Iles RK, Shepherd J, Chard T. Serum concentrations of cancer antigen 125, placental alkaline phosphatase, cancer-associated serum antigen and free beta human chorionic gonadotropin as prognostic markers for epithelial ovarian cancer. Br J Obstet Gynaecol 1997;104:1024–9.
[48] Vartiainen J, Lassus H, Lehtovirta P, Finne P, Alfthan H, Butzow R, et al. Combination of serum hCG beta and p53 tissue expression defines distinct subgroups of serous ovarian carcinoma. Int J Cancer 2008;122:2125–9.
[49] Guo X, Liu G, Schaver IG, Yang G, Mercado-Uribe I, Yang F, et al. Over expression of the B subunit of human chorionic gonadotropin promotes the transformation of human ovarian epithelial cells and ovarian tumorigenesis. Am J Pathol 2011;179 (3):1385–93.
[50] Kubiczak M, Walkowiak GP, Nowak-Markwitz E, Jankowska A. Human chorionic gonadotropin beta subunit genes CGB1 and CGB2 are transcriptionally active in ovarian cancer. Int J Mol Sci 2013;14:12650–60.
[51] Syrigos KN, Fyssas I, Konstandoulakis MM, Harrington KJ, Papadopoulos S, Milingos N. Beta human chorionic gonadotropin concentrations in serum of patients with pancreatic adenocarcinoma. Gut 1998;42:88–91.
[52] Louhimo J, Nordling S, Alfthan H, von Boguslawski K, Stenman UH, Haglund C. Specific staining of human chorionic gonadotropin β in benign and malignant gastrointestinal tissues with monoclonal antibodies. Histopathology 2001;38:418–24.
[53] Louhimo J, Alfthan H, Stenman UH, Haglund C. Serum hCG b and CA 72–4 are stronger prognostic factors than CEA, CA 19-9 and CA 242 in pancreatic cancer. Oncology 2004;66:126–31.
[54] Sheaff MT, Martin JE, Badenoch DF, Baithun SI. βhCG as a prognostic marker in adenocarcinoma of the prostate. J Clin Pathol 1996;49:329–32.
[55] Otite U, Baithun S, Chinegwundoh F, Nargund VH, Iles RK. Detection of human chorionic gonadotropin-beta in serum or urine of prostate cancer patients is of no clinical significance. Tumor Biol 2006;27:181–6.

[56] Carter PG, Iles RK, Neven P, Ind TEJ, Shepherd JH, Chard T. Measurement of urinary β core fragment of human chorionic gonadotropin in women with vulvovaginal malignancy and its prognostic significance. Br J Cancer 1995;71:350–3.

[57] de Bruijn HW, ten Hoor KA, Krans M, van der Zee AG. Rising serum values of beta-subunit human chorionic gonadotropin (hCG) in patients with progressive vulvar carcinomas. Br J Cancer 1997;75:1217–18.

[58] Hussa RO. The clinical marker hCG. New York, NY: Praeger Publishers; 1987.

[59] Iles RK, Czepulkowski BH, Young BD, Chard T. Amplification and re-arrangement of the β-human chorionic gonadotropin (β-hCG)-human LH gene cluster is not responsible for the ectopic production of β-hCG by bladder tumor cells. J Mol Endocrinol 1989;2:113–17.

[60] Bo M, Boime I. Identification of the transcriptionally active genes of the chorionic gonadotropin β gene *in vivo*. J Biol Chem 1992;267:3179–84.

[61] Rull K, Laan M. Expression of β-subunit of human chorionic gonadotropin genes during the normal and failed pregnancy. Hum Reprod 2005;20:3360–8.

[62] Giovangrandi Y, Parfait B, Asheuer M, Olivi M, Lidereau R, Vidaud M, et al. Analysis of the human CGB/LHB gene cluster in breast tumors by real-time quantitative RT-PCR assays. Cancer Lett 2001;168:93–100.

[63] Span PN, Manders P, Heuvel JJ, Thomas CM, Bosch RR, Beex LV, et al. Molecular beacon reverse transcription-PCR of human chorionic gonadotropin-beta-3, -5, and -8 mRNAs has prognostic value in breast cancer. Clin Chem 2003;49:1074–80.

[64] Boorstein WR, Vamvakopoulos NC, Fiddes JC. Human chorionic gonadotropin beta subunit is encoded by at least eight gene3s arranged in tandem and inversted pairs. Nature 1982;300:419–22.

[65] Burczynska BB, Kobrouly L, Butler SA, Naase M, Iles RK. Novel insights into the expression of CGB1 & 2 genes by epithelial cancer cell lines secreting ectopic free hCGβ. Anticancer Res 2014;34(5):2239–48.

[66] Braunstein GD, Vaitukaitis JL, Carbone PP, Ross GT. Ectopic production of human chorionic gonadotropin by neoplasms. Ann Intern Med 1973;78:39–45.

[67] Dosogne-Geurin M, Stolarczyk A, Borkowski A. Prospective study of the alpha and beta subunits of hCG in the blood of patients with various benign and malignant conditions. Eur J Cancer 1978;14:525–32.

[68] Braunstein GD. hCG expression in trophoblastic and nontrophoblastic tumors. In: Fishman WH, editor. Oncodevelopmental markers: biologic, diagnostic and monitoring aspects. New York, NY: Academic Press; 1983. p. 35.

[69] Rodenburg CJ, Nieuwenhuyzen-Kruseman AC, DeMaaker HA, Fieuren EJ, Van Oosterom AT. Immunohistochemical localization and chromatographic characterization of human chorionic gonadotropin in a bladder carcinoma. Arch Path Lab Med 1985;109:1046–8.

[70] Iles RK, Oliver RTD, Kitau M, Walker C, Chard T. *In vitro* secretion of human chorionic gonadotropin by bladder tumor cells. Br J Cancer 1987;55:623–6.

[71] Hagen C, Gilby ED, McNeilly AS, Olgaard K, Bondy PK, Rees LH. Comparison of circulating glycoprotein hormones and their subunits in patients with oat cell carcinoma of the lung and uraemic patients on chronic dialysis. Acta Endocrinol 1976;83:26–35.

[72] Kahn CR, Rosen SW, Weintraub BD, Fajans SS, Gorden P. Ectopic production of chorionic gonadotropin and its subunits by islet-cell tumors. A specific marker for malignancy. N Engl J Med 1977;297:565–9.

[73] Adejuwon CA, Segal SJ. Apparent chorionic gonadotropin immunoreactivity in human non-placental tissues: reduction by protease inhibitors or heat treatment. Clin Chim Acta 1984;137:245–53.
[74] Monteiro JC, Ferguson KM, McKinna JA, Greening WP, Neville AM. Ectopic production of human chorionic gonadotropin-like material by breast cancer. Cancer 1984;53:957–62.
[75] Cowley G, Smith JA, Ellison M, Gusterson B. Production of beta-human chorionic gonadotropin by human squamous carcinoma cell lines. Int J Cancer 1985;35:575–9.
[76] Heitz PU, von Herbay G, Kloppel G, Komminoth P, Kasper M, Hofler H, et al. The expression of subunits of human chorionic gonadotropin (hCG) by nontrophoblastic, nonendocrine, and endocrine tumors. Am J Clin Pathol 1987;88:467–72.
[77] Cole LA, Wang Y, Elliot M, Latif M, Chambers JT, Chambers SK, et al. Urinary human chorionic gonadotropin free β-subunit and core fragment: a new marker of gynecological cancers. Cancer Res 1988;48:1356–60.
[78] Yamaguchi A, Ishida T, Nishimura G, Kumaki T, Katoh M, Kosaka T, et al. Human chorionic gonadotropin in colorectal cancer and its relationship to prognosis. Br J Cancer 1989;60:382–4.
[79] Marcillac I, Troalen F, Bidart J-M, Ghillani P, Ribrag V, Escudier B, et al. Free human chorionic gonadotropin β subunit in gonadal and nongonadal neoplasms. Cancer Res 1992;52:3901–7.
[80] Marcillac I, Cottu P, Theodore C, Terrier-Lacombe MJ, Bellet D, Droz JP. Free hCG-β subunit as tumor marker in urothelial cancer. Lancet 1993;341:1354–5.
[81] Iles RK, Chard T. Molecular insights into structure and function of human chorionic gonadotropin. J Mol Endocrinol 1993;10:217–34.
[82] Vaitukaitis JL, Braunstein GD, Ross GT. A radioimmunoassay which specifically measures human chorionic gonadotropin in the presence of human luteinizing hormone. Am J Obstet Gynecol 1972;113:751–8.
[83] Iles RK, Delves PJ, Butler SA. Does hCG or hCGβ play a role in cancer cell biology? Mol Cell Endocrinol 2010;329(1–2):62–70.
[84] Gillott DJ, Iles RK, Chard T. The effects of β-human chorionic gonadotropin on the *in vitro* growth of bladder cancer cell lines. Br J Cancer 1996;73:323–6.
[85] Bagshawe KD, Springer CJ, Searle F, Antoniw P, Sharma SK, Melton RG, et al. A cytotoxic agent can be generated selectively at cancer sites. Br J Cancer 1988;58:700–3.
[86] Butler SA, Staite EM, Iles RK. Reduction of bladder cancer cell growth in response to hCGβ CTP37 vaccinated mouse serum. Oncol Res 2003;14:93–100.
[87] Devi GR, Oldenkamp JR, London CA, Iversen PL. Inhibition of human chorionic gonadotropin beta-subunit modulates the mitogenic effect of c-myc in human prostate cancer cells. Prostate 2002;53:200–10.
[88] Jankowska A, Gunderson SI, Andrusiewicz M, Burczynska B, Szczerba A, Jarmolowski A, et al. Reduction of human chorionic gonadotropin beta subunit expression by modified U1 snRNA caused apoptosis in cervical cancer cells. Mol Cancer 2008;7:26.
[89] Burczynska B, Booth MJ, Iles RK, Shah A, Shiled A, Butler SA. Stable knockdown of hCGβ mRNA expression in bladder cancer cells results in significant growth inhibition. Anticancer Res 2013;33:3611–14.
[90] Butler SA, Ikram MS, Mathieu S, Iles RK. The increase in bladder carcinoma cell population induced by the free β subunit of human chorionic gonadotropin is a result of an anti-apoptosis effect and not cell proliferation. Br J Cancer 2000;82:1553–6.

[91] Lapthorn AJ, Harris DC, Littlejohn A, Lustbader JW, Canfield RE, Machin KJ, et al. Crystal structure of human chorionic gonadotropin. Nature 1994;369:455–61.
[92] Wu H, Lustbader JW, Liu Y, Canfield RE, Hendrickson WA. Structure of human chorionic gonadotropin at 2.6 Å resolution from MAD analysis of the selenomethionyl protein. Structure 1994;2:545–8.
[93] Butler SA, Laidler P, Porter JR, Kicman AT, Chard T, Cowan DA, et al. The beta subunit of human chorionic gonadotropin exists as a homodimer. J Mol Endocrinol 1999;22:185–92.
[94] Butler SA, Iles RK. Monomeric human chorionic gonadotropin beta is responsible for autocrine growth effects. Tumor Biol 2004;25:18–23.
[95] Muller YA, Li B, Christinger HW, Wells JA, Cunningham BC, De Vos AM. Vascular endothelial growth factor: crystal structure and functional mapping of the kinase domain receptor binding site. Proc Natl Acad Sci USA 1997;94:7192–7.
[96] Herr F, Baal N, Reisinger K, Lorenz A, McKinnon T, Preissner KT, et al. HCG in the regulation of placental angiogenesis. Results of an *in vitro* study. Placenta 2007; S85–93.
[97] Reisinger K, Baal N, McKinnon T, Münstedt K, Zygmunt M. The gonadotropins: tissue-specific angiogenic factors? Mol Cell Endocrinol 2007;269:65–80.
[98] Zygmunt M, Herr F, Keller-Schoenwetter S, Kunzi-Rapp K, Münstedt K, Rao CV, et al. Characterization of human chorionic gonadotropin as a novel angiogenic factor. J Clin Endocrinol Metab 2002;87:5290–6.
[99] Demir R, Kayisli UA, Seval Y, Celik-Ozenci C, Korgun ET, Demir-Weusten AY, et al. Sequential expression of VEGF and its receptors in human placental villi during very early pregnancy: differences between placental vasculogenesis and angiogenesis. Placenta 2004;25:560–72.
[100] Arrieta O, Michel Ortega RM, Angeles-Sánchez J, Villarreal-Garza C, Avilés-Salas A, Chanona-Vilchis JG, et al. Serum human chorionic gonadotropin is associated with angiogenesis in germ cell testicular tumors. J Exp Clin Cancer Res 2009;28:120.
[101] Coombs LM, Pigott DA, Eydmann ME, Proctor AJ, Knowles MA. Reduced expression of TGFβ3 is associated with advanced disease in transitional cell carcinoma. Br J Cancer 1993;67:578–84.
[102] Eder IE, Stenzl A, Hobisch A, Cronauer MV, Bartsch G, Klocker H. Transforming growth factors-β1 and β2 in serum and urine from patients with bladder carcinoma. J Urol 1996;157:953–7.
[103] Eder IE, Stenzl A, Hobisch A, Cronauer MV, Bartsch G, Klocker H. Expression of transforming growth factors-β1, β2 and β3 in human bladder carcinomas. Br J Cancer 1997;75:1753–60.
[104] Fujii D. Transforming growth factor β gene maps to human chromosome 19 long arm and to mouse chromosome 7. Somat Cell Mol Genet 1986;12:281–8.
[105] Vitt UA, Hsu SY, Hsueh AJW. Evolution and classification of cystine knot-containing hormones and related extracellular signaling molecules. Mol Endocrinol 2001;15:681–94.
[106] Levitsky HI, Mongomery J, Ahmadzadeh M, Staveley-O'Carroll K, Guarnieri F, Longo DL, et al. Immunization with granulocyte-macrophage colony-stimulating factor-transduced, but not B7–1 transduced, lymphoma cells primes idiotype-specific T cells and generates potent systemic antitumor immunity. J Immunol 1996;156:3858–65.
[107] Allison JP, Hurwitz AA, Leach DR. Manipulation of costimulatory signals to enhance antitumor T-cell responses. Curr Opin Immunol 1995;7:682–6.
[108] Boon T, Cerottini J-C, Van den Eynde B, van der Bruggen P, Van Pel A. Tumour antigens recognized by T lymphocytes. Annu Rev Immunol 1994;12:337–65.

[109] Maeurer MJ, Storjus WJ, Kirkwood JM, Lotze MT. New treatment options for patients with melanoma: review of melanoma-derived epitope-based peptide vaccines. Melanoma Res 1996;6:11–24.
[110] Strauss HJ. Mutant ras proteins and peptides: bad news for tumors. J Natl Cancer Inst 1995;87:1820–2.
[111] Fisk B, Chesak B, Pollack MS, Wharton JT, Ioannides CG. Oligopeptide induction of a cytotoxic T lymphocyte response to HER-2/Neu prot-oncogene *in vitro*. Cell Immunol 1994;157:415–27.
[112] Humphrey PA, Wong AJ, Vogelstein B, Zalutsky MR, Fuller GN, Archer G, et al. Antisynthetic peptide antibody reacting at the fusion junction of deletion-mutant epidermal growth factor receptors in human glioblastoma. Proc Natl Acad Sci USA 1990;87:4207–11.
[113] Burchell J, Taylor-Papdimitriou J, Boshell M, Gendler S, Duhig T. A short sequence within the amino acid tandem repeat of a cancer-associated mucin, contains immunodominant epitopes. Int J Cancer 1989;44:691–6.
[114] Renard V, Leach DR. Perspectives on the development of a therapeutic HER-2 cancer vaccine. Vaccine 2007;25S:B17–23.
[115] Delves PJ, Iles RK, Roitt IM, Lund T. Designing a new generation of anti-hCG vaccines for cancer therapy. Mol Cell Endocrinol 2007;260–262:276–81.
[116] Brennan FR, Jones TD, Hamilton WDO. Cowpea mosaic virus as a vaccine carrier of heterologous antigens. Mol Biotechnol 2001;17:15–26.
[117] Talwar GP, Sharma NC, Dubey SK, Salahuddin M, Das C, Ramakrishnan S, et al. Isoimmunization against human chorionic gonadotropin with conjugates of processed 3-subunit of the hormone and tetanus toxoid. Proc Nat Acad Sci USA 1976;73:218–22.
[118] Griffin PD. Immunisation against hCG. Hum Reprod 1994;9:267–72.
[119] Stevens VC. Progress in the development of human chorionic gonadotropin antifertility vaccine. Am J Reprod Immunol 1996;35:148–55.
[120] Stevens VC, Cinader B, Powell JE, Lee AC, Koh SW. Preparation and formulation of a human chrionic gonadotropin antifertility vaccine: selection of adjuvant and vehicle. Am J Reprod Immunol 1981;1:315–21.
[121] Stevens VC, Chou WS, Powell JE, Lee AC, Smoot J. The identification of peptide sequences of human chorionic gonadotropin containing a conformational epitope. Immunol Lett 1986;12:11–18.
[122] Triozzi PL, Stevens VC. Human chorionic gonadotropin as a target for cancer vaccines. Oncol Rep 1999;6:7–17.
[123] Talwar GP. Fertility regulating and immunotherapeutic vaccines reaching trials stage. Hum Reprod Update 1997;3(4):301–10.
[124] Dirnhofer S, Klieber R, De Leeuw R, Bidart JM, Merz WE, Wick G, et al. Functional and immunological relevance of the COOH-terminal extension of human chorionic gonadotropin beta: implications for the WHO birth control vaccine. FASEB J 1993;7:1381–5.
[125] Berger P, Sturgeon C, Bidart JM, Paus E, Gerth R, Niang M. . The ISOBM TD-7 workshop on hCG and related molecules. Towards user-oriented standardization of pregnancy and tumor diagnosis: assignment of epitopes to the three-dimensional structure of diagnostically and commercially relevant monoclonal antibodies directed against human chorionic gonadotropin and derivatives. Tumour Biol 2002;23:1–38.
[126] Talwar GP. Vaccines and passive immunological approaches for the control of fertility and hormone-dependent cancers. Immunol Rev 1999;171:173–92.

[127] Porakishvili N, Chiesa MD, Chikadze N, Martensen P, Justesen J, Lund T, et al. Elimination of luteinizing hormone cross-reactive epitopes from human chorionic gonadotropin. Vaccine 2002;20:2053–9.
[128] Stevens VC. Future perspectives for vaccine development. Scand J Immunol Suppl 1992;11:137–43.
[129] Triozzi PL, Martin EW, Gouchnour D, Aldritch W. Phase 1b trial of a synthetic beta human chorionic gonadotropin vaccine in patients with metastatic cancer. Ann NY Acad Sci 1993;630:358–9.
[130] Triozzi PL. Immunological effects of a synthetic beta-hCG vaccine. Int J Oncol 1994;5:1447–53.
[131] Moulton HM, Yoshihara PH, Mason DH, Iversen PL, Triozzi PL. Active specific immunotherapy with a beta-human chorionic gonadotropin peptide vaccine in patients with metastatic colorectal cancer: antibody response is associated with improved survival. Clin Cancer Res 2002;8:2044–51.
[132] Morse MA, Chapman R, Powderly J, Blackwell K, Keler T, Green J, et al. Phase I study utilizing a novel antigen-presenting cell-targeted vaccine with Toll-like receptor stimulation to induce immunity to self-antigens in cancer patients. Clin Cancer Res 2011;17(14):4844–53.
[133] Baselga J. Clinical trials of Herceptin® (trastuzumab). Eur J Cancer 2001;37:S18–24.

hCG and breast cancer conundrum 24

Laurence A. Cole
USA hCG Reference Service, Angel Fire, NM, USA

A major puzzle burdens science today. One group of cancer scientists has shown that administration of hCG variants [1] or hCG production during pregnancy [2] avoids or inhibits breast cancer. This was first shown by Jose and Irma Russo [1,2]. This was confirmed by five other independent groups [3–7]. In contrast, seven separate independent groups throughout the world have shown, confirmed, and double-confirmed that three hCG variants, hCG free β-subunit, hyperglycosylated hCG free β-subunit, and hyperglycosylated hCG, are the principle promoters promoting most cancers and molecules that enhance cancer risk (see Chapters 21–23).

We have recently shown that 15 of 15 separate malignancies, including breast cancer, are primarily promoted to grow and invade by hyperglycosylated hCG, hyperglycosylated hCG free β-subunit, or hCG free β-subunit driven mechanisms [8] (see Chapter 23). Many others, at least seven independent groups, have reported similar hCG-driven malignancy findings with hCG-related molecules as the promoters of malignancy [9–15] (see Chapters 22 and 23).

Each of these two large groups of scientists is satisfied with their findings and solidly agrees with them. They have trouble believing the other results. I am part of the second group and have had multiple papers rejected because of the confusion of this puzzle. We ask, can both separate groups be correct? Can hCG and its variants both prevent and block breast cancer and at the same time cause and promote cancers?

A possible explanation is proposed here for this conflicting promotion/blocking conundrum. hCG is produced by placental cells and acts as a hormone binding a luteinizing hormone (LH)/hCG receptor. hCG functions at this receptor by promoting intracellular cyclic AMP (cAMP), which promotes hCG function [16]. As shown, hyperglycosylated hCG, hyperglycosylated hCG free β-subunit, and hCG free β-subunit are actually autocrines that bind a TGFβ type II receptor on cytotrophoblast and cancer cells [10,17,18]. TGFβ functions at this receptor by activating SMADs (a family of intracellular TGFβ signaling proteins [17]) through promotion of cAMP [19]. These SMADs promote cell growth, apoptosis, and invasion through promoting collagenases and metalloproteinases [20] or malignancy. It is reasoned that TGFβ receptor antagonism as occurs with hyperglycosylated hCG, hyperglycosylated hCG free β-subunit, and hCG free β-subunit [10,18] blocks cAMP and blocks SMADs. hCG through the LH/hCG receptor promotes cAMP, or each does the opposite.

Few cells express both the LH/hCG receptor and the TGFβ receptor, which perform opposite cAMP-driven functions. Exceptions are the cytotrophoblast cells,

Human Chorionic Gonadotropin (hCG). DOI: http://dx.doi.org/10.1016/B978-0-12-800749-5.00024-9

testicular germ cells, choriocarcinoma, and some breast cancer cells. It is postulated that hCG may act on these cells, enhancing intracellular cAMP levels and surmounting the antagonism and cAMP suppression caused by the hCG variants at the TGFβ receptor. It is thought that through this cAMP mechanism, hCG can surmount an hCG variant – induced cancer induction and cancer promotion process, reducing cAMP, as proposed.

The problem is that hCG is widely used today as a dietary supplement, following the questionable 1954 *Lancet* claim of Simeons [21]. The big question today is whether hCG variants are dangerous and promote cancer or are safe and block cancer. Which side of the conundrum is most valid? Currently, the many hCG-promoting sites on the Internet and newspaper articles support hCG dietary use, claiming that hCG has been shown to inhibit breast cancer and avoiding the mass of publications showing that forms of hCG are cancer promoters. Is this correct? Are people using hCG as a dietary aid doing the right thing?

Most hCG preparations sold for dietary purposes are hCG nose repellents and hCG tablets that are produced from pregnancy urine. These have questionable hCG hormonal value and high contents of hyperglycosylated hCG and hCG free β-subunit. Many other products sold are extremely crude pregnancy urine extract that also have a high content of hyperglycosylated hCG and hCG free β-subunit (Table 24.1). These hCG variants clearly need to be avoided.

We think that there is only one answer. There is one product manufactured by Serono Inc. comprising Chinese hamster ovary cell recombinant hCG (Table 24.1) [22]. It is mostly used by physicians for promoting ovulation. It is pure hCG with 0% hyperglycosylated hCG. Yes, it can very slowly dissociate releasing hCG free β-subunit, but it is clearly the safest form of hCG. It is unfortunately the

Table 24.1 **Commercial preparation of hCG and its variants. CHO cells are Chinese hamster ovary cells.**

Preparation	Origin	Hyper-glycosylated hCG	hCG free β-subunit	Biological activity calibrated with WHO 3rd I.S.
Serono Ovidrel	CHO cell Recombinant	<0.1%	<0.1%	11,900 IU/mg
Serono Profasi	Pregnancy Urine	6.0%	1.6%	10,000 IU/mg
Scripps >99% hCG	Pregnancy Urine	2.2%	0.8%	11,000 IU/mg
Scripps >80% hCG	Pregnancy Urine	6.8%	5%	9,000 IU/mg
Ferring Choragon	Pregnancy Urine	4%	5%	5,000 IU/mg
Organon Pregnyl	Pregnancy Urine	7%	12%	3,000 IU/mg
Scripps >15% hCG	Pregnancy Urine	13%	13%	2,000 IU/mg
Sigma C5297	Pregnancy Urine	12%	14%	3,000 IU/mg

most expensive hCG preparation. Those determined to use hCG for dietary purposes, however, should invest in this expensive form of pure hCG, Serono Ovidrel; otherwise, they should avoid using those seemingly highly carcinogenic urinary hCG mixtures.

The same applies to athletes who use hCG for its anabolic actions, and to doctors who use hCG to initiate *in vitro* fertilization and promote ovulation. All commercial and dietary urinary hCG preparations should be avoided because of their content of these cancer promoters, hyperglycosylated hCG and hCG free β-subunit.

References

[1] Russo IH, Russo J. Use of human chorionic gonadotropin in the prevention of breast cancer. Women's Health 2008;4:1–5.

[2] Russo IH, Russo J. Pregnancy induced changes in breast cancer risk. J Mamm Gland Biol Neopl 2011;16:221–33.

[3] Lopez D, Sekharam M, Coppola D, Carter WB. Purified human chorionic gonadotropin induces apoptosis in breast cancer. Mol Cancer Ther 2008;7:2837–44.

[4] Kocdor H, Kocdor MA, Russo J, Snider KE, Vanegas JE, Russo IH, et al. Human chorionic gonadotropin (hCG) prevents the transformed phenotypes induced by 17 beta-estradiol in human breast epithelial cells. Cell Biol Intl 2009;33:1135–43.

[5] Tuma R. Mimicking pregnancy to reduce breast cancer risk. J Natl Cancer Inst 2010;102:517–18.

[6] Toniolo P, Grankvist K, Wulff M, Chen T, Johansson R, Schock H, et al. Human chorionic gonadotropin in pregnancy and maternal risk of breast cancer. Cancer Res 2010;70:6779–86.

[7] Boukaidi SA, Cooley A, Hardy A, Matthews L, Zelivianski S, Jeruss JS. Impact of infertility regimens on breast cancer cells: follicle-stimulating hormone and luteinizing hormone lack a direct effect on breast cell proliferation *in vitro*. Fertil Steril 2012; 97:440–4.

[8] Cole LA, Iles RK, Butler SA. Hyperglycosylated hCG and its β-subunit: pervasive promoters of invasion and growth in cancer. Mol Cell Endocrinol 2013.

[9] Bellet D, Lazar V, Bleche I, Paradis V, Giovangrandi Y, Paterliru P. Malignant transformation of nontrophoblastic cells in association with the expression of chorionic gonadotropin β genes normally transcribed in trophoblastic cells. Cancer Res 1997; 57:516–23.

[10] Butler SA, Ikram MS, Mathieu S, Iles RK. The increase in bladder carcinoma cell population induced by the free beta subunit of hCG is a result of an anti-apoptosis effect and not cell proliferation. Br J Cancer 2000;82:1553–6.

[11] Cole LA, Dai D, Leslie KK, Butler SA, Kohorn EI. Gestational trophoblastic diseases. 1. Pathophysiology of hyperglycosylated hCG-regulated neoplasia. Gynecol Oncol 2006;102:144–9.

[12] Carter WB, Sekharem M, Coppola D. Purified hCG induces apoptosis in breast cancer. Breast Cancer Res Treat 2006;100:S243–4.

[13] Iles RK. Ectopic hCGβ expression by epithelial cancer: malignant behavior metastasis and inhibition of tumor cell apoptosis. Mol Cell Endocrinol 2007;260:264–70.

[14] Li D, Wen X, Ghali L, Al-Shalabi FM, Docherty SM, Purkis P, et al. hCG beta expression by cervical squamous carcinoma-*in vivo* histological association with tumour invasion and apoptosis. Histopathology 2008;53:147–55.

[15] Jankowska A, Andrusiewicz M, Grabowski J, Nowak-Markwitz E, Warchol JB. Coexpression of human chorionic gonadotropin beta subunit and its receptor in nontrophoblastic gynecological cancer. Int J Gynecol Cancer 2008;18:1102–7.

[16] Strott CA, Yoshimi T, Ross GT, Lipsett MB. Ovarian physiology: relationship between plasma LH and steroidogenesis by the follicle and corpus luteum; effect of HCG. J Clin Endocrinol Metab 1969;29:1157–67.

[17] Berndt S, Blacher S, Munuat C, Detilleux J, Evain-Brion D, Noel A, et al. Hyperglycosylated human chorionic gonadotropin stimulates angiogenesis through TGF-β receptor activation. FASEB J 2013; Abstract 12-213686.

[18] Cole LA, Butler SA. Hyperglycosylated hCG, hCGß and hyperglycosylated hCGβ: interchangeable cancer promoters. Mol Cell Endocrinol 2012;349:232–8.

[19] Schiller M, Dennler S, Anderegg U, Kokot A, Simon JC, Luger TA, et al. Increased cAMP levels modulate transforming growth factor-beta/Smad-induced expression of extracellular matrix components and other key fibroblast effector functions. J Biol Chem 2010;285:409–21.

[20] Murphy G, Reynolds JJ, Whitham SE, Docherty AJ, Angel P, Heath JK. Transforming growth factor beta modulates the expression of collagenase and metalloproteinase inhibitor. Euro Mol Biol Org J 1987;6:1899–904.

[21] Simeons ATW. The action of chorionic gonadotropin in the obese. Lancet 1954; 2:946–7.

[22] Cole LA, Sutton JM, Higgins TN, Cembrowski GS. Between-method variation in hCG test results. Clin Chem 2004;50:874–82.

Part I

Clinical Applications

Use of hCG in reproductive dysfunction

25

Francis W. Byrn
Obstetrics and Gynecology, University of New Mexico, Albuquerque, NM, USA

Human chorionic gonadotropin (hCG) is a heterodimeric glycoprotein. The primary structures of the α-subunits and β-subunits were determined in 1972. The hormone is approximately 30% carbohydrate [1]. Its similarity to luteinizing hormone (LH) and its capacity to bind to its receptors have fostered interest in the clinical utility of hCG in human reproductive medicine. Studies were performed that clarified the disappearance rates of hCG after delivery of a pregnancy [2], after intravenous (IV) and intramuscular (IM) injections to healthy males and females [3], and after subcutaneous or IM injections in healthy males [4], and have calculated metabolic and renal clearance rates with continuous-infusion and single-injection techniques [5]. These studies similarly confirmed that hCG disappears in a rapid phase and a slower phase, taking approximately 6 h and 36 h [5], respectively. Using an injection of 10,000 IU, Midgeley and Jaffe's early data regarding hCG disappearance after delivery also estimated a fast half-life of 8.9 h and a longer, trailing half-life of 37.2 h. They discussed the fact that the rate of disappearance at a given time after delivery did not appear to be related to the concentration of hCG, nor to the duration of pregnancy, nor to the route of delivery [2]. With IV rates, Rizkallah demonstrated an initial fast half-life of 3.7–5.9 h and a slower half-life of 22–26.7 h [3].

Studies examining the pharmacokinetics of hCG when administered to humans have been published [3–5]. In males, the peak hCG concentrations are reached 6–8 h after IM injection [4]. When administered intramuscularly to three subjects, the half-life of the tail was longer (27.5, 31.2, and 31.6 h) compared with clearance rates after IV injection, which averaged approximately 26 h. There were no appreciable differences in the disappearance curves between males and females [3]. In this study, the researchers also made other observations, including that the metabolic clearance rate of hCG was more prolonged than that of human LH. The maximal blood concentrations achieved with IV versus IM injections of 10,000 IU of hCG were approximately 5.8 and 2.2 IU/ml, respectively. The peak of hCG was delayed 6 h and somewhat lower compared with IV administration. Using the second international reference preparation – human menopausal gonadotropins (IRP-hMG) standards, the authors noted that LH in human ovulation reaches 40–200 mIU during a 24-h period. They commented on the fact that the maximum concentration of hCG after an IM injection in one subject was equivalent to 4000 mIU of the second IRP-hMG standard, noting that this was an approximately 20-fold increase over the maximum values of the LH peak (in terms of standard reference) during

Human Chorionic Gonadotropin (hCG). DOI: http://dx.doi.org/10.1016/B978-0-12-800749-5.00025-0

spontaneous ovulation. The slower disappearance rate of hCG relative to LH was characterized as a disadvantage in that slower metabolism could be an aggravating factor in multiple ovulation and an independent factor for ovarian hyperstimulation syndrome (OHSS) [3].

A study by Fischer et al. investigated the time interval of ovulation in women undergoing ovulation management with clomiphene after either IM injection of 10,000 or 500 IU provided intravascularly on an alternating protocol for two cycles [6]. Ultrasound assessment and serum samples were obtained every 2 h beginning 30 h after IV administration and 31 h after IM administration. Given the apparent variability of ovulation intervals demonstrated by IM hCG administration, the authors hypothesized that an ovulation-inducing threshold of hCG provided by IV administration would result in less variance in the endpoint of ovulation. Twenty-five women provided data by completing both induction protocols. The mean time to ovulation with IM injection of hCG was 40.4 h, and was 38.3 h with IV administration of hCG. The 500-IU dosage utilized in this study had previously been noted to exceed the peak LH concentration by three-fold [6]. Ultimately, 48 of 55 (87.3%) studied cycles provided physical or endocrinologic evidence of ovulation.

Patterns of serum hCG levels were monitored in four random patients. In comparison with previously mentioned pharmacokinetic studies [1,5], the study by Fischer et al. [6] obtained clinical trial data more practically consistent with actual patient management strategies. Thirty-six hours after IM injection, hCG concentrations had increased to a mean level of 311 IU/l (range 201–359 IU/l) from pre-administration hCG concentrations of <2 IU/l. Twenty-one hours after IV administration of 500 IU of hCG, serum levels were 45 IU/l (range of 36–55 IU/l). The patterns and rates of change of estradiol and progesterone levels from the time of hCG administration until documented ovulation were reported not to be different as a consequence of the routes of administration of the two different doses of hCG.

These studies used urinary-derived chorionic gonadotropin to augment or substitute for LH. Early data suggest that hCG has a longer half-life than LH, and that IM administration of 5000 IU (and certainly 10,000 IU) significantly exceeds the maximal concentrations of LH during the peak of the surge during spontaneous ovulation [3]. Although not specifically studied, the authors discussed possible implications for multiple ovulations and/or sustaining supraphysiologic estrogen and progesterone secretion from ovaries stimulated with hMG.

25.1 Historical overview and perspective

The indications for hCG administration within clinical reproductive medicine most often involve ovulation management. One must keep in mind that this is a very broad category. The most common clinical indication for an hCG injection is to secure ovulation (as a completing event of natural or induced folliculogenesis). For example, this could be an eventual recommendation for an otherwise anovulatory woman who wishes to conceive. A second hCG use, however, is to prospectively promote the release of one or more oocytes at a time to coincide with an ancillary procedure such as timed intercourse or artificial insemination. In this circumstance, anovulation might not be the primary problem.

Although definitions cannot always be uniformly applied in discussions of medical protocols that involve an injection of hCG, it will be helpful to think of protocols as a continuum in three stages. Ovulation induction (OI) is focused on helping a woman ovulate if, without assistance, she might ovulate infrequently or not at all. Controlled ovarian stimulation (COS) is a more comprehensive regimen of either oral or parenteral medications that promote ovulation. However, these medical regimens might result in the development of more than one ovarian follicle. They are generally more dependent on the use of hCG and, because of this, necessitate more clinical surveillance to avoid overstimulation of the ovary. The most aggressive protocols, termed controlled ovarian hyperstimulation (COH), are variations of COS, but with a more focused intent of multiple follicle recruitment. Monitoring of COH protocols is quite necessary. These three protocols are the foundation for assisted reproduction protocols such as *in vitro* fertilization (IVF).

The progress of clinical OI over the past 45 years has revolved around protocols that utilize oral agents [7–12] and parenteral agents [11,13,14]. When successful, the former (i.e., clomiphene citrate) do not necessarily depend on the adjunctive IM injection of hCG; however, reviews of standard treatment plans with clomiphene citrate may include a recommendation for the use of hCG [15] in select instances.

Specific clinical indications for hCG use in managing infertility can be gleaned from previous reviews [15–20]. The oldest and most consistently discussed indications are OI (including both clomiphene citrate-managed and gonadotropin-managed patients) [7,15,19] as an adjunct for timing artificial insemination procedures, whether using the husband's or donor sperm samples [21–25]. Less commonly, secondary indications for use of hCG-component protocols have been for women with unexplained infertility [26] or who have not successfully conceived after surgery or focused medical management of pelvic endometriosis [27]. These secondary categories might involve women who do not necessarily exhibit ovulatory dysfunction and a partner whose semen samples might be entirely normal when empirical data suggest that they might otherwise benefit.

It should be mentioned that hCG is also used for luteal support in OI protocols, particularly as a component of cycles controlled with hMG [28–30]. Soules et al. noted that it is probably important to maintain threshold levels of plasma LH in the luteal phase of a cycle to adequately support luteal function. Physiologically, this is associated with maintaining a concentration of progesterone in the postovulation or postembryo transfer interval of clinical protocols.

hCG derived from urinary sources has been variably used for these indications for more than 30 years [13,31,32]. Publications relating to its earliest use were not reviewed, but it is evident that by the mid 1970s, hCG was being administered in doses of 10,000 IU, 5000 IU, or less [14]. When use of clomiphene citrate is not successful, it is recommended that the woman should consider use of gonadotropin therapy. A general definition of clomiphene resistance implies a failure to ovulate with conventional multiday clomiphene citrate doses (up to 150 or 200 mg/day), or a failure to conceive with an ovulatory dose of clomiphene citrate during the three to six previous cycles [33]. Thus, clomiphene resistance is a common diagnosis that leads to COS.

This transition typically requires a timed hCG injection as an integral component of the induction protocol utilizing menopausal gonadotropins [34–37]. The conventional assumption is that hCG acts directly and independently as the surrogate LH signal; however, studies examining the extent to which an endogenous LH surge is suppressed by gonadotropin induction suggest that it is not entirely suppressed. Given that the LH surge dynamic remains but is otherwise truncated or altered, other studies have examined possible clinical consequences that might result when an ancillary hCG signal is provided in addition to an endogenous LH signal [38].

To summarize, hCG administration is a common adjunct in the contemporary management of human infertility. It can be included in differing protocols involving oral agents; if this first-line therapy is unsuccessful, then patients can proceed to gonadotropin therapy. Protocols utilizing exogenous follicle-stimulating hormone (FSH) customarily require hCG administration.

As hCG injections became more specifically integrated into treatment algorithms, increasing numbers of clinical studies have been published. Concomitantly, as steroid and glycoprotein immunoassays became more practically available, and as ultrasound technology improved, studies ascertaining the indications for and the utility and practical effectiveness of hCG administration also have become more informative.

For example, the practical endpoint of determining if ovulation has occurred has been [39] and remains a considerable challenge, whether it is the result of normal hormonal dynamics or the goal after hCG administration [6,26,40]. Historically, clomiphene OI has been monitored with clinical assessment of hormones [8,19], cervical mucus [41–43], basal body temperature graphing [8,44,45], timed assessment of serum progesterone [15], and eventually transabdominal [39,46,47] and transvaginal ultrasound [28,48]. OI with gonadotropin protocols have been monitored with cervical mucus [49,50], vaginal cytology, cervical mucus changes [49], estrogen levels, and ultrasound [51].

Progressing from imprecise clinical markers like cervical mucus changes [31,49,50] to progressively more accurate methods like timed [39,46] or serial high-resolution transvaginal ultrasound series [52] to determine ovulation has been transformative. As more sophisticated modalities for monitoring OI were developed and incorporated into clinical use, positive and negative predictive accuracy for the event of ovulation improved. As an evolving process, eventually studies were published that could focus on clinical endpoints such as pregnancy rates consequent to variations in stimulation protocols rather than efficacy studies of OI [27,29,53,54].

The following section is a brief review of the use of hCG in the clinical management of reproductive dysfunction. It devotes more attention to the most common indications of OI and COS and use in artificial insemination protocols.

25.2 Considerations of hCG administration within clinical protocols

25.2.1 OI—oral agents

Ovulatory dysfunction is the first or second most common problem of clinical infertility [55]. A pivotal report was published in 1961 noting effective ovulation with the chemical agent clomiphene citrate, called MRL/41 [13]. This agent, first synthesized in 1956 and approved for use in the United States in 1967 [32], was the basis of more than 2000 publications by 1976. Consensus algorithms using this agent have been published and periodically revised [11,12,15,19,20]. By 1976, the general side effects, risks, and benefits of this agent had been documented and discussed. They included low pregnancy rates, compromise of cervical mucus, follicular luteinization (pseudoovulation), elevated miscarriage rates, and a multiple birth rate of 6–12%. Clomiphene citrate remains the most common oral medication utilized in OI [7,8,12,15,19,36,56]. Collectively, this mixed estrogen agonist/antagonist has been used in various protocols, over a varying number of days [49], and at sequentially higher dosages in nonresponsive patients. More recently, other oral agents, including insulin sensitizing agents [57–59] and aromatase inhibitors [32,60,61], have been used, either independently or in combination with clomiphene citrate. This chapter does not focus specifically on these agents.

By 2003, a Practice Committee Report from the American Society for Reproductive Medicine listed anovulation, luteal phase deficiency, and unexplained infertility as indications for management with clomiphene alone or in combination with other treatment regimens. Because of the increased cost of monitoring, hCG use was cautionary, with the consideration that it be limited to women who require intrauterine insemination (IUI) or in whom a mid-cycle LH surge could not be reliably detected [56]. The most recent publication providing consensus recommendations for OI still lists clomiphene citrate as the initial medication [36].

25.2.2 OI—parenteral agents with hCG

The history of nonoral, parenteral agents for ovulation management was initiated in 1959, when hMG was first used for OI in Europe. Four years later, gonadotropins were approved for clinical trials in the United States. hMG was approved for sale in the United States in 1975 [32]. One of the common indications for second-line OI therapy cited in earlier publications was nonresponse or unacceptable side effects of clomiphene citrate [7,20]. hMG were used when patients failed to respond to OI with clomiphene citrate, which resulted in standard protocols covering conventional OI [33,34,37].

Although urinary-derived menotropins had been available since the 1960s, they contained variable activities of FSH, LH, and quantities of potentially allergenic urinary proteins [33]. In the 1980s, more uniformly manufactured products, such as Pergonal®, Humegon®, Urofollitropin®, and Metrodin®, became available. Pergonal and Humegon contained fixed activities of FSH and LH; Urofollitropin and Metrodin contained standard dosages of FSH, but with reduced or essentially nil LH activity [62]. They still had a very small amount of urinary proteins. In another 10 years, recombinant FSH (rFSH) products were developed, providing increased purity, absence of urinary proteins, and production consistency. Two preparations of rFSH were developed in the 1990s; Gonal F® (follitropin alpha) became available in 1995 and Puregon® (follitropin beta) became available in 1997. rFSH was first approved for use in the United States the same year. rFSH (1988) was purer, contained no LH or co-purified proteins, had high specific bioactivity, was available independent of urine collection, and boasted absolute source control and batch-to-batch consistency [33,63].

The era of assisted reproduction began with the first IVF births in England (1978) and in the United States (1981) [32]. In retrospect, these conceptions were more exceptional than clinically routine. Subsequently, all factors contributing to success with IVF were actively scrutinized. This included adapting the extant induction protocols that relied on clomiphene or hMG. The need for reliable controlled stimulation protocols for assisted reproductive technology (ART) was a strong impetus for the development of the increasingly pure FSH products mentioned previously. Rather than limited follicle recruitment, the new paradigm was to develop efficient controlled stimulation protocols that would result in consistently high numbers of periovulatory follicles for assisted reproduction [64,65].

As those same FSH preparations supplanted hMG, they were, out of necessity, incorporated into conventional OI protocols. This resulted in another set of clinical concerns for patients considering more conventional stimulation protocols, because of the increased risks of high-order multiple pregnancies and OHSS [66–72] that were associated with the more aggressive protocols being developed.

Some of the information that applies to the timing of hCG injection was developed in the late 1980s as clinical efficacy protocols for controlled stimulation for *IVF*. These protocols were being clarified and understood [73–75].

25.2.3 OI protocols and FSH preparations

Publications appeared that examined and compared the clinical efficacy of the many FSH-containing preparations that were then available (including hMG, urinary FSH [uFSH] products, and rFSH products). In 1984, one publication provided the first study of COS with IUI as an alternative to IVF [76]. The clinical reproductive conditions addressed with the use of these newer formulations were historically the same: polycystic ovary syndrome (PCOS), clomiphene-resistant OI, adjuncts for artificial insemination, and controlled stimulation or hyperstimulation for unexplained infertility or IVF programs. Entirely new models of OI, however, had to be re-examined in light of the new products.

The reader is reminded that the basic formula of controlled stimulation—FSH administration culminating in timed hCG injection—remained the prototypical sequence for clinical studies assessing novel formulations for managing clinical infertility; however, older clinical concerns remained and new concerns were added.

For example, did FSH-only products work as well as products containing both FSH and LH? Did rFSH work differently than urinary-sourced pure FSH? Two representative studies can be mentioned. The availability of pure FSH formulations was expected to benefit the management of patients with PCOS, who often had abnormally high levels of LH. The lack of LH activity had no negative influence on stimulation of PCOS patients [77,78]. Also, a Cochrane Review in 2000 noted that urinary-derived FSH products did not seem to improve pregnancy rates when compared with less expensive hMG in PCOS patients. The FSH products did, however, reduce OHSS [79].

Another review examined the efficacy and safety of rFSH versus purified uFSH in patients with PCOS, as well as outcome assessment of either of these preparations compared with more conventional hMG products [33]. Six randomized, controlled trials involving clomiphene-resistant anovulatory women were sufficient for review. The primary outcome of interest was a live birth or ongoing pregnancy per woman. The secondary outcomes included ovulation rate per cycle, ongoing pregnancy rate, live birth, miscarriage rate, incidence of OHSS, incidence of multiple pregnancy, total gonadotropin dose, total duration of stimulation, single-follicle development, and cancelation rate per woman. Chronic low-dose regimens appear to be less likely to cause OHSS when compared with conventional regimens using either uFSH or rFSH. This meta-analysis concluded, "in summary, there is as yet insufficient evidence to conclude that rFSH is more effective than uFSH for OI in women with clomiphene citrate-resistant PCOS."

What must be acknowledged is that hCG was an integral aspect of almost all of the stimulation protocols then being developed that utilized FSH activity. There was no consensus about what dosage of hCG to use, when to administer the medication, or whether more than one injection of hCG was necessary. A sampling of protocol details from the references reviewed for this chapter reinforces this impression. Examples derived from both oral and parenteral protocols include the following synopses.

In a relatively early publication on induction of ovulation with menotropins, it was mentioned that up to 25,000 IU of hCG was administered over a 3-day to 4-day interval to induce ovulation after assuming sufficient stimulation by serial determinations of total urinary estrogen [80]. One investigator specifically mentioned that 6000–10,000 IU of hCG could be administered at mid-cycle to augment the LH surge for luteinized unruptured follicle (LUF) syndrome [19].

Another study examined ovulation response rates in a sample of former clomiphene nonresponders (at 100 mg) when they were stratified to receive either a recommended daily micro-dose of 200 IU of hCG administered intramuscularly after the largest (ultrasound-monitored) follicle reached 12 mm in women who repeated a 100-mg dose of clomiphene. Their responses were compared with ovulation rates in a control group of women who also had not responded to 100 mg of clomiphene, and who repeated a trial at an increased dosage of 150 mg. The ovulation rates were 57% for the Clomid/hCG group and 18% for the Clomid-only group [81]. Individualized hMG doses from 75–225 IU/day provided stimulation, and hCG was administered at 5000–10,000 IU/day for 2 or 3 days, based on cervical mucus changes [50].

In a clinical trial of rFSH compared with pure uFSH, 10,000 IU of hCG was administered when one follicle of 218 mm or two to three follicles of 215 mm were seen [82]. Yarali et al. administered 10,000 IU of hCG when at least one stimulated follicle at least 17 mm in diameter was noted [83]. Using an induction protocol that began with 150 IU FSH daily but that was individually adjusted after ultrasound noting ovarian ultrasound characteristics on cycle day 8, Guzick et al. [25] administered 10,000 IU of hCG when two or more follicles were 218 mm and estradiol levels varied from 500 to 3000 pg/ml. Similarly, Hedon et al. administered 5000 IU of hCG when a follicle of at least 16 mm was seen [84]. Investigating a rFSH preparation compared with an FSH/LH preparation in a step-up induction regimen, Balasch et al. [85] administered 10,000 IU of hCG when a follicle >17 mm developed.

In a study by Fisch et al. [26], patients in two of four randomization arms received either a 100-mg clomiphene citrate cycle on days 5–9 and 5000 IU of hCG administered intramuscularly on cycle days 19, 22, 25, and 28 or placebo tablets on days 5–9 but hCG injections on the same schedule.

One report from 1988 noted that among 52 pregnancies conceived during a specific hMG/hCG protocol, there was a 27% incidence of multiple gestations when the pregnancy continued beyond the first trimester. The majority of these multiple pregnancies (9 of 14 or 62%) were twins. The indications for treatment in that study were male factor infertility (sperm density <20 million per milliliter and/or motility <40%), abnormalities in cervical mucus or a poor postcoital test, or management of patients receiving donor insemination [86]. When assessed, the data only suggested an intrauterine inseminant density of more than 20 million sperm as correlating with increased multiple pregnancy risk. Parameters that did not significantly correlate included total hMG dose, type or total hCG dose, day of hCG administration, or peak estradiol level. It is interesting to note that, in this study, the timing of hCG administration, usually 10,000 IU, was based on serial estradiol assessment only. Ultrasound information was not performed per the described protocol. The hCG dose was administered 24 h after the estradiol level attained 400 pg/ml but was <1000 pg/ml.

Continuing with other examples of hCG inclusion in protocols used in ovulation management for PCOS and unexplained infertility, Tadakoro administered 3000, 6000, or 10,000 IU of hCG and provided luteal support with 1500 IU of hCG on days +3, +6, and +9, depending on hormone surveillance [37].

A variation of mid-cycle stimulation was investigated. The authors utilized hMG at 150 IU/day, monitored ovarian response with transvaginal ultrasound until at least one follicle was at least 16 mm in diameter, and randomized patients to receive either 10,000 IU of hCG as a single or divided injection, with the second 5000-IU injection provided 1 week after the initial 5000 IU injection. Data confirmed no significant clinical differences except length of the luteal phase, which was longer in the divided-dose protocol [30].

Luteal support is a variable management step in the OI protocols [59–61]. Some studies mention specific protocols, such as 5000 IU of hCG when at least one follicle was >17 mm, and serum estradiol consistency (noting on day of determined hCG injection that mean numbers of follicles >10 mm were 4.7 ± 2.2 and 410 pg/ml ± 272 pg/ml). The luteal phase was supported with 2500 IU of hCG 4–7 days after the hCG ovulatory injection [68].

In a review of approaches to OI from the early 1980s derived from the Royal Women's Hospital, use of hCG for gonadotropin protocols was highlighted [20], emphasizing an incremental dosing paradigm for ovulation dose of hCG. In a first induction cycle, 3000 IU of hCG was administered 48 h after last FSH dose. Subsequently, the cycle was monitored by daily estradiol and progesterone values to validate ovulation. If ovulation was determined to have occurred, then supplementary doses of hCG (1000 IU on days 9 and 12) after the ovulating dose of hCG were administered. If ovulation did not occur during the first cycle, then stimulation was repeated; if necessary, the initial hCG dose was incrementally increased to 5000, 7500, or 10,000 IU by the fourth cycle. If the 10,000-IU dose was necessary, then supplementary doses were withheld. To summarize, the historical and continuing use of hCG results in its inclusion in a variety of ovulation management protocols. This heterogeneity has occurred because when hCG is being considered as a component of clinical infertility treatment, many interrelated factors have to be taken into account. What is the indication being treated? What is the clinical goal? Is it ovulation of the lowest achievable number of follicles or maximal stimulation? What medication or combination of medications is being used as the stimulation component? What modalities are available for monitoring and surveillance? What dose will be selected? Will more than one dose be necessary? What are the expected benefits of the management protocol, and how will they be determined? What are the inherent risks associated with the protocol, and how will they be minimized or monitored? What criteria will be used to determine when to administer hCG?

Given the confounding variables inherent in clinical studies, the timing of hCG administration has been based on one or more of the following parameters: consideration of preovulatory serum progesterone; suspicion or early documentation of an incipient LH surge; ovarian follicle size or number; and estradiol levels.

25.3 Timing administration of hCG—ultrasound monitoring, progesterone patterns, and endogenous LH surge patterns

When ultrasound technology emerged in the early 1980s, it was incorporated into clinical protocols [39,46,87]. A 1981 publication stated that although (transabdominal) ultrasound scan does not supplant estrogen monitoring, it can assist in detection of hyperstimulation, in the more accurate timing of the ovulatory dose of hCG, and in the withholding of hCG. It was recommended that an ultrasound examination should be performed on all patients prior to hCG administration [87].

As ultrasound technology advanced, other relevant studies were published. An article published in 1991 [40] provided data regarding a critical assumption about hCG administration. The purpose was to derive an equation for predicting ovulation based on data for correlating ovulation with follicle size on the day of hCG administration in hMG-stimulated cycles. The primary question was, what size follicle (in millimeters) ovulates in response to a timed injection of 10,000 IU of hCG after COH with hMG? In their discussion, the study authors remarked on the then-current lack of consensus and the confounding issues related to the understanding that follicles reaching 20–25 mm in spontaneous development were likely to ovulate after LH surge induction. None of the previous studies had compared ovulation percentages using induced follicle size as a parameter; it was remarked that oocytes obtained from follicles as small as 16 mm could be fertilized in IVF circumstances. These authors also listed related uncertainties, such as whether a follicle of a certain size would guarantee ovulation, a lack of understanding regarding the number of ovulated oocytes that was correlated with a chance of pregnancy, and the fact that no optimal E2 range correlating with increased fecundity had been determined (recall that these questions were being raised more than a decade after the use of hCG in many OI protocols; see Dickey 2009 for a comprehensive review) [32].

Patients were stimulated with 150 IU of hMG that was initiated daily and adjusted by protocol per changes in estradiol values and the number of follicles reaching critical sizes. After the hCG injection, ultrasound surveillance was performed to confirm ovulation. No exogenous luteal support was provided. Among 49 patients in 122 cycles, 1344 follicles were studied. Follicles measuring 17 mm or larger on the day of hCG administration ovulated significantly more often than follicles 16 mm or smaller. Follicles measuring 14 mm or smaller rarely ovulated. The average estradiol value on the day of hCG administration was 947 pg/ml. No significant difference in the estrogen value was correlated with cycles resulting in ongoing pregnancy, spontaneous miscarriage, or failed conception. The data from this article reflect a current-use paradigm for OI with controlled stimulation.

25.4 Risks of ovulation management with hCG

Controlled ovulation is the goal, but it must be understood that the preceding induction protocols also carry inherent risks of superovulation, ovarian enlargement, multiple follicles, OHSS, and high-order multiple pregnancies and births [32]. More recent publications are focusing on strategies to minimize these risks. These options include reducing gonadotropin exposure by combining FSH with clomiphene [11] and limiting the dose or the number of days of stimulation via withholding hCG or other maneuvers.

One strategy to minimize or avoid OHSS in patients with PCOS was to use low-dose FSH protocols. Formerly, it was unclear if a step-up, step-down, or combination protocol would provide better efficacy and lower OHSS risk. In one well-conducted study, Christin-Maitre and Hugues [35] examined the benefit of either a step-up or a step-down protocol with rFSH in clomiphene citrate – resistant patients with PCOS. The study noted similar cumulative rates of gestation: 38.6% versus

30.8% in the step-up versus the step-down protocols, respectively. The latter protocol, however, was more efficient in promoting monofollicular development and ovulation and had a lower rate of hyperstimulation, but a longer duration of treatment seemed necessary.

Details of the protocol included administering 5000 IU of hCG as a single dose when the leading follicle was more than 18 mm; hCG was withheld if four or more follicles >16 mm were present and/or the estradiol value reached 1000 pg/ml or higher.

The protocol also stated specifically that the luteal phase was not supported. The data from this study reported that ovarian follicles exceeding 16 mm were noted in 83% of the inductions with the step-up protocol, compared with only 56% in the step-down protocol. Also, multiple follicles (more than three in number) reaching more than 16 mm were noted in only 4.7% of the step-up cycles, compared with 36% of the step-down cycles. These favorable comparisons resulted in the administration of hCG in 84.6% of step-up cycles, compared with only 61.8% of step-down cycles. These results suggest that continuing to modify induction protocols can result in risk reduction for multiple follicle recruitment.

It was noted that most clinical induction protocols rely on estrogen and ultrasound follicle characteristics to help timed hCG administration. Other studies also examined the confounding variable of endogenous LH secretion. There is debate over whether superovulation protocols inhibit or modify the endogenous LH surge. During the 1980s, when controlled hyperstimulation protocols for IVF programs were studied, the problems of premature luteinization and/or progesterone secretion both compromised success rates. The data derived from clinical ART studies help contribute to an understanding of the larger issue of timing of hCG for non-ART induction protocols; specifically, do endogenous LH levels or progesterone levels influence a patient's response to an hCG injection?

The relationship of hCG administration relative to the LH surge has not been extensively studied. However, an interesting retrospective data analysis of pregnancy rates after IUI in women who underwent the procedure in 49 unstimulated cycles and in 856 consecutive hMG-stimulated cycles is relevant. Group A had an endogenous LH surge and was not given hCG. Group B was given hCG after an LH surge. Group C received hCG before an LH surge. Pregnancy rates provided were 16% per cycle; however, the lowest rate was in group A and the highest was in group B. The authors concluded that the better pregnancy rate occurs when hCG is given after an LH surge. Their protocol involved hMG at 75 IU of FSH, adjusted as necessary until one to two follicles >16 mm in diameter were detected. With increased hormonal surveillance, LH, and progesterone, 5000 IU of hCG was given under selective conditions per protocol. This related to presence or absence of LH and/or a progesterone shift. Timing of IUI was contingent on specific administration of hCG. Luteal support was provided: 1000 IU every 3 days for three doses. The analysis of the results suggested that timed administration of hCG after an LH surge could statistically improve cycle pregnancy rate, particularly if hCG was given up to 20 h after detection of the surge and if IUI could be completed within 48 h of the hCG injection [38].

The suggestion that hCG superimposed on a spontaneously manifested LH surge improved the chances of pregnancy was noted in another publication by Mitwally et al. [61]. In this review of retrospective data, the authors examined the influence of hCG administration in relationship to detected LH surge as a component of pregnancy success in cycles involving ovarian stimulation. First, they concluded that hCG administration was associated with a favorable outcome during stimulation in general. Higher clinical pregnancy rates associated with hCG administration were noted in a variety of stimulation protocols, sperm exposure models (timed intercourse or IUI) of presumed causes of infertility, specifically unexplained infertility, or PCOS. In addition, they reported a practical distinction regarding the timing of the hCG injection being pertinent to the stimulating protocol. If clomiphene citrate were used, then better pregnancy results were achieved when the LH surge was monitored and hCG was given subsequent to the surge. In FSH-driven protocols, monitoring and using an LH surge parameter for hCG administration was not as successful as administering hCG based on more common parameters involving estradiol and ultrasound criteria.

25.5 Efficacy of LUF syndrome

The central question regarding LUF syndrome is whether luteinization of a follicle cyst can occur in the absence of a released oocyte from the mature Graafian follicle [88]. The phenomenon was first considered in 1935 and then examined again after 1966 [89,90]. Components of the diagnosis presupposed that the patient showed evidence of failure to ovulate after maturing a follicle, yet exhibited manifestations of corpus luteum function (primarily progesterone secretion) with variable clinical characterization of the luteal phase. The prevalence of this diagnosis remains controversial [89,91,92]. Issues relate to the accuracy with which the condition might be diagnosed, its persistence, and whether it is associated with specific conditions such as endometriosis, unexplained infertility, or OI [89,93]. Relevant points in the discussion about the possible etiologies of LUF also consider deficient LH surge dynamics and concomitant variations in the FSH surge associated with the LH surge [92,94] and its treatment [92,93].

Studies have surveyed the possible occurrence of this syndrome with technology available at the time. Early incidence studies used combinations of assessments of peritoneal fluid concentrations of ovarian steroids, basal body temperature graphing, and serum progesterone levels [88,89]. Hamilton et al. published a study in which they prospectively monitored 600 cycles among 270 infertile patients every other day from cycle day 8 onward, and then daily ovarian ultrasound was performed when an ovarian follicle exceeded 15 mm in diameter. Ultrasound scanning continued until follicle dynamics met criteria consistent with ovulation or persisted in size to 30 mm or more for 3 consecutive days. A progesterone measurement was obtained on each day of ultrasound. The data from this study suggested that (using ultrasound) approximately 10% of the women and more than 6% of cycles exhibited a follicle growth pattern consistent with failure to rupture but, with suggestion of luteinization, manifested as shifts in basal temperature patterns and increases in luteal phase progesterone. This pattern was noted in 14 of the 27 women who received medical management during the index cycle producing the data.

The medications used for cycles with apparent ovulation failure included clomiphene citrate alone or in combination with hCG, hMG, or LH-releasing hormone [94]. This well-executed study was the first to characterize the follicular growth pattern of LUF cycles compared with those in normal ovulatory cycles. The growth rates were comparable until the serum-detected LH surge, at which time follicle rupture occurred in the 45 control cycles. In contrast, the LUF pattern was marked by a rapid increase in follicle diameter (reaching 35–36 mm on average) and persisted through the remainder of the cycle. Additionally, LH levels from LUF cycles surged 2 days later and were ultimately lower than LH values compared with values from normal cycles (despite the fact that the clinical luteal phases were similar in length). The mid-luteal progesterone levels were significantly lower when obtained from cycles exhibiting a LUF pattern.

Ovulation documentation has been as direct as laparoscopic assessment of ovarian stigmata [88], with many related studies substituting persistent or enlarging ovarian cyst formation beyond the window of assumed ovulation [88,93]. Based on data suggesting that concentrations were significantly higher in women who actually ovulated versus those women who did not apparently ovulate, one study proposed substituting estradiol and progesterone concentration values derived from peritoneal fluid obtained within 5 days of suspected ovulation. The data presented suggested overlap in concentrations in peritoneal hormone levels, fluid volume, and protein concentrations from women with and without laparoscopically determined ovulation stigmata, as well as from some women with no evidence of persistent ovarian cysts [88].

In an early study, Check et al. [93] monitored ovarian folliculogenesis via ultrasound in 333 patients. Of these, 89 failed to release an ovum by ultrasound criteria in two consecutive cycles. Relevant criteria included documenting when a mature follicle attained 18–24 mm in diameter, monitoring whether the index follicle decreased in size by 5 mm, free fluid appearing in the cul-de-sac, and cervical mucus regressing and serum progesterone increasing at least 3 ng/ml. Free fluid was variable, and it was not clear if all patients had progesterone assessments. The 89 patients considered to have LUF on two occasions were derived from three groups. Group 1 was not using any medications ($n = 39$). Group 2 was given clomiphene citrate ($n = 17$). Group 3 was treated with hMG ($n = 33$). They were subsequently managed per protocol to induce successful ovulation. They were first treated with 10,000 IU of hCG alone. If unsuccessful, then a repeat cycle was managed with 15,000 IU of hCG and 150 IU of hMG. In group 1, 35 of 39 patients provided ultrasonographic evidence of ovulation. In group 2, 13 of 16 patients provided ultrasonographic evidence of ovulation. In group 3, 31 of 33 patients ovulated when specifically administered the higher dose of hCG and hMG after failing to ovulate with hCG alone. No control groups were provided, and the authors acknowledged the lack of a placebo arm.

In a later publication from the same author [92], the assessment and treatment of LUF were again explored with more comprehensively monitored patients. LUF was characterized by two patterns, both involving prospective monitoring of estradiol, progesterone, and LH starting 17 days before expected menses and continuing until the serum progesterone was >5 ng/ml. Daily ultrasounds were obtained on the same days. LUF was considered when a lead follicle failed to demonstrate a 5-mm reduction in average diameter within 72 h of maturity and corroborated with LH surge, a decrease in estrogen, and an increase in progesterone to more than 2.5 ng/ml in two consecutive cycles.

Two patterns of LUF were described: a mature follicle type and a premature luteinization type. The former type manifested a follicle achieving average diameter of 18–24 mm and estrogen of more than 200 pg/ml, whereas the progesterone remained less than 2.5 ng/ml. Subsequently, the follicle size did not diminish and the mid-luteal phase progesterone reached a minimum of 8 ng/ml. In the latter type, a follicle attained a diameter of more than 18 mm, but progesterone increased to more than 2.5 ng/ml and the estradiol level never reached 200 pg/ml. As in the former group, the follicle size did not decrease and the mid-luteal phase progesterone value exceeded 8 ng/ml. Patients in both ovulation and non-OI protocols with LUF syndrome were then treated, in sequential cycles if not responding, with triggering doses of 10,000 IU of hCG, 15,000 IU of hCG, 15,000 IU of hCG, and 150 IU of hMG, or a divided-dose treatment of 1000 IU of hCG followed by 9000 IU of hCG 24 h later. Failure to respond resulted in transfer of patients to an active OI arm, sequencing variations of combinations of clomiphene citrate, and then serial hMG and the hCG permutations were used in the non-OI arm. Again, there were no controlled cycles. It was noted that there were no conceptions in the 168 cycles in which LUF was noted, as opposed to 24 conceptions in the 198 (12.1%) cycles that appeared to be corrected with the strategies used in this study [92].

A related study focused on whether the release efficacy of ova by hCG in hMG treated cycles was critically related to the late follicular phase progesterone level. In the discussion, the authors pointed out that one of the proposed functions of early progesterone production is to enhance the activity of certain prostaglandins associated with follicular rupture. Because hCG is administered in menotropin OI cycles based on estrogen levels and follicle size criteria, the authors wondered if critical levels of progesterone had to be evident before administration of hCG. Based on data assessment of progesterone levels in discrete intervals of 0.1–0.5, 0.6–0.9, 1.0–1.5, and >1.6 ng/ml, they concluded that rates of ovulation did not show significant differences after hCG administration of 10,000 IU of hCG (related to the existing pre-injection progesterone level) [92].

The incidence of LUF from different studies varies from 6.7% [94] to 10.7% in spontaneous cycles [95] and to more than 25% in some earlier studies [89]. Without being able to control for a variety of factors, the discussions about LUF remained focused on etiologies, that is, whether this condition is repetitive and whether it has higher associations of occurrence with conditions such as unexplained infertility, endometriosis, or pelvic inflammatory disease. In studies that did not use serial ultrasound methods to prospectively monitor follicle dynamics, indirect ascertainment of ovulation (particularly in studies utilizing ovulation management) might provide confounding data on efficacy of response and, indirectly, pregnancy rates [89,92]. The reasons for including a brief review of this syndrome are, first, that it is germane to patients who might need to be treated with OI protocols and, second, because it might be more important to incorporate postinjection surveillance for documenting the timing and occurrence of ovulation.

25.6 Considerations of hCG administration for timing IUI

IUI is probably the least standardized infertility treatment. Variables reported to be important for clinical management include different treatment groups and population composition [22], age, clinical etiologies and indications for the procedure [50], previous diagnostic workup, previous treatments, suspicion of cervical mucus abnormalities [96], cycle management [97], ovarian stimulation, ovarian monitoring, IUI timing, number of inseminations per cycle [98–101], associated treatments in the male and female partner, and number of treatment cycles to be performed [23,101–105]. The specific methods for sperm collection, storage, and laboratory preparation for IUI [106–109] additionally confound purported endpoints, independent from the stimulation protocol, and the timing determinants for the insemination itself.

Given these clinical variables and the acknowledged heterogeneity of most studies [23,101–110], the focus of an insemination procedure still remains the identification of the day of suspected ovulation [111]. Earlier in this chapter we reviewed the use of methods for urinary LH determination, as well as the elective use of injections of hCG associated with OI and COS. Both of these modalities were seamlessly applied to insemination scheduling. Urinary LH monitoring was rapidly embraced as an adjunct to timing of artificial insemination procedures [53,112]. Elective injection of hCG was used in gonadotropin-controlled stimulation protocols for timing IUI [25,38,54,60,113]. The indications for IUI as first-line therapy include unexplained infertility, male infertility, and ovulatory dysfunctions [114]. Whether an ovulation management protocol is utilized as an adjunct to the timed procedure is variable. These protocols have included both clomiphene citrate [27,52,114,115] and controlled stimulation with gonadotropins [38,54,85,113]. Clomiphene citrate administration has been an initial step in many management strategies for IUI. Clomiphene use results in ovulation during a window of time that cannot be precisely estimated.

Elsewhere in this chapter we have commented on the variables of folliculogenesis, preovulatory follicle diameters, and spontaneous LH surge chronology, each of which results in the estimation that the actual event of ovulation can occur 24–56 h after a detected LH surge [53]. It is well-understood that detection of the LH surge by urinary detection kits is an independent confounder in identifying the time interval for insemination. Components of this difficulty include variability of the pituitary secretion pattern in a given 24-h interval, rate of urine production (which affects concentrations of LH in urine on which a semi-quantitative enzyme-linked immuno-spectrometric assay [ELISA] test depends), the inconsistency of detection between early morning, mid-day, or evening assessment among available detection kits, and the interference of augmented LH release when certain patients utilize clomiphene citrate.

The use of hCG as a timing agent addresses some of the variance associated with endogenous LH release patterns but also introduces other considerations. One important consideration is the need for ovarian follicle surveillance to determine the timing of the injection. Although an injection can be empirically timed, one or more ultrasound procedures usually provide a better estimate of when to administer hCG. Follicles have predictable growth trajectories, supporting the possibility of empiric timing. Many carefully performed studies, however, have noted ranges in the numbers of follicles starting and completing development (endpoint diameters associated with inducing spontaneous LH surges) that can critically influence when to proceed [38,48,114]. Once the injection is given, the expected time of ovulation has less variance than that associated with spontaneous LH surges, perhaps 36–48 h [114]. In a study of IM versus IV hCG administration after clomiphene OI, Fischer et al. monitored the time interval for ovulation, estrogen, and progesterone values. They noted a mean ovulation time of 40.4 h after IM and 38.3 h after IV administration. The range among responses in both groups was less than 36 h and more than 48 h [6]. Anderson et al. noted a mean time to ovulation of 38.3 h, which they derived from a range of 34–46 h among 37 cycles. They also reported that in 66% of cases, the largest follicle was the first to rupture. Three similarly designed studies were published in 1997, 1999 [53], and 2006 [52]. They examined the pregnancy rates in women receiving clomiphene citrate induction who subsequently underwent insemination based on either urinary LH determination or hCG injection. The protocols were similar, and the two later studies are reviewed here in detail.

Deaton et al. published a retrospective review of results for patients with unexplained, anovulatory, or male factor infertility. Among all three clinical categories, 250 were conducted with LH timing and 182 were conducted with hCG timing of a single insemination. There was no significant difference in the clinical pregnancy rates when each clinical category was compared.

In 1999, Zreik et al. published the results of a prospective trial in which women with unexplained infertility, ovulatory dysfunction, or male factor infertility and those who were being treated with clomiphene citrate underwent IUI after being randomized to two different protocols to facilitate insemination timing. The two protocols were urinary LH determination versus hCG injection. Data were derived from 141 cycles performed in 54 couples. Clomiphene was administered on cycle days 3–7. Daily LH monitoring began on day 10, and two daily inseminations were performed once a urinary LH surge was detected. An ultrasound was performed once on the day of the first insemination in this group. In the other study group, ultrasound monitoring was conducted. When a lead follicle was 218 mm, 10,000 IU of hCG was administered. Two inseminations were completed beginning the following day. If a pregnancy did not occur, then the patient crossed over to the alternate protocol for the next cycle. This continued for up to four cycles.

In this study, unexplained infertility included patients who had undergone laparoscopy or had minimal endometriosis. The two protocols used for timing IUI yielded similar pregnancy rates for the three diagnostic groups [53]. Follicle sizes were similar between the LH groups: 22.5 mm compared with 21.1 mm on the day women in the second group were given hCG. The authors stated that 12% of normal ovulatory women might have had a urinary LH peak of <20 IU/ml, which is below the threshold of many urinary LH detection kits.

In 2006, Lewis et al. [52] also published a prospective, randomized trial comparing pregnancy rates with two different methods of timing IUI in women who were utilizing clomiphene citrate. One group timed the insemination procedure the day after detecting a positive urinary LH surge. The other group received 10,000 IU of hCG when criteria were met for endometrial development (at least 8 mm thickness with at least partial layering), acceptable follicle size (220 mm mean diameter), and acceptable follicle number (less than four follicles larger than 20 mm in diameter). The injection was given in the evening and the insemination procedure was completed 33–42 h later. Patients acceptable for study utilized 100 mg oral clomiphene citrate on days 5–9. All subjects underwent vaginal ultrasound each study cycle before starting clomiphene. Women randomized to the LH detection group self-monitored urinary LH beginning on day 12.

Women in the hCG group were seen on day 12 for transvaginal ultrasound. Repeat scans were scheduled according to initial follicle dimensions. Patients not meeting criteria for hCG administration in a specific cycle self-monitored urinary LH. Patients remained in their treatment group assignment for three cycles. Subjects in the surge group were removed from the study if a surge was not detected. Subjects who did not meet hCG injection criteria and those who did not have an LH surge were removed from the study.

This study is noteworthy because it is clinically relevant, and the protocols within the study are similar to common current office practices. The major indication for treatment was unexplained infertility, with a minor proportion of male factor infertility. Of the randomized 150 patients, 129 completed at least one cycle of treatment (58 in the LH surge group; 71 in the hCG group). This resulted in 17 and 23 pregnancies, respectively. A preponderance of pregnancies occurred in the first treatment cycle. By intention-to-treat and cumulative pregnancy rate analyses, there was no significant difference in pregnancy rates. Study dropout rates were higher for the LH group: 31% compared with 11% for the hCG group. Failure to detect an LH surge was the most common contributor to noncompletion of a cycle in the home-monitoring group.

The authors concluded that ultrasound monitoring with hCG administration did not seem to result in a higher pregnancy rate when compared with home urinary LH monitoring; however, the failure to detect a surge (commented on in other studies) might be one reason to consider hCG administration [52,56]. Thus, the data from retrospective and prospective controlled studies [52,53] suggest that hCG timing does not appear to increase clinical pregnancy rates as a result of timed insemination in clomiphene-induced ovulation when compared with LH monitoring. Alternatively, Deaton et al. noted that ultrasound-timed injection of hCG after clomiphene induction, followed by IUI, was more efficacious than cycles concluding with clomiphene and hCG-timed intercourse [27].

A more recent meta-analysis [114] of seven studies meeting criteria for clomiphene treatment, IUI, and use of urinary versus hCG-timed IUI procedures was conducted. The data were reviewed separately for different indications: male factor; unexplained infertility; ovulatory dysfunction (the same clinical triad treated in previously mentioned studies). Overall, 1461 patients received hCG and 1162 served as controls. Preparation steps for the semen samples were variable; clomiphene citrate dosages varied from 50 to 150 mg/day among studies. The dose of hCG administered was consistently 10,000 IU among studies. Inseminations were conducted at 18–24 h in the LH monitored groups and at 36–42 h in the hCG group. Criteria for hCG administration (requiring a lead follicle of >20 mm) were mentioned in four studies.

From this meta-analysis, the authors reported that there was a significant difference in pregnancy rates between the hCG-administered group and the LH-recording group that favored LH monitoring [114]. This interpretation was qualified and it pertained only to the groups treated for male factor and unexplained infertility. An opposite, improved trend was described for pregnancy rates resulting from hCG administration rather than LH timing for the clinical indication of ovulatory dysfunction.

Given that hCG can be used for induction protocols, one must keep in mind the relative benefits, uncertainties, costs, and risks of its use in contrast to the same considerations for urinary LH monitoring. The advantages of selectively timed administration of hCG include a proven biological basis, ease of planning, and clinical predictability [114]. The disadvantages include the cost of the medication and the need for ultrasound monitoring. If IUI is being considered as the first-line approach for couples with unexplained infertility, ovulatory dysfunction, or male factor infertility, then the use of LH kits to time IUI could produce important cost-savings over the use of ultrasonography [53].

The specific advantages of using urinary LH detection for timing include its cost-effectiveness and the fact that it can be used in a private setting [114]. The disadvantages are that the associated false positive and negative rates lead to suboptimally timed procedures or skipped or canceled cycles because of nondetection in a truly ovulatory cycle [52].

Because IUI might be an isolated treatment recommendation or an adjunct to both OI and COS protocols [103,113], a derivative set of clinical risks as well as anticipated benefits are encountered when multiple follicle development occurs followed by hCG-timed IUI. The primary benefit is a higher pregnancy rate per treatment cycle. Some studies did not confirm this advantage, however. In 1991, Martinez et al. studied 48 patients with male factor or unexplained infertility; they underwent either timed intercourse or timed IUI after hMG stimulation with hCG injection. No significant difference in pregnancy rates between the two options of sperm exposure was detected [113]. Other studies and reviews confirmed an improved chance of pregnancy [37,66].

The derivative risks of controlled stimulation over OI include persistent ovarian cyst formation [40,116], ovarian hyperstimulation [71], and creation of high-order multiple pregnancies [32,37,86]. One study [40] utilized estradiol assays and transvaginal sonography to monitor ovarian follicle size, number, and response to 10,000 IU of hCG administration after controlled stimulation with hMG. The primary purpose of the study was to develop a prediction model for the likelihood of conception. Data derived from the study, however, suggest a high rate of ovarian cyst persistence after controlled stimulation. Fifty-four previously ovulating but nonconceiving patients presented for a consecutive stimulation cycle. It was noted that residual ovarian cysts were found in 28 (51.9%) of those returning patients. Interestingly, data from this study suggested that the likelihood of conception did not correlate with the number of ovulatory follicles, the total number of follicles present at time of hCG administration, or the total hMG dose administered. There was, however, a correlation with serum E2 levels on the day of hCG administration.

The continuum of problems associated with hMG/hCG protocols, ranging from persistent cyst formation to iatrogenic OHSS, is an important risk in controlled stimulation protocols but is not independently addressed in this chapter.

25.7 High-order multiple pregnancies

The third relevant risk in controlled stimulation or timed insemination is that of multiple follicle recruitment and high-order multiple pregnancy creation [32]. Use of hCG is now an integral component of various fertility-promoting strategies. A brief chronologic progression synopsis of the development and incorporation of gonadotropin preparations into clinical protocols has been published [32]. The review provides a cogent discussion about the contribution of non-ART to high-order multiple pregnancies and births that occur as a result of the modern era of OI. For example, it was estimated that approximately 230,000 OI cycles were conducted in the United States in 2003, and that 22% of all twin, 40% of all triplet, and 71% of all higher-order pregnancies in 2004 were the result of controlled stimulation protocols. As a specific example, one publication reported a 27% incidence of multiple gestations (14 of 52) conceived within an hMG/FSH protocol. The majority of conceptions (62%) were twins [86].

Dickey articulated possible factors contributing to the disproportionate increase in multiple pregnancies over the natural background rate. In his opinion, this increase is the result of more clinicians trained to use gonadotropins, more potent gonadotropins, and the adaptation of dosing regimens developed for IVF for more conventional OI-IUI protocols. Publications identify the conundrum of how gonadotropin doses associated with assisted reproduction are being used in empiric controlled stimulation protocols with IUI, possibly because of higher pregnancy rates [25,54,97,105] relative to clomiphene citrate-based protocols [37,53], adding to the risk of high-order multiple pregnancies.

Patient and cycle factors that affect the risk of multiple pregnancies have been identified [32,69,86]. Women younger than age 32 years have an increased risk of multiple conceptions, and women older 38 years of age have a significantly reduced risk [32]. Physiological responses to gonadotropins (manifested as elevated periovulatory estradiol levels and increased number of follicles) have been suggested based on univariate and multivariate analyses in prospectively modeled studies [69,70].

Another set of variables (possibly promoting increased conception numbers) is introduced when IUI is combined with OI [101]. These confounding variables include the indication for treatment (mild male infertility, cervical factor, or unexplained infertility), the stimulation protocol, the criteria for follicle development (number desired, cycles canceled, converted to IVF), quality of sperm preparation, number of cycles per patient, and outcome focusing on final live birth rate [110].

Other studies relevant to the contribution of the inseminant itself have been published. Only the study by Shelden et al. [86] was positive in concluding that increased sperm density might be associated with increased risk of multiple pregnancies. Others focusing on fresh versus frozen semen samples, motile count [106,107], and motility longevity [108] did not conclude that these particular factors significantly increased the risk of multiple conception. There is a paucity of controlled trials that comprehensively account for this large set of confounders [110].

Given this information, a consensus goal still remains: to reduce the incidence of multiple pregnancies as a result of the clinical protocol for IUI [32,86]. Strategies to reduce clinical risk of multiple pregnancies have been offered. One review listed three points of focus to help reduce multiple births due to OI. Maximizing the potential effectiveness of first-step treatment algorithms using oral medications (including antiestrogens and aromatase inhibitors) was paramount. When gonadotropin protocols became necessary, it was proposed that the lowest effective doses should be used [32,35,61,66,69,70,74]. Dickey reiterated that, historically, early use of hMG relied on daily doses of 75 IU of FSH activity; these met with good success. He particularly emphasized clinical data that reinforced the safety and efficacy of low FSH dose strategies (based on a review of the pregnancy rate outcomes associated with different protocol doses, estrogen levels, duration of stimulation, number of cycles conducted) and the effect of maternal age (often most pertinent to women <32 years of age). A second recommendation was to have contingencies in place to deal with unacceptable multiple follicle development, including cycle cancelation or conversion to IVF. A third recommendation, in sequence, involved selective pregnancy reduction if available [3].

References

[1] O'Connor JF, Birken S, Lustbader JW, Krichevsky A, Chen Y, Canfield RE. Recent advances in the chemistry and immunochemistry of human chorionic gonadotropin: impact on clinical measurements. Endocr Rev 1994;15:650–83.

[2] Midgeley AR, Jaffe RB. Regulation of human gonadotropins. II. Disappearance of human chorionic gonadotropin following delivery. J Clin Endocrinol Metab 1968;28:1712–18.

[3] Rizkallah T, Gurpide E, Vande Wiele RI. Metabolism of HCG in man. J Clin Endocrinol Metab 1969;29:92–100.

[4] Saal W, Glowania H-J, Hengst W, Happ J. Pharmacodynamics and pharmacokinetics after subcutaneous and intramuscular injection of human chorionic gonadotropin. Fertil Steril 1991;56:225–9.
[5] Wehmann RE, Nisula BC. Metabolic and renal clearance rates of purified human chorionic gonadotropin. J Clin Invest 1981;68:184–94.
[6] Fischer RA, Nakajima ST, Gibson M, Brumsted JR. Ovulation after intravenous and intramuscular human chorionic gonadotropin. Fertil Steril 1993;60:418–22.
[7] Wolf LJ. Ovulation induction. Clin Obstet Gynecol 2000;43:902–15.
[8] Wu CH. Monitoring of ovulation induction. Fertil Steril 1978;30:617–30.
[9] Weseley AC, Melnick H. Tamoxifen in clomiphene-resistant hypothalamic anovulation. Int J Fertil 1987;32:226–8.
[10] Tajima C. Endocrine profiles in tamoxifen-induced conception cycles. Fertil Steril 1984;42:548–53.
[11] Kemmann E, Jones JR. Sequential clomiphene citrate-menotropin therapy for induction or enhancement of ovulation. Fertil Steril 1983;39:772–9.
[12] Lobo RA, Gysler M, March CM, Goebelsmann U, Mishell DR. Clinical and laboratory predictors of clomiphene response. Fertil Steril 1982;37:168–73.
[13] Greenblatt RB, Barfield WE, Jungck ED, Ray AW. Induction of ovulation with MRL/41: preliminary report. J Am Med Assoc 1961;178:101–4.
[14] Schwartz M, Jewelewicz R. The use of gonadotropins for induction of ovulation. Fertil Steril 1981;35:3–12.
[15] Hammond MG. Monitoring techniques for improved pregnancy rates during clomiphene ovulation induction. Fertil Steril 1984;42:499–509.
[16] Keye Jr WR, Chang RJ, Rebar RW, Soules MR. Infertility evaluation and treatment. Philadelphia, PA: WB Saunders Co; 1995. p. 58, 140–41, 130.
[17] Bruno L, Vaclav I, Glezerman M. Diagnosis and treatment of functional infertility. 3rd revised ed. Berlin: Blackwell Wissenschaft; 1993.
[18] Lawrence A, Healy D. Management of women with chronic anovulation. In: Kovacs G, editor. The subfertility handbook. A clinician's guide. Cambridge UK: Cambridge University Press; 1997. p. 69–82 [Chapter 6].
[19] Wu CH. A rational and practical approach to clomiphene therapy. Clin Obstet Gynecol 1984;27:953–65.
[20] Pepperell RJ. A rational approach to ovulation induction. Fertil Steril 1983;40:1–14.
[21] Andersen AG, Als-Nielsen B, Hornes PJ, Andersen LF. Time interval from human chorionic gonadotropin (HCG) injection to follicular rupture. Hum Reprod 1995;10:3202–5.
[22] Shenfield F, Doyle P, Valentine A, Steele SJ, Tan S-L. Effects of age, gravidity and male infertility status on cumulative conception rates following artificial insemination with cryopreserved donor semen: analysis of 2998 cycles of treatment in one centre over 10 years. Hum Reprod 1993;8:60–4.
[23] Bensdorp AJ, Cohlen BJ, Heineman MJ, Vandekerckhove P. Intra-uterine insemination for male subfertility. Cochrane Database Syst Rev 2007;4:CD000360.
[24] Cantineau AE, Cohlen BJ, Heineman MJ. Ovarian stimulation protocols (anti-oestrogens, gonadotropins with and without GnRH agonists/antagonists) for intrauterine insemination (IUI) in women with subfertility. Cochrane Database Syst Rev 2007;18:CD005356.
[25] Guzick DS, Sullivan MW, Adamson GD, Cedars MI, Falk RJ, Peterson EP, et al. Efficacy of treatment for unexplained infertility. Fertil Steril 1998;70:207–13.
[26] Fisch P, Casper RF, Brown SE, Wrixon W, Collins JA, Reid RL, et al. Unexplained infertility: evaluation of treatment with clomiphene citrate and human chorionic gonadotropin. Fertil Steril 1989;51:828–33.

[27] Deaton JL, Gibson M, Blackmer KM, Nakamima ST, Badger GJ, Brumsted JR. A randomized, controlled trial of clomiphene citrate and intrauterine insemination in couples with unexplained infertility or surgically corrected endometriosis. Fertil Steril 1990;54:1083–8.
[28] Erdem A, Erdem M, Atmaca S, Guler I. Impact of luteal phase support on pregnancy rates in intrauterine insemination cycles: a prospective randomized study. Fertil Steril 2009;91:2508–13.
[29] Pritts EA, Atwood AK. Luteal phase support in infertility treatment: a meta-analysis of randomized trials. Hum Reprod 2002;17:2287–99.
[30] Grazi RV, Taney FH, Gagliardi CL, Von Hagen S, Weiss G, Schmidt CL. The luteal phase during gonadotropin therapy: effects of two human chorionic gonadotropin regimens. Fertil Steril 1991;55:1088–92.
[31] Marshall JR, Jacobson A, Hammond CB. Dose response relationships of ovulation induction with human menopausal gonadotropin. J Clin Endocrinol Metab 1969;29:106–10.
[32] Dickey RP. Strategies to reduce multiple pregnancies due to ovulation stimulation. Fertil Steril 2009;91:1–17.
[33] van Wely M, Bayram N, van der Veen F. Recombinant FSH in alternative doses or versus urinary gonadotrophins for ovulation induction in subfertilituy associated with polycystic ovary syndrome: a systematic review based on a cochrane review. Hum Reprod 2003;18:1143–9.
[34] van Santbrink EJP, Fauser BCJM. Urinary follicle-stimulating hormone for normogonadotropic clomiphene-resistant anovulatory infertility: prospective, randomized comparison between low dose step-up and step-down dose regimens. J Clin Endocrinol Metab 1997;82:3597–602.
[35] Christin-Maitre S, Hugues JN. A comparative randomized multicentric study comparing the step-up versus step-down protocol in polycystic ovary syndrome. Hum Reprod 2003;18:1626–31.
[36] Thessaloniki ESRM/ASRM-sponsored PCOS Consensus Workshop Group. Consensus on infertility treatment related to polycystic ovary syndrome. Hum Reprod 2008;23:462–77.
[37] Tadakoro N, Vollenhoven B, Clark S, Baker G, Kovacs G, Burger H, et al. Cumulative pregnancy rates in couples with anovulatory compared with unexplained infertility in an ovulation induction programme. Hum Reprod 1997;12:1939–44.
[38] Fuh KW, Wang X, Tai A, Wong I, Norman RJ. Intrauterine insemination: effect of the temporal relationship between the luteinizing hormone surge, human chorionic gonadotropin administration and insemination on pregnancy rates. Hum Reprod 1997;12:2162–6.
[39] Seibel MM, McArdle CR, Thompson IE, Berger MJ, Taymor ML. The role of ultrasound in ovulation induction: a critical appraisal. Fertil Steril 1981;36:573–7.
[40] Silverberg KM, Olive DL, Burns WN, Johnson JV, Groff TR, Schenken RS. Follicular size at the time of human chorionic gonadotropin administration predicts ovulation outcome in human menopausal gonadotropin-stimulated cycles. Fertil Steril 1991;56:296–300.
[41] Bateman BG, Nunley Jr WC, Kulp LA. Exogenous estrogen therapy for treatment of clomiphene-citrate-induced cervical mucus abnormalities: is it effective? Fertil Steril 1990;54:577–9.
[42] Moghissi KS. Cervical mucus changes and ovulation prediction and detection. J Reprod Med 1986;31:748–53.
[43] Maxson WS, Pittaway DE, Herbert CM, Garner CH, Wentz AC. Antiestrogenic effect of clomiphene citrate: correlation with serum estradiol concentrations. Fertil Steril 1984;42:356–9.
[44] de Mouzon J, Testart J, Lefevre B, Pouly J-L, Frydman R. Time relationships between basal body temperature and ovulation or plasma progestins. Fertil Steril 1984;41:254–8.

[45] McCarthy JJ, Rockette HE. Prediction of ovulation with basal body temperature. J Reprod Med 1986;31:742–7.
[46] Queenan JT, O'Brien GD, Bains LM, Simpson J, Collins WP, Campbell S. Ultrasound scanning of ovaries to detect ovulation in women. Fertil Steril 1980;34:99–105.
[47] Batzer FR. Ultrasonic indices of ovulation. J Reprod Med 1986;31:764–9.
[48] Baerwald AR, Walker RA, Pierson RA. Growth rates of ovarian follicles during natural menstrual cycles, oral contraception cycles, and ovarian stimulation cycles. Fertil Steril 2009;91:440–9.
[49] Marrs RP, March CM, Mishell DR. A comparison of clinical and laboratory methods in monitoring human menopausal gonadotropin therapy. Fertil Steril 1980;34:542–7.
[50] Kurachi K, Aono T, Suzuki M, Hirano M, Kobayashi T, Kaibara M. Results of HMG (Humegon)-HCG therapy in 6096 treatment cycles in 2166 Japanese women with anovulatory infertility. Eur J Obstet Gynecol Reprod Biol 1985;19:43–51.
[51] Marrs RP, Vargyas JM, March CM. Correlation of ultrasonic and endocrinologic measurements in human menopausal gonadotropin therapy. Am J Obstet Gynecol 1983;145:417–21.
[52] Lewis V, Queenan Jr J, Hoeger K, Stevens J, Guzick DS. Clomiphene citrate monitoring for intrauterine insemination timing: a randomized trial. Fertil Steril 2006;85:401–6.
[53] Zreik TG, Garcia-Velasco JA, Habboosh MS, Olive DL, Arici A. Prospective, randomized, crossover study to evaluate the benefit of human chorionic gonadotropin-timed versus urinary luteinizing hormone-timed intrauterine inseminations in clomiphene citrate-stimulated cycles. Fertil Steril 1999;71:1070–4.
[54] Karlström P, Bergh T, Lundkvist O. A prospective randomized trial of artificial insemination versus intercourse in cycles stimulated with human menopausal gonadotropin or clomiphene citrate. Fertil Steril 1993;59:554–9.
[55] Jaffe SB, Jewelewicz R. The basic infertility investigation. Fertil Steril 1991;56: 599–610.
[56] Practice Committee Report. Use of clomiphene citrate in women. Fertil Steril 2003;80: 1302–8.
[57] de Leo V, La Marca A, Petraglia F. Insulin lowering agents in the management of poly-cystic ovary syndrome. Endocr Rev 2003;24:633–67.
[58] Baillargeon J-P, Iuorno MJ, Nestler JE. Insulin sensitizers for polycystic ovary syndrome. Clin Obstet Gynecol 2003;46:325–40.
[59] Legro RS, Barnhart HX, Schlaff WD, Carr BR, Diamond MP, Carson SA, et al. Clomiphene, metformin, or both for infertility in the polycystic ovary syndrome. N Engl J Med 2007;356:551–66.
[60] de Ziegler D. The dawning on the non-cancer uses of aromatase inhibitors in gynaecology. Hum Reprod 2003;18:1598–602.
[61] Mitwally MF, Abdel-Razeq S, Casper RF. Human chorionic gonadotropin administration is associated with high pregnancy rates during ovarian stimulation and timed intercourse or intrauterine insemination. Reprod Biol Endocrinol 2004;2:55.
[62] Diczfalusy E, Harlin J. Clinical–pharmacological studies on human menopausal gonadotropin. Hum Reprod 1988;3:21–7.
[63] Howles CM. Genetic engineering of human FSH (Gonal-F). Hum Reprod Update 1996;2:172–91.
[64] Cohen JJ, Debache C, Pigueau F, Mandelbaum J, Plachot M, de Brux J. Sequential use of clomiphene citrate, human menopausal gonadotropin, and human chorionic

gonadotropin in human *in vitro* fertilization. II. Study of luteal phase adequacy following aspiration of the preovulatory follicles. Fertil Steril 1984;42:360–5.
[65] Stone BA, Vargyas JM, Marrs RP, Quinn PJ, Batzofin JH, Tan T, et al. Levels of steroid and protein hormones in antral fluids of women treated with different combinations of gonadotropins, clomiphene citrate, and a gonadotropin-releasing hormone analog. Fertil Steril 1988;49:249–57.
[66] Eijkemans MJC, Imani B, Mulders AGMGJ, Habbema JDF, Fauser BCJM. High singleton live birth rate following classical ovulation induction in normogonadotrophic anovulatory infertility (WHO 2). Hum Reprod 2003;18:2357–62.
[67] White DM, Polson DW, Kiddy D, Sagle P, Watson H, Gilling-Smith C, et al. Induction of ovulation with low-dose gonadotropins in polycystic ovary syndrome: an analysis of 109 pregnancies in 225 women. J Clin Endocrinol Metab 1996;81:3821–4.
[68] Balasch J, Fabregues F, Penarrubia J, Creus M, Vidal R, Casamitjana R, et al. Follicular development and hormonal levels following highly purified or recombinant follicle-stimulating hormone administration in ovulatory women and WHO Group II anovulatory infertile patients. J Assist Reprod Genet 1998;15:552–9.
[69] Tur R, Barri PN, Coroleu B, Buxaderas R, Parera N, Balasch J. Use of a prediction model for high-order multiple implantation after ovarian stimulation with gonadotropins. Fertil Steril 2005;83:116–21.
[70] Imani B, Eijkemans MJC, Faessen GH, Bouchard P, Guidice LC, Fauser BCJM. Prediction of the individual follicle-stimulating hormone threshold for gonadotropin induction of ovulation in normogonadotropic anovulatory infertility: an approach to increase safety and efficiency. Fertil Steril 2002;77:83–90.
[71] Dourrons NE, Williams DB. Prevention and treatment of ovarian hyperstimulation syndrome. Sem Reprod Endocrinol 1996;14:355–65.
[72] Golan A, Ron-El R, Herman A, Soffer Y, Weinraub Z, Caspi E. Ovarian hyperstimulation syndrome: an update review. Obstet Gynecol Surv 1989;44:430–40.
[73] Yee B, Vargyas JM. Multiple follicle development utilizing combinations of clomiphene citrate and human menopausal gonadotropins. Clin Obstet Gynecol 1986;29:141–7.
[74] Rosenwaks Z, Mausher SJ, Acosta AA. Use of HMG and/or FSH for multiple follicle development. Clin Obstet Gynecol 1986;29:148–57.
[75] Bernardus RE, Seegar Jones G, Acosta AA, Garcia JE, Liu HC, Jones DL, et al. The significance of the ratio of follicle-stimulating hormone and luteinizing hormone in induction of multiple follicular growth. Fertil Steril 1985;43:373–8.
[76] Sher G, Knutzen VK, Stratton CJ, Montakhab MM, Allenson SG. *In vitro* sperm capacitation and transcervical intrauterine insemination for the treatment of refractory infertility: phase I. Fertil Steril 1984;41:260–4.
[77] Hayden CJ, Balen AH, Rutherford AJ. Recombinant gonadotrophins. Br J Obstet Gynaecol 1999;106:188–96.
[78] van Weissenbruch MM, Schoemaker HC, Drexhage HA, Schoemaker J. Pharmacodynamics of human menopausal gonadotrophin (HMG) and follicle-stimulating hormone (FSH): the importance of the FSH concentration in initiating follicular growth in polycystic ovary-like disease. Hum Reprod 1993;8:813–21.
[79] Cohlen BJ, Vandekerckhove P, te Velde ER, Habbena JDF. Timed intercourse versus intra-uterine insemination with or without ovarian hyperstimulation for subfertility in men. Cochrane Database Syst Rev 2000;2:CD000360.
[80] Oelsner F, Serr DM, Mashiach S, Blankstein J, Lunenfeld B, Snyder M. The study of induction of ovulation with menotropins: analysis of results of 1897 treatment cycles. Fertil Steril 1978;30:538–44.

[81] Branigan EF, Estes A. Use of micro-dose human chorionic gonadotropin (hCG) after clomiphene citrate (CC) to complete folliculogenesis in previously CC-resistant anovulation. Am J Obstet Gynecol 2005;192:1890–6.
[82] Coelingh Bennink HJ, Fauser BC, Out HJ. Recombinant follicle-stimulating hormone (FSH; Puregon) is more efficient than urinary FSH (Metrodin) in women with clomiphene citrate-resistant normogonadotropic, chronic anovulation: a prospective, multicenter assessor-blind, randomized clinical trial. European Puregon Collaborative Anovulation Study Group. Fertil Steril 1998;69:19–25.
[83] Yarali H, Bukulmez O, Gurgan T. Urinary follicle-stimulating hormone (FSH) versus recombinant FSH in clomiphene citrate-resistant normogonadotropic, chronic anovulation: a prospective, randomized study. Fertil Steril 1999;72:276–81.
[84] Hedon B, Hugues JN, Emperaire JC, Chabaud JJ, Barbereau D, Boujenah A, et al. A comparative prospective study of a chronic low dose versus a conventional ovulation stimulation regimen using recombinant human follicle stimulating hormone in anovulatory infertile women. Hum Reprod 1998;13:2688–92.
[85] Balasch J, Fabregues F, Creus M, Casamitjina R, Puerto B, Vanrell JA. Recombinant human follicle stimulating hormone for ovulation induction in polycystic ovary syndrome: a prospective, randomized trial of two starting doses in a chronic low-dose step-up protocol. J Assist Reprod Genet 2000;17:561–5.
[86] Sheldon R, Kemmann E, Bohrer M, Pasquale S. Multiple gestation is associated with the use of high sperm numbers in intrauterine insemination specimen in women undergoing gonadotropin stimulation. Fertil Steril 1988;49:607–10.
[87] O'Herlihy C, Evans JH, Brown JB, Ch de Crespigny LJ, Robinson HP. Use of ultrasound in monitoring ovulation induction with human pituitary gonadotropins. Obstet Gynecol 1982;60:577–82.
[88] Koninckx PR, De Moor P, Bosens IA. Diagnosis of the luteinized unruptured follicle syndrome by steroid hormone assays on peritoneal fluid. Br J Obstet Gynaecol 1980;87:929–34.
[89] Katz E. The luteinized unruptured follicle and other ovulatory dysfunctions. Fertil Steril 1988;50:839–48.
[90] Marik J, Hulka J. Luteinized unruptured follicle syndrome: a subtle cause of infertility. Fertil Steril 1978;29:270–4.
[91] Kerin JF, Kirby C, Morris D, McEvoy M, Ward B, Cox LW. Incidence of the luteinized unruptured follicle phenomenon in cycling women. Fertil Steril 1983;40:620–6.
[92] Check JH, Dietterich C, Nowroozi K, Wu CH. Comparison of various therapies for the luteinized unruptured follicle syndrome. Int J Fertil 1992;37:33–7.
[93] Check JH, Chase JS, Adelson HG, Dietterich C. New approaches to diagnosis and therapy of the luteinized unruptured follicle syndrome. Int J Fertil 1986;30:29–32.
[94] Hamilton CJCM, Wetzels LCG, Evers JLH, et al. Follicle growth curves and hormonal patterns in patients with luteinized unruptured follicle syndrome. Fertil Steril 1985;43:541–8.
[95] Killick S, Elstein M. Pharmacologic production of luteinized unruptured follicles by prostaglandin synthetase inhibitors. Fertil Steril 1987;47:773–7.
[96] Helmerhorst FM, van Vlier HAAM, Gornas T, Finken MJ, Grimes DA. Intrauterine insemination versus timed intercourse for cervical hostility in subfertile couples. Obstet Gynecol Surv 2006;61:402–14.
[97] Hughes EG. "Effective treatment" or "not a natural choice"? Hum Reprod 2003;18:912–14.
[98] Cantineau AEP, Heineman MJ, Cohlen BJ. Single versus double intrauterine insemination in stimulated cycles for subfertile couples: a systematic review based on a Cochrane review. Hum Reprod 2003;18:941–6.

[99] Randall GW, Gantt PA. Double vs. single intrauterine insemination per cycle. Use in gonadotropin cycles and in diagnostic categories of ovulatory dysfunction and male factor infertility. J Reprod Med 2008;53:196–202.
[100] Guzick DS. For now, one well-timed intrauterine insemination is the way to go. Fertil Steril 2004;82:30–1.
[101] Osuna C, Matorras R, Pijoan JI, Rodriguez-Escudero FJ. One versus two inseminations per cycle in intrauterine insemination with sperm from patients' husbands: a systematic review of the literature. Fertil Steril 2004;82:17–24.
[102] Martinez AR, Bernadus RE, Vermeiden JPW, Shoemaker J. Basic questions on intrauterine insemination: an update. Obstet Gynecol Surv 1993;46:811–28.
[103] Ombelet W, Puttemans P, Bosmans E. Intrauterine insemination: a first-step procedure in the algorithm of male subfertility treatment. Hum Reprod 1995;10: 90–102.
[104] Stewart JA. Stimulated intra-uterine insemination is not a natural choice for the treatment of unexplained infertility. Should the guidelines be changed? Hum Reprod 2003;18:903–14.
[105] Botchan A, Hauser R, Gamzu R, Yogev L, Paz G, Yavetz H. Results of 6139 artificial insemination cycles with donor spermatozoa. Hum Reprod 2001;16: 2298–304.
[106] Horvath PM, Bohrer M, Shelden RM, Kemmann E. The relationship of sperm parameters to cycle fecundity in superovulated women undergoing intrauterine insemination. Fertil Steril 1989;52:288–94.
[107] Berg U, Brucker C, Berg FD. Effect of motile sperm count after swim-up on outcome of intrauterine insemination. Fertil Steril 1997;67:747–50.
[108] Denil J, Ohal DA, Hurd WW, Menge AC, Hiner MR. Motility longevity of sperm samples processed for intrauterine insemination. Fertil Steril 1992;56:436–8.
[109] Bordson BL, Ricci E, Dickey RP, Dunaway H, Taylor SN, Curole DN. Comparison of fecundibility with fresh and frozen semen in therapeutic donor insemination. Fertil Steril 1986;46:466–9.
[110] Cooke ID. Randomized studies in intrauterine insemination. Fertil Steril 2004;82:27–9.
[111] Ragni G, Somigliana E, Vegetti W. Timing of intrauterine insemination: where are we? Fertil Steril 2004;82:25–6.
[112] Federman CA, Dumesic DA, Boone WR, Shapiro SS. Relative efficiency of therapeutic donor insemination using a luteinizing hormone monitor. Fertil Steril 1990;54: 489–92.
[113] Martinez AR, Bernadus RE, Voorhjorst FJ, Vermeiden JPW, Shoemaker J. A controlled study of human chorionic gonadotropin induced ovulation versus urinary luteinizing hormone surge for time of intrauterine insemination. Hum Reprod 1991;6:1247–51.
[114] Kosmas IP, Tatsioni A, Fatemi HM, Kolibianakis EM, Tournaye H, Devroey P. Human chorionic gonadotropin administration vs. luteinizing monitoring for intrauterine insemination timing, after administration of clomiphene citrate: a meta-analysis. Fertil Steril 2007;87:607–12.
[115] Martinez AR, Bernadus RE, Voorhjorst FJ, Vermeiden JPW, Shoemaker J. Intrauterine insemination does and clomiphene citrate does not improve fecundity in couples with infertility due to male or idiopathic factors: a prospective, randomized, controlled study. Fertil Steril 1990;53:847–53.
[116] Tummon IS, Henig I, Radwanska E, Binor Z, Rawlins R, Dimowski WP. Persistent ovarian cysts following administration of human menopausal and chorionic gonadotropins: an attenuated form of ovarian hyperstimulation syndrome. Fertil Steril 1988;49:244–8.

hCG in assisted reproduction

26

Ervin E. Jones
Genetics and IVF Institute, Fairfax, VA, USA

Both human chorionic gonadotropin (hCG) and luteinizing hormone (LH) are heterodimeric glycoprotein hormones [1]. Both are composed of common α-subunits and β-subunits that provide their specificity. Both hormones bind to the same LH/hCG receptor. Classic expression of the LH/hCG receptor has been well-established in cells of testis, ovarian theca, granulose, and luteal cells. LH regulates both steroid hormone synthesis and gametogenesis in both males and females. hCG is secreted by the syncytiotrophoblast cells of the early blastocyst and later by the placenta [2]. Many other tissues are also known to produce hCG, including the human pituitary.

The circulating half-life of hCG is approximately 20–37 h, whereas the circulating half-life of LH is only 25 min to 2 h. This difference in half-life and stronger binding affinity account for the approximate 50-fold difference in potency of hCG compared with LH [3]. Although LH is present in all mammalian species, CG is produced only in humans and advanced primates. The biological and immunological activity of hCG is derived from the structural differences in the β-subunit. The carboxyl terminal end 24 amino acids unique to hCG provide both the biological and immunological specificity of the molecule. This difference allows modern immunoassays to distinguish between the two molecules [4].

More recently, it has become clear that many tissues produce hCG. Recently, it has been shown that menopausal urine contains hCG of pituitary origin. Odell and Griffin [5] were able to show that hCG is secreted in a pulsatile fashion in adult humans of both sexes. The same authors were able to show that cultured human pituitary cells secrete hCG [6]. These observations are supported by the findings of Muller and Simoni, who showed that mRNA for CG is expressed in sections of marmoset pituitary; sections of marmoset pituitary gland also stained positive for the hormone using immunohistochemical techniques [7].

hCG is widely used in assisted reproduction as an agonist for LH. hCG is easily obtained from urine of pregnant women and placental tissue, and it is produced in recombinant Chinese hamster ovary cells. hCG is also less costly when compared with LH. Human menopausal gonadotropin (hMG) is widely used in assisted reproduction and is now known to contain hCG [8].

Human Chorionic Gonadotropin (hCG). DOI: http://dx.doi.org/10.1016/B978-0-12-800749-5.00026-2

26.1 The ovarian cycle and hCG use in assisted reproduction

To describe the use of hCG in assisted reproduction, it is necessary to briefly review the normal ovarian cycle and describe how hCG is used in both natural cycles and in controlled ovarian stimulation cycles for intrauterine insemination or *in vitro* fertilization (IVF). The role of hCG in the luteal – follicular transition and its potential impact on implantation is discussed. The use of hCG in the diagnosis and management of abnormal pregnancy is also reviewed.

26.2 The follicular phase: the role of LH

During the natural ovarian cycle, LH/hCG receptors are present in granulosa, theca, and luteal cells of ovarian follicles. Before ovulation, LH stimulates the production of estrogen by the production of androgens, which are acted on by the enzyme aromatase to produce estrogens. LH levels are low during the early follicular phase and begin to increase just before ovulation, culminating in the LH surge. Increasing LH levels facilitate meiosis in and maturation of oocytes and cause rupture of the follicle. After ovulation, LH induces luteinization of the remaining follicle cells to become a corpus luteum. The corpus luteum comprises large luteal cells (derived from the follicular granulosa cells) and small granulosa cells (derived from the theca cells). Follicle-stimulating hormone (FSH) receptors are only present on granulosa cells. FSH stimulates follicle growth and selection of the dominant follicle in natural cycles and increases aromatase activity, which results in increased production of estrogens from androgens. FSH promotes the production of LH receptors on granulosa cells, which, along with cells from the theca compartment, produce the corpus luteum.

hCG is often used in ovarian stimulation cycles for IVF. Low levels of LH are secreted during the early follicular phase. Low doses of hCG might be substituted for LH in controlled ovarian stimulation cycles. Low-dose hCG (50–200 mIU/ml) provides a simplified, cost-effective protocol for ovarian stimulation when used in combination with FSH. Several of the combined commercial preparations contain hCG of pituitary origin. Filicori and coworkers studied the impact of low-dose hCG alone and its efficiency in clinical practice to replace LH-containing gonadotropins in controlled ovarian stimulation cycles using a controlled, prospective, and randomized study [8]. They found that the duration of FSH administration was reduced, preovulatory steroid levels were higher, the amount of FSH used was lower, and the number of small (but not large) follicles was reduced. Fertilization rates were higher and the pregnancy outcome did not differ. hCG-stimulated follicle growth and maturation in the presence of FSH administration did not cause premature luteinization and resulted in a more estrogenic intrafollicular environment. They concluded that in low doses in the latter stages of controlled ovarian hyperstimulation, hCG alone reduced recombinant FSH consumption, and that the outcomes were comparable with traditional regimens (Figure 26.1).

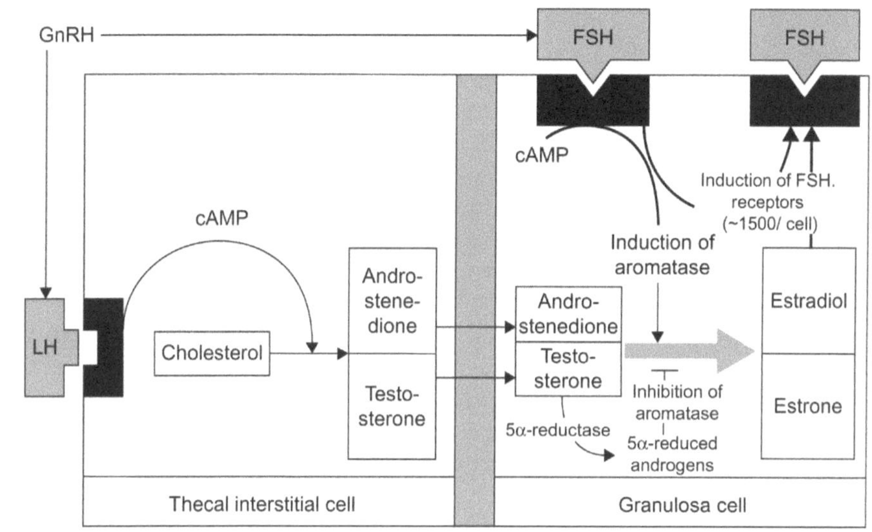

Figure 26.1 Schema depicting the two-cell, two-gonadotropin hypothesis.
Source: Modified with permission from Ref. [9].

Other studies have shown that addition of hCG in cycles of women who seemed to produce high numbers of immature oocytes is a useful maneuver. Huddleston et al. described a subgroup of patients with a high rate of oocyte immaturity during IVF cycles stimulated with recombinant FSH alone. Addition of LH in a subsequent cycle significantly increased the number of mature oocytes. The study was a matched pair design in which each patient served as their own control. Better-quality embryos were also obtained compared with FSH-only cycles [10].

26.3 The periovulatory phase and the mid-cycle LH surge

In natural cycles, estradiol secreted by the developing follicle sensitizes the anterior pituitary to gonadotropin releasing hormone (GnRH), and the LH surge is triggered. During the preovulatory phase increasing LH levels, which eventually culminate in a surge, cause nuclear maturation of the oocyte, softening, and expansion of the oocyte—cumulus complex and ovulation. The endogenous LH surge mechanism is undesirable in cycles stimulated for IVF and embryo transfer because timing of oocyte collection would be unpredictable. GnRH agonists and antagonists are widely used in assisted reproduction to prevent the endogeneous LH surge. In cycles stimulated for IVF, gonadotropins are administered to increase the number of follicles. In IVF cycles, hCG is administered when certain monitoring criteria are met. Such criteria consist of a certain number of preovulatory follicles of predetermined size. Once such criteria are met, hCG is usually administered as a single injection of 5000–10,000 IU to simulate the LH surge. In this instance, hCG injection substitutes for LH and allows for controlled timing of oocyte collection. The hCG injection might be thought of as a "surrogate" LH surge. If hCG is administered too late, then one or more of the most advanced follicles yield postmature (fragmented) eggs of low viability; however, if hCG is injected too soon, then the eggs might be immature. Ovulation will generally occur 36–39 h after hCG administration. Therefore, retrieval of oocytes is timed for 34–36 h after injection of hCG.

In vitro maturation of oocytes is currently a subject of intense investigation. The ability to harvest immature oocytes and mature them *in vitro* would contribute immensely to the field of reproductive medicine. For example, patients with malignant disease would be able to bank oocytes for future use without the need for undergoing potentially risky stimulation protocols that are commonly used to stimulate multiple follicle development for oocyte retrieval. As described in this chapter, hCG is most often used prior to retrieval of oocytes for IVF. Similarly, hCG is also used to induce maturation of immature oocytes *in vivo* and *in vitro* [11]. *In vivo*, priming with hCG results in more oocytes than in nonprimed and FSH-primed cycles, and *in vivo* matured oocytes result in better embryo development and higher pregnancy rates. Their findings illustrate the importance of the use of hCG in the induction of oocyte maturation in modern reproductive medicine.

In controlled ovarian stimulation cycles for timed intercourse or intrauterine insemination, hCG is also used to trigger ovulation. In such cycles, follicle growth is monitored using serial transvaginal ultrasound scanning along with measurements of serum estradiol levels. Timed administration of hCG helps to assure ovulation and facilitates timing of intercourse or insemination.

26.4 The follicular – luteal transition

After natural ovulation or after oocyte retrieval in IVF cycles, the granulosa and theca cells that remain in the follicle are transformed (under the influence of LH) into luteal cells that comprise the corpus luteum. The granulosa cells become the large luteal cells of the corpus luteum, and the theca cells become the small luteal cells. The large cells decline in number in the late luteal phase in concert with a decline in steroidogenesis. The small cells also decrease their ability to respond to LH. Importantly, the large luteal cells that remain seem to increase their ability to respond to hCG. The follicular – luteal transition begins. The corpus luteum primarily secretes progesterone, although it also synthesizes and secretes estradiol.

LH is believed to play a key role in the follicular – luteal transition. The follicular – luteal transition is primarily driven by LH. The LH surge induces the formation of the corpus luteum in natural cycles, and low levels of LH are continuously secreted to support the early development of the corpus luteum. hCG binds to LH receptors and facilitates the follicular – luteal transition. In controlled ovarian stimulation cycles for IVF, exogenous hCG is often used to supplement the luteal phase. In conception cycles, hCG produced by the developing blastocyst stimulates prolongation of the life of the corpus luteum.

26.5 The luteal – placental shift

The corpus luteum generally lives for 11–12 days in nonconception cycles; progesterone levels decrease, menses follows, and the next menstrual cycle ensues. In conception cycles, the placenta must eventually take over progesterone production from the corpus luteum to prevent menses and to permit the pregnancy to succeed. hCG is secreted exponentially and the long half-life extends the life span of the corpus luteum, maintaining progesterone secretion until the placenta begins to produce significant progesterone at approximately 7 weeks of gestation [12]. hCG can be detected in the maternal circulation around the time of implantation; it peaks around 9–11 weeks of gestation. hCG then decreases to approximately one-tenth the peak level. The secretion of hCG is crucial for maintenance of pregnancy until approximately the seventh week of gestation. Between 7 and 10 weeks of gestation, the corpus luteum is replaced by the placenta in terms of progesterone production. This phase of luteal function has been referred to as the "luteal – placental shift." In conception cycles, the life of the corpus luteum is extended until the luteal – placental shift occurs.

In a study of trinucleate embryos obtained from IVF, the eight-cell stage embryo expresses mRNA [13]. hCG can be detected in maternal serum 1–2 days after implantation. Locate et al. studied the differential distribution of mRNA for the α-subunits and β-subunits of CG in the implantation stage blastocyst of the marmoset monkey. CG mRNA was detected mainly in polar and mural trophoblasts 13–15 days postconception [14]. Thus, early in the first trimester, hCG that is manufactured by the syncytiotrophoblast "rescues" the corpus luteum, which is the major source of estrogens and progesterone. This function of the corpus luteum continues well into early pregnancy. By itself, however, the corpus luteum is not capable of generating the very high steroid levels that are characteristic of late pregnancy. The placenta augments the production of progesterone and estrogens such that by 8 weeks of gestation, the placenta has become the major source of these steroids (Figure 26.2).

The timing of the luteal – placental shift is crucial, particularly in IVF cycles. At approximately 8 weeks of gestation, most centers used either progesterone or hCG to supplement the luteal phase. Lower doses of hCG are usually administered (1250–2500 IU) daily after pregnancy has been confirmed. Some centers combine progesterone and hCG in the luteal phase.

26.6 The potential role of hCG in implantation

Precise synchronization between the developing embryo and the maternal endometrium is necessary for successful implantation. hCG is one of the earliest molecules produced by the embryo and is the most specific marker of its presence [16]. A dynamic interchange occurs at the maternal–fetal interface during embryo implantation. Studies using primate models have shown that CG is the major

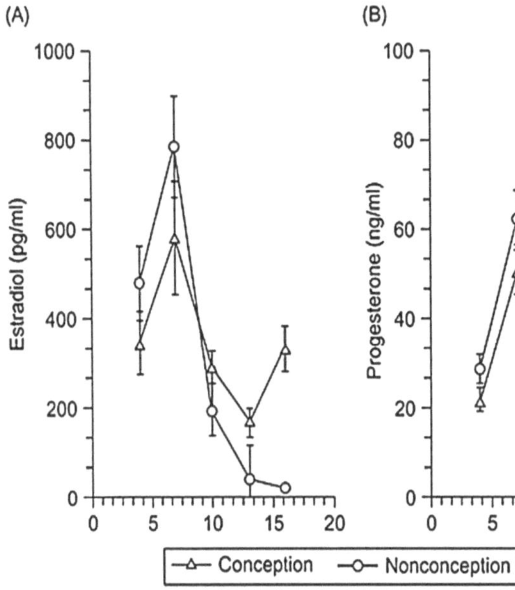

Figure 26.2 Conception and nonconception from IVF menstrual cycles. Mean (±SEM) serum concentrations of estradiol (A) and progesterone (B) in conception and nonconception cycles from 28 IVF cycles without luteal support. Note luteal rescue as reflected by increasing estradiol and progesterone levels under the influence of endogenous hCG in conception cycles.
Source: Reproduced with permission from Ref. [15].

trophoblast signal; it not only rescues the corpus luteum but also modulates the uterine environment in preparation for implantation [12]. This response is characterized by alteration in both the morphological and biochemical characteristics in the three major cell types: luminal and glandular epithelium and stromal fibroblasts. Local effects of CG that influenced the immune system might permit survival of the fetal allograft and prevent endometrial cell death.

The hCG/LH receptor has received considerable attention for its role as a luteotropic factor. This role not only relays the inadequate support provided by the reduced rates of LH secretion but also influences the pregnancy in a paracrine fashion, probably by binding with the LH/hCG receptor on the endometrial epithelium. The embryo actively participates in its own implantation, tolerance, and placentation [17] (Figure 26.3).

hCG functions as a long-range signal and sustains the corpus luteum in the face of precipitously declining LH levels of maternal origin. hCG is one of the most significant molecules secreted by the developing blastocyst, both before and after implantation. In addition to rescuing the corpus luteum, hCG also acts as an immunosuppressive agent and as a growth factor that promotes trophoblast growth and placental development. hCG might also have a role in the adhesion of the trophoblasts to the maternal endometrium; hCG has protease activity, and hCG levels are

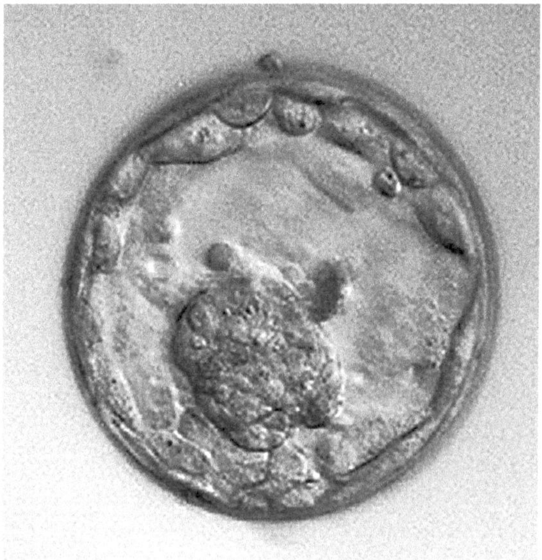

Figure 26.3 A fully expanded 5-day human blastocyst in culture. The trophectoderm and prominent inner cell mass are visible.
Source: Courtesy of Andy Dorfmann, Director Embryology Laboratory, Genetics and IVF Institute, Fairfax, VA.

high in the area where the trophoblasts face and make contact with the endometrium [17]. hCG appears to have a role in uterine angiogenesis and vascular remodeling, both of which are crucial to embryo implantation and normal development of the placenta [18].

26.7 hCG in the management of normal pregnancy

In addition to its therapeutic role in assisted reproduction, hCG is highly useful as a diagnostic tool. Normal serum hCG levels are approximately 100 mIU/ml at the time of missed menses in natural cycles and reach approximately 100,000 mIU/ml by 8–10 weeks of gestation. The first significant increase in plasma hCG occurs 9–12 days after ovulation. There is a rapid increase in the daily output of hCG by *in vitro* cultured blastocyst up to approximately day 11 after fertilization. Gonadotropin production then tends to plateau to day 14 after fertilization. A similar observation was reported using blastocyst recovered from the uterine cavity of the baboon [19]. Catt et al. found that hCG does not appear in the maternal circulation until after implantation of the blastocyst [20]. The serum concentrations of hCG then increase sharply and, with a doubling time of approximately 2.3 days, reach levels of 10,000–100,000 mIU/ml near the end of the first trimester of pregnancy. hCG levels begin to decline after the tenth week of pregnancy.

hCG detected in maternal blood is by far the most reliable marker for biochemical diagnosis of early pregnancy. The development of monoclonal antibody assays for hCG provided increased sensitivity and also decreased the risk of cross-reactivity substances such as LH. hCG is ubiquitous even in the nonpregnant state (at low concentrations). Thus, the risks of a false-positive test have decreased with the advent of monoclonal antibody assays for hCG. A single estimation for hCG should only be considered indicative of pregnancy if it is in excess of 25 mIU/ml; pregnancy can be definitively identified if the level of hCG is seen to double in concentration in serial samples obtained at 2- to 3-day intervals. Knowledge of the chemistry of hCG has led to the development of sensitive and specific immunoassays that are used in assisted reproduction. hCG diagnostics are used in the management of normal uterine pregnancy, abnormal intrauterine pregnancy (IUP), ectopic pregnancy, and cancers of the reproductive tract. For each laboratory, the detection limit for hCG assays should be determined to establish the levels of hCG above which the diagnosis can be confidently made in relation to assay background. hCG measurements should be postponed until the hCG injected before ovum retrieval has been cleared from the patient's blood.

Clinical assays for beta hCG measure between 1 and 5 mIU/ml; levels less than 5 mIU/ml are considered negative for pregnancy; levels between 5 and 25 mIU/ml might still be considered as spurious for pregnancy and should be repeated depending on the clinical circumstance. All pregnancy tests are forms of immunoassays of some sort; home, slide, and dipstick tests detect 20–50 mIU/ml. Developments of highly specific monoclonal antibodies to the β-subunit of hCG have resulted in assays of greater sensitivity and specificity. The β-subunit of hCG provides a sensitive test for hCG. Using this method, the diagnosis of normal pregnancy can be made as early as 8–9 days after ovulation.

26.8 Hyperstimulation syndrome

Ovarian hyperstimulation syndrome occurs after luteinization of ovarian follicles. Hyperstimulation syndrome only occurs after the administration of hCG. Clinical signs of hyperstimulation syndrome usually occur 5–10 days after hCG administration or after the LH surge. Hyperstimulation syndrome is rare in women with a spontaneous LH surge. hCG is thought to be causal in initiation of hyperstimulation syndrome. hCG administration provokes profound effects on paracrine parameters of differentiation at implantation—insulin growth factor binding protein-1 and prolactin and implantation. Leukemia inhibitory factor (LIF), macrophage colony-stimulating factor (MCSF), and vascular endothelial growth factor (VEGF), which are cytokines important for neoangiogenesis, were significantly stimulated by hCG, suggesting that hCG has a role in controlling endometrial vascularization and placentation [16]. It is believed that ovarian secretion causes increased capillary permeability and increased peritoneal fluid shift. Increased capillary permeability results in massive ascites and hypovolemia.

Patients at increased risk for hyperstimulation syndrome should not be treated with luteal phase hCG. Most programs use progesterone supplementation instead of hCG. A major preventive measure is to withhold hCG in the face of impending hyperstimulation syndrome.

Interestingly, hyperstimulation syndrome rarely occurs after a spontaneous LH surge induced by increasing estrogen levels. Recent advances in the control of ovarian cycles stimulated for IVF have led to the use of gonadotropin hormone-releasing hormone agonists to trigger an LH surge. Patients managed in this manner rarely progress to hyperstimulation syndrome. This observation suggests that characteristic actions of the hCG molecule cause hyperstimulation syndrome.

26.9 hCG in the management of ectopic pregnancy

Ectopic pregnancy is a major clinical problem occurring in more than 75,000 cases per year in the United States. Ectopic pregnancy can be an unintended consequence of assisted reproduction. Ectopic pregnancy occurs in approximately 2% of IVF cycles. When compared with normal pregnancies, hCG levels increase at different rates in ectopic pregnancies [21]. The rate of increase of hCG in ectopic pregnancy is lower than that of normal pregnancies [21]. Similarly, the rate of decrease of hCG is lower for ectopic pregnancies than for completed spontaneous abortions. Thus, the earliest manifestation of ectopic pregnancy might be low, or slowly increasing, titers of hCG. Most low or poorly increasing hCG titers are due to abnormal intrauterine pregnancies. hCG levels are measured at approximately 9 days after presumed implantation. For example, most pregnancy tests are obtained 7 days after a day-5 transfer, or 10 days after a day-3 transfer. Ectopic pregnancy must be suspected if the first titer is low or when titers fail to double appropriately.

The measurement of hCG by radioimmunoassay shows the clinical usefulness of hCG. It is a specific indicator of presence of a viable trophoblastic tissue. The gold standard for monitoring suspected ectopic pregnancy is a combination of serial hCG titers and transvaginal ultrasound. When hCG titers reach 1500–2000 mIU/ml, a gestational sac should be visible on transvaginal ultrasound. Nyberg et al. examined 150 women with early pregnancy, 76 intrauterine pregnancies, and 74 ectopic pregnancies [22]. When the hCG titer was 1800 mIU/ml or more in all women with IUP, a gestational sac was seen in all patients. In contrast, in patients with ectopic pregnancy and hCG titer more than 1800 mIU/ml, no gestational sac was seen. The authors concluded that when hCG levels exceed 1800 mIU/ml, an intrauterine gestational sac is normally detected and its absence is evidence for ectopic pregnancy.

hCG doubles every 2 days during the 2–4 weeks after ovulation. An obstetrician should see a sac if hCG is >1500 mIU/ml. When hCG doubles every 2 days, the situation is reassuring but does not rule out abnormal pregnancy. Failure of hCG titers to double must raise the suspicion of ectopic pregnancy. hCG doubles every 2 days if <1200 mIU/ml, every 3 days between 1200 and 6000 mIU/ml, and every 4 days if >6000 mIU/ml. Clinicians must be aware that the discriminatory zone varies with the assay and with the reference preparation. Mixed-effects regression models have also been used to estimate the hCG trajectory in its variability and relationship to pregnancy outcomes [19]. The authors observed a three-fold increase in hCG between the day of detection and the next day. They described the average profile of hCG increase and its variability during the 7 days after estimated implantation in a population of naturally conceived pregnancies. The relative rate of increase decreased thereafter, reaching 1.6-fold between days 6 and 7. hCG levels followed a log-quadratic trajectory, and the patterns of increase were unrelated to number of fetuses or risk of spontaneous abortion for up to six babies.

hCG titers are also used to monitor treatment of ectopic pregnancy, regardless of the treatment modality (i.e., medical or surgical). Methotrexate is a foliate antagonist and is widely used in assisted reproduction for ectopic pregnancy. In all cases, titers must be followed until they become negative. Agostini et al. examined the change in hCG levels after methotrexate injection and outcome in 129 consecutive patients [23]. A 20% decline in hCG levels between days 1 and 4 of methotrexate treatment had a positive predictive value of 97%.

With the evolution of conservative surgery for ectopic pregnancy, the risk of persistent ectopic pregnancy increased [24]. The authors discussed a series of 11 patients, 10 of whom had repeat surgeries for either symptoms or plateauing hCG levels. While monitoring symptoms, serial measurements of hCG levels constitute optimum treatment for this group of patients. Mock et al. also studied the disappearance rate of hCG in patients treated for ectopic pregnancy by laparoscopic salpingostomy. The disappearance of hCG from the circulation followed biexponential decay with a rapid decline followed by slow disappearance. The disappearance kinetics is characterized by two half-lives: one early and one late. The authors concluded that the late half-life should be used as a parameter to follow ectopic pregnancies after treatment [25].

Clearly, low serum levels of hCG might indicate early pregnancy. Persistently low serum hCG levels might indicate abnormal placentation associated with either ectopic pregnancy or IUP. Patients with suspected ectopic pregnancy often have persistently low serum hCG concentrations (50–500 mIU/ml). Undetectable levels of hCG (<5 mIU/ml) rule out the existence of viable trophoblastic tissue and render a medical or surgical procedure unnecessary.

26.10 Conclusions

The use of hCG occupies an extraordinarily prominent place in the field of reproductive medicine. The use of hCG can be partitioned into two major categories: diagnostic and therapeutic. Diagnostic use of hCG includes detection, monitoring, and management of normal pregnancy. hCG is also widely used in the diagnosis, management, and monitoring of abnormal pregnancies, including ectopic pregnancies and persistent trophoblastic disease. Therapeutic uses of hCG in reproductive medicine have found greatest utility in controlled ovarian stimulation protocols for IVF (including *in vitro* maturation of oocytes) and in protocols for ovulation induction for intrauterine insemination.

Several isoforms of hCG resulting from posttranslational modification of the native molecule have been identified. These will find utility in the diagnosis and treatment of both disease states and in ovarian stimulation protocols used in assisted reproduction.

References

[1] Lapthorn AJ, Harris DC. Crystal structure of human chorionic gonadotropin. Nature 1994;369:455–61.

[2] Pierce JG, Parsons TF. Glycoprotein hormones: structure and function. Annu Rev Biochem 1981;50:465–95.

[3] Rao CV, Estergreen VL, Carman Jr FR, Moss GE. Receptors for gonadotrophin and prostaglandin F2 alpha in bovine corpora lutea of early, mid and late luteal phase. Acta Endocrinol 1979;91:529–37.

[4] Filicori M. Use of luteinizing hormone in the treatment of infertility: time for reassessment? Fertil Steril 2003;79:253–5.

[5] Odell WD, Griffin J. Pulsatile secretion of human chorionic gonadotropin in normal adults. N Engl J Med 1987;317:1688–91.

[6] Odell WD, Griffin J. Secretion of chorionic gonadotropin by cultured human pituitary cells. J Clin Endocrinol Metab 1990;71:1318–21.

[7] Muller TM, Simoni M. Chorionic gonadotropin beta subunit mRNA but not luteinizing hormone beta subunit mRNA is expressed in the pituitary of the common marmoset (*Callithrix jacchus*). J Mol Endocrinol 2004;32:115–18.

[8] Filicori M, Fazleabas AT. Novel concepts of human chorionic gonadotropin: reproductive system interactions and potential in the management of infertility. Fertil Steril 2005;84:275–84.

[9] Graunwald, et al. Manual of assisted reproduction. 2nd ed. Berlin: Springer Verlag; 2000.

[10] Huddleston HG, Jackson KV. hMG increases the yield of mature oocytes and excellent-quality embryos in patients with a previous cycle having a high incidence of oocyte immaturity. Fertil Steril 2009;92:946–9.

[11] Son WY, Chung JT, Chian R-C, Herrero B, Demirtas E, Elizur S, et al. A 38 h interval between hCG priming and oocyte retrieval increases *in vivo* and *in vitro* oocyte maturation rate in programmed IVM cycles. Hum Reprod 2008;23:2010–16.

[12] Srisuparp S, Strakova Z. The role of chorionic gonadotropin (CG) in blastocyst implantation. Arch Med Res 2001;32:627–34.

[13] Bonduelle ML, Dodd R. Chorionic gonadotrophin-beta mRNA, a trophoblast marker, is expressed in human 8-cell embryos derived from tripronucleate zygotes. Hum Reprod 1988;3:909–14.

[14] Lopata A, Berka J. Differential distribution of mRNA for the alpha- and beta-subunits of chorionic gonadotrophin in the implantation stage blastocyst of the marmoset monkey. Placenta 1995;16:335–46.

[15] Hutchinson-Williams, et al. Fertil Steril 1990;53:495.
[16] Licht P, Russu V. On the role of human chorionic gonadotropin (hCG) in the embryo-endometrial microenvironment: implications for differentiation and implantation. Semin Reprod Med 2001;19:37–47.
[17] Perrier d'Hauterive S, Berndt S. Dialogue between blastocyst hCG and endometrial LH/ hCG receptor: which role in implantation? Gynecol Obstet Invest 2007;64: 156–60.
[18] Zygmunt M, Herr F, Keller-Schoenwetter S, Kunzi-Rapp K, Munsteadt K, Rao CV, et al. Characterization of human chorionic gonadotropin as a novel angiogenic factor. J Clin Endocrinol Metab 2002;87:5290–6.
[19] Nepomnaschy PA, Weinberg CR. Urinary hCG patterns during the week following implantation. Hum Reprod 2008;23:271–7.
[20] Catt KJ, Dufau ML. Appearance of hCG in pregnancy plasma following the initiation of implantation of the blastocyst. J Clin Endocrinol Metab 1975;40:537–40.
[21] Silva C, Sammel MD. Human chorionic gonadotropin profile for women with ectopic pregnancy. Obstet Gynecol 2006;107:605–10.
[22] Nyberg DA, Filly RA. Ectopic pregnancy. Diagnosis by sonography correlated with quantitative HCG levels. J Ultrasound Med 1987;6:145–50.
[23] Agostini A, Blanc K. Prognostic value of human chorionic gonadotropin changes after methotrexate injection for ectopic pregnancy. Fertil Steril 2007;88:504–6.
[24] Seifer DB, Gutmann JN, Doyle MB, Jones EE, Diamond MP, DeCherney AH. Persistent ectopic pregnancy following laparoscopic linear salpingostomy. Obstet Gynecol 1990;76:1121–5.
[25] Mock P, Chardonnens D, Stamm P, Campana A, Bischof P. The apparent late half-life of human chorionic gonadotropin (hCG) after surgical treatment for ectopic pregnancy. A new approach to diagnose persistent trophoblastic activity. Eur J Obstet Gynecol Reprod Biol 1998;78:99–102.

Illicit use of hCG in dietary programs and to promote anabolism

27

Laurence A. Cole
USA hCG Reference Service, Angel Fire, NM, USA

The most common therapeutic use of hCG is not to promote ovulation or to promote progesterone production and maintain fertility. It is the illicit use of hCG in dietary programs. There are thousands of websites, books, newspaper advertisements, and TV advertisements selling diets with these ridiculous claims. Some athletes even give themselves regular shots of hCG to promote testosterone and anabolism for muscle and bone growth. These uses inspire us to ask, why? This chapter discusses the pros and cons of illicit hCG applications and uses and misuses of hCG administration. Finally, this chapter considers the dangers of any hCG variant administration.

27.1 Dietary programs

Advertisements trying to sell prescription hCG and novel diets come from medical clinics, doctors' offices, and pharmacies all over the world. These advertisements are in newspapers and even presented by medical correspondents on major TV shows (KSL5 Utah, Fox Morning Show, Mike & Juliet Show, and others). This story starts with one very small study performed in England in the early 1950s. Dr. Simeons conducted a nonblinded, noncontrolled study in which participants were restricted to 500 kcal/day and administered 125 IU supplements of urine-origin hCG [1]. The study had amazing results and led to weight loss. According to Dr. Simeons [1], hCG must mobilize stored fat all over the body and suppresses the appetite. If you search the Internet, there are literally thousands of advertisements for the hCG-based diet of Dr. Simeons.

In 1959, Sohar [2] argued that it was solely the 500-kcal diet that led to the weight loss and that the hCG supplements do absolutely nothing. In the 1960s, Craig et al. [3] and Frank [4] each conducted their own study regarding the Simeons diet. The two separate experiments both agreed that there is no possible or even conceivable relationship between hCG administration and loss of weight or hunger. In the years that followed, physicians profiting from the dietary administration of hCG began performing double-blind studies. These small, inappropriately controlled studies claimed to confirm that the Simeons diet worked [5,6].

Human Chorionic Gonadotropin (hCG). DOI: http://dx.doi.org/10.1016/B978-0-12-800749-5.00027-4

Then, in the 1970s, several double-blind studies were independently performed by Young et al. [7], Stein et al. [8], Greenway and Bray [9], and Shetty and Kalkoff [10]. They all reached the same conclusion as Sohar did in 1959—that hCG has absolutely nothing to do with hunger and fat mobilization and does not promote weight loss. In the 1980s, Richer and Runnebaum [11] reached the same conclusion. In the 1990s, the high-standard controlled studies of Bosch et al. [12] and Lijesen et al. [13] once again confirmed that an hCG diet simply does not and could not work. The amount of medical evidence that began accumulating was overwhelming. Countless studies used clean, double-blind, and carefully controlled data to disprove the claims of Simeons' hCG-based diet. It simply does not work [14,15]. It should be noted that in 2007, the USA Federal Trade Commission charged Kevin Trudeau with misrepresentation for writing a book praising the hCG diet [16].

Today, the CV Rao Laboratory [17–19], world experts on hCG biological function, has identified hCG receptors in numerous sites associated with pregnancy, the uterus, the placenta, the fetus, and the brain. The laboratory has not, however, ever found evidence for a receptor in the digestive tract, in the liver, or in adipose tissues that could explain how hCG mobilizes stored fat and suppresses hunger. Thus, Simeons's 1954 results are seemingly false and were apparently contrived to support a new enterprise. Nevertheless, Simeons started a dietary fad, and even after 59 years his diet is still sold throughout the world. Dr. Simeons even published a support book for followers of his diet. Somehow, Simeons became a public and scientific hero for this nonsense diet. After the accumulation of all this evidence, how and why there are still clinics, doctors' offices, and pharmacies completely sold on the hCG diet defies all logic, unless they are falsely led by money. Yes, hCG will promote emesis or nausea and vomiting. Is this the secret to the diet's claims of exceptional weight loss—extreme hyperemesis or nausea and constant vomiting?

Today, clinics and pharmacies give patients a choice of injectable hCG, hCG drops that one places under the tongue, or hCG–green tea pills [16,20–23]. They claim that each works as well as the other, but there is no evidence to support that hCG can mobilize fats, suppress hunger, or induce euphoria, as claimed. Surely, the digestive tract destroys the large glycoprotein hormone before it is absorbed into the circulation. Pills comprising hCG are a fraud. Injectable hCG preparations range from partially purified human pregnancy urine extracts called Profasi, Pregnyl, Novarel, Chorex, and Follotein to super-pure Chinese hamster ovary cell-line recombinant hCG called Ovidrel. All are misused for dietary purposes with claims of great weight loss. It is one thing for a woman to be sold by crazy advertisements about some miracle diet, but it is another to consider the consequence of hCG administration, cessation of menstrual periods, infertility, and hyperemesis gravidarum.

27.2 hCG and anabolism promotion

Unquestionably, hCG promotes testicular testosterone that acts on muscles and bones to promote growth and anabolism in men. Interestingly, injections of hCG do not promote significant testosterone production in women [24], but rather promote progesterone and androstenedione production [24]. As such, the other illicit use of hCG is in athletics, particularly in major professional sports and world sports such as the Olympic Games. The hormone is also sold on the Internet and in magazine advertisements, purporting to make people strong like Mr. Universe by aiding muscle growth.

Agencies such as the World Anti-Doping Agency (WADA) and the U.S. Anti-Doping Agency (USADA) started to test the urine of Olympic and other international athletes. In the United States, all major athletic associations perform random urine tests that detect testosterone, growth hormone, and the hormone hCG, including the National College Athletic Association, National Football League, National Hockey League, National Basketball League, National League Baseball, and the American League Baseball. As shown by Stenman et al. [25], hCG can circulate for 7–11 days after injection. The USA hCG Reference Service has shown WADA and USADA that during this period, hCG is gradually degraded to free subunit, to nicked hCG, to nicked molecules missing the β-subunit C-terminal peptide, and finally to β-core fragment. Thus, it is important to measure each of these molecules in urine tests.

Today, most sporting agencies test urine using the Siemens Immulite assay, which is the only automated hCG test that detects all of these degradation products [26,27]. Laidler et al. [28] measured hCG in 1400 men and statistically determined that a 5-mIU/ml cutoff was very acceptable in terms of total hCG concentration present in urine. Delbenke et al. [29] examined 5663 men and found background hCG in men reaching 2.28 mIU/ml. They also supported the 5-mIU/ml sensitivity limit. This is now used as the cutoff in doping studies by WADA and USADA.

27.3 hCG variants as dangerous substances

Research presented in this book clearly demonstrates and confirms that hCG variants, most notably hyperglycosylated hCG, hyperglycosylated hCG free β-subunit, and hCG free β-subunit, are the principal drivers of most human cancers (see Chapters 10, 18–23). These promoters, but not regular hCG, function by binding and antagonizing a TGFβ type II receptor [30–32]. Expression of the molecules and their TGFβ pathways appear to be a major part of carcinogenesis or human cancer transformation.

These three molecules, hyperglycosylated hCG, hyperglycosylated hCG free β-subunit, and hCG free β-subunit, appear to transfer cells into malignant cells by driving growth, blocking apoptosis, and driving invasion (see Chapters 10, 18–23). As such, if a person has damaged tissues, pre-cancerous tissue, or immune-suppressed cancer tissue in their body, then that person is likely to have that tissue transformed into cancer tissue by the presence of these molecules. As shown in Table 27.1, most common prescription commercial hCG preparations are contaminated with these molecules. This is probably particularly true regarding hCG pills

Table 27.1 **Commercial preparation of hCG and its variants**

Preparation	Origin	Hypergly-cosylated hCG	hCG free β-subunit	Biological activity calibrated with WHO 3rd I.S.
Serono Ovidrel	CHO cell recombinant	<0.1%	<0.1%	11,900 IU/mg
Serono Profasi	Pregnancy urine	6.0%	1.6%	10,000 IU/mg
Scripps >99% hCG	Pregnancy urine	2.2%	0.8%	11,000 IU/mg
Scripps >80% hCG	Pregnancy urine	6.8%	5%	9000 IU/mg
Ferring Choragon	Pregnancy urine	4%	5%	5000 IU/mg
Organon Pregnyl	Pregnancy urine	7%	12%	3000 IU/mg
Scripps >15% hCG	Pregnancy urine	13%	13%	2000 IU/mg
Sigma C5297	Pregnancy urine	12%	14%	3000 IU/mg

CHO cells are Chinese hamster ovary cells.

and hCG nasal drops sold on the Internet. It is inferred that these urinary-derived hCG preparations are highly carcinogenic or very dangerous substances.

One hCG preparation, Serono Ovidrel, is a recombinant form of hCG made with Chinese hamster ovary cells. It is an absolutely pure hCG containing no variants. This is seemingly the safest form of hCG to use. It still is very slowly or partially dissociated into hCG free β-subunit, one of the cancer promoters [33], so it is not completely harmless. If one has to administer hCG to oneself, then this expensive preparation, Serono Ovidrel, is clearly the only form to use. Ideally, however, one should stay away from all forms of hCG.

References

[1] Simeons ATW. The action of chorionic gonadotropin in the obese. Lancet 1954;2: 946–7.

[2] Sohar EA. A forty-day 550 calorie diet in the treatment of obese outpatients. Am J Clin Nutr 1959;5:514–19.

[3] Craig LS, Ray RE, Waxler SH, Madigan HM. Chorionic gonadotropin in the treatment of obese women. J Clin Nutr 1963;12:230–4.

[4] Frank BW. The use of chorionic gonadotropin hormone in the treatment of obesity. Am J Clin Nutr 1964;14:133–6.

[5] Gusman HA. Chorionic gonadotropin in obesity. Further clinical observations. Am J Clin Nutr 1969;22:686–95.

[6] Asher WL, Harper HW. Effect of human chorionic gonadotropin on weight loss, hunger and feeling of well being. Am J Clin Nutr 1973;26:211–18.

[7] Young RL, Fuchs RJ, Woinjen MJ. Chorionic gonadotropin in weight control. A double blind crossover study. J Am Med Assoc 1976;236:2495–7.

[8] Stein MR, Julis RE, Peck CC, Hinshaw W, Sawicki JE, Deller JJ. Ineffectiveness of human chorionic gonadotropin in weight reduction: a double-blind study. Am J Clin Nutr 1976;29:940–8.
[9] Greenway FL, Bray GA. Human chorionic gonadotropin (hCG) in the treatment of obesity. West J Med 1977;127:461–3.
[10] Shetty KR, Kalkoff RK. Human chorionic gonadotropin (hCG) treatment of obesity. Arch Intern Med 1977;137:151–5.
[11] Richer RT, Runnebaum B. Risiko-Nutzen risk–benefit analysis of a hCG-500 kcal reducing diet (Cura Romana) in females. Geburtshitfe und Frauenhellkunde 1987;47: 297–305.
[12] Bosch B, Venter I, Stewart RI, Bertram SR. Human chorionic gonadotropin and weight loss. A double-blind placebo-controlled trial. S Afr Med J 1990;77:185–9.
[13] Lijesen GK, Theeuwen I, Assendelft WJ, Van Der Wal G. The effect of human chorionic gonadotropin (hCG) in the treatment of obesity by means of the Simeon therapy: a criteria-based meta-analysis. Br J Clin Pharmacol 1995;40:237–43.
[14] Barrett S. hCG worthless as a weight-loss aid. Diet Scam Watch 2007, <http://www.dietscam.org/reports/hcg.shtml>; 2010.
[15] Peyman T. The hCG myth, <http://www.drtarapeyman.com/hCGArticle.pdf>.
[16] FTC: Marketer Kevin Trudeau violated prior court order, charges him with misrepresenting contents of book, <http://www.casewatch.org/ftc/news/2007/trudeau.shtml>; 2007.
[17] Lei ZM, Reshef E, Rao CV. The expression of human chorionic gonadotropin/luteinizing hormone receptors in human endometrial and myometrial blood vessels. J Clin Endocrinol Metab 1992;75:651–9.
[18] Lei ZM, Rao CV, Kornyei J, Licht P, Hiatt ES. Novel expression of human chorionic gonadotropin/luteinizing hormone receptor gene in brain. Endocrinology 1992;132: 2262–70.
[19] Lei Z, Rao CV. Gonadotropin receptors in human fetoplacental unit: implications for hCG as an intracrine, paracrine and endocrine regulator of human fetoplacental function. Placenta 1992;13:213–24.
[20] hCG Diet, <http://www.hcgdietdirect.com/?gclid = CIGSp_60z50CFRESawod9HZFrg>; 2010.
[21] Your hCG.com, <http://www.yourhcg.com/purchase.php>; 2010.
[22] hCG Medical. Advanced Wellness and Hormone Therapy, <http://www.hcgmedical.com/inthenews/index.html>; 2008.
[23] OraThin oral hCG alternative, <http://www.naturalcuresstore.com/product/ORAL_HCG?gclid = CPG_5-q4z50CFShSagodGjqdrw>; 2010.
[24] Handelsman DJ. Clinical review: the rationale for banning human chorionic gonadotropin and estrogen blocker in sport. J Clin Endocrinol Metab 2006;91:1646–53.
[25] Stenman UH, UnkilaKallio L, Korhonen J, Alfhan H. Immunoprocedures for detecting human chorionic gonadotropin: clinical aspects and doping control. Clin Chem 1997; 43:1293–8.
[26] Cole LA, Sutton JM, Higgins TN, Cembrowski GS. Between-method variation in hCG test results. Clin Chem 2004;50:874–82.
[27] Cole LA, Shahabi S, Butler S, Mitchell H, Newlands ES, Behrman HR, et al. Utility of commonly used commercial hCG immunoassays in the diagnosis and management of trophoblastic diseases. Clin Chem 2001;47:308–15.
[28] Laidler P, Cowan DA, Hider RC, Kickman AT. New decision limits and quality control material for detecting human chorionic gonadotropin misuse in sports. Clin Chem 1994;40:1306–11.

[29] Delbenke FT, Van Eenoo P, De Backer F. Detection of human chorionic gonadotropin misuse in sports. Int J Sports Med 1998;19:287–90.
[30] Butler SA, Ikram MS, Mathieu S, Iles RK. The increase in bladder carcinoma cell population induced by the free beta subunit of hCG is a result of an anti-apoptosis effect and not cell proliferation. Br J Cancer 2000;82:1553–6.
[31] Berndt S, Blacher S, Munuat C, Detilleux J, Evain-Brion D, Noel A, et al. Hyperglycosylated human chorionic gonadotropin stimulates angiogenesis through TGF-β receptor activation. FASEB J 2013; Abstract 12-213686.
[32] Cole LA, Butler SA. Hyperglycosylated hCG hCGβ and hyperglycosylated hCGβ: interchangeable cancer promoters. Mol Cell Endocrinol 2012;349:232–8.
[33] Norman RJ, Buchholtz MW, Somogyl AA, Amato F. hCG-β core fragment is a metabolite of hCG: evidence from the infusion of recombinant hCG. J Endocrinol 2000;164: 299–305.

Positive hCG tests: Causes other than pregnancy

28

Laurence A. Cole
USA hCG Reference Service, Angel Fire, NM, USA

The hCG test is considered by most physicians to be a pregnancy test and nothing else. Unfortunately, today when a woman has a positive result on an hCG test and is proven not to be pregnant and not to have an ectopic or tubal pregnancy, chaos occurs. I say chaos in that generally most physicians have no idea what to do or what to test. Medical students and obstetrics and gynecology textbooks teach and are taught that an hCG test can only be positive in pregnancy. If a patient has a positive result on an hCG test and is not pregnant, then cancer must be considered along with choriocarcinoma or gestational trophoblastic neoplasms. As such, most patients will then be sent to an oncologist or gynecologic oncologist, and chemotherapy will be started or hysterectomy or salipingooophrectomy will be considered. Unfortunately, it is only when these procedures fail that the physician will consider referring the patient to the USA hCG Reference Service or reading one of the many publications published by them in the *New England Journal of Medicine* [1], *Gynecology Oncology* [2], or a wide mixture of other medical journals. The problem today is that nobody seems to know what to do.

The USA hCG Reference Service was started in 1998. It specializes in this problem and in complications of gestational trophoblastic diseases. To date, it has consulted on 529 cases positive for hCG but not involving pregnancy. With this experience it has become a world expert on patients with positive hCG test results but who are not pregnant. The USA hCG Reference Service has found eight reasons why individuals can have a positive result on an hCG test but not be pregnant. These reasons are described in detail in this chapter.

The USA hCG Reference Service was started in 1998 during a problematic time with false-positive hCG test results. In their first 4 years, most cases had false-positive results. The only other reasons known at this time for hCG test being positive but the person not being pregnant was pituitary hCG, cancer, and choriocarcinoma/gestational trophoblastic disease. It was the discovery of women positive for hCG on urine and serum tests with a history of miscarried pregnancy or molar pregnancy that led to the USA hCG Reference Service discovery of quiescent gestational trophoblastic disease [3,4]. As we now know today, quiescent gestational trophoblastic disease is the principal explanation for a woman having a positive result on an hCG test but without being pregnant. With time, the USA hCG Reference Service discovered cases that were due Munchausen's syndrome and to women administering hCG as a diet drug [5].

Human Chorionic Gonadotropin (hCG). DOI: http://dx.doi.org/10.1016/B978-0-12-800749-5.00028-6

In 2010, the USA hCG Reference Service discovered cases that could not be explained by false-positive hCG, quiescent gestational trophoblastic disease, pituitary hCG, cancer, or choriocarcinoma/gestational trophoblastic neoplasm. This is when they discovered incidentally familial hCG syndrome, a new genetic reason for producing hCG outside of pregnancy.

Looking back at the USA hCG Reference Service's 15 years of experience with 529 cases. A total of 210 cases (40%) were diagnosed as quiescent gestational trophoblastic disease, 129 were diagnosed as producing pituitary hCG (24%), 103 (19%) had a false-positive hCG test, 44 (8.3%) were diagnosed as having an undetermined cancer, 23 (4.3%) were diagnosed as having choriocarcinoma/gestational trophoblastic neoplasm, 15 (2.8%) were diagnosed as having familial hCG syndrome, 3 were diagnosed with Munchausen's syndrome (0.56%), and 2 (3.8%) were self-administrating hCG. The eight causes of positive hCG test are presented here. No other causes of hCG production or positive hCG test results are known.

28.1 False-positive hCG test results

False-positive hCG results have been a major issue regarding quantitative serum hCG assays. Throughout their history, both immunometric and radioimmunoassays have used animal antibodies to bind hCG [1–3]. Humans produce heterophilic antibodies against human immunoglobulins. These can cross-react with animal immunoglobulins. Humans also produce antibodies against the animal antibodies used in vaccinations and other therapies, such as human anti-mouse antibodies (HAMA). Most modern assays use mouse monoclonal antibodies and goat polyclonal antibodies. The presence of HAMA, human anti-animal antibodies (HAAA), and heterophilic antibodies in blood samples can interfere with assays and cause false-positive results.

Hormones like hydrocortisone, luteinizing hormone (LH), estradiol, and progesterone are always present in serum; therefore, tests are always positive. Levels become slightly higher than normal when blood contains interfering HAMA, HAAA, or heterophilic antibodies, and these interfere with assay and falsely indicate raised values.

In contrast, hCG is normally only present in serum during pregnancy. In terms of becoming false-positive, hCG is a unique hormone. When HAMA/HAAA or heterophilic antibodies interfere with an estradiol, progesterone, LH, or hydrocortisone test, slightly elevated results occur, which is a minor issue. In contrast, if hCG is falsely positive due to heterophilic antibodies/HAAA/HAMA, it suggests pregnancy; however, if the person is not pregnant, then it can create an emergency.

Heterophilic antibodies and false-positive test results can occur because of a family history of immunoglobulin-A deficiency syndrome or a history of a relatively common infection like mononucleosis [4]. False-positive hCG test results are an old problem. Heterophilic antibodies first became a major problem with the old radioimmunoassay of the 1970s [1,2]. All too often, women were mistakenly identified as pregnant because of interfering antibody false-positive radioimmunoassays.

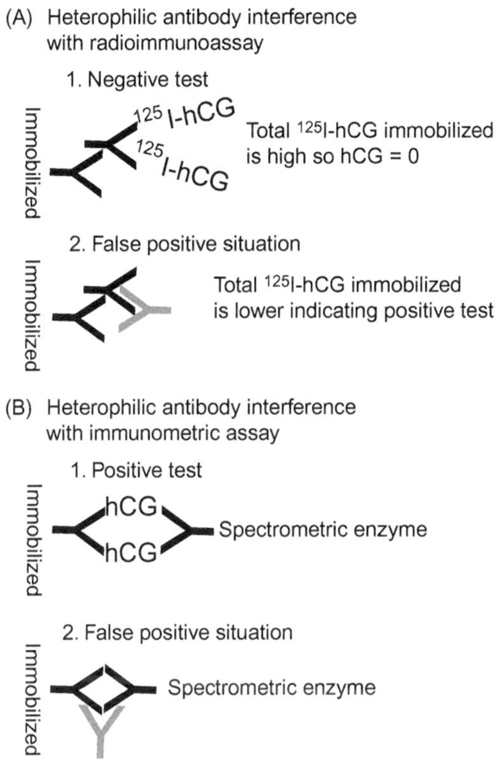

Figure 28.1 HAMA/HAAA/heterophilic antibody interference in the (A) radioimmunoassay and (B) immunometric assay.

Radioimmunoassays are read as being negative when all the radioactive hCG in the assay is bound by the limiting animal antibody. When HAMA/HAAA or a heterophilic antibody is present in the serum samples, it may block the ability of the limiting animal antibody to bind the radioactive hCG (Figure 28.1). This will reduce the amount of radioactive hCG bound to antibody in the assay. This will falsely make the assay to appear positive. The assay will read the lesser amount of radioactive hCG bound to the antibody. This falsely indicates a positive test result (Figure 28.1).

The multi-antibody immunometric test was invented in the early 1980s. It was hoped that these tests would prevent false-positive problems in hCG assays [3]. There were no reported troubles until 1999, when false-positive cases began to appear throughout the United States [5–10]. Using immunometric assays, a sandwich is formed between a capture antibody, hCG, and a second site tracer antibody. The concentration of sandwiches is measured. The interfering HAMA/HAAA or a heterophilic antibody can link the capture and tracer antibodies like hCG and give false-positive results (Figure 28.1).

These problems created a demand for a laboratory to resolve false-positive hCG problems, and the USA hCG Reference Service was born. They identified five cases as false-positive in 1999, 12 at the end of 2000, and an alarming 52 cases between 1999 and 2004. Of the 52 cases identified, 34 (65%) underwent major chemotherapy or surgery for assumed cancer. Of these 34 cases, 10 young women underwent hysterectomy and 3 received a bilateral salpingooophorectomy that left them either sterile or unable to carry a pregnancy [1–6].

The USA hCG Reference Service has established three clear criteria from its experience for identifying false-positive hCG [5–10]. HAMA/HAAA/heterophilic antibodies are very large molecules that do not cross the kidney and enter the urine. As such, the first criterion that shows antibody interference is that there is negative urine hCG in the presence of positive serum hCG [5–10]. Urine was measured by the USA hCG Reference Service in 71 cases. Urine testing identified 71 of 71 (100%) false-positive cases and was by far the most sensitive method for detecting false-positive hCG. Urine was measured using a sensitive automated immunoassay designed for serum (Siemens Immulite, sensitivity 1 mIU/ml) and not an insensitive point-of-care test as commonly used in the United States (sensitivity >20 mIU/ml).

The second criterion is the observation of different or negative results between the Service's assay (the Siemens Immulite 1000) and the referring institute's assay. This identified and confirmed 83% of cases. The final criterion is the fact that urine β-core fragment is made in the kidney and is only present in human urine. Positive β-core fragment in serum indicated HAMA/HAAA/heterophilic antibodies or false-positive hCG. This was useful in 57% of cases. We did not find that examining diluted samples was of much use. Laboratories today have a test for heterophilic antibodies in serum. We find that this test is only partially useful in identifying false-positive hCG cases. We propose that, to demonstrate a false-positive, at least two of these three methods described here need to show the false-positive hCG problem. This was achieved in all USA hCG Reference Service false-positive cases. These methods and the use of heterophilic antibody blocking agents have been described previously [11,12].

Investigations have since shown that the Abbott AxSym test was not appropriately protected against HAMA/HAAA/heterophilic antibodies [13–19]. Abbott added animal serum to its sample diluting buffer to protect against HAMA/heterophilic antibodies but did not add anything to the antibody preparations. As such, the test was protected using diluted samples, but not when testing undiluted samples with lower hCG concentrations [5–10]. This was the source of their false-positive problem. Subsequent to the USA hCG Reference Service's early publications [5–10], most women receiving needless therapy sued Abbott in individual and class-action lawsuits. This brought the false-positive problem to the attention of the media to physicians [16,18,19]. With the lawsuits, Abbott AxSym moved from being the number one selling assay to a minor selling assay in the United States. With the highly publicized false-positive problem, laboratories wanted to avoid using the Abbott AxSym hCG test. Abbott eventually fixed their test. Since 2004, the USA hCG Reference Service has not heard about a single false-positive result with the Abbott AxSym test or with Abbott's new platform, the Abbott Architect hCG assay.

With all the attention from the media and lawsuits, physicians became weary of false-positive cases and started using laboratory tests to check women for HAMA, HAAA, and heterophilic antibodies. As such, the USA hCG Reference Service had only 37 false-positive cases referred between 2004 and 2013. It appeared that the crisis was over. The USA hCG Reference Service is aware of false-positive results with all major laboratory hCG assays in extreme HAMA/HAAA/heterophilic antibody circumstances. Most quantitative serum hCG tests are protected by animal serum in the antibody preparations, but some HAMA/HAAA/heterophilic antibody cases have high levels of anti-animal antibody, which can still influence otherwise appropriately protected assays and cause false-positive results. This was the case with the Siemens Centaur, Siemens ACS180, Siemens Dimension, Ortho Vitros, and Beckman Access automated assays. Considering the 103 false-positive cases, 49 underwent needless chemotherapy or hysterectomy for an assumed cancer or gestational trophoblastic disease. Physicians need to rethink how to treat cases positive for hCG outside of pregnancy.

28.2 Pituitary hCG

As described in detail in Chapter 16, menopausal pituitary hCG is a common cause of false-positive pregnancy hCG test results. As described and illustrated, the hypothalamus of the brain produces gonadotropin-releasing hormone (GnRH). This acts on gonadotrope cells promoting LH and FSH production. The LH β-subunit gene is on chromosome 19. A single copy of LH β-subunit is found in the middle of eight parallel hCG β-subunit genes. It appears, based on the low concentration of hCG produced (<1–33.6 mIU/ml), that GnRH, while trying to promote LH, incidentally also promotes one hCG β-subunit gene. The gonadotrope cells of the pituitary produce excess common α-subunit, which combines with hCG β-subunit to form hCG.

Normally, GnRH production is regulated by feedback from estrogen. When estrogen feedback decline and halts in perimenopause (age >38) and menopause (age >43), GnRH pulse maximizes, leading to normal high production of LH, FSH, and hCG (<1–33.6 mIU/ml) [20–23]. The easiest way to demonstrate hCG production by the pituitary is to show increased FSH (FSH >30 mIU/ml) [24]. Other methods include suppressing hCG and showing its reduction to <1 mIU/ml by treating for 3 weeks with a high-estrogen contraceptive pill [23].

28.3 Quiescent gestational trophoblastic disease

The USA hCG Reference Service has now consulted on 210 cases of quiescent gestational trophoblastic disease. In the Reference Service's experience, it is probably the most common cause for persistent low hCG levels outside of pregnancy in a woman of menstrual age. Each year, the USA hCG Reference Service is referred approximately 90 cases of persistent low hCG levels with non-increasing hCG levels outside of pregnancy. Approximately 35 are pituitary hCG in menopausal women, 3 are familial hCG syndrome, 2 are cancer, 5 are false-positive hCG (quite rare today), and 45 (50%) are quiescent gestational trophoblastic disease.

A total of 97 of these 210 patients (46%) consulting with the USA hCG Reference Service had cases of quiescent gestational trophoblastic disease followed by the evacuation of complete hydatidiform mole or the termination of a complete mole pregnancy. A total of 53 patients (25%) had genetically undefined miscarriage of pregnancy, 28 patients (13%) had a history of choriocarcinoma/gestational trophoblastic neoplasm, 10 patients (4.8%) had partial hydatidiform mole or partial molar miscarriage, and 8 patients (3.8%) had termination of pregnancy.

In each case, after miscarriage, dilation and curettage, or chemotherapy, hCG levels never diminished to 0; they plateaued between 1.1 and 207 mIU/ml and only declined with the death or apoptosis of residual quiescent tissue. Two patients that we have consulted underwent transabdominal hysterectomy. Examining the histology, a very small amount of highly differentiated syncytiotrophoblast tissue was observed [25]. Quiescent gestational trophoblastic disease seemingly represents highly differentiated syncytiotrophoblast cells with no significant cytotrophoblast cells. Because there are no significant cytotrophoblast cells, there is no detectable hyperglycosylated hCG or no invasion promoter [25–29]. To date, we have observed cases with hCG ranging from 1.1 to 207 mIU/ml. Higher levels have not yet been observed. The median hCG level in those patients consulted was 16.05 mIU/ml. In every case, quiescent gestational trophoblastic disease was marked by the absence of hyperglycosylated hCG or the low presence of hyperglycosylated hCG (>10% of total hCG).

If your center does not have access to the hyperglycosylated hCG test to diagnose quiescent disease (available from Quest Diagnostics Inc., order test 11303X), it is still easy to identify. Three months of constant low hCG levels (1.1–207 mIU/ml) with no more than two-fold oscillation in hCG results can be observed in all cases of quiescent disease.

Quiescent gestational trophoblastic disease has no hyperglycosylated hCG and only slow-growing non-invasive syncytiotrophoblast cells. We have never observed a case of quiescent gestational trophoblastic disease that responds to any kind of chemotherapy. In the USA hCG Reference Service's experience, quiescent gestational trophoblastic disease disappears within 6 months in most cases. We have observed rare exceptions where prolonged low hCG levels lasted as long as 9 years. In patients with a history of hydatidiform mole or spontaneously aborted pregnancy, 10% of cases led to invasive trophoblastic disease, with hCG levels doubling and then doubling again over a period of time. In patients with a history of gestational trophoblastic neoplasm or choriocarcinoma, the incidence of recurrence of disease is notably higher (25%).

The USA hCG Reference Service first identified quiescent gestational trophoblastic disease in patients with no hyperglycosylated hCG or no invasion promoter. Their findings have been confirmed in England and by Yale University [30,31]. If quiescent gestational trophoblastic disease is identified, then hCG levels should be followed closely over the ensuing months until hCG disappears.

28.4 Cancer

Unfortunately, a small but low proportion of quiescent gestational trophoblastic disease cases are actually cancer (8.3% of 529 cases). Commonly, this is an early-stage cancer that is otherwise asymptomatic. In nine of 41 cancer cases (22%) diagnosed by the USA hCG Reference Service in which histology was confirmed later by a referring center, the pathology was a rarer cancer, ovarian dysgerminoma, which is an ovarian germ cell malignancy (Table 28.1). Among the 44 cases diagnosed by the USA hCG Reference Service, the serum total hCG was generally quite low (median 32.5 mIU/ml, range 0.48–683 mIU/ml). Most of the serum total hCG was hCG free β-subunit (84 ± 17% of total hCG) (Table 28.1). Eight of the 44 cases were found in men. These were mostly prostate malignancies (Table 28.1).

It can be difficult to identify an hCG source as cancer. Generally, procedures start by excluding quiescent gestational trophoblastic disease, pituitary hCG, and false-positive hCG results. The only unique attribute of cancer cases is that they mostly secrete the free β-subunit of hCG, with 84 ± 17% of total hCG as this free subunit (Table 28.1). Detection of low levels of hCG (<683 mIU/ml) that are mostly free β-subunit is clearly suggestive of cancer [32]. The ultimate test is a brain and pelvis MRI and a chest CT scan.

28.5 Choriocarcinoma/gestational trophoblastic neoplasm

Choriocarcinoma is pathology-confirmed malignancy of the trophoblast cell of the placenta. Commonly, symptoms or test results can indicate choriocarcinoma. Such cases are called gestational trophoblastic neoplasm until pathology is confirmed. Choriocarcinoma occurs after pregnancy in 1 in 30,000 pregnancies in the United States [33]. Choriocarcinoma also occurs after complete hydatidiform mole. Complete hydatidiform mole occurs in 1 in 900 pregnancies in the United States. It affects approximately 1 in 50 complete hydatidiform mole cases [33].

As published, choriocarcinoma is totally driven by the TGFβ antagonist and autocrine hyperglycosylated hCG [25,34]. As described in Chapter 22, through this mechanism hyperglycosylated hCG drives choriocarcinoma cell growth, blockage of apoptosis, and production of invasive enzymes, collagenases, and metalloproteinases [25,35,36].

Similarly, the detection of the presence of hyperglycosylated hCG in serum (>40% of total hCG) is a solid indicator of invasive gestational trophoblastic disease or choriocarcinoma (gestational trophoblastic neoplasm before pathology) (Table 28.2), and thus is a marker of choriocarcinoma. Serum and urine hCG titers in untreated choriocarcinoma cases can be as high as 596,000 mIU/ml.

Table 28.1 **Women not pregnant with positive hCG, diagnosis cancer, $n = 44$**

Case #	Age	Total hCG serum	Intact hCG	Free β serum	Free β%	Total hCG urine	β-core urine	β-core%	Ultimate cancer histology
67	26	206	30.9	145	70%	160	387	>100%	Adenosquamous cervical
27D	31	6.6	<1.0	6.6	100%	37	46.8	>100%	Adenosquamous cervical
9	41	8	<1.0	82.8	>100%	155	149	96%	Brain glioblastoma
364	52	274	38.4	198	72%	302	272	90%	Clear cell renal carcinoma
27E	22	6.4	<1.0	10.4	>100%	4.4	2.1	48%	Ductal breast cancer
195	40	27.9	3.2	21.6	77%	83.8	40	48%	Ductal breast cancer
114	17	0.48	<1.0	0.54	>100%	<1.0	<0.8	ND	Endometrial adenocarcinoma
277	21	1.5	<1.0	8.3	>100%	1.5	ND	ND	Endometrial adenocarcinoma
121	34	160	<1.0	172	>100%	49	120	>100%	Endometrial adenocarcinoma
28	33	19	2.1	14.9	78%	3.3	4	>100%	Hepatocellular carcinoma
365	27	209	36.6	136	65%	220	136	62%	Lobular breast cancer
362	20	235	25.9	183	78%	140	56	40%	Lobular breast cancer
361	27	73.4	7.0	59.4	81%	ND	ND	ND	Multiple myeloma
57	37	3.1	<1.0	5	>100%	3.6	ND	ND	Non small cell lung
179	18	45.9	5.7	34.2	75%	15	56	>100%	Ovarian dysgerminoma
175	14	4.9	0.9	3.06	62%	3.4	16	>100%	Ovarian dysgerminoma

185	21	320	2.3	329	>100%	230	840	>100%	Ovarian dysgerminoma
515	17	110	19.3	72	65%	129	80	62%	Ovarian dysgerminoma
27C	26	12	0.5	10.9	91%	42	37.2	89%	Ovarian dysgerminoma
126	19	21	<1.0	25.2	>100%	24	33.2	>100%	Ovarian dysgerminoma
278	21	150	<1.0	153	>100%	213	86.8	41%	Ovarian dysgerminoma
27A	18	10	2.4	5.22	52%	30	20.8	69%	Ovarian dysgerminoma
76	25	16.5	2.3	11.8	72%	4.4	<0.8	ND	Ovarian dysgerminoma
29	50	41	<1.0	41.4	>100%	23	<0.8	ND	Ovarian serous
279	37	2.3	<1.0	3.4	>100%	<1.0	ND	ND	Ovarian serous
280	31	3.5	<1.0	19.4	>100%	4.6	ND	ND	Ovarian serous
196	41	474	126	221	47%	5000	19600	>100%	Pancreatic adenocarcinoma
250	45	43	<1.0	49.1	>100%	ND	ND	ND	Papillary renal carcinoma
27B	29	6.8	<1.0	6.8	100%	5.0	5.2	>100%	Prostate adenocarcinoma
266	37	13.3	0.1	13.1	98%	10	36	>100%	Prostate adenocarcinoma
145	51	31.1	3.0	25.2	81%	28.9	14.1	48%	Prostate adenocarcinoma
281	34	1.9	<1.0	19.6	>100%	2.4	ND	ND	Small cell lung
101	46	11	0.2	10.6	96%	<1.0	<0.8	ND	Squamous cell hepatic

(*Continued*)

Table 28.1 **(Continued)**

Case #	Age	Total hCG serum	Intact hCG	Free β serum	Free β%	Total hCG urine	β-core urine	β-core%	Ultimate cancer histology
177	26	24.4	3.7	17.2	70%	25	12	48%	Squamous vulvar
155	39	119	5.4	108	91%	252	904	>100%	Urothelial bladder
403	50	129	<1.0	504	>100%	2323	9100	>100%	Urothelial bladder
577	40	223	27.9	167	75%	480	483	100%	Urothelial bladder
363	26	249	24.9	198	80%	857	797	93%	Urothelial bladder
482	54	564	67.7	428	76%	4012	6680	>100%	Urothelial bladder
133	25	683	212	281	41%	9	1360	>100%	Urothelial bladder
255	29	47.9	10.8	35.2	55%	26.4	ND	ND	Uterine mixed mullerian
569	31	147	57.8	88.2	60%	170	680	>100%	Histology not available
571	24	34	2.4	30	88%	23	18	78%	Histology not available
543	29	6.4	<1.0	6.6	>100%	5.5	4.2	76%	Histology not available

Table 28.2 Women not pregnant with positive hCG, diagnosis choriocarcinoma, $n = 23$

Case	Age	Total hCG serum	Total hCG urine	hCG-H serum	hCG-H% Total hCG	Free β-subunit	Final diagnosis
522	31	18.1	13.6	7.37	41%	<0.36	Choriocarcinoma
82B	53	23	18	17.27	75%	5.58	Choriocarcinoma
1	30	30	13	52.8	>100%	<0.36	Choriocarcinoma
302	34	98.6	52	101.2	>100%	1.08	Choriocarcinoma
258	40	330	678	275	83%	ND	Choriocarcinoma
259	21	670	450	825	>100%	ND	Choriocarcinoma
552	20	821	872	638	78%	16.02	Choriocarcinoma
408	34	901	253	539	60%	43.2	Choriocarcinoma
214	28	1504	847	1540	>100%	156.6	Choriocarcinoma
574	25	1890	<1.0	851	45%	16.2	Choriocarcinoma
371	39	1929	1499	1991	>100%	273.6	Choriocarcinoma
211	35	1993	2234	1870	94%	252	Choriocarcinoma
260	25	2000	3400	2530	>100%	ND	Choriocarcinoma
212	27	2550	1256	2211	87%	158.4	Choriocarcinoma
208	33	2608	4356	1430	55%	174.6	Choriocarcinoma
198C	21	3578	5000	1628	46%	88.2	Choriocarcinoma
569	21	4975	1532	2607	52%	54	Choriocarcinoma
264	30	37,500	34,670	34,210	91%	<0.38	Choriocarcinoma
213	35	40,644	67,320	33,132	82%	2322	Choriocarcinoma
382	32	80,400	181,900	88,550	>100%	216	Choriocarcinoma
134B	33	95,100	ND	40,568	43%	20,394	Choriocarcinoma
134A	33	158,000	ND	89,980	57%	37,386	Choriocarcinoma
118	21	596,000	233,545	414,095	69%	15,120	Choriocarcinoma

hCG-H is hyperglycosylated hCG.

28.6 Munchausen's syndrome

In 15 years of operation, the USA hCG Reference Service has been referred three Munchausen syndrome cases. These three cases were referred to us by oncologists and gynecologic oncologists because of consistent elevated hCG blood levels in the absence of any pregnancy. In these three cases we carefully considered, and then considered again, quiescent gestational trophoblastic disease, false-positive hCG, pituitary hCG cancer, choriocarcinoma/gestational trophoblastic disease, and familial hCG syndrome, After all these diagnoses had been considered and excluded, we started talking to physicians and patients for further information. For every case, we collected additional or serial serum and urine samples.

What was noteworthy was that serial serum and urine samples showed oscillation, levels declined slowly to zero, and then levels increased and started to decline again. It was also noteworthy that hCG in blood seemed unusually clean; there were extremely low levels of hCG free β-subunit (<0.1% of total hCG), of nicked hCG (<0.1% of total hCG), and of hyperglycosylated hCG (<0.1% of total hCG). One of the main selling preparations of hCG is Ovidrel. This is recombinant hCG, 100% intact hCG dimer, prepared by Serono Inc. After an arduous and confusing consultation, it was concluded in these three cases that they were using this ultrapure hCG, Ovidrel, and Munchausen syndrome was diagnosed.

Two of the patients were, incidentally, nurses and one was a physician. It was inferred that they were self-administering Ovidrel from their clinic's stocks and claiming to be sick. The three cases very much confused the oncologists and gynecologic oncologists. The two nurses admitted that this was the case and were referred to psychiatrists; the physician denied using Ovidrel and the case continues. If they had been using other hCG preparations like Ovidrel or Profasi, which are urinary extracts, then the administration of hCG never would have been discovered.

28.7 Familial hCG syndrome

In 2003, the USA hCG Reference Service received serum and urine sample from two sisters in Utah, claiming that they had false-positive hCG and wanting to join the ongoing class-action law suit against Abbott Laboratories. They had been tested in the Abbott Axsym test, a test that, at that time, was having false-positive test problems. The problem was that they were both positive for hCG in serum and urine, which excludes false-positive hCG. Neither had elevated FSH, which excludes the possibility of pituitary hCG, and neither had elevated hyperglycosylated hCG, which excludes gestational trophoblastic disease and choriocarcinoma, or a history of miscarriage, which excluded quiescent disease [32]. It appeared that we had a new hCG problem that we could not resolve. When one sister told us that her mother was also similarly hCG-positive, we started questioning the idea of a genetic disorder affecting the mother and her two daughters. We tested the mother's blood and urine and, true enough, her test was positive, just like that of her daughters. In all three women, total hCG (Siemens Immulite total hCG) was 79%; 50% and 30% were accounted for by free β-subunit and 72%, 72%, and 65% of molecules were missing the β-subunit C-terminal peptide (Table 1, case 1). The three women had no other symptoms or disease of any kind other than that they were all producing an inherited hCG.

After finding multiple other cases, we called this strange familial production "familial hCG syndrome" [37,38]. It is seemingly a strange inherited genetic problem only affecting the hCG α-subunit and β-subunit genes. We had no idea what tissue in the normal healthy body was producing hCG, or whether it had any possible complications. Since 2003, we have seen 15 cases of familial hCG syndrome, eight in women and seven in men (Table 28.3) [38,39]. All cases were confirmed by demonstration of a familial or inherited link (Table 28.3) [38,39].

One of the strange findings associated with this genetic disease is the detection of hCG missing the β-subunit C-terminal. In the world today, there are 12 automated total hCG immunoassays used by clinical laboratories. Of these assays, 11 use an hCG free β-subunit C-terminal-specific antibody. These are Abbott AxSyn, Abbott Architect, Beckman Access, Beckman DxI 800, Ortho Vitros EciQ, Roche Elecsis hCG+β, Perkin-Elmer Delfia, Siemens ACS180, Siemens Centaur, Siemens Dimension, and Tosoh A1A total hCG assays [37]. The only exception is the Siemens Immulite series of assay, which uses two antibodies to the core of

Table 28.3 Men and women not pregnant with positive serum hCG, diagnosis familial hCG syndrome, $n = 15$

Case	Sex	Total hCG mIU/ml	Total hCG, C-terminal peptide determinant, mIU/ml	Missing C-terminal peptide, estimated mIU/ml	hCG free β, ng/ml, mIU/ml hCG calculated	Confirmation (positive hCG)
1	F	43	12	31 (72%)	34 (79%)	Mother
2	M	1.0	0.5	0.5 (50%)	0.6 (60%)	Father
3	M	2.0	0.4	1.6 (80%)	1.3 (65%)	Father
4	M	2.8	<1.0	2.8 (100%)	1.7 (61%)	Father
5	F	201	29	172 (86%)	113 (56%)	Father
6	M	1.8	0.7	1.1 (61%)	1.0 (55%)	Mother
7	F	216	24	192 (89%)	121 (56%)	Father
8	F	32	16	16 (50%)	17 (53%)	Son
9	F	9.4	2.5	6.9 (73%)	7.6 (81%)	Daughter
10	M	2.0	0.8	1.2 (60%)	1.2 (60%)	Father
11	M	2.8	<1.0	2.8 (100%)	1.8 (64%)	Father
12	M	6.0	3.1	2.9 (48%)	3.1 (52%)	Father
13	F	17	<1.0	17 (100%)	13 (76%)	Father
14	F	168	50	118 (70%)	168 (100%)	Mother
15	F	136	39	97 (71%)	146 (100%)	Mother

β-subunit. The end result is that 11 of 12 assays do not detect molecules missing the β-subunit C-terminal peptide [37]. It is speculated that the reason that this genetic disorder was only discovered recently by the USA hCG Reference Service is that most total hCG tests used by clinical laboratories throughout the world do not detect molecules missing the β-subunit C-terminal peptide, so they did not detect familial hCG syndrome cases.

Strange history accompanied the referrals of the 15 cases of familial hCG syndrome; it is unusual that 15 of the referred cases included 7 men. Men are tested throughout the world for hCG as part of routine anabolic drug screening for those involved in professional and amateur athletics. If a man uses hCG, then it will promote testicular testosterone production; the testosterone will, in turn, promote muscle and bone growth, providing an advantage in sports. Interestingly, five of the seven male cases were major athletes: case 2 (Table 28.3) was a National College Athletic Association player accused of doping with hCG; case 3 (Table 28.3) was an Olympic silver medal winner accused of doping with hCG; case 6 (Table 28.3) is a National Football League (NFL) player accused of doping with hCG; case 11 (Table 28.3) is another NFL player accused of doping with hCG; and case 12 (Table 28.3) was a professional volleyball player in Greece accused of hCG doping.

The USA hCG Reference Service has been involved in several fights with those accused of doping in sports. Although some sports authorities were able to accept the familial hCG syndrome explanation, others were not. Much of the confusion came from the oscillation in hCG production that occurs among familial hCG syndrome cases (Figure 28.2). hCG production can be zero for several months, as illustrated in Figure 28.2C, and then increase to 6 mIU/ml. Many of the arguments with sports authorities arose because the hCG was 10 mIU/ml but then was 30 mIU/ml at the time of the doping testing, so the athlete must have been doping.

All 15 cases of familial hCG syndrome produced a primary molecule seemingly missing the β-subunit C-terminal peptide (48–100% of immunoreactive hCG is missing the C-terminal peptide) or highly degraded, in that they were not bound by the C-terminal antibody used in the Siemens Centaur assay. Molecules missing the C-terminal peptide in such high proportion have never been described in any other disorder [38,39]. Why do these familial hCG syndrome cases stand out as being so extreme and so unusual? One possibility is that the β-subunit is an alternative structure molecule produced by an alternative hCG β-subunit gene to that expressed in pregnancy. There are eight hCG β-subunit genes on chromosome 19 and only four (CGβ, CGβ5, CGβ7, and CGβ8) are expressed in pregnancy. The alternative β-subunit could lead to greater degradation or to lack of recognition by C-terminal peptide antibody. Another option that has to be considered is expression of a mutant hCG β-subunit in familial hCG syndrome. This also may not be recognized by different total hCG assays.

Ulf Stenman, PhD, has studied many of our cases of familial hCG syndrome in Finland. He claims that gel filtration shows that the size of the β-subunit is that of the normal β-subunit (Ulf Stenman, PhD, personal communication) and not the

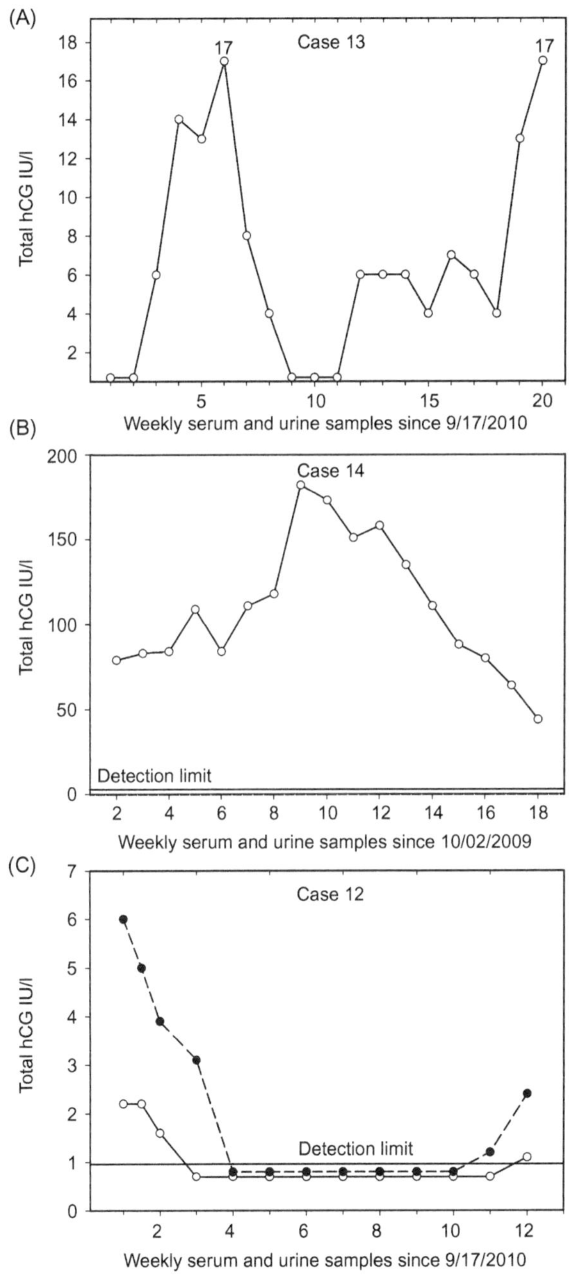

Figure 28.2 Oscillating hCG production in familial hCG syndrome cases.

smaller (such as the size of hCG of β-subunit missing the C-terminal peptide). It is inferred that familial hCG syndrome may not be exceptionally degraded as indicated, but rather it has mutations blocking recognition by a C-terminal peptide antibody. Mutant hCG may be the explanation for everything, such as non-recognition by the Siemens Centaur C-terminal peptide assay and variable hCG production, as indicated in Figure 28.2. Variable production could be attributable to variable recognition of mutant hCG.

Ongoing studies by Stephen Butler, PhD, involving a DNA sample from a single case of familial hCG syndrome show the presence of mutant hCG β-subunit in this syndrome. This explains the variable detection by the Siemens Immulite and other hCG assays. Further genetic studies involving familial hCG syndrome cases are urgently needed.

28.8 Administering hCG

The final category of individuals who are not pregnant and have a positive result on an hCG test involves those who administer hCG for dietary reasons [40]. Today, hCG is a very popular dietary aid that is taken as injections, nasal drops, and even as tablets. This is described in detail in Chapter 27. The USA hCG Reference Service has had 2 cases (out of 529 cases of not being pregnant but positive for hCG) that were attributable to women not revealing the dietary drug that they were using and their false positivity on a pregnancy test. There is no test in the USA hCG Reference Service profile that can identify this, beyond the finding of serial hCG levels that slowly decrease and then suddenly increase. The only useful method is a confidential discussion with the patient and having the patient admit to using hCG as a dietary supplement.

28.9 Managing non-pregnant individuals positive for hCG

Table 28.4 lists all the methodologies commonly used by the USA hCG Reference Service. The total hCG Immulite assay (or the only total hCG assay that will detect hCG missing the β-subunit C-terminal peptide [37]) can be ordered from Tricore Reference Laboratories, Albuquerque, New Mexico. The hyperglycosylated hCG test can be ordered from Quest Diagnostics Inc. (order #11303X), and the free β-subunit test can be ordered from Hereditary Disease Laboratory, University of Alberta Hospital, Edmonton, Canada (not available in the USA). One can run the USA hCG Reference Service – proven program oneself according to the procedures listed in Table 28.4.

Table 28.4 **Detecting positive hCG cases outside of pregnancy**

Diagnosis	Supporting data
Quiescent gestational trophoblastic disease	1. Has history of molar pregnancy or miscarriage 2. Negative (<10% total hCG) in hyperglycosylated hCG test
Pituitary hCG	1. FSH test >30 mIU/ml 2. Hyperglycosylated hCG not detected 3. Three weeks treatment with high estrogen pill suppresses hCG
False positive hCG test	1. In quantitative hCG test, hCG detected in serum but not urine 2. Multiple serum hCG tests give widely varying results
Cancer	1. Data shows all hCG to be free β-subunit 2. Variable results in Immulite and other automated hCG tests
Choriocarcinoma/GTN	1. Hyperglycosylated hCG more that 50% of total hCG 2. Uterine mass
Familial hCG syndrome	1. Other automated assay give <50% of Immulite hCG result 2. Total hCG is >50% free β-subunit 3. Family members positive
Munchausen's syndrome	1. Exclude all other causes of positive hCG 2. Serum lacks hyperglycosylated hCG and hCG free β-subunit
Administering hCG	1. hCG levels that slowly over a month falls and then rises 2. Need confidential conversation with patient

GTN is gestational trophoblastic neoplasm.

References

[1] Vladutiu AO, Sulewski JM, Pudlak KA, Stull CG. Heterophilic antibodies interfering with radioimmunoassay. J Am Med Assoc 1982;248:2489–90.

[2] Hussa RO, Rinke ML, Schweitzer PG. Discordant human chorionic gonadotropin results: causes and solutions. Obstet Gynecol 1985;65:211–19.

[3] Armstrong EG, Ehrlich PH, Birken S, Canfield R. Use of a highly sensitive and specific immunoradiometric assay for detection of human chorionic gonadotropin in urine of normal, nonpregnant, and pregnant individuals. J Clin Endocrinol Metab 1984;59:867–74.

[4] Knight AK, Bingemann T, Cole L, Cunningham-Rundles C. Frequent false positive beta human chorionic gonadotropin in immunoglobulin a deficiency. Clin Exp Immunol 2005;141:333–7.
[5] Cole LA, Rinne KM, Shahabi S, Omrani A. False positive hCG levels leading to unnecessary surgery and chemotherapy, and needless occurrences of diabetes and coma. Clin Chem 1999;45:313–14.
[6] Butler SA, Cole LA. Falsely elevated hCG leading to unnecessary therapy. Obstet Gynecol 2002;99:515–16.
[7] Khanlian SA, Smith HO, Cole LA. Persistent low levels of hCG: a pre-malignant gestational trophoblastic disease. Am J Obstet Gynecol 2003;188:1254–9.
[8] Cole LA. Problems with hCG measurements and interpretation of hCG results in gestational trophoblastic diseases. Jpn J Obstet Gynecol 2003;20:11–14.
[9] Cole LA. Phantom hCG and phantom choriocarcinoma. Gynecol Oncol 1998;71:325–9.
[10] Rotmensch S, Cole LA. False diagnosis and needless therapy of presumed malignant disease in women with false positive human chorionic gonadotropin concentrations. Lancet 2000;355:712–15.
[11] Olsen TG, Hubert PR, Nycum LR. Falsely elevated human chorionic gonadotropin leading to unnecessary therapy. Obstet Gynecol 2001;98:843–5.
[12] Flam F, Hambraeus-Jonzon K, Hansson L, Kjaeldgaard A. Hydatidiform mole with non-metastatic pulmonary complication and false low level of hCG. Eur J Obstet Gynecol Reprod Biol 1998;77:235–7.
[13] Chu J-W, Golman JO. False elevation of serum hCG. Laboratory updates: Detroit Medical Center University Labs; 2003jau_janice.htm.
[14] Rode L, Daugaard G, Fenger M, Hilsted L, Møller LK, Raaberg L, et al. Serum hCG: still a problematic marker. Acta Obstet Gynecol Scand 2003;82:199–202.
[15] Trojan A, Joller-Jemelka H, Stahel RA, Jacky E, Hersberger M. False positive hCG in a patient with germ cell cancer. Oncology 2004;66:336–8.
[16] Pesce MA. False elevated hCG results with the AxSym assay. Clin Chem 2003;49:A92.
[17] Esfandiari N, Goldberg JM. Heterophile antibody blocking agent to confirm false positive serum human chorionic gonadotropin assay. Obstet Gynecol 2003;101:1144–6.
[18] Giannopoulos P, Jones G, Lim E-M. A simple protocol for identification of heterophilic antibody interferences in the Abbott AxSym hCG assay. Aust Assoc Clin Chem 2003; Gold Coast, Queensland. Abstract.
[19] Cole LA, Khanlian SA. Easy fix for clinical laboratories for the false positive defect with the Abbott AxSym total β-hCG test. Clin Biochem 2004;37:344–9.
[20] Birken S, Maydelman Y, Gawinowicz MA, Pound A, Liu Y, Hartree AS. Isolation and characterization of human pituitary chorionic gonadotropin. Endocrinology 1996;137:1402–11.
[21] Odell WD, Griffin J. Pulsatile secretion of human chorionic gonadotropin in normal adults. N Engl J Med 1987;317:1688–91.
[22] Odell WD, Griffin J. Pulsatile secretion of chorionic gonadotropin during the normal menstrual cycle. J Clin Endocrinol Metab 1989;69:528–32.
[23] Cole LA, Khanlian SA, Sasaki Y, Muller CY. Normal production of hCG in menopause: a medical management Dilemma. N Eng J Med 2007;356:1184–6.
[24] Gronowski AM, Fantz CR, Parvin CA, Sokoll LJ, Wiley CL, Wener MH, et al. Use of serum FSH to identify perimenopausal women with pituitary hCG. Clin Chem 2008;54:652–6.

[25] Cole LA, Butler SA, Khanlian SA, Giddings A, Muller CY, Seckl MJ, et al. Gestational trophoblastic diseases. 2. Hyperglycosylated hCG as a reliable marker of active neoplasia. Gynecol Oncol 2006;102:150–8.
[26] Muller CY, Cole LA. The quagmire of hCG and hCG testing in gynecologic oncology. Gynecol Oncol 2009;112:663–72.
[27] Cole LA. Use of hCG tests for evaluating trophoblastic diseases: choosing an appropriate hCG assay, false detection of hCG, unexplained elevated hCG, and quiescent trophoblastic disease. In: Hancock BW, Newland ES, Berkowitz RS, Cole LA, editors. Gestational trophoblastic disease. 2nd ed. London: Chapman and Hall; 2002.
[28] Cole LA, Butler SA. Hyperglycosylated hCG and its free β-subunit, tumor markers and tumor promoters: a review. J Reprod Med 2008;53:499–510.
[29] Cole LA, Khanlian SA, Giddings A, Butler SA, Muller CY, Hammond C, et al. Gestational trophoblastic diseases. 4. Presentation with persistent low positive human chorionic gonadotropin. Gynecol Oncol 2006;102(116):3–9.
[30] Cole LA, Muller Y. hCG in the management of quiescent and chemorefractory gestational trophoblastic diseases. Gynecol Oncol 2010;116:3–9.
[31] Hancock BW, Tidy JA. Clinical management of persistent low-level hCG elevation. Trophobl Dis Update 2003;4:5–6.
[32] Cole LA, Cole LA, Laidler L, Muller C. USA hCG reference service, 10 year report. Clin Biochem 2010;43:1013–22.
[33] Hancock BW, Berkowitz RS, Cole LA. Gestational trophoblastic disease. 2nd ed. London: Chapman and Hall; 450 pages.
[34] Cole LA, Khanlian SA, Riley JM, Butler SA. Hyperglycosylated hCG (hCG-H) in gestational implantation, and in choriocarcinoma and testicular germ cell malignancy tumorigenesis. J Reprod Med 2006;51:919–29.
[35] Hamade AL, Nakabayashi K, Sato A, Kiyoshi K, Takamatsu Y, Laoag-Fernandez JB, et al. Transfection of antisense chorionic gonadotropin β gene into choriocarcinoma cells suppresses the cell proliferation and induces apoptosis. J Clin Endocrinol Metab 2005;90:4873–9.
[36] Cole LA, Butler SA. Hyperglycosylated hCG, hCGβ and hyperglycosylated hCGβ: interchangeable cancer promoters. Mol Cell Endocrinol 2012;349:232–8.
[37] Cole LA, Sutton JM, Higgins TN, Cembrowski GS. Between-method variation in hCG test results. Clin Chem 2004;50:874–82.
[38] Cole LA, Laidler LL. Inherited hCG. J Reprod Med 2010;55:99–102.
[39] Cole LA. Familial hCG. J Reprod Immunol 2012;93:52–7.
[40] Simeons ATW. The action of chorionic gonadotropin in the obese. Lancet 1954;2:946–7.

Part J

Assays and Antibodies

Antibodies and hCG tests

29

Laurence A. Cole
USA hCG Reference Service, Angel Fire, NM, USA

hCG assays, or pregnancy tests, have a long and interesting history. With this history came great improvement in assay specificity and improvements (radioimmunoassay [RIA] over previous assays) and declinations (immunometric assays vs. RIA) in assay specificity. In 1927, Aschheim and Zondek discovered that the blood and urine of pregnant women contained a gonad-stimulating substance. With their findings, the first hCG/pregnancy tests were born [1–3]. These tests primarily used urine to promote ovulation in mice and rabbits [1–7] and were commonly referred to as the "rabbit" or "Friedman" test.

Over the next four decades, bioassays like the rabbit test were cumbersome for testing hCG or pregnancy [1–7]. The first antibody-based tests were launched in 1960, and they examined hemagglutination inhibition in pregnancy testing [8,9]. In 1964, the competitive hCG RIA was invented [10–15]. It revolutionized pregnancy testing by detecting as little as 5 mIU/ml. The problem with the first hCG RIAs is that the polyclonal antibody was raised against hCG dimer, and it detected both hCG and LH. In 1973, Vaitukaitis et al. introduced the hCGβ test [15], a RIA pregnancy test using an antibody against the β-subunit of hCG. The hCGβ test was the first hCG-specific RIA not detecting LH. Unlike its predecessor, the hCGβ test measured hCG variants alone. The discovery of monoclonal antibodies in 1975 was paramount to the development of modern immunometric hCG tests [16]. Modern two- or three-antibody immunometric hCG assays were developed in 1981–1990. These largely replaced the RIA. With these assays came the concept of antibody enzyme labeling and high-sensitivity fluorimetric and spectrometric and chemiluminescent detection [17–21].

29.1 Antibody sites on hCG

The big problem with hCG or pregnancy tests is that hCG is not the only molecule produced; a wide variety of hCG-related molecules and hCG variants are present in serum and urine (the detection fluids) (Figure 29.1). First, normal pregnancy must produce hCG and hyperglycosylated hCG; hCG is made by syncytiotrophoblast cells and hyperglycosylated hCG is made by cytotrophoblast cells [22–24]. In addition, cancer cells make hyperglycosylated hCG free β-subunit and the pituitary gland makes sulfated hCG.

The heterogeneity issue does not stop at that point. As illustrated in Figure 29.1, there are macrophages and monocytes attached to the placental trophoblast plasma

Human Chorionic Gonadotropin (hCG). DOI: http://dx.doi.org/10.1016/B978-0-12-800749-5.00029-8

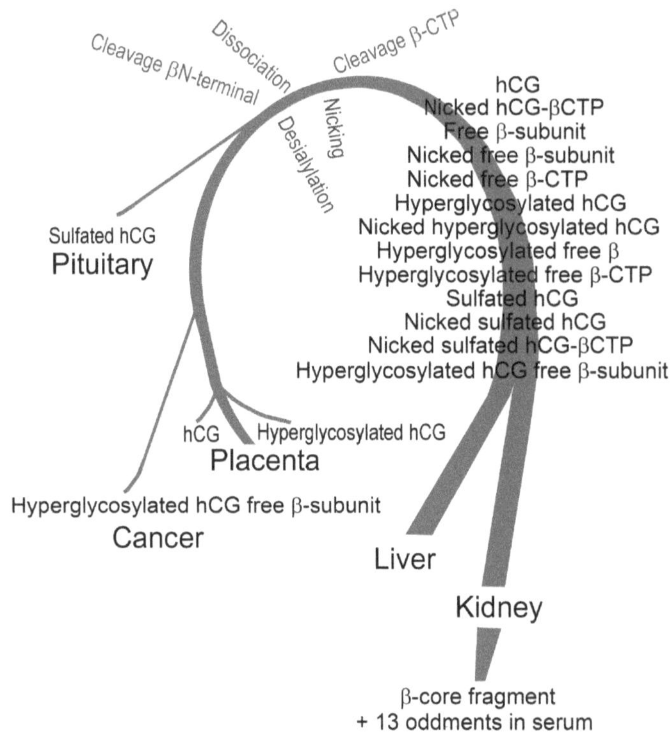

Figure 29.1 The continuous dissociation and degradation of hCG and hyperglycosylated hCG on secretion. β-CTP indicates molecules missing the β-subunit C-terminal peptide.

membranes and to cancer and pituitary membranes. The second that hCG, hyperglycosylated hCG, sulfated hCG, and hyperglycosylated hCG free β-subunit are secreted, the diversification starts, with nicking (cleavages at β47–48) by macrophage and monocyte leukocyte elastase and other enzymes, dissociation into free subunits, desialylation by a macrophage neuraminidase, and a cleavage of the β-subunit C-terminal peptide, starting immediately on secretion (Figure 29.1). The cleavage and the heterogeneity issue does not stop at that point.

The cleavage and dissociation are an ongoing process in the blood, continuously degrading hCG, dissociating hCG subunits, and dismantling its oligosaccharides (Figure 29.1). These hCG dissociation and degradation products are rapidly cleared from the circulation by the liver and kidney. In the kidney, hCG products are further degraded to the terminal degradation product β-core fragment detected only in urine.

With a good hCG assay, the following mixture of intact, dissociation, and degradation heterogeneous products of hCG can to be detected in serum at any one time: hCG; nicked hCG; nicked hCG missing the β-subunit C-terminal peptide; free β-subunit; nicked free β-subunit; nicked free β-subunit missing the β-subunit C-terminal peptide; hyperglycosylated hCG; nicked hyperglycosylated hCG; nicked hyperglycosylated hCG missing the β-subunit C-terminal peptide; hyperglycosylated hCG free

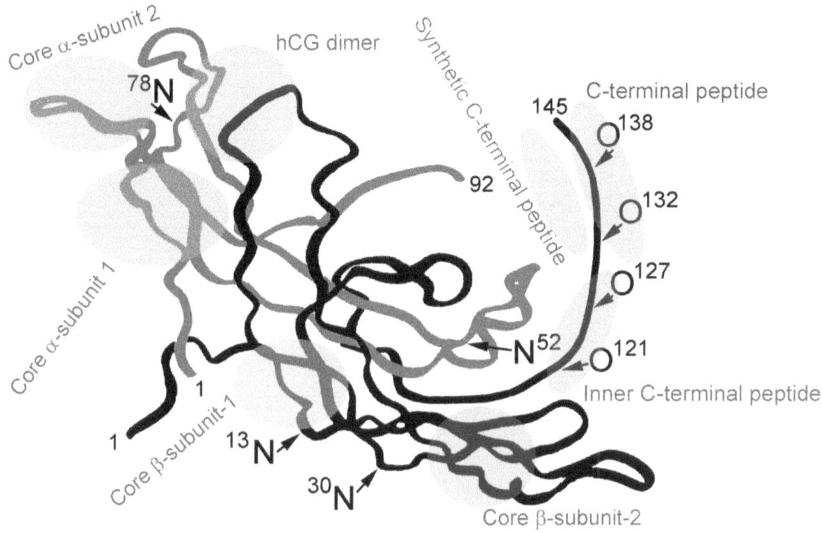

Figure 29.2 Monoclonal and polyclonal antibody epitopes for hCG.

β-subunit; sulfated hCG; nicked sulfated hCG; and nicked sulfated hCG missing the β-subunit C-terminal peptide. In urine, the same mixture of molecules plus β-core fragment, a urinary terminal degradation product, may be present [25,26].

The problem with an hCG or pregnancy test is that it needs to detect this wide variety of 13 molecules and still be absolutely specific for hCG. An occasional person with pregnancy, or pregnancy problem, for instance, may be producing just one of these 13 variants. Thus, detection of all of these dissociation and degradation products as well as the secretagogues hCG, hyperglycosylated hCG, sulfated hCG, and hyperglycosylated hCG free β-subunit is critical.

As shown in Figure 29.2 there are at least nine major monoclonal and polyclonal antibody epitopes on hCG. If intact hCG is injected in a mouse, goat, sheep, or rabbit, then the principal antibody will be an antibody to the hCG dimer site and core α-subunit-1 and α-subunit-2 sites (Figure 29.2). If the separated hCG β-subunit is injected in an animal, then the principal antibody will be an antibody to the core β-subunit-1 and α-subunit-2 sites or the C-terminal peptide or inner C-terminal peptide sites (Figure 29.2). Insert a synthetic 15-amino-acid peptide linked to conjugant into an animal and the result may be a C-terminal peptide specific antibody (Figure 29.2). Other antibodies not shown are possible, including antibodies to free α-subunit and β-subunit and antibodies to carbohydrate variants of hCG, hyperglycosylated hCG, and β-core fragment.

In the next chapter, Chapter 30, the opinion that modern immunometric assays are a major regression or non-improvement over the 1970s or 1980s RIA is discussed. This is because of poor detection of the 13 hCG variants by 10 of 11 automated immunometric assays sold today. This limitation in detection causes clear failures in modern laboratory medicine, with failure to appropriately detect hCG variants in pregnancy, in gestational trophoblastic disease cases, and after clearance of hCG post-failure or post-delivery. It is for these reasons that it is probably best to start by discussing the last appropriate hCG test or the RIA.

29.2 Radioimmunoassay

The RIA was introduced in 1964 and remained the principal hCG or pregnancy tests until approximately 1990, when the immunometric hCG assay was introduced [10–18]. The hCG RIA works on a simple principal. Each tube contains a set low and limiting concentration of antibody to hCG (approximately 0.5 ng), a set concentration of radioiodinated hCG (approximately 0.5 ng or 45,000 cpm), and an unknown concentration of serum or urine hCG (total volume 0.5–1.0 ml). The radioactive hCG is allowed to compete with the unknown concentration of hCG in the sample for 12–16 h in competitively binding the limiting amount of antibody at 4°C. A second antibody is then added to precipitate the first antibody, like goat anti-rabbit gamma globulin. The precipitated first antibody is centrifuged to a pellet. The pellet is isolated and the radioactivity is measured. The higher the counts for radioactive iodine, or radioiodinated hCG, the lesser the amount of unknown serum hCG that has won the competitive battle and is in the pellet. A standard curve is created; the higher the concentration of unknown hCG, the lower the radioactivity count in the pellet for each assay. The standard curve is read and concentration of hCG in the unknown sample is calculated.

The original hCG assay used antibodies to hCG dimer and α-subunit and detected both hCG and LH. It was in 1973 when Vaitukaitis et al. introduced the hCGβ test; the hCG RIA used hCG β-subunit antisera [15]. This assay detected only hCG and its β-subunit variants. This was the first perfect hCG test, and it probably remains the most perfect hCG pregnancy test today. It was perfect in that it distinguished LH and hCG 100%, and it is the most perfect test in that it equally recognized all 13 molecules listed in Figure 29.1, which all contain an hCG β-subunit core structure.

The big advantage of the RIA was that it uses a single antibody, usually a polyclonal, against hCG β-subunit core (core-1 or core-2; Figure 29.2). Figure 29.2 shows the epitopes on hCG and hyperglycosylated hCG; the core is indicated as core β-subunit-1 or core β-subunit-2. Usually an antibody against this site will bind all hCG variants present in serum, with all having a common core site.

The RIA is still used today by several clinical laboratories demanding a high specificity assay. It is the choice of the Charing Cross Hospital Trophoblast Disease Center in London, England, to accurately distinguish hydatidiform mole and choriocarcinoma cases in referred cases. It is also the choice, so we are told, of the Obstetrics and Gynecology laboratory at Yale University School of Medicine. The end of the RIA was caused by the need to radioiodinate afresh hCG every month and the strict rules regarding 125-I handling. Yes, industry tried enzyme labeling in the RIA, but it had its competitive limitations. The 1990s saw the enzyme-labeled immunometric assay largely replace the RIA for hCG.

29.3 Laboratory immunometric assays

The 1990s saw the development of non-competitive immunometric assays [16–21]. Basically, one antibody is immobilized on beads or plasticware. This is the immobilized antibody. This grabs the hCG and its variants and immobilizes them. A second antibody to hCG and to a distant site on the molecule is labeled with an enzyme. This is the tracer antibody. This binds the immobilized hCG variants, forms a sandwich (immobilized antibody–hCG–tracer antibody), and labels them. The mix is washed and the amount of tracer is measured. The tracer is enzyme-labeled so that an enzyme substrate is added and, after a short incubation, the concentration of product is measured by chemilumenescence or fluorimetry. The concentration of product should be directly proportional to the concentration of tracer-bound enzyme or the concentration of the unknown hCG sample.

The inherent problem with immunometric assays is that they must have antibodies to at least two distant sites on the hCG molecule, the immobilization antibody or antibodies, and one of the tracer antibody or antibodies. The RIA needs just one antibody to core β-subunit-1 or β-subunit-2 and can measure all variants of hCG. The immunometric assay with its requirement of distant antibodies can only measure variants that have the two antibody epitopes. This very much limits what can be detected and measured.

If you inject hCG β-subunit into animals, the most common antibodies generated are antibodies to the β-subunit core (sites 1 or 2) and antibodies to the β-subunit C-terminal peptide. This combination, β-subunit core antibodies, and β-subunit C-terminal peptide antibodies are the antibody of choice out of most laboratory immunometric assays used today. Unfortunately, these antibodies are a poor antibody choice and limit hCG variant detection. Assays designed this way cannot detect any of the 4 of 13 hCG variants missing the β-subunit C-terminal peptide (Figure 29.1). The β-subunit C-terminal peptide is a region rich in carbohydrates, so assays using this antibody combination poorly detect hyperglycosylated hCG and related molecules and other oligosaccharide variants. As described in the next chapter, Chapter 30, β-subunit C-terminal peptide antibodies are a major problem with most laboratory immunometric assay used today.

29.4 Point-of-care hCG immunoassays

Point-of-care (POC) pregnancy tests are available at physician's offices and prenatal care clinics around the world. They are also used by clinical laboratories as urine hCG tests. Most of today's POC tests work on the dual antibody immunometric principle.

Basically, antibody to hCG site 1 is bound to a dye tracer (tracer antibody). This combines with the hCG, which is concentrated as it emerges through a nitrocellulose pad. The hCG–antibody 1–dye complex then migrates through the nitrocellulose coated device. When it comes in contact with a line site, a line of antibody to hCG site 2, it is immobilized at this site (immobilized antibody) and it forms a

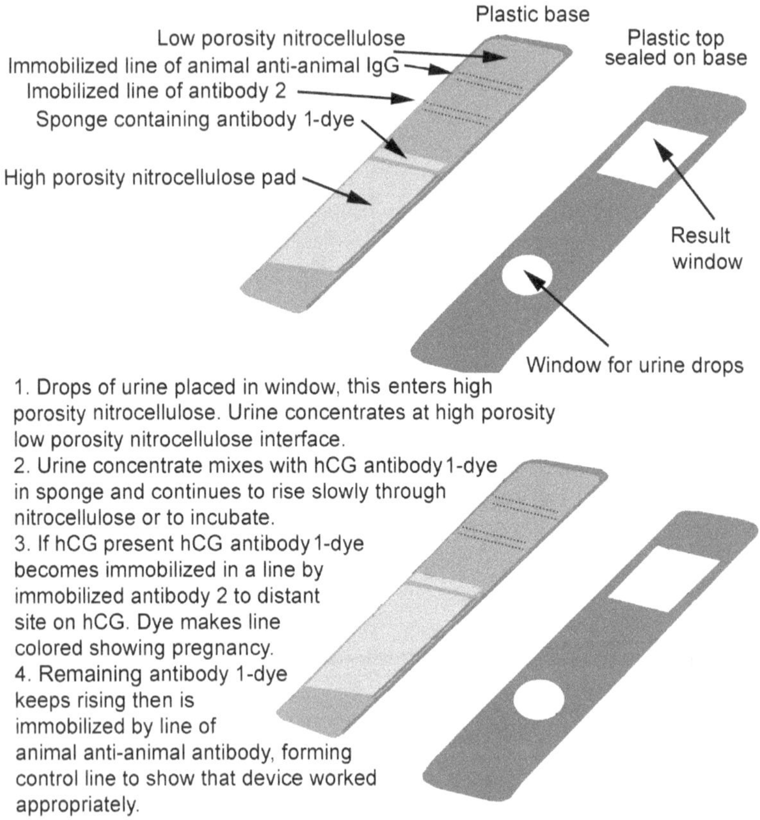

Figure 29.3 Principle operation of POC hCG assay.

sandwich called the antibody 2–hCG–antibody 1–dye complex. The result is that dye now appears in the line site. This line site is uncovered and seen on the outside of the device indicating a positive test result (Figure 29.3).

POC devices can be used with urine or serum samples. They are strictly qualitative assays with sensitivities ranging from 20 to 100 mIU/ml hCG. These tests range from urine-drop devices to dipstick and digital devices (Figure 29.3). The principal manufacturers of multiple POC products include Quidel (San Diego, CA), Beckman-Coulter (Fullerton, CA), and Inverness Medical Innovations (Waltham, MA). Other manufacturers include Mainline (Ann Arbor, MI), Siemens Healthcare Diagnostic (New York, NY), Genzyme (Cambridge, MA), Abbott Rapid Diagnostics (Abbott Park, IL), SCI International (Frederick, MD), and multiple others.

29.5 Over-the-counter hCG immunoassays

In 1976, the FDA approved the first home, or over-the-counter (OTC), pregnancy test. The product, called e.p.t.® (abbreviation of "early pregnancy test"), claimed to

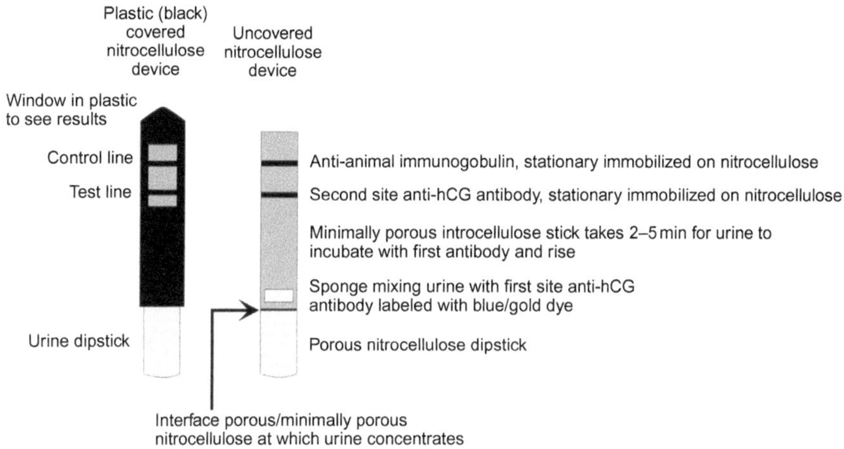

Figure 29.4 Principal of operation of OTC immunometric pregnancy test.

detect pregnancy 4 days before missing the menstrual period [27]. Since then, OTC devices have undergone many changes, including the adaption of the immunometric assay format [28], which all OTC tests now use (Figure 29.4). Many of today's OTC tests are digital. Instead of relying on the presence or absence of a faint test line to determine pregnancy (see Figure 29.4), a digital test clearly indicates the words "Yes" or "No" or "Pregnant" or "Not pregnant."

OTC tests work on a similar principal as the POC test (Figures 29.3 and 29.4). Briefly, antibody to hCG site 1 is bound to a dye tracer (tracer antibody). This combines with the hCG, which is concentrated as it emerges through a porous nitrocellulose urine stick. The hCG–antibody 1–dye complex then migrates through the nitrocellulose-coated device. When it comes in contact with a line site, a line of antibody to hCG site 2, it is immobilized at this site (immobilized antibody) and it forms a sandwich, antibody 2–hCG–antibody 1–dye complex. The result is that dye now shows in the line site. This line site is uncovered and seen on the outside of the device, indicating a positive test (Figure 29.4). Some of the hCG–antibody 1–dye keeps migrating. It reaches a second line containing immobilized anti-animal IgG. It forms an antibody 2–hCG–anti-animal IgG complex. A second line is formed at this second line window, indicating that the device worked properly. If no hCG is present, then a line will form at this second window, indicating that the device worked.

29.6 Specific hCG free subunit, fragment, and carbohydrate variant assays

Specific antibodies and immunometric assays have been generated to hCG free α-subunit (34P2C2) [29], to hCG free β-subunit FBT11 [30], 1E5 [31], B204 [32], and to β-core fragment (B210) [33]. Further antibodies and assays have been generated to asialo-hCG (R141) [34], nicked hCG (B151), and hyperglycosylated hCG (B152) [35]. Although no antibodies are yet available for pituitary sulfated hCG or fetal hCG, it appears that antibodies can be obtained for most hCG-related molecules.

References

[1] Aschheim S, Zondek B. Das Hormon des hypophysenvorderlappens: Testobjekt zum Nachweis des hormons. Klin Wochenschr 1927;6:248–52.

[2] Zondek B, Aschheim S. The Zondek–Ascheim pregnancy test. Can Med Assoc J 1930;22:251–3.

[3] Friedman MH, Lapham ME. A simple, rapid procedure for the laboratory diagnosis of early pregnancies. Am J Obstet Gynecol 1931;21:405–10.

[4] Shapiro HA, Zwarenstein H. Metabolic changes associated with endocrine activity and the reproductive cycle in *Xenopus laevis*: IV. The effects of injection of ovarian and pituitary extracts on the serum calcium in normal, ovariectomised and hypophysectomised toads. J Exp Biol 1934;11:267–72.

[5] Delfs E. An assay method for human chorionic gonadotropin. Endocrinology 1941;28:196–202.

[6] Kupperman HS, Greenblatt RB, Noback CR. A two and six-hour pregnancy test. J Clin Endocrinol 1943;3:548–50.

[7] Galli-Mainini C. Pregnancy test using the male batrachia. J Am Med Assoc 1948;138:121–5.

[8] Wide L, Gemzell CA. An immunological pregnancy test. Acta Endocrinol 1960;35:261–7.

[9] Wide L. An immunological method for the assay of human chorionic gonadotropin. Acta Endocrinol (Kobenhavn) 1962;4:1–111.

[10] Paul WE, Odell WD. Radiation inactivation of the immunological and biological activities of human chorionic gonadotropin. Nature 1964;203:979–80.

[11] Wilde CE, Orr A, Bagshaw K. A radioimmunoassay for human chorionic gonadotropin. Nature 1965;205:191–2.

[12] Lunenfeld B, Eshkol A. Immunology of human chorionic gonadotropin (HCG). Vit Hor 1967;25:137–90.

[13] Aono T, Goldstein DP, Taymor ML, Dolch K. A radioimmunoassay method for human pituitary luteinizing hormone (LH) and human chorionic gonadotropin (HCG) using 125-I-labeled LH. Am J Obstet Gynecol 1967;98:996–1001.

[14] Rushworth AG, Orr AH, Bagshawe KD. The concentration of HCG in the plasma and spinal fluid of patients with trophoblastic tumours in the central nervous system. Br J Cancer 1968;22:253–7.

[15] Vaitukaitis JL, Braunstein GD, Ross GT. A radioimmunoassay which specifically measures human chorionic gonadotropin in the presence of human luteinizing hormone. Am J Obstet Gynecol 1972;113:751–8.

[16] Kohler G, Milstein C. Continuous cultures of fused cells secreting antibody of predefined specificity. Nature 1975;256:495–7.

[17] Maggio ET, Nakamura RM. Biomedical advances: 1. Clinical assays employing enzymes in innovative ways. Ligand Rev 1981;3:16–24.

[18] Ishikawa E, Kawai T, Miyai K. Enzyme immunoassay. Tokyo: Igaku-Shoin; 19811–280.
[19] Hussa RO, Hudson EN. A two-site immunometric assay in evaluation of low levels of serum hCG. Am Clin Prod Rev 1984;3:12–17.
[20] Sekiya T, Furuhashi Y, Goto S, Kaseki S, Tomoda Y, Kato K. Sandwich-type enzyme immunoassay for human chorionic gonadotropin. J Endocrinol Invest 1981;4:275–9.
[21] Batzer FR. Hormonal evaluation of early pregnancy. Fertil Steril 1980;34:1–13.
[22] Cole LA, Dai D, Butler SA, Leslie KK, Kohorn EI. Gestational trophoblastic diseases: 1. Patho-physiology of hyperglycosylated hCG-regulated neoplasia. Gynecol Oncol 2006;102:144–9.
[23] Kovalevskaya G, Genbacev O, Fisher SJ, Cacere E, O'Connor JF. Trophoblast origin of hCG isoforms: cytotrophoblasts are the primary source of choriocarcinoma-like hCG. Mol Cell Endocrinol 2002;194:147–55.
[24] Cole LA, Khanlian SA, Sutton JM, Davies S, Stephens N. Hyperglycosylated hCG (invasive trophoblast antigen, ITA), a key antigen for early pregnancy detection. Clin Biochem 2003;36:647–55.
[25] Cole LA, Shahabi S, Butler SA, Mitchell H, Newlands ES, Behrman HR, et al. Utility of commercially used human chorionic gonadotropin immunoassays in the diagnosis and management of trophoblastic diseases. Clin Chem 2001;47:308–15.
[26] Cole LA, Sutton JM, Higgins TN, Cembrowski GS. Between-method variation in hCG test results. Clin Chem 2004;50:874–82.
[27] Anonymous. Home pregnancy test simple to use, reasonably accurate, especially if result is positive. Fam Plann Persp 1979;11:190–1.
[28] Joshi UM, Roy R, Sheth AR, Shah HP. A simple and sensitive color test for the detection of human chorionic gonadotropin. Obstet Gynecol 1981;57:252–4.
[29] Moodley D, Moodley J, Buck R, Haneef R, Payne A. Free alpha-subunits of human chorionic gonadotropin in preeclampsia. Int J Gynaecol Obstet 1995;49:283–7.
[30] Puisieux A, Bellet D, Troalen F, Razafindratsita A, Lhomme C, Bohuon C, et al. Occurrence of fragmentation of free and combined forms of the beta-subunit of human chorionic gonadotropin. Endocrinology 1990;126:687–94.
[31] Khazaeli MB, Buchina ES, Pattillo RA, Soong SJ, Hatch KD. Radioimmunoassay of free beta-subunit of human chorionic gonadotropin in diagnosis of high-risk and low-risk gestational trophoblastic disease. Am J Obstet Gynecol 1989;160:444–9.
[32] Kardana A, Cole LA. Polypeptide nicks cause erroneous results in assays of human chorionic gonadotropin free β-subunit. Clin Chem 1992;38:26–33.
[33] Okamoto T, Matsuo K, Niu R, Osawa M, Suzuki H. Human chorionic gonadotropin (hCG) β-core fragment is produced by degradation of hCG or free hCG in gestational trophoblastictumors: a possible marker for early detection of persistent postmolar gestational trophoblastic disease. J Endocrinol 2001;171:435–43.
[34] Amr S, Rosa C, Birken S, Canfield R, Nisula B. Carboxyterminal peptide fragments of the beta subunit are urinary products of the metabolism of desialylated human choriogonadotropin. J Clin Invest 1985;76:350–6.
[35] Birken S, Krichevsky A, O'Connor J, Schlatterer J, Cole L, Kardana A, et al. Development and characterization of antibodies to a nicked and hyperglycosylated form of hCG from a choriocarcinoma patient: generation of antibodies that differentiate between pregnancy hCG and choriocarcinoma hCG. Endocrine 1999;10:137–44.

Problems with today's hCG pregnancy tests

30

Laurence A. Cole
USA hCG Reference Service, Angel Fire, NM, USA

Any discussion of hCG assays continues much more than a discussion of hCG antibodies and a description of the mechanism of modern hCG assays (Chapter 29).

30.1 Automated immunometric assays

Today, most laboratories use automated immunometric assays. These are assays that use machines to do all the work; machines make dilution, mix all solutions, and incubate samples. Machines separate samples and take fluorimetric or chemiluminescent readings. Machines then calculate results against a new or an inherent standard curve, and machines print a final result or feeds it into a computer after 15–40 min. No longer does a laboratory technician have the job of mixing everything and doing everything. The job is just to line up serum samples inside a machine and to read and accept the final results.

In laboratories today, automated immunometric assay machines include the Abbott AxSym platform, the Abbott Architect platform, the Beckman Access platform, the Beckman Dxi 800 platform, the Ortho Vitros EciQ platform, the Roche Elecsys platform, the Tosoh A1A platform, the Perkin-Elmer Delfia platform, the Siemens Dimension platform, the Siemens ADVIA Centaur platform, the Siemens ACS180 platform, and the Siemens Immulite platform. Each platform comes in multiple models designed for low and high throughput and different levels of automation. The Siemens Immulite system, for instance, is available as the Immulite 1000 with minimal automation and low throughput, and Immulite 2000 has superior automation and medium – high throughput (200 samples per hour).

Each automated platform may run 30–50 different immunometric assays. The Abbott AxSym platform, for instance, may run tests for acetaminophen, active-B12 (HoloTC), AFP, amphetamine/methamphetamine, anti-CCP, anti-HAV IgM, anti-HBc, anti-HBc IgM, anti-HBe, anti-HBs, anti-HCV, anti-HIV-1/HIV-2, anti-Tg, anti-TPO, B12, barbiturates, benzodiazepines, BNP, CA 125, CA 15–3, cannabinoids, carbamazepine, CEA, CK-MB, CMV IgG, CMV IgM, cocaine, cortisol, cyclosporine, D-dimer, digoxin, estradiol, ethanol, ferritin, folate, free PSA, free T3, free T4, FSH, gentamicin, HBsAg, HBsAg confirmatory, hCG (total B-hCG), hemoglobin A1c, homocysteine, insulin, LH, methadone, myoglobin, opiates, phencyclidine (PCP), phenobarbital, phenytoin,

Human Chorionic Gonadotropin (hCG). DOI: http://dx.doi.org/10.1016/B978-0-12-800749-5.00030-4

progesterone, prolactin, rubella IgG, rubella IgM, salicylate, testosterone, theophylline, tobramycin, total PSA, total T3, total T4, toxo IgG, toxo IgM, tricyclic antidepressants, troponin-I, T-uptake, valproic acid, and vancomycin. There is a wide variation in composition and molecular weight (estradiol MW = 272, IgM subunit a MW = 184,000). Estradiol is a lipid and IgM is a sugar–amino acid complex.

Each assay uses specific plug-in cartridges containing the antibody and other ingredients. Each assay also comes with a CD containing any special calculation software. Standard curves and diluents may be available. Injection needles draw required volumes of racked samples and conveyer belts process drawn samples.

The problem with automated laboratory hCG tests is that 11 of 12 total hCG tests using these automated platforms use an antibody to hCG β-subunit C-terminal peptide as the capture antibody or tracer antibody. This antibody can severely restrict the specificity of the test (see specificity and detection described in this chapter). Requiring an antibody to this site prevents detection of molecules missing this C-terminal peptide site, a major component of circulating hCG variants. Then, because the site is mostly carbohydrate (50–60% by molecular weight), it may generate an assay that is carbohydrate specific, poorly detects hyperglycosylated molecules, and has variable sialylated or variable glycosylated molecules, for instance.

30.2 Problems with automated immunometric hCG assays and specificity

Specificity very much emphasizes the problem with 11 of 12 automated hCG immunometric assays using a β-subunit C-terminal peptide antibody (Table 30.1). Only 11 of the 12 assays are available in the United States. The twelfth assay is the Perkin-Elmer Delfia hCGβ, which also uses a β-subunit C-terminal peptide antibody, so it may be represented by the results in Table 30.1, but it is not included in the table.

Table 30.1 shows assay specificity blindly determined at multiple clinical laboratories in the United States, Canada, and New Zealand (one to seven laboratories blindly detected each standard) [1–3]. Each of 21 laboratories involved in this study were blindly (no identifiers) sent standards in normal male serum. All assays except the Siemens Immulite extremely poorly detected molecules missing the β-subunit C-terminal peptide, detecting only 0.0–42% of the original amino acid analyses-calibrated concentration [1–3].

The Siemens Immulite is one assay exception that does not use this kind of antibody. It uses two antibodies to the core β-subunit site 1 and site 2 but not a C-terminal peptide antibody. As also shown in Table 30.1, all assays except the Siemens Immulite invariably, and in many cases very poorly, detect nicked hCG, hyperglycosylated hCG, nicked hyperglycosylated hCG, asialo hCG, hCG β-subunit, and nicked hCG β-subunit, or most hCG degradation and dissociation products [1–3].

Table 30.1 The properties of 11 automated laboratory total hCG tests used in commercial laboratories

	Abbott Architect	Abbott AxSYM	Beckman Access 2	Beckman DxI 800	Ortho Vitros ECiQ	Roche Elecsys hCG + β	Siemens ACS180	Siemens Centaur	Siemens Dimension	Siemens Immulite	Tosoh A1A
Assay mechanism	Chemi	Micro	Chemi	Chemi	Chemi	Electro	Chemi	Chemi	Spectro	Chemi	Fluori
Sensitivity (mIU/ml)	1.2	2	0.5	0.5	0.7	0.5	2	2	1	1	0.5
Upper limit (mIU/ml)	1000	1000	1000	1000	1000	1000	1000	1000	1000	5000	400
Specificity											
hCG	98% 1X	103% 5X	103% 1X	100% 1X	98% 3X	109% 1X	106% 5X	105% 2X	93% 3X	101% 3X	100% 3X
Nicked hCG	70% 1X	101% 7X	92% 3X	71% 1X	75% 2X	69% 2X	82% 4X	71% 2X	85% 3X	111% 2X	101% 2X
Nicked hCG missing CTP	0.0% 1X	0.0% 1X	1.0% 1X	1.0% 1X	17% 1X	12% 1X	0.0% 1X	0.0% 1X	35% 1X	110% 2X	42% 1X
Hyperglycosylated hCG	85% 1X	85% 7X	127% 3X	97% 1X	75% 3X	136% 3X	102% 5X	98% 3X	55% 3X	102% 4X	96% 2X
Nicked hyper-glycosylated hCG	41% 1X	46% 3X	47% 1X	52% 1X	82% 1X	102% 1X	70% 1X	40% 1X	81% 1X	125% 2X	71% 1X
asialo hCG	35% 1X	69% 1X	48% 1X	46% 1X	85% 1X	46% 1X	81% 1X	39% 1X	65% 1X	113% 2X	59% 1X
hCG β-subunit	88% 1X	103% 7X	142% 2X	137% 1X	62% 2X	83% 2X	105% 3X	55% 2X	56% 2X	105% 2X	66% 2X
Nicked hCG β-subunit	33% 1X	33% 1X	56% 1X	64% 1X	19% 1X	53% 1X	72% 1X	19% 1X	41% 1X	131% 2X	60% 1X
Urine β-core fragment	1.0% 1X	1.0% 1X	1.0% 1X	1.0% 1X	1.0% 1X	16% 1X	0.0% 1X	1.0% 1X	1.0% 1X	38% 2X	1.0% 1X
Average specificity (mean specificity)	50%	60%	69%	63%	57%	70%	69%	48%	57%	104%	66%

βCTP is β-subunit C-terminal peptide. For sensitivity, standards were blindly tested at one to seven independent laboratories across the United States, Canada, and New Zealand [1–3], values are average (mean) percentage of actual concentrations determined by amino acid analysis. Values 1X, 2X, 3X, 4X, 5X, 6X, and 7X are number of independent laboratory results averaged. Underlined values are extremely poor detection, <70%. Mechanisms include enzyme chemiluminescent immunometric assay (abbreviated: Chemi); microparticle enzyme immunometric assay (abbreviated: Micro); electrochemiluminescent enzyme immunometric assay (abbreviated: Electro); spectrometric enzyme immunometric assay (abbreviated: Spectro); and flourimetric enzyme assay (abbreviated: Flouri).

Urine β-core fragment is a terminally degraded carbohydrate and peptide product of the β-subunit only present in urine (MW = 9500). It is barely detected by the Siemens Immulite (38% detection) and poorly or not detected by any other assay (0–16%).

Overall, the mean sensitivities of the β-subunit C-terminal peptide antibody assays were just 50%, 60%, 69%, 63%, 57%, 70%, 69%, 48%, 57%, and 66% of hCG-related variants (Table 30.1) [1–3]. The exception is the Siemens Immulite total hCG assay, which detected an average of 104% of variants. This assay has similar specificity as the core β-subunit-directed hCG β-subunit radioimmunoassay [1–3].

This major specificity problem would end here as an interesting detail about assays except for the fact that the 48–70% specificity problem with 11 of 12 automated hCG immunometric assays goes much further and disturbs clinical practice.

It is my understanding that the Siemens ADVIA Centaur total hCG test with just 48% detection of hCG variants, the poorest of the poor tests, as shown here, is the US market leader. It is the test used by Quest Diagnostics, which is the largest laboratory complex and has branches in every major city in the United States. Unfortunately, the Siemens Immulite is a relatively poor contender in the hCG market in the United States and the world.

Over a period of 1.5 years, the USA hCG Reference Service tested the serum of 37 cases with the Siemens Immulite hCG test (run by Tricore Reference Laboratories), and the serum was also tested blindly at other laboratories (mostly Tricore Reference Laboratories and Quest Diagnostics and also referring laboratories) with the other 10 C-terminal peptide hCG immunometric assays available in the United States (Table 30.2). As found, the results were significantly much lower when tested with these 10 other assays.

As shown in Table 30.2, the median result from 37 serum samples was 130 mIU/ml when measured with the Siemens Immulite assay and just 48 mIU/ml, or a mean 55% of the Immulite results, when measured with the other 10 assays. This is not acceptable. The specificity study indicates that the Immulite test detected 104% of hCG variables and the other 10 assays only detected an average of 48–70% of hCG variants. The serum studies representative of normal pregnancy, failing pregnancy, and other hCG-producing disorders show that these assays only detect 55% of what they seemingly should detect (the Siemens Immulite result).

This is not acceptable. It is not acceptable that 10 of 11 hCG assays that cover most of the world's market are only detecting 55% of virtually half of what they should be detecting. I considered the possibility that the Immulite assay result could be wrong or exaggerated. I sent the serum standards used in the sensitivity study (Table 30.1) to Charing Cross Hospital in England, where they measured the samples by hCG β-subunit RIA. Summarizing concentration results, their result was that the RIA detected 99% of the amino acid analysis concentrations (hCG, 99%; hyperglycosylated hCG, 92%; nicked hCG, 98%; nicked hCG missing β-subunit C-terminal peptide, 95%; nicked hyperglycosylated hCG, 70%; asialo hCG, 126%; hCG β-subunit, 86%; nicked hCG, 80%; β-core fragment, 76%) versus 104% detected by the Siemens Immulite assay [1–3]. This is a very similar result.

I have spoken to Siemens about adopting the antibodies used in the Immulite assay in their Dimension, Centaur, and ACS180 assays. I am not sure if they are

Table 30.2 Detection of serum total hCG by the Siemens Immulite and 10 other automated assays

Diagnosis	Total hCG, Immulite assay (mIU/ml)	Total hCG, other assay (mIU/ml)	Other assay, % Immulite result	Other assay used
Normal term pregnancy	441,624	292,136	66%	Siemens Dimension
Normal term pregnancy	9358	16,621	56%	Siemens Dimension
Normal term pregnancy	7370	5743	78%	Siemens Dimension
Normal term pregnancy	275	263	96%	Siemens Dimension
Spontaneous abortion	6	5	83%	Siemens Dimension
Spontaneous abortion	11	11	100%	Siemens Dimension
Spontaneous abortion	46	38	83%	Siemens Dimension
Spontaneous abortion	2	1	50%	Siemens Dimension
Spontaneous abortion	2.4	<0.5	<21%	Beckman Access 2
Spontaneous abortion	140	66	47%	Ortho Vitros Eci
Spontaneous abortion	444	103	23%	Siemens Centaur
Spontaneous abortion	148	38	26%	Siemens Centaur
Spontaneous abortion	140	66	47%	Siemens Centaur
Spontaneous abortion	16	8	50%	Siemens Centaur
Spontaneous abortion	156	103	66%	Siemens Centaur
Spontaneous abortion	137	93	68%	Tosoh A1A
hCG doping	80,400	78,000	97%	Siemens Centaur
hCG doping	106	75	71%	Abbott AxSym
Menopausal hCG	10.0	9.0	90%	Siemens Centaur
Choriocarcinoma	2315	1875	81%	Abbott Architect
Choriocarcinoma	717	651	91%	Siemens Dimension

(Continued)

Table 30.2 (Continued)

Diagnosis	Total hCG, Immulite assay (mIU/ml)	Total hCG, other assay (mIU/ml)	Other assay, % Immulite result	Other assay used
Choriocarcinoma	3.0	2.0	67%	Siemens Dimension
Trophoblastic neoplasm	204	101	50%	Siemens Centaur
Trophoblastic neoplasm	107	88	83%	Siemens Centaur
Familial hCG syndrome	216	24	11%	Beckman Dxi
Familial hCG syndrome	201	29	14%	Siemens Centaur
Familial hCG syndrome	168	50	30%	Siemens Centaur
Familial hCG syndrome	130	39	30%	Siemens Centaur
Familial hCG syndrome	43	12	28%	Siemens Centaur
Familial hCG syndrome	17	<2.0	<12%	Siemens Centaur
Familial hCG syndrome	2.8	<2.0	<71%	Siemens Centaur
Molar pregnancy	47	38	81%	Siemens Dimension
Molar pregnancy	235	48	20%	Abbott AxSym
Molar pregnancy	118	70	59%	Abbott AxSym
Molar pregnancy	118	66	56%	Beckman Access 2
Molar pregnancy	9.2	7.1	77%	Roche Elecsys hCG + β
Molar pregnancy	5.9	4.6	78%	Siemens ACS180
Average	**Median 130**	**Median 48**	**Mean 55%**	

commercially interested. One of the biggest problems is the lack of availability of standards. There is no official standard for hyperglycosylated hCG and no official standard for hCG missing the β-subunit C-terminal peptide. Two of the most critical problem detection molecules are shown here. In part, this lack of available standards caused the problems present here because the manufacturers are unaware that the problem exists. I have written to Siemens and Abbott about this problem. I have heard no response from Abbott. I am very open to helping any manufacturer fix this very unacceptable problem.

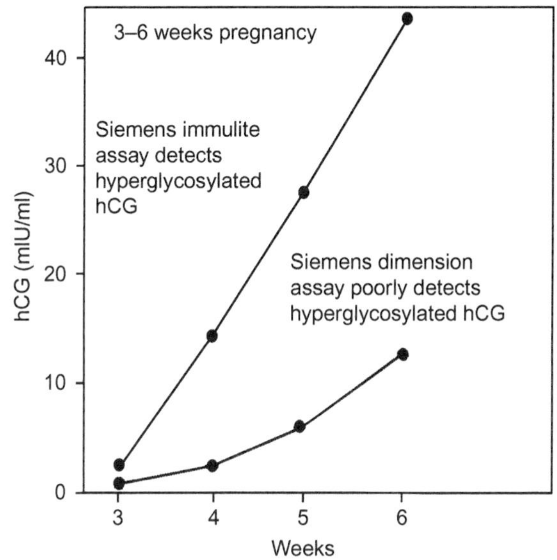

Figure 30.1 Example of problems encountered with assays using C-terminal peptide antibody.

Four figures are presented in this chapter (Figures 30.1–30.3). They are self-explanatory examples that very much emphasize the problem with the 10 assays using β-subunit C-terminal antibodies.

30.3 Sensitivity and specificity of point-of-care hCG immunoassays

Physician office, clinic, clinical laboratory urine hCG, or point-of-care (POC) pregnancy tests are inexpensive dip-stick tests sold in bulk. These tests are mostly designed for urine pregnancy testing but also may work with blood pin pricks. They all work on the multi-antibody immunometric assay principal as used in automated serum assays. As manager (stayed independent from testing devices), I employed a team of dip-stick device readers (three other PhD scientists and laboratory technicians) and blindly (everything coded) examined the sensitivity and specificity of six representative POC hCG tests (Table 30.3).

As shown in Table 30.3, most tests are strangely just intact hCG tests and not total hCG tests [4] that do not detect or poorly detect hCG free β-subunit and its multiple variants, including β-core fragment, which is the principal hCG variant in urine.

Detection of early pregnancy hCG by a total hCG assay requires at least appropriate detection of hyperglycosylated hCG, the early pregnancy marker. As shown in Table 30.3 [4], only two of the six tests evaluated appropriately detected hyperglycosylated hCG. These were the Quidel QuickVue and Beckman–Coulter

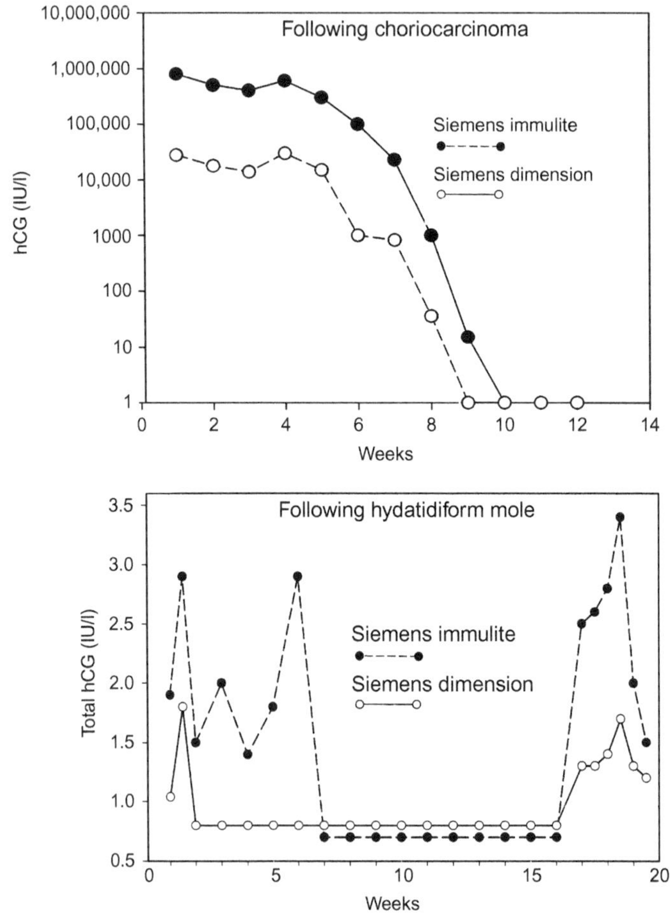

Figure 30.2 Example of problems encountered with assays using C-terminal peptide antibody.

One-Step test. The Beckman–Coulter One-Step was the most sensitive test investigated, with a sensitivity of 12 mIU/ml.

The POC test seems to be an insensitive mixture of intact hCG tests of very limited application.

30.4 Sensitivity and specificity of over-the-counter immunoassays

Home or over-the-counter (OTC) immunoassays are another pregnancy testing option. There are manual OTC hCG tests, which are basically a sophisticated

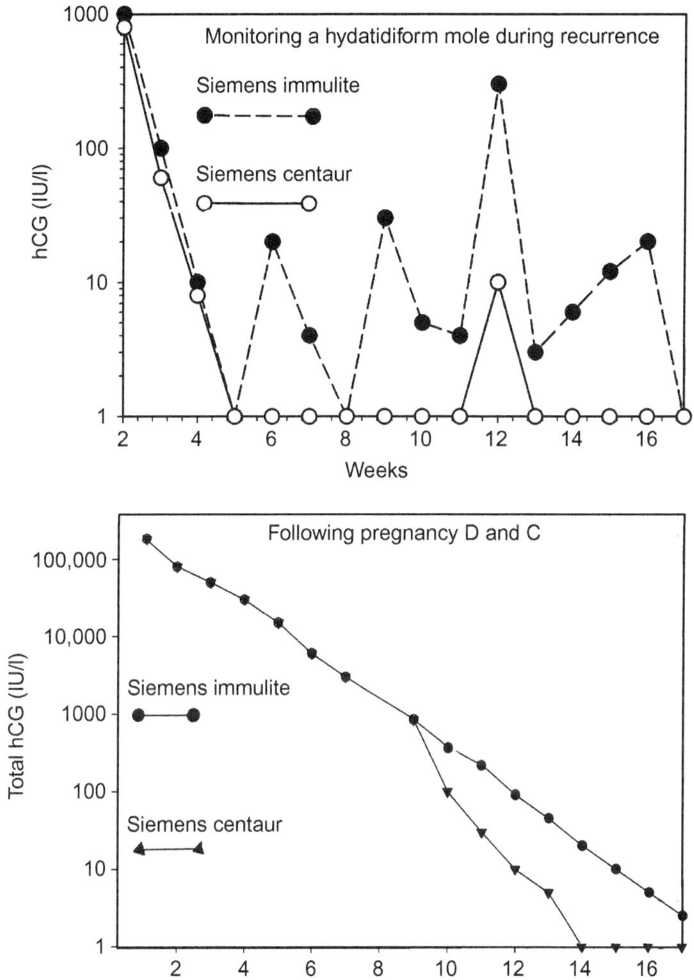

Figure 30.3 Example of problems encountered with assays using C-terminal peptide antibody.

version of a POC test designed for easy home use, and there are digital tests, which instead of presenting a dye line or symbol that has to be carefully read, present a digital message "Yes or "No" or "Pregnant" or "Not Pregnant" to avoid all home confusion. The digital devices are generally less sensitive than manual devices. As manager (stayed independent from testing devices), I employed a team of dip-stick device readers (three other PhD scientists and laboratory technicians) and blindly (everything coded) examined the sensitivity and specificity of three of the most common digital and three of the most common manual OTC devices sold.

To conduct the sensitivity/specificity studies, the USA hCG Reference Service purchased and tested OTC devices from 12 pharmacies. Purchased devices included

Table 30.3 Sensitivity of POC hCG assays, values are calculated on a molar basis

POC test	hCG standard Sensitivity (mIU/ml)	Hyperglycosylated hCG standard Sensitivity (mIU/ml)	Free β-subunit standard Sensitivity (mIU/ml)
Quidel QuickVue One-Step	24	24	88
Beckman–Coulter Icon 25	12	12	88
Siemens Clinitest, Digital Analyzer	50	100	>500
Inverness Acceava hCG Basic II	18	50	>500
Mainline Maxine hCG	24	50	>500
Mainline Confirms hCG	24	50	>500

Six devices were tested at each of six concentrations of hCG (first Reference Reagent): 6, 12, 24, 50, 100, and 500 IU/l. Lowest concentrations leading to six of six positive results are reported [4].

First Response, First Response Gold digital, ClearBlue Easy manual, and ClearBlue Easy digital, EPT manual, and EPT Certainty digital [4–7].

Specificity studies showed that the First Response and First Response Gold devices have a sensitivity for hCG and hyperglycosylated hCG of 3.3 and 5.5 mIU/ml. The First Response devices were total hCG assays detecting 9.0 mIU hCG free β-subunit. The EPT manual and digital devices had an hCG sensitivity of 11 mIU/ml and hyperglycosylated hCG sensitivity of 22 mIU/ml, and the similar ClearBlue Easy manual and digital devices also had hCG sensitivity of 11 mIU/ml and hyperglycosylated hCG sensitivity of 22 mIU/ml [6]. The very similar EPT and ClearBlue Easy devices were intact hCG devices that did not detect hCG free β-subunit. The First Response devices were the most sensitive devices and the only devices that equally detected hCG and hyperglycosylated hCG.

Basically, daily urine samples were collected from 220 women eager to achieve pregnancy. Eighty of these 220 women achieved normal term outcome pregnancies with normal deliveries [4–7]. Table 30.4 shows detection by OTC tests in daily urine; data are normalized to day of missing the established due date of menstrual period (−6, −5, −4, −3, −2, −1, 0, +1, + , +3, and +4 days from the day of the missed period). The "achieved pregnancy" column shows the proportion of the 80 who achieved pregnancy implantation by this time. This was shown by first appearance of urine hCG. hCG median is the median hCG concentration at the specified time of pregnancy. Proportions indicate the percentage of tests that were positive at the specified time of pregnancy [3–6].

As shown in Table 30.4, on the day of missing the menstrual period (day 0), 99% of pregnancies had implanted. The First Response manual device and the First

Table 30.4 Evaluation of OTC tests with pregnancy urines [4–7]

Day relative to day of missed period	Achieved pregnancy	hCG median mIU/ml	First Response manual	First Response Gold digital	EPT manual	EPT digital	ClearBlue Easy manual	ClearBlue Easy digital
−6	29%	2.1	25%	25%	0%	0%	0%	0%
−5	40%	2.9	33%	25%	5.00%	5.00%	5.00%	5.00%
−4	76%	5.2	58%	42%	6.30%	6.30%	8.80%	3.80%
−3	88%	12	74%	68%	14%	18%	27%	12%
−2	92%	21	76%	81%	29%	31%	29%	28%
−1	99%	40	93%	91%	42%	55%	57%	51%
0	99%	70	96%	96%	53%	68%	67%	51%
1	100%	143	100%	96%	64%	71%	74%	69%
2	100%	227	100%	99%	77%	79%	81%	77%
3	100%	302	100%	100%	80%	86%	87%	84%
4	100%	534	100%	100%	100%	100%	100%	100%

Response Gold digital test detected 96% of these pregnancies. The EPT manual device detected just 53% of these pregnancies, and the EPT digital device detected 68% of the pregnancies. The ClearBlue Easy manual test detected 67% of these pregnancies, and the ClearBlue digital device detected 51% of these pregnancies. Relatively comparable results favoring the First Response devices were seen on other dates (−6, −5, −4, −3, −2, −1, +1, +2, and +3 days from the missed period date) [4–7]. These appear to be the most sensitive devices for home testing for pregnancy.

Manufactures advertise claims such as "Detect pregnancies 3–5 days before the day of missing the menstrual period." In my opinion, these statements are invalid because only a small proportion of pregnancies have implanted at this time. Detection statements should be limited to the day of missing the menstrual cycle or afterward. As shown in Table 30.4, at 4 days before missing the menstrual cycle, 58% of pregnancies can be detected with the First Response manual test, 42% with the First Response Gold digital device, a meaningless 6.3% with EPT manual and digital tests, 8.8% with ClearBlue manual, and just 3.8% with the ClearBlue digital test. It is noted that manufacturers date their dates of missed menstrual period somewhat differently than that used here and claim somewhat better results.

References

[1] Cole LA, Shahabi S, Butler S, Mitchell H, Newlands ES, Behrman HR, et al. Utility of commonly used commercial hCG immunoassays in the diagnosis and management of trophoblastic diseases. Clin Chem 2001;47:308–15.

[2] Cole LA, Sutton JM, Higgins TN, Cembrowski GS. Between-method variation in hCG test results. Clin Chem 2004;50:874–82.

[3] Cole LA, DuToit S, Higgins TN. Total hCG tests. Clin Chim Acta 2011;412:2216–22.

[4] Cole LA, Ladner DG. Background hCG in non-pregnant individuals: need for more sensitive point-of-care and over-the-counter pregnancy tests. Clin Biochem 2009;42:168–75.

[5] Cole LA, Sutton-Riley JM, Khanlian SA, Borkovskaya M, Rayburn BR, Rayburn WF. Sensitivity of over-the-counter pregnancy tests: comparison of utility and marketing messages. J Am Pharm Assoc 2005;45:608–15.

[6] Cole LA. The utility of 6 over-the-counter (home) pregnancy tests. Clin Chem Lab Med 2011;49:1317–22.

[7] Cole LA, Khanlian SA, Sutton JM, Davies S, Rayburn WF. Accuracy of home pregnancy tests at the time of missed menses. Am J Obstet Gynecol 2004;190:100–5.

The future and a new generation of pregnancy tests

31

Laurence A. Cole
USA hCG Reference Service, Angel Fire, NM, USA

The biggest problem today with pregnancy tests is that they detect pregnancy, but they do not tell whether the pregnancy will be a successful pregnancy that will proceed to term and result in delivery of a baby, or whether the pregnancy may be a complete failure. This is essential information. Today, women or their doctors may use an OTC, POC or laboratory pregnancy test. The test may be positive, but they do not know whether to cry, think positively, or celebrate. Unfortunately, too many families celebrate a pregnancy and then cry later when the pregnancy fails.

In the United States today, 25% of all pregnancies are biochemical pregnancies and terminate in the early weeks of pregnancy [1,2], 18% of pregnancies are first- or second-trimester spontaneous abortions or miscarriages of pregnancies [3,4], and 2.0% of pregnancies are fallopian tube or ectopic pregnancies [5,6]. As such, 45% (25% + 18% + 2.0%) of pregnancies have no chance or are destined failures. The remainder, 55% of pregnancies, are destined to head to term or likely yield a baby. In some other countries, much higher proportions of all these pregnancy failures occur [7,8]. Higher incidences of failures can occur among *in vitro* fertilized (IVF) pregnancies and other fertility clinic groups [9,10].

One might ask the question, how can a pregnancy test determine if a pregnancy is going to fail? This is a good question. Current pregnancy tests measure hCG, the hormone produced by fused syncytiotrophoblast cells of the placenta during pregnancy. This hormone provides no consistent information about pregnancy failure.

What if instead we measure the molecule that controls implantation of pregnancy, hyperglycosylated hCG, an autocrine that binds a TGFβ receptor [11,12]? What if shortage of this molecule marks ectopic pregnancy, marks biochemical pregnancies, and marks spontaneous abortion pregnancies [13,14]? As shown, gross deficiency of hyperglycosylated hCG seems to be the principal cause of most failing pregnancies, biochemical pregnancies, and spontaneous abortions [13,14], and it clearly marks measurement of this autocrine [13,14]. Low hyperglycosylated hCG also seems to mark other failures like ectopic pregnancies [15]. What if we measure hyperglycosylated hCG as a pregnancy test and mark failures as low-positive hyperglycosylated hCG tests?

The first problem is that positive hyperglycosylated hCG concentrations start at the time of implantation of pregnancy, and then increases exponentially in the weeks that follow [1,13,14]. Concentrations decline after reaching a peak a 10-week pregnancy. Hyperglycosylated hCG increases and peaks parallel to the

Human Chorionic Gonadotropin (hCG). DOI: http://dx.doi.org/10.1016/B978-0-12-800749-5.00031-6

hormone hCG [1]; what if hyperglycosylated hCG results are expressed as a proportion of total hCG (hCG + hyperglycosylated hCG + hCG free β-subunit) or normalized? This is what we did, and we propose proportions of hyperglycosylated hCG as the future pregnancy test.

We enrolled 247 women eager to achieve pregnancy into a clinical trial. We first suggested the daily collection of serum from more than five menstrual cycles until pregnancy was achieved, and until 7 weeks of gestation thereafter. This required the collection of up to 170 daily serum samples. Nobody was interested in volunteering to perform extensive daily phlebotomies. We changed the program to daily urine collection, and every one of the 247 couples volunteered.

A total of 93 women achieved normal term pregnancies (53%), 38 had spontaneously aborting pregnancies (22%), 40 had early biochemical pregnancies (23%), and 2 had ectopic pregnancies (2.3%). Figure 31.1 shows proportion of hyperglycosylated hCG on the day of pregnancy implantation in these 174 pregnancies. As shown by Wilcox and colleagues [1], the day of implantation is the day of first urine total hCG appearance (>1 mIU/ml).

As shown (Figure 31.1), proportion of hyperglycosylated hCG marks most pregnancy losses. As found by Norwitz et al. [16], Semprini and Simon [17], and Wilcox et al. [1], 70% of spontaneous abortion and 70–75% of biochemical pregnancies are due to incomplete implantation of pregnancy. The balance of failures is attributable to genetic abnormality leading to properly implanted pregnancies that fail later. As found by my group, 70–75% of failures are

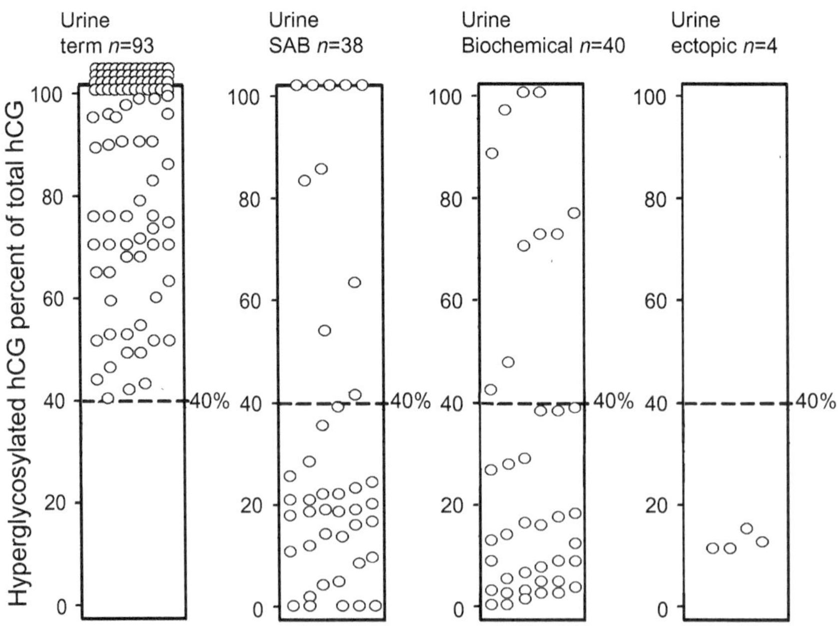

Figure 31.1 Proportion hyperglycosylated hCG on day of pregnancy implantation.

attributable to an insufficient supply of hyperglycosylated hCG [14,15]. This is confirmed by Figure 31.1. As shown in Figure 31.1, 100% of term outcome pregnancies produce >40% hyperglycosylated hCG at the time of implantation, 28 of 38 (74%) spontaneous abortion outcome pregnancies produce deficient hyperglycosylated hCG (<40%) at the time of implantation, and 30 of 40 (75%) biochemical pregnancies produce deficient hyperglycosylated hCG (<40%) at the time of implantation.

I considered whether proportion of hyperglycosylated hCG can be used as an early pregnancy test. As described in Figure 31.1, 93 of 93 term outcome pregnancies were detected on the day of implantation of pregnancy, the earliest possible time to detect pregnancy. These pregnancies ranged from 40.3% of hyperglycosylated hCG to 100% of hyperglycosylated hCG; 48 of the 93 produced 100% hyperglycosylated hCG. Using the data in Figure 31.1, 100% of pregnancies producing <40% hyperglycosylated hCG failed because of ectopic pregnancy, spontaneous abortion, or biochemical pregnancy. It has been concluded that measurement of the proportion of hyperglycosylated hCG is a pregnancy test and predictor of pregnancy outcome.

As calculated on the day of pregnancy implantation, a positive proportion of hyperglycosylated hCG test (>40%) indicates 92% likelihood of term pregnancy. A positive hCG test and negative proportion hyperglycosylated hCG (>40%) test indicate failure with 100% certainty. I consider the use of these tests during the 2 weeks that follow the day of implantation (3 weeks after last menstrual period and 5 weeks of gestation). As shown in Table 31.1, prediction statistics quickly decline from the 92% and 100% with time.

Although a negative test indicated 100% certainty of a failure on the day of implantation, that prediction diminishes to 85% 3 days later, to 81% 7 days later, to 79% 10 days later, and to 68% 14 days later (Table 31.1). Similarly, the prediction of a term pregnancy or a delivery with 92% certainty on the day of implantation diminishes to 83% at 3 days and 7 days, to 70% at 7 days, and to 64% at 14 days. It should be noted that 64% prediction of term pregnancy and 68% prediction of failure at 14 days is still a lot better that no prediction, and there is a 45% chance of failure and 55% chance of term outcome [1–6].

Clearly, this new pregnancy test is best used close to the day of implantation. This can be performed with IVF pregnancies 6 days after embryo transfer. This test can also be performed in those attending a fertility clinic measuring LH peak; testing 7 days after LH peak should be performed. This test may also be possible with women noting the first day of their last menstrual period, or when their next menstrual period is due. In these women, the test could be performed 3–4 weeks after the start of the period. It will tell women for the first time if their pregnancy is "real" (resulting in a baby) or if their pregnancy will be a failure.

The group of 254 women volunteering for the clinical trial were willing to provide daily urine samples but not daily serum samples. For this reason, the studies and their statistics are limited to spot urine pregnancy testing. The statistics of this testing can be used immediately. The test will not be able to be switched to serum testing until similar statistics can be determined for serum.

Table 31.1 Proportions hyperglycosylated hCG in the days following pregnancy implantation

	Term pregnancy outcome			Biochemical pregnancy outcome			Spontaneous abortion outcome			Ectopic pregnancy outcome		Accuracy of test	Predictions		
Time	Count	>40%	%	Count	>40%	%	Count	>40%	%	Count	>40%		1	2	3
Implantation	93	93	100%	40	10	25%	38	10	26%	4	0	89%	92%	98%	100%
+1 day	83	72	87%	20	2	10%	31	6	19%	4	0	86%	91%	97%	90%
+2 days	83	71	86%	28	2	7%	30	9	30%	4	1	84%	89%	97%	89%
+3 days	83	65	78%	36	6	17%	31	14	45%	2	1	74%	83%	96%	85%
+4 days	83	61	73%	36	2	6%	31	13	42%	4	1	75%	83%	95%	84%
+5 days	81	62	76%	36	6	17%	28	11	39%	4	1	75%	83%	95%	85%
+6 days	84	63	75%	28	2	7%	30	8	27%	4	0	79%	85%	95%	84%
+7 days	84	63	75%	26	4	15%	30	14	47%	4	0	73%	83%	95%	81%
+8 days	84	63	75%	22	2	9%	31	17	55%	4	0	71%	82%	95%	80%
+9 days	84	65	77%	22	2	9%	31	16	52%	4	1	73%	84%	95%	82%
+10 days	84	62	74%	20	4	20%	31	16	52%	4	0	70%	82%	95%	79%
+11 days	84	57	68%	18	2	11%	31	19	61%	4	1	65%	80%	94%	75%
+12 days	83	58	70%	10	0	0%	31	19	61%	4	1	65%	81%	94%	73%
+13 days	83	48	58%	8	0	0%	31	13	42%	2	1	61%	78%	92%	69%
+14 days	83	58	70%	2	0	0%	31	15	48%	4	3	64%	82%	94%	69%

The count is measured exceeding >40% hyperglycosylated hCG (% total hCG). Accuracy is (total failures <40%, plus total term outcome >40%) ÷ total count. Predictions column 1 are the statistical odds if proportions are >40% that a pregnancy will be normal and will go to term ((Term >40%/Count Term + Spontaneous abortion >40% + Biochemical >40% + Ectopic >40%) x 45%) + 55%. Prediction 2 follows prediction 1 after 19 days pregnancy when biochemical pregnancies are excluded. Prediction 3 are the statistical odds if a test is <40% of a pregnancy failure occurring ((Spontaneous abortion <40% + Biochemical <40% + Ectopic <40%/Total <40%) x 55%) + 45%. Decreasing supply of urine as pregnancy advances due to lessening volunteer cooperation.

References

[1] Wilcox AJ, Weinberg CR, O'Connor JF, Baird DD, Schlatterer JP, Canfield RE, et al. Incidence of early loss of pregnancy. N Engl J Med 1988;319:189–94.

[2] Wilcox AJ, Weinberg CR, Baird DD. Risk factors for early pregnancy loss. Epidemiology 1990;1:382–5.

[3] Wilcox AJ, Treloar AE, Sandler DP. Spontaneous abortion over time: comparing occurrence in two cohorts of women a generation apart. Am J Epidemiol 1981;114:548–53.

[4] Katz VL. Spontaneous and recurrent abortion: etiology, diagnosis, treatment. In: Katz VL, Lentz GM, Lobo RA, Gershenson DM, editors. Comprehensive gynecology. 6th ed. Philadelphia, PA: Elsevier Mosby; 2012 [chapter 16].

[5] Berg CJ, Callaghan WM, Syverson C, Henderson Z. Pregnancy-related mortality in the United States, 1998 to 2005. Obstet Gynecol 2010;116:1302–9.

[6] Tulandi T. Incidence, risk factors, and pathology of ectopic pregnancy, <http://www.uptodate.com/contents/incidence-risk-factors-and-pathology-of-ectopic-pregnancy>.

[7] Miscarriages in different countries, statistic of miscarriages, <http://miscarriagesindifferentcountries.wordpress.com/miscarriages/statistics-of-miscarriages/>.

[8] Buyukbayrak EE. Worldwide incidence of ectopic pregnancy, a protocol of systematic review. Postgraduate training course in reproductive health, <http://www.gfmer.ch/Medical_education_En/PGC_RH_2004/buyukbarak_review.htm>; 2004.

[9] IVF and ectopic pregnancy, <http://www.womens-health.co.uk/ivf-ectopic-pregnancy.html>.

[10] Danielsson K. Does conceiving via *in vitro* fertilization mean higher risk of miscarriage? <http://miscarriage.about.com/od/riskfactors/f/invitro.htm>.

[11] Cole LA, Butler SA. Hyperglycosylated hCG hCGβ and hyperglycosylated hCGβ: interchangeable cancer promoters. Mol Cell Endocrinol 2012;349:232–8.

[12] Berndt S, Blacher S, Munuat C, Detilleux J, Evain-Brion D, Noel A, et al. Hyperglycosylated human chorionic gonadotropin stimulates angiogenesis through TGF-β receptor activation. J Fed Am Soc Exp Biol 2013;27:1309–21.

[13] Cole LA. Hyperglycosylated hCG and pregnancy failures. J Reprod Immunol 2012;93:119–22.

[14] Sasaki Y, Ladner DG, Cole LA. Hyperglycosylated hCG the source of pregnancy failures. Fertil Steril 2008;89:1781–6.

[15] Cole LA, Khanlian SA, Sutton JM, Davies S, Stephens N. Hyperglycosylated hCG (invasive trophoblast antigen, ITA) a key antigen for early pregnancy detection. Clin Biochem 2003;36:647–55.

[16] Norwitz ER, Schust DJ, Fisher SJ. Implantation and the survival of early pregnancy. N Engl J Med 2001;345:1400–8.

[17] Semprini AE, Simon G. Not so efficient reproduction. Lancet 2000;356:257–8.

Part K

Test Applications and Standards

Detecting Down syndrome pregnancies and preeclampsia

32

Laurence A. Cole
USA hCG Reference Service, Angel Fire, NM, USA

In 1995, with the new discovery of hyperglycosylated hCG, I presented the story of a discovery by the Obstetrics and Gynecology faculty at Yale University. I was approached by a member of the Maternal Fetal Medicine Division at Yale University to test a library of second-trimester Down syndrome pregnancy samples and controls. Screening pregnancies for Down syndrome ended up being the first application of many found for hyperglycosylated hCG [1]. Why should hyperglycosylated hCG be a marker of Down syndrome pregnancy? hCG and hyperglycosylated hCG take on different profiles in Down syndrome pregnancies. The limited fusion of trophoblast cells in Down syndrome causes an unduly high proportion of cytotrophoblast cells to remain unfused. This leads to unduly high production of hyperglycosylated hCG in Down syndrome pregnancies [2–4]. In this chapter, we look at the procedures involved in screening for Down syndrome pregnancies. We then examine screening for preeclampsia and pregnancy-induced hypertension (PIH), a new application desperately needing a good marker.

32.1 Down syndrome screening

Trisomy 21, also referred to as Down syndrome, is the most common genetic abnormality that complicates pregnancies. Trisomy 21 and other genetic monosomies, trisomies, and triploidies can be detected through a cytogenetic analysis procedure. Amniocentesis is performed at 14–26 weeks of gestation and can lead to late pregnancy termination between 16 and 30 weeks of gestation (for those choosing to abort). Amniocentesis is complicated by a failure rate of approximately 1 in 300, with the fetus being aborted by the procedure. Chorionic villous sampling can determine cytogenetics at 9–11 weeks of gestation. The prenatal risk for analysis by chorionic villous sampling is approximately two-times higher at approximately 1 in 150 failures, depending on the expertise of the center performing the procedure. The advantage of chorionic villous sampling is that abortion can be performed if a genetic abnormality is found early, at 12–16 weeks of gestation.

As age advances, risk for Down syndrome pregnancies increases significantly, occurring in 1 in 105 pregnancies in women age 40 years or older [5]. Women younger than age 30 years have a 1 in 1000 Down syndrome risk [5]. Because more women achieve pregnancy before age 30, more cases of fetuses with Down syndrome occur in

Human Chorionic Gonadotropin (hCG). DOI: http://dx.doi.org/10.1016/B978-0-12-800749-5.00032-8

women younger than age 30. Thus, it is statistically more appropriate not to take the higher risk of undergoing amniocentesis (1 in 300 failures) or chorionic villous sampling (1 in 150 failures) to determine their cytogenetic outcomes. New tests and supplemental procedures are needed for younger women to determine whether amniocentesis or chorionic villous samples are statistically justified.

In 1984, Chard and colleagues [6] showed that α-fetoprotein was an approximate indicator of Down syndrome pregnancy. In 1987, Bogart and colleagues [7] showed that Down syndrome pregnancies were associated with two-times higher hCG levels in the second trimester of pregnancy. Wald and colleagues [8] then showed that unconjugated estriol was also a useful marker. This led to the development of the "triple test," which used a combination of hCG, α-fetoprotein, and unconjugated estriol as markers for predicting those at high risk for Down syndrome pregnancies. This triple test normally had a 60% detection rate of Down syndrome pregnancy and a 5% false-positive rate. Even though there was a low false-positive rate and minimal risks for women younger than 30 years old, a woman with a Down syndrome – positive result in a triple test still had an approximately 1 in 100 chance of having a fetus with Down syndrome. The chance was enough to justify amniocentesis, which has a failure rate of 1 in 300 pregnancies.

The addition of inhibin A as a fourth marker in the quadruple test improved the sensitivity for identifying Down syndrome pregnancies (68% detection rate with a 5% chance of false-positive results vs. 60% and 5%) [9]. Major improvements also arose from using higher-sensitivity hyperglycosylated hCG in place of regular hCG (80% detection rate with a 5% chance of false-positive results) [10]. However, all these biochemical tests are still limited by the 5% false-positive results. There is still a 1 in 50 to 1 in 100 chance of a positive test resulting in a Down syndrome pregnancy, and each test is limited to the second trimester of pregnancy.

In the late 1990s, serum markers were identified for first-trimester Down syndrome screening; hCG free β-subunit was preferred over regular hCG [11]. These serum markers required high-risk chorionic villous sampling, which was required to determine cytogenetics in those indicated as being at risk for positive Down syndrome test results. A woman with positive results on first-trimester biochemical testing had a 1 in 100 chance of having a Down syndrome pregnancy, and chorionic villous sampling had a 1 in 150 chance of destroying the pregnancy. For women with genetic predispositions to Down syndrome, it was a gamble.

In 1992, Nicolaides and colleagues [12] found that fetal nuchal translucency or neck translucency (measured using ultrasound) was a powerful indicator of Down syndrome pregnancy. The combination of ultrasound nuchal translucency with laboratory measurement of free β-subunit and pregnancy-associated plasma protein-A became the standard combination of tests to assess risk for a Down syndrome fetus. In 2003 [13], free β-subunit was replaced with hCG-H and the sensitivity of the Nicolaides test combination improved. Today, the nuchal translucency combinations with pregnancy-associated plasma protein-A (PAPP-A) and free β-subunit have a 94% detection rate with a 5% chance of false-positive results. Combinations using hyperglycosylated hCG in place of free β-subunit have a 95% accurate detection rate with a 5% chance of false-positive results. Regardless of whether the test is performed during the first or second trimester, the hCG variant (hCG, hCG-H, or free β-subunit) is a major part of the test. Both first-trimester and second-trimester combinations are used throughout the world to predict Down syndrome.

In terms of predicting a Down syndrome pregnancy, the hyperglycosylated hCG, PAPP-A, and nuchal translucency combination is the best method available today [13–15]. Although this combination of markers can yield a positive result for Down syndrome pregnancy, it still only has a limited chance of indicating a Down syndrome fetus. Furthermore, these markers require chorionic villous sampling to give a definitive prediction, and chorionic villous sampling is a high-risk procedure. New tests that examine fetal chromosomes in separate maternal blood are on the horizon [16]. Such complex methodologies will eventually replace biochemical testing, ultrasound methods, and amniocentesis, or chorionic villous sampling will be improved and the associated risks will be removed.

32.2 Preeclampsia screening

Hypertensive disorders, PIH, preeclampsia, eclampsia, and hemolysis, elevated liver enzymes, and low platelet count (HELP syndrome) are the most deadly complications of pregnancy. Each can lead to hemorrhage and death. They are most common in nulliparous, or first-time, pregnancies. They occur in approximately 7% of all pregnancies.

As discussed in Chapter 14, hypertensive disorders might be a complication derived from adequate hyperglycosylated hCG production, appropriate invasion, and establishment and growth of efficient hemochorial placentation during pregnancy [17–21]. As such, impending hypertensive disorders can be predicted

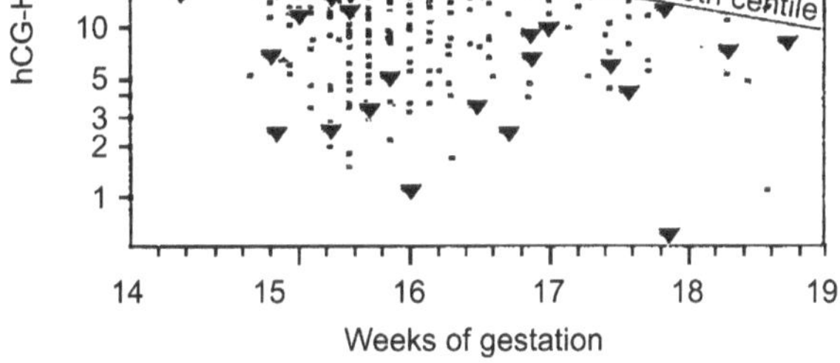

Figure 32.1 Hyperglycosylated hCG as a marker of pregnancy hypertensive disorders. Multiple of median (MoM) plot. Small black dots indicate normal term pregnancies and large black triangles indicate pregnancies with proven Down syndrome outcome.

by measuring hyperglycosylated hCG in serum or urine during the first and second trimesters of pregnancy [20,21]. Bahado-Singh and colleagues tested 27 samples from women genetically destined for development of hypertensive disorders and 510 controls [20]. As illustrated in Figure 32.1, using MoM algorithms, hyperglycosylated hCG is an outstanding indicator of preeclampsia and hypertensive diseases. A progressive increase in risk is noted with decreasing MoM. As such, hyperglycosylated hCG presents itself as an ideal marker for predicting pregnancy hypertensive disorders.

References

[1] Cole LA, Cermik D, Bahado-Singh R. Oligosaccharide variants of hCG-related molecules: potential screening markers for Down syndrome. Prenat Diagn 1997;17:1188–90.

[2] Kato Y, Braunstein GD. Discordant secretion of placental protein hormones in differentiating trophoblasts *in vitro*. J Clin Endocrinol Metab 1989;68:814–20.

[3] Knöfler M. Regulation of hCG during normal gestation and in pregnancies affected by Down's syndrome? Mol Hum Reprod 1999;5:895–7.

[4] Frendo JL, Vidaud M, Guibourdenche J, Luton D, Muller F, Bellet D, et al. Defect of villous cytotrophoblast differentiation into syncytiotrophoblast in Down's syndrome. J Clin Endocrinol Metab 2000;85:3700–7.

[5] Newberger DS. Down syndrome: prenatal risk assessment and diagnosis. Am Fam Physician 2000;62:837–8.

[6] Chard T, Lowings C, Kitau MJ. Alpha fetoprotein and chorionic gonadotropin levels in relation to Down's syndrome. Lancet 1984;29:750–1.

[7] Bogart MH, Pandian MR, Jones OW. Abnormal maternal serum chorionic gonadotropin levels in pregnancies with fetal chromosome abnormalities. Prenat Diagn 1987;7:623–30.

[8] Wald NJ, Cuckle HS, Densem JW, Nanchahal K, Canick JA, Haddow JE, et al. Maternal serum unconjugated oestriol as an antenatal screening test for Down's syndrome. Br J Obstet Gynaecol. 1988;95:334–41.

[9] Wald NJ, Densem JW, Muttukvishna S, Knight PG. Prenatal screening for Down syndrome using inhibin A as a serum marker. Prenat Diagn 1996;16:143–53.

[10] Cole LA, Shahabi S, Oz UA, Bahado-Singh RO, Mahoney MJ. Hyperglycosylated hCG (invasive trophoblast antigen) immunoassay: a new basis for gestational Down syndrome screening. Clin Chem 1999;45:2109–19.

[11] Macri JN, Kasturi RV, Krantz DA, Cook EJ, Moore ND, Young JA, et al. Maternal serum Down syndrome screening: free beta protein is a more effective marker than human chorionic gonadotropin. Am J Obstet Gynecol 1990;163:1248–53.

[12] Nicolaides KH, Azar G, Byrne D, Mansur C, Marks K. Fetal nuchal translucency: ultra sound screening for chromosomal defects in the first trimester of pregnancy. Br Med J 1992;304:867–89.

[13] Palomaki GE, Knight GJ, Roberson MM, Cunningham GC, Lee JE, Strom CM, et al. Invasive trophoblast antigen (hyperglycosylated human chorionic gonadotropin) in second-trimester maternal urine as a marker for Down syndrome: preliminary results of an observational study on fresh samples. Clin Chem 2003;50:182–9.

[14] Palomaki GE, Knight GJ, Neveux LM, Pandian R, Haddow JE. Maternal serum invasive trophoblast antigen and first-trimester Down syndrome screening. Clin Chem 2005;51:1499–504.
[15] Quest Diagnostics, Maternal Serum Screen, First Trimester. Quest Diagnostics Incorporated, <http://www.questdiagnostics.com/hcp/topics/genetictesting/mss_t1.html>; 2000–2010.
[16] Poon L, Leung T, Lau T, Lo Y. Prenatal detection of fetal Down's syndrome from maternal plasma. Lancet 2000;356:1819–20.
[17] Robillard PY, Hulset TC, Deka CG. Preeclampsia and human reproduction. An essay of a long term reflection. J Reprod Immunol 2003;59:93–100.
[18] Burton GJ. Placental oxidative stress: from miscarriage to preeclampsia. J Soc Gynecol Invest 2004;11:342–52.
[19] Salas SP. What causes pre-eclampsia? Clin Obstet Gynecol 1999;13:141–57.
[20] Bahado-Singh RO, Oz AU, Kingston JM, Shahabi S, Hsu CD, Cole LA. The role of hyperglycosylated hCG in trophoblast invasion and the prediction of subsequent preeclampsia. Prenat Diagn 2002;22:478–83.
[21] Brennan MC, Wolfe MD, Murray-Krezan CM, Cole LA, Rayburn WF. First trimester hyperglycosylated human chorionic gonadotropin and development of hypertension. Prenat Diagn 2013.

The biology of gestational trophoblastic neoplasms

33

Laurence A. Cole
USA hCG Reference Service, Angel Fire, NM, USA

Gestational trophoblastic diseases include a bizarre mixture of reproductive diseases that define many of the oddities of biological sciences. Gestational trophoblastic diseases explain the biology of a fertilized ovum comprising chromosomes from two sperm and no oocyte. Gestational trophoblastic diseases explain what happens when an unfertilized oocyte lacks chromosomes, and what happens when an ovum is fertilized by more than one sperm. Gestational trophoblastic diseases also include a root malignancy, a disease that may predate or define most human cancers.

Let me start by defining four very different kinds of gestational trophoblastic diseases: complete hydatidiform mole, partial hydatidiform mole, choriocarcinoma, and placental site trophoblastic disease.

33.1 Complete hydatidiform mole

A complete hydatidiform mole is a pregnancy comprising only placenta tissue (villous placenta tissue). No fetal elements of any kind may be present. A complete hydatidiform mole takes on a unique anatomy of its own. It presents as hundreds of balls or bloody cysts (hygromatous cysts) [1–3]. Each ball arises by distention of the placental callous tissue by fluids. At hysterectomy, the complete hydatidiform mole may appear like a "bunch of red grapes" or a "bowl of cherries." Figure 33.1 illustrates the appearance of a complete hydatidiform mole at hysterectomy.

The complete hydatidiform mole pregnancy comprises only hygromatous cysts (Figure 33.1). Normally, the woman will look unduly large for the gestation age of her pregnancy. The human chorionic gonadotropin (hCG) levels may be 2-fold to 20-fold higher than that seen in a pregnancy with the sheer mass of trophoblastic tissue. A complete hydatidiform mole is a consideration in any pregnancy case with total hCG greater than 100,000 mIU/ml. Performing an ultrasound and looking for fetal components comprises the definitive test [1–3].

Normally, a complete mole is first discovered by ultrasound (Figure 33.2). It yields a snowflake-like pattern from cystic trophoblast tissue on ultrasound and no hint of a fetus. A complete hydatidiform mole is removed by dilation and curettage, normally at 8–16 weeks of gestation.

Human Chorionic Gonadotropin (hCG). DOI: http://dx.doi.org/10.1016/B978-0-12-800749-5.00033-X

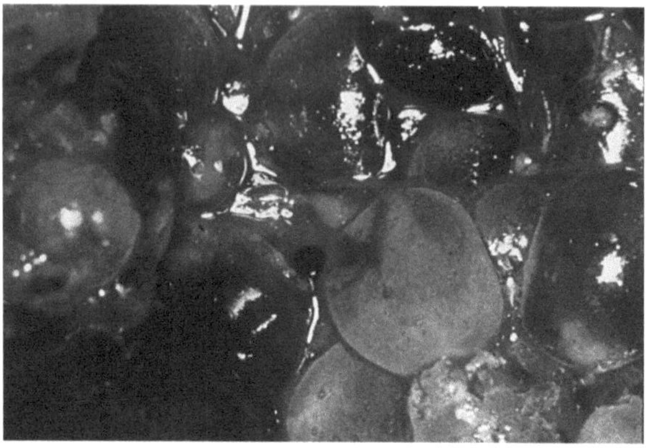

Figure 33.1 A complete hydatidiform mole at hysterectomy.

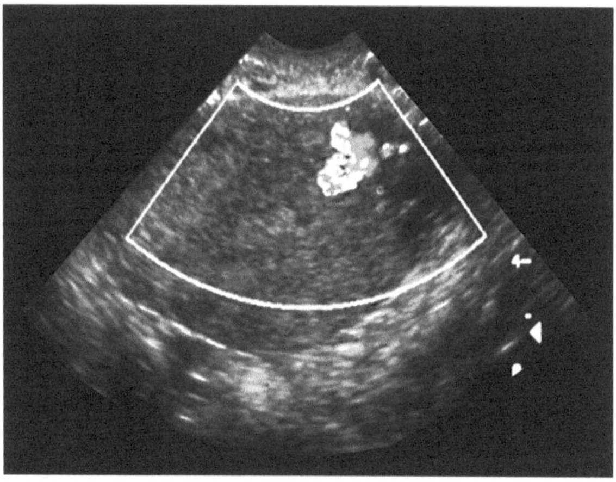

Figure 33.2 A complete hydatidiform mole at ultrasound. Note the classic snowflake appearance without any fetal component.

What is a complete hydatidiform mole? How does biology create a being appearing like a bunch of grapes? Multiple genetic experts have examined complete hydatidiform mole. It was first found that hydatidiform mole may be unusual in that it contains Barr bodies in the interphase of cell division [4]. Other studies noted that most complete hydatidiform moles were female or had 46, XX cytogenetics [5,6]. Amazingly, research showed that complete hydatidiform moles were always diploid. They comprised two haploid sets of 100% androgenic or paternal origin chromosomes [7–12]. No chromosomal element of this living being was of maternal origin.

Because all diploid chromosomes in a complete hydatidiform mole were of paternal origin, it is inferred that a mole is derived from an anucleate ovum [7–12].

Interestingly, complete hydatidiform mole has strictly paternal chromosomal DNA but maternal mitochondrial DNA, suggesting one maternal element [13–15].

It has been speculated that the anucleate oocyte may be a result of both meiosis-generated haploid sets being discharged as polar bodies, or caused by destruction in either the first stage or the second stage of meiosis [13–16]. Other possibilities include a complete failure of the initial stage of meiosis in the maternal ovary, or failure of luteinizing hormone (LH) and pituitary hCG to appropriately stimulate the first stage of meiosis, as they should, as part of the multistep ovarian ovulation procedure.

Figure 33.3 illustrates ovarian fertilization possibilities considering anucleate ovum, which could lead to a diploid complete hydatidiform mole. As illustrated, a single sperm may fertilize an anucleate ovum. The haploid chromosomes are then duplicated. They form two pronucleuses, as the haploid sets would normally with a sperm and an ovum, and then form a diploid fertilized being (Figure 33.3). Alternatively, an ovum may be fertilized by two sperm, with each forming a pronucleus and together forming a diploid being (Figure 33.3).

Normally, the crosslinking of the ovum shell or zona pellucida by enzymes made by the cortical granules will prevent more than one sperm from penetrating the ovum. It is possible that this process may fail in an anucleate ovum, or this is the reason why dispermic fertilization is less commonly the cause of complete

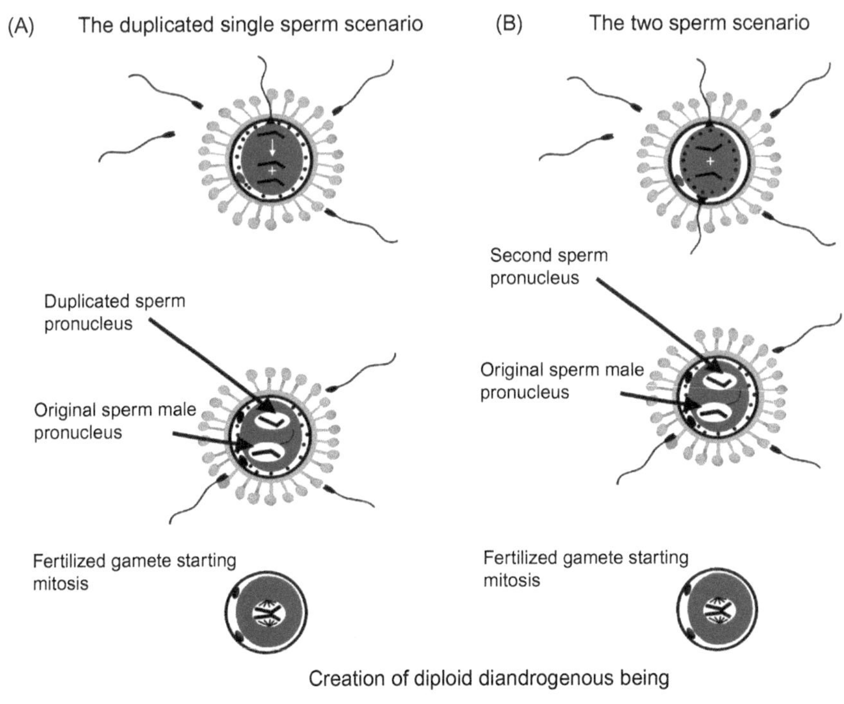

Figure 33.3 Creation of complete hydatidiform mole, a diploid diandrogenous being.

hydatidiform mole. Dispermic fertilization must occur, however, because it is the only possible explanation for 46, XY complete hydatidiform mole. It is estimated that most complete hydatidiform moles are 46, XX and most likely have a duplicated sperm haploid as their origin. Approximately 20–25% are 46, XY, so they are dispermic in origin [17–19]. Imagine a life form resulting from the combination and fusing of the haploid sets from two sperms.

One of the greatest questions is, what happens next? A viable being has been created (an androgenic diploid being). Cells divide to two cells, four cells, eight cells, 16 cells, and beyond, but how do we end up with a being that is only placenta tissue? The zygote divides, forming two identical cells, four identical cells, eight identical cells, and 16 identical cells in the normal way. Then, a morula is normally formed, which normally divides cells into an amnion (fetal tissue) and a chorion (placental tissue). Clearly, in the diandrogenous complete hydatidiform mole case, the morula fails to divide the tissue into amnion and chorion and, by default, becomes 100% chorion. It is though that possibly the maternal set of chromosomes or possibly the maternal X chromosome signals this division. Clearly, there is no signal in the complete mole case. The chorionic being divides further to form a chorion-only blastocyst with no fetal mass and then implants.

I have wondered many times whether it is possible that some diandrogenous zygotes make it through the morula division, differentiate appropriately, and end up with normal babies. Nobody has investigated this possibility. Homosexual men might be very pleased if it was possible to make a baby from two men's sperm. It is clearly by this process and failure of morula division that the complete hydatidiform mole ends up as a mass of villous placental tissue. Further differentiation beyond the morula must occur to generate the villous structures and the hygromatous cysts.

Complete hydatidiform mole or androgenic diploid pregnancies occur in approximately 1 in every 900 pregnancies to 1 in every 1500 pregnancies in the United States and Europe [20–22]. Occurrence is much more common in Southeast Asia and Africa, particularly among tribal people. These people seemingly have an inherited racial precedence for this disorder. The independently assessed incidence in Southeast Asia and Africa is approximately 1 in 400 pregnancies [23].

Rarely, familial occurrences of hydatidiform mole are detected. One Indian family was detected as having two complete hydatidiform mole pregnancies [24]. Two twin sisters were detected as having molar pregnancies in Italy [25]. A Lebanese family was identified in which two sisters and a cousin each had a molar pregnancy [26]. A German family was found in which three sisters each had a molar pregnancy [27]. Once a person has had a molar pregnancy, she is at increased risk (1 in 60) for having a second molar pregnancy [28].

The major complication of complete hydatidiform mole is the risk of persistent or invasive disease. This is a malignant gestational trophoblastic disease that invades tissues and grows. It is our understanding that complete hydatidiform mole adopts the pregnancy implantation mechanism driven by hyperglycosylated hCG. This is the signal that drives growth and enzymatic invasion in pregnancy implantation [29–31]. As now proven by multiple groups, hyperglycosylated hCG binds a transforming growth factor-β (TGFβ) receptor, not an LH–hCG receptor, promoting autocrine cell growth and invasion through metalloproteinases and collagenases [29–31].

Our own studies (Table 33.1) show a minimal presence of the hyperglycosylated hCG invasion signal in complete hydatidiform mole cases not requiring chemotherapy (mean 7.2% ± 4.7% hyperglycosylated hCG of total hCG). A slightly higher proportion of hyperglycosylated hCG is detected in cases that eventually become persistent disease requiring chemotherapy (mean 10.0 ± 3.1% hyperglycosylated hCG of total hCG). A much clearer difference is noted in those with active persistent disease needing chemotherapy (22.1 ± 5.5% hyperglycosylated hCG of total hCG) (Table 33.1).

It is believed that when the hyperglycosylated hCG signal is low (as in the mean 7.2 ± 4.7%), significant invasion does not occur. However, when the hyperglycosylated hCG or invasion and growth signal reaches the 20% range, the hyperglycosylated hCG drives persistent disease needing chemotherapy (Table 33.1).

Reports vary widely regarding the proportion of complete hydatidiform mole cases that will develop persistent gestational trophoblastic disease and need chemotherapy. A study by Seckl [28] indicates that just 16% of cases will need chemotherapy (28%). Other studies by Goldstein and Berkowitz [20] indicate that 18.7% of complete mole cases will need chemotherapy. A study by Kohorn [32] indicates, in contrast, that as many as 29.1% of cases will need chemotherapy.

As a general rule, any complete hydatidiform mole cases with decreasing total hCG levels after dilation and curettage needs to be investigated carefully. If hCG levels plateau or cease to decline for three consecutive measurements, then persistent gestational trophoblastic disease should be assumed and chemotherapy should be started.

Choriocarcinoma is a possible complication of complete hydatidiform mole. Normally, choriocarcinoma occurs in 1 in every 30,000 pregnancies; it is extremely rare. As demonstrated, choriocarcinoma can result from 3.0% of complete mole cases [20,28]. If complete hydatidiform mole has an incidence of 1 in 1000 pregnancies [20–22], then the incidence of choriocarcinoma as a complication of a complete mole is approximately 1 in 33,000 pregnancies or is similar to the incidence of choriocarcinoma as a pregnancy complication. Approximately 50% of choriocarcinoma cases occur as a complication of complete hydatidiform mole.

Rarely, a complete hydatidiform mole exists as a conjoined twin with a normal pregnancy. In such cases, the complete hydatidiform mole tends to be an aggressive mole with a high incidence of persistent gestational trophoblastic disease. Statistics show that only 40% of such conjoined twins make it to term and to delivery.

A complete hydatidiform mole case needs to be examined with extreme care. The ultrasound showing a snowflake pattern is not sufficient to make the final diagnosis. After dilation and curettage, a sample needs to be examined by pathology and cytogenetics needs to be determined. If a diploid 46, XX or 46, XY karyotype is determined, then a complete mole can be confirmed. It is also essential to measure total hCG at least weekly until the concentration has become undetectable. Levels should be monitored for another 3 months to make sure that any chance of persistent gestational trophoblastic disease is eliminated. If help is needed in

Table 33.1 **hCG and hyperglycosylated hCG (hCG-H) in complete hydatidiform mole**

hCG mIU/ml	hCG ng/ml	hCG-H ng/ml	%hCG-H
Complete mole, no need for chemotherapy			
619	56	0.3	0.5%
1160	105	0.59	0.6%
869	79	1.44	1.8%
746	68	1.92	2.8%
735	67	3.31	4.9%
22,540	2049	125	6.1%
455	41	2.67	6.4%
4080	371	26	7.1%
2416	220	16	7.4%
386	35	2.67	7.6%
4750	432	42	9.8%
16,317	1483	160	10.8%
30,255	2750	355	12.9%
15,358	1396	187	13.4%
4890	445	71	16.1%
		Mean ± SD	7.2% ± 4.7%
Complete mole, will need chemotherapy			
3316	301	12	4.0%
2662	242	14	5.6%
882	80	5.33	6.7%
55,504	5046	435	8.6%
3577	325	29	8.9%
642	58	5.23	9.0%
5720	520	47	9.0%
700	64	6.61	10.4%
6149	559	69	12.3%
36,818	3347	419	12.5%
15,170	1379	173	12.6%
9580	871	110	12.7%
11,490	1045	145	13.9%
15,092	1372	192	14.0%
		Mean ± SD	10.0% ± 3.1%
Active persistent gestational trophoblastic disease cases, receiving chemotherapy			
2364	215	49	22.6%
12,710	1155	283	24.5%
109,056	9914	1227	12.4%
1067	97	17.5	18.0%
1164	106	28	26.5%
1596	145	42	28.9%
204,400	18,582	4025	21.7%
		Mean ± SD	22.1% ± 5.5%

managing the case, then advice should be sought from a gynecologic oncologist or gestational trophoblastic disease specialty clinic.

33.2 Partial hydatidiform mole

Partial hydatidiform mole appears to be an ailment separate from complete hydatidiform mole. Partial hydatidiform mole presents as a mixture of villous tissue, hygromatous balls or cysts, and some fetal tissues (no intact fetus). Figure 33.4 illustrates a partial hydatidiform mole at hysterectomy.

A partial hydatidiform mole is a triploid pregnancy and another oddball complication of the reproductive process. Fertilization is an intricate process. Part of this process is the cortical granule action. As illustrated in Figure 33.5 on single sperm penetration of the zona pellucida. The cortical granules enter the periviteline space and secrete enzymes that cross-link the zona pellucida, making it impenetrable to further sperm. This limits the ovum to penetration by one sperm.

Partial hydatidiform mole is the direct result of what happens when this cortical granule process fails, and an egg can be penetrated by multiple sperm (Figure 33.5). Classically, a partial hydatidiform mole is triploid. It could be said that a complete hydatidiform mole is a consequence of an anucleate ovum, and a partial hydatidiform mole is the consequence of failure of the cortical granules to prevent dispermic penetration.

Partial hydatidiform mole can be identified by ultrasound showing placenta tissue, some hygromatous cysts, some fetal oddities, but no clear fetus structure. Ultrasound is strictly an indicator. After spontaneous abortion or dilation and curettage, tissue needs to be examined by pathology and cytogenetics needs to be determined. If a triploid karyotype is determined as 69, XXX, 69XXY, or 69 XYY [6], then a partial mole can be confirmed.

Unlike the complete hydatidiform mole, where morula differentiation to a chorion and amnion fails, a partial hydatidiform mole has some fetal elements. As speculated with complete hydatidiform mole, the presence of the maternal X

Figure 33.4 Partial hydatidiform mole at hysterectomy.

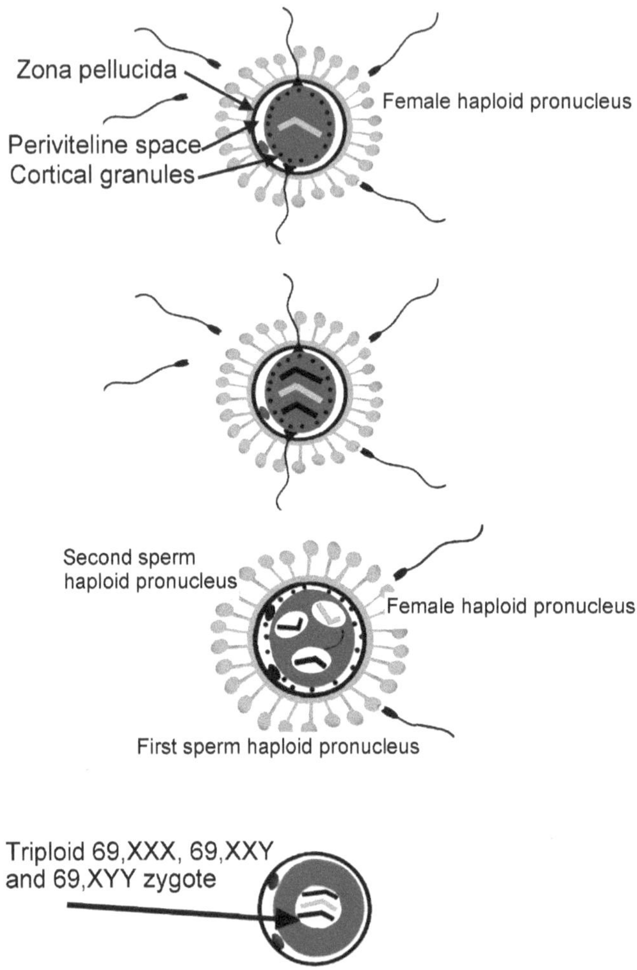

Figure 33.5 Dispermic penetration of ovum. Creation of a partial hydatidiform mole.

chromosome may signal the differentiation. In a partial hydatidiform mole, this signal may be present.

However, development of a complete fetus does not happen, suggesting later problems with fetal differentiation. I have personally dissected a few partial hydatidiform mole masses. It can be quite squeamish. One can dissect one's way through cystic and placenta tissues to suddenly come across an intact fully formed finger or an odd eye. It is haunting.

Partial hydatidiform mole rarely becomes invasive, developing persistent gestational trophoblastic disease like complete partial moles. A large study by Seckl indicates that 0.5% of partial mole cases become persistent gestational trophoblastic disease [28].

Very commonly, partial hydatidiform moles cause miscarriage of pregnancy. In some countries, dilation and curettage is not performed when a partial hydatidiform mole is indicated by ultrasound. The mole is allowed to remain until the person has a miscarriage; however, problems can arise if this is done and the mole can become persistent disease.

It is important to differentiate a complete and partial hydatidiform mole correctly using cytogenetics on the dilation and curettage or spontaneous abortion results. A complete mole is traditionally diploid, whereas a partial mole is triploid. A complete hydatidiform mole must be distinguished from a partial hydatidiform mole because of the invasive nature of the complete mole. In cases of both complete and partial moles, weekly placental hCG measurements are suggested until the concentration reaches <1 mIU/ml. At that point, placental hCG should be measured for up to 3 months to ensure that recurrence or persistence is not a possibility.

33.3 Choriocarcinoma

Choriocarcinoma was first discovered in Germany by Hans Chiari in 1877 [33], then in 1889 by Max Saenger [34], and in 1898 by Felix Marchand [35]. Hans Chiari, Max Saenger, Felix Marchand, nor any physician or scientist since has ever demonstrated that choriocarcinoma is actually a cancer by showing a transformation or any major change between pregnancy and choriocarcinoma cells [36]. It has always been assumed that choriocarcinoma is a cancer because its cells are rapidly growing or are highly invasive disease. Again, the invasive cytotrophoblast cells of pregnancy implantation are rapidly growing and highly invasive. However, nobody has demonstrated that choriocarcinoma cytotrophoblast cells differ in any way from normal pregnancy cytotrophoblast cells, which are naturally extremely invasive cells. Furthermore, nobody has demonstrated by biochemistry, cell culture, histology, or other methods that choriocarcinoma cells appear or function differently than normal pregnancy cytotrophoblast cells.

As shown, choriocarcinoma and normal pregnancy cytotrophoblast cells are very similar. Pregnancy cytotrophoblast cells have multiple glucose transporter proteins, as do choriocarcinoma cells [37,38]. Pregnancy-specific β1-glycoprotein is active in normal pregnancy cells and similarly active in choriocarcinoma cells [39,40]. Pregnancy cytotrophoblast cells have receptors for cyclic AMP (gonadotropin-releasing hormone) for estradiol, TGFβ, and progesterone, as do choriocarcinoma cells [40–42]. Cytotrophoblast cells have the enzyme aromatase and estrogen synthetase, as do choriocarcinoma cells [41,43]. As reported, interleukin, activin, insulin, and glucocorticoid receptors have all been shown to be active in pregnancy cytotrophoblast cells and in choriocarcinoma cells [43–46]. I could go on for many pages about completed studies that have shown that no difference can be found between choriocarcinoma cells and normal pregnancy cytotrophoblast cells.

Cytotrophoblast cells are a new tissue to human biology. It was only in 1983 that this type of cell was mentioned for the first time in biology texts or medical journals. The physiology of these new cells is still being dissected [30,47–49]. It was only in 2006 that I discovered that these cells produce hyperglycosylated hCG, and that this molecule drove cytotrophoblast growth and invasion [30,48].

In 2000, Butler and colleagues showed that cancer cells produce an hCG free β-subunit, and that this hCG variant functioned as an autocrine through a TGFβ receptor [50]. In 2012, Butler and I showed that hyperglycosylated hCG and hCG free β-subunit were interchangeable and that both seemingly functioned as autocrines through a TGFβ receptor [51]. In 2012, it was unquestionably confirmed by Berndt et al. in France that hyperglycosylated hCG functions through an autocrine cytotrophoblast cell TGFβ receptor [47].

As we now understand, cytotrophoblast cells are the natural invasive cells of pregnancy implantation. The cells produce hyperglycosylated hCG, which is an autocrine that function through the TGFβ receptor [30,47,48,51], promoting cell growth and extreme cell invasion by production of degradative enzymes, metalloproteinases, and collagenases [52,53]. Cytotrophoblast cells naturally rapidly invade the uterus to a depth of 30–40% of uterine thickness in pregnancy.

It has now been demonstrated by six independent groups that human cancer cells are driven to grow and invade by hyperglycosylated hCG and its free β-subunit [50,51,54–57]. Seemingly, in transformation, most cancer cells steal the only known normal human invasion process, the process that drives cytotrophoblast cells in pregnancy implantation, taking the autocrine hyperglycosylated hCG and its free β-subunit and the TGFβ receptor to drive the malignancy.

Cells called choriocarcinoma cells predominantly comprise cytotrophoblast cells. To a lesser extent, they contain fused syncytiotrophoblast cells [28,30]. Just like the cytotrophoblast cells of implantation, they primarily produce hyperglycosylated hCG. Research shows that choriocarcinoma cells are (similar to implantation cells) driven to grow and invade solely by hyperglycosylated hCG functioning through the TGFβ receptor [30,47]. In this respect they are similar to implantation cells.

Every human cancer has to transform (carcinogenesis) to express the stolen hCG variants and to express the TGFβ receptor. Choriocarcinoma cells or cytotrophoblast cells naturally produce these key molecules. Choriocarcinoma is not transformed, so it is not cancer. Research presented here shows that choriocarcinoma cells and the pregnancy cytotrophoblast cells are very similar (Table 33.2). Both primarily produce hyperglycosylated hCG. In both cells, invasion is promoted by hyperglycosylated hCG, assumingly through the TGFβ pathway [30,47,48].

Choriocarcinoma cells, unlike all other cancers, are not transformed. The cells that comprise choriocarcinoma are naturally highly invasive like choriocarcinoma. It is inferred that choriocarcinoma cells are not a cancer, but rather a variant of normal pregnancy implantation cytotrophoblast cells.

Normally during pregnancy, the villous placenta of hemochorial placentation is bonded to the uterus through a ball (approximately 5–7 balls of 100 cells) of cytotrophoblast cells, called the extravillous trophoblast cells (Figure 33.6).

Typically, a pregnant mother enters parturition. A normal healthy baby is delivered. The placenta is delivered. As indicated in Figure 33.6, delivery of the placenta

Table 33.2 Comparison of normal placenta cytotrophoblast cells and choriocarcinoma cells

Cells	hCG production	Total hCG production	Hyperglycosylated hCG production
Production			
Term normal placenta cytotrophoblast cells	2.3 ng/5000 cells	0.46 ng/1000 cells	0.46 ng/1000 cells
JEG-3 Choriocarcinoma cells	13.6 ng/30,000 cells	0.45 ng/1000 cells	0.50 ng/1000 cells
JaR Choriocarcinoma cells	12.7 ng/30,000 cells	0.42 ng/1000 cells	0.38 ng/1000 cells
Matrigel invasion chamber studies			
Term normal placenta cytotrophoblast cells	No additive	40 ± 10% invasion	
Term normal placenta cytotrophoblast cells	Hyperglycosylated hCG 3 ng/1000 cells/ml	66 ± 13% invasion	
JEG-3 Choriocarcinoma cells	No additive	48 ± 11% invasion	
JEG-3 Choriocarcinoma cells	Hyperglycosylated hCG 3 ng/1000 cells/ml	88 ± 6.0% invasion	

Cells are culture in Matrigel chambers. These comprises a disk of mouse sarcoma cells. Placental cells and choriocarcinoma cells are culture over the mouse cells, and the extent of cell penetration or invasion is measured. Matrigel studies were performed in quadruplicate.

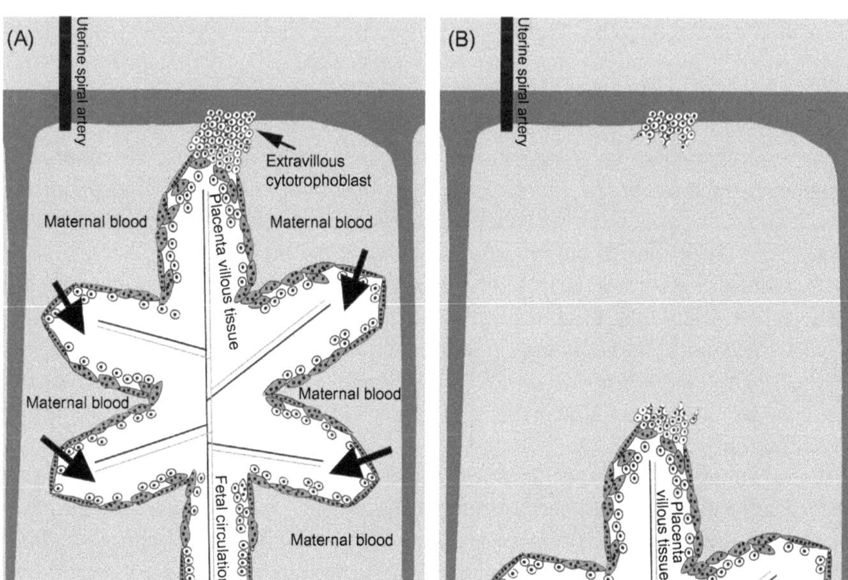

Figure 33.6 Placentation during pregnancy (A) and at parturition (B).

involves tearing those balls of cytotrophoblast cells at the top of each hemochorial placentation chamber. These balls of cells tie together the uterus and the villous placental tissue. It is concluded that the residual cytotrophoblast cells from the torn balls of cells deeply implanted in the uterus (approximately 5–7 half balls of 50 cells) become invasive and propagate in a choriocarcinoma case.

Choriocarcinoma occurs in the United States and Europe at a small incidence of 1 in 30,000 pregnancies in the Western world [20,28]. In 2006, I studied total hCG and hyperglycosylated hCG in pregnancy serum. Choriocarcinoma occurs as a complication in 1 in 33 complete hydatidiform mole cases [28]. If complete mole has an incidence of 1 in 1000 pregnancies, then choriocarcinoma as a complication of pregnancy (1 in 30,000 pregnancies) and as a complication of complete mole (1 in 33,000 pregnancies) have similar incidences.

In 2009, we examined 696 serum samples from the length of pregnancy. This included 49 third-trimester pregnancies [58]. The third trimester total hCG ranged from 121 to 93,000 mIU/ml among the 49 individuals, and hyperglycosylated hCG ranged from 3.3 to 134 mIU/ml. In 2011, we examined 158 third-trimester pregnancy serum samples. The hyperglycosylated hCG concentration ranged much further, from 1.4 to 1282 mIU/ml [59]. As a proportion of total hCG, this was 0.0015% hyperglycosylated hCG to 24.8% hyperglycosylated hCG. It is believed that if we tested a much larger group, like 50,000 pregnancies, we would detect the extreme case with as much as 30% hyperglycosylated hCG. Just as was indicated by research in this section regarding complete hydatidiform mole, that complete hydatidiform mole producing more than 20% hyperglycosylated hCG was activated and became persistent gestational trophoblastic disease. We claim here that, similarly, the 1 in 30,000 extreme pregnancies producing >30% hyperglycosylated hCG will likely be activated at parturition, becoming choriocarcinoma. Publications show that >30% hyperglycosylated hCG is needed for choriocarcinoma to be active disease responsive to chemotherapy [60,61]. Here, it is believed that >30% hyperglycosylated hCG in term pregnancy may activate choriocarcinoma.

Choriocarcinoma is an invasive condition seemingly involving reactivation of pregnancy implantation. In choriocarcinoma, placental tissue can invade through the myometrium and into the lymphatic system, metastasizing possibly to the pelvis, to the liver, then to the lungs, and finally to the brain.

If histology is not possible from choriocarcinoma surgery or biopsy, then the disease is called gestational trophoblastic neoplasm. Choriocarcinoma as a non-cancer is best treated with chemotherapy. Choriocarcinoma as a rapidly growing disease responds particularly well to combination chemotherapy, EMA-CO, etoposide, methotrexate, and actinomycin D cycled weekly with cyclophosphamide and vincristine (onvocin).

As shown by my laboratory, when choriocarcinoma cells are grown in nude mice (Figure 33.7), cells are completely blocked, growing and invading by monoclonal antibody to hyperglycosylated hCG [62]. This confirms that this is a hyperglycosylated hCG-dependent disease. In conclusion, choriocarcinoma is not a cancer, but rather a variant of normal pregnancy invasive cytotrophoblast cells.

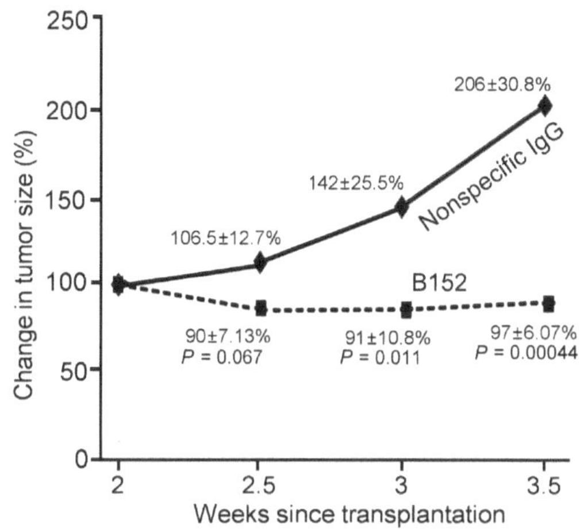

Figure 33.7 JEG-3 choriocarcinoma cells were transplanted into eight nude mice. Two weeks were allowed for the tumor to establish and metastasize. Mice were given either B152 monoclonal antibody to hyperglycosylated hCG (dashed line) or nonspecific IgG (solid line) biweekly.

33.4 Placental site trophoblastic tumor

There are just two kinds of placental trophoblast cells, cytotrophoblast cells and fused syncytiotrophoblast cells [30,63]. Yes, these cells are in villous cytotrophoblast and extravillous cytotrophoblast cells, which are functionally related cells [30,64]. All cytotrophoblast cells produce hyperglycosylated hCG [30,48,49,63], and all syncytiotrophoblast cell produce hCG [30,48,63]. Persistent gestational trophoblastic disease tissue and villous tissue both produce hCG and hyperglycosylated hCG. Choriocarcinoma cytotrophoblast cells produce hyperglycosylated hCG. Is there a general rule that all trophoblast cells make a form of hCG dimer? Placental site trophoblastic tumor (PSTT) is the exception. These cells produce a small amount of hCG free β-subunit and possibly a small amount of hCG dimer [65,66].

Why is PSTT not a big hCG producer like all other trophoblastic cells? It has long been speculated that PSTT derives from intermediate trophoblast cells, a third kind of trophoblast cell. As a twenty-first century placentologist I ask, what are intermediate trophoblast cells? I have never seen any such cells. Cytotrophoblast cells are root placental cells. These fuse, forming syncytiotrophoblast cells containing 5–50 nuclei. Is an intermediate trophoblast cell something halfway between cytotrophoblast cells and syncytiotrophoblast cells? Is it possibly a cell with 2.5–25 nuclei? This does not exist.

Realistically, it appears the PSTT cells are transformed trophoblastic cells. They have lost their production of hCG or hyperglycosylated hCG altogether as part of transformation. Like other non-trophoblastic malignancies, they have gained the expression of the hCG β-subunit gene. Like other non-trophoblastic malignancies, they do not have the enzymes needed to complete β-subunit disulfide bonds (lack placental isomerases needed to complete the last two disulfide bridges) [67,68], and β-subunit will not combine with α-subunit. Thus, cells produce low levels of hCG free β-subunit [65,66].

The hCG free β-subunit test can be used to diagnose a case of PSTT and differentiate the disease from choriocarcinoma [65,66]. PSTT should be treated very differently from choriocarcinoma. Uterine tumors should be removed by hysterectomy, and combination chemotherapy should be used to stop progression of malignant disease. PSTT appears to be the only trophoblastic true cancer.

References

[1] Takeuchi S. Immunology of spontaneous abortion and hydatidiform mole. Am J Reprod Immunol 1980;1:23–8.
[2] Yapar EG, Ayhan A, Ergeneli MH. Pregnancy outcome after hydatidiform mole, initial and recurrent. J Reprod Med 1994;39:297–9.
[3] Baxton CL. Management of patients with hydatidiform mole. Ann NY Acad Sci 1959;80:121–4.
[4] Park WW. The occurrence of sex chromatin in chorionepitheliomas and hydatidiform moles. J Pathol Bacteriol 1957;74:197–206.
[5] Sasaki M, Fukuschima T, Makino S. Some aspects of the chromosome constitution of hydatidiform moles and normal chorionic villi. Gann 1962;53:101–6.
[6] Lawler SD, Fisher RA, Pickthall VJ, Povey S, Evans MW. Genetic studies on hydatidiform moles. I. The origin of partial moles. Cancer Genet Cytogenet 1982;5:309–20.
[7] Kajii T, Ohama K. Androgenetic origin of hydatidiform mole. Nature 1977;268:633–4.
[8] Wake N, Takagi N, Sasaki M. Androgenesis as a cause of hydatidiform mole. J Natl Cancer Inst 1978;60:51–7.
[9] Jacobs PA, Wilson CM, Sprenkle JA, Rosenshein NB, Migeon BR. Mechanism of origin of complete hydatidiform moles. Nature 1980;286:714–16.
[10] Ohama K, Kajii T, Okamoto E, Fukada Y, Imaizumi K, Tsukahara M, et al. Dispermic origin of XY hydatidiform moles. Nature 1981;29:551–2.
[11] Pattillo RA, Sasaki S, Katayama KP, Roesler M, Mattingly RF. Genesis of 46, XY hydatidiform mole. Am J Obstet Gynecol 1981;141:104–5.
[12] Surti U, Szulman AE, O'Brien S. Dispermic origin and clinical outcome of three complete hydatidiform moles with 46, XY karyotype. Am J Obstet Gynecol 1982;144:84–7.
[13] Wallace DC, Surti U, Adams CW, Szulman AE. Complete moles have paternal chromosomes but maternal mitochondrial DNA. Hum Genet 1982;61:145–7.
[14] Edwards YH, Jeremiah SJ, McMillan SL, Povey S, Fisher RA, Lawler SD. Complete hydatidiform moles combine maternal mitochondria with a paternal nuclear genome. Ann Hum Genet 1984;48:119–27.
[15] Azuma C, Saji F, Tokugawa Y, Kimura T, Nobunaga T, Takemura M, et al. Application of gene amplification by polymerase chain reaction to genetic analysis of molar mitochondrial DNA: the detection of anuclear empty ovum as the cause of complete mole. Gynecol Oncol 1991;40:29–33.

[16] van Blerkom J, Henry G. Cytogenetic analysis of living oocytes: cellular basis and developmental consequences of perturbations in chromosomal organization and complement. Hum Reprod 1988;3:777–90.
[17] Fisher RA, Povey S, Jeffreys AJ, Martin CA, Patel I, Lawler SD. Frequency of heterozygous complete hydatidiform moles, estimated by locus-specific minisatellite and Y chromosome-specific probes. Hum Genet 1989;82:259–63.
[18] Kovaks BW, Shahbahrami B, Tast DE, Curtin JP. Molecular genetic analysis of complete hydatidiform moles. Cancer Genet Cytogenet 1991;54:143–52.
[19] Lawler SD, Fisher RA, Dent J. A prospective study of hydatidiform mole. Am J Obstet Gynecol 1991;164:1270–7.
[20] Goldstein DP, Berkowitz RS. The diagnosis and management of molar pregnancy. Gestational trophoblastic neoplasms: clinical principles of diagnosis and management. Philadelphia, PA: Saunders; 1982Ch. 7, p. 143–175.
[21] Schuler SJ, Beckett JH, Gettings S. Incidence of hydatidiform mole. 2008; <www.freemd.com/hydatidiform-mole.incidence.htm>.
[22] Gestational trophoblastic disease. 2012; <www.cancer.org/cancer/gestationaltrophoblasticdisease/detailedguide/gestational-trophoblastic-disease-key-statistics>.
[23] Bracken MB. Incidence and aetiology of hydatidiform mole: an epidemiological review. Br J Obstet Gynaecol 1987;94:1123–35.
[24] Ambani LM, Vaidya RA, Rao CS, Daftary SD, Motashaw ND. Familial occurrence of trophoblastic disease-report of recurrent molar pregnancies in sisters in three families. Clin Genet 1980;18:27–9.
[25] La Vecchia C, Franceschi S, Fasoli M, Mangioni C. Gestational trophoblastic neoplasias in homozygous twins. Obstet Gynecol 1982;60:250–2.
[26] Seoud M, Khalil A, Frangich A, Zahled A, Azar G, Nuwayri-Salti N. Recurrent molar pregnancies in a family with extensive intermarriage: report of a family and reviw of the literature. Obstet Gynecol 1995;86:692–5.
[27] Kircheisen R, Ried T. Hydatidiform moles. Hum Reprod 1994;9:1783–5.
[28] Seckl M. Presentation and management of persistent gestational trophoblastic disease (GTD) and gestational trophoblastic tumours/neoplasia (GTT/GTN) in the United Kingdom. In: Gestational trophoblastic disease, 3rd ed., 2009; Hancock BW, SecklMS, Berkowitx R, Cole LA. editors. <http://isstd.org/isstd/chapter10.html>; [Chapter 10].
[29] Cole LA. Hyperglycosylated hCG and pregnancy failures. J Reprod Immunol 2012;93:119–22.
[30] Cole LA, Dai D, Butler SA, Leslie KK, Kohorn EI. Gestational trophoblastic diseases: 1. Pathophysiology of hyperglycosylated hCG-regulated neoplasia. Gynecol Oncol 2006;102:144–9.
[31] Handschuh K, Guibourdenche J, Tsatsaris V, Guesnon M, Laurendeau I, Evain-Brion D, et al. Human chorionic gonadotropin expression in human trophoblasts from early placenta, comparative study between villous and extravillous trophoblastic cells. Placenta 2007;28:175–84.
[32] Szulman AE, Surti U. The syndromes of hydatidiform mole. 1. Cytogenetic and morphologic correlations. Am J Obstet Gyncol 1978;131:665–71.
[33] Chiari H. Uber drei Falle von primarem kacino in findus und corpus des uterus. Med Jahrb 1877;7:364–7.
[34] Saenger M. Deciduoma malignum. Zbl Gyak 1889;167:537.
[35] Marchand FJ. Uber die sogenannten "decidualen" geshwulskeim im anshluss an normale geburt, abort, blasenmole und extrauterineschwanggerahaft. Monatsschr Geburtshilfe Gynakol 1895;1:419–38.

[36] Ober WB, Fass RO. The early history of choriocarcinoma. J Hist Med Allied Sci 1961;16:49–73.
[37] Oguru K, Sakata M, Okamota Y, Yasui Y, Tadokoro C, Yoshimoto Y, et al. 8 Bromo-cyclic AMP stimulates glucose transporter-1 expression in a human choriocarcinoma cell line. J Endocrinol 2000;164:171–8.
[38] Takata K, Kasahara T, Kasahara M, Ezaki O, Hirano H. Immunolocalization of glucose transporter GLUT1 and the gap junction in the transport of glucose across the placental barrier. Cell Tissue Res 1994;3:411–18.
[39] Chou JY. Effects of retinoic acid on differentiation of choriocarcinoma cells *in vitro*. J Clin Endocrinol Metab 1982;54:1174–80.
[40] Olgilvie S, Kvello-Stenstrom AG, Hammond G, Buhl WC, Larkin LH, Shiverick KT. Identification of proteins immunochemically related to human pregnancy-specific beta 1-glycoprotein in rat placenta. Endocrinology 1989;125:287–94.
[41] Ackermann K, Fauss J, Pyerin W. Inhibition of cyclic AMP-triggered aromatase gene expression in human choriocarcinoma cells by antisense oligodeoxynucleotide. Cancer Res 1994;54:4940–6.
[42] Bellino FS, Hussa RO, Osawa Y. Estrogen synthetase in choriocarcinoma cell culture. Stimulation by bibutyryl cyclic adenosine monophosphate and theophylline. Steroids 1978;32:37–44.
[43] King A, Loke YW. Human trophoblast and JEG choriocarcinoma cells are sensitive to lysis by IL-2 stimulated decidual natural killer cells. Cell Immunol 1990;129:435–48.
[44] Chambers SK, Ivins CM, Kacinski BM, Hochberg RB. An unexpected effect of glucocorticoids on stimulation of c-fms proto-oncogene expression in choriocarcinoma cells that express little glucocorticoid receptor. Am J Obstet Gynecol 2004;190:974–82.
[45] Ni X, Luo S, Minegishi T, Pung C. Activin A in JEG-3 cells: potential role as an autocrine regulator of steroidogenesis in humans. Bio Reprod 2000;62:1224–30.
[46] Deal CL, Guyda HJ. Insulin receptors on human term placental cells and choriocarcinoma (JEG-3) cells: characteristics and regulation. Endocrinology 1983;112:1512–23.
[47] Berndt S, Blacher S, Munuat C, Detilleux J, Evain-Brion D, Noel A, et al. Hyperglycosylated human chorionic gonadotropin stimulates angiogenesis through TGF-β receptor activation. FASEB J 2013; Abstract 12-213686.
[48] Guibourdenche J, Handschuh K, Tsatsaris V, Gerbaud MC, Legul F, Muller D, et al. Hyperglycosylated hCG is a marker of early human trophoblast invasion. J Clin Endocrinol Metab 2010;95:E240–4.
[49] Handschuh K, Guibourdenche J, Tsatsari V, Guesnon M, Laurendeau I, Evain-Brion D, et al. Human chorionic gonadotropin produced by the invasive trophoblast but not the villous trophoblast promotes cell invasion and is down-regulated by peroxisome proliferator-activated receptor-gamma. Endocrinology 2007;148:5011–19.
[50] Butler SA, Ikram MS, Mathieu S, Iles RK. The increase in bladder carcinoma cell population induced by the free beta subunit of hCG is a result of an anti-apoptosis effect and not cell proliferation. Brit J Cancer 2000;82:1553–6.
[51] Cole LA, Butler SA. Hyperglycosylated hCG, hCGβ and hyperglycosylated hCGβ: interchangeable cancer promoters. Mol Cell Endocrinol 2012;349:232–8.
[52] Karmakar S, Das C. Regulation of trophoblast invasion by IL-1β and TGF-β1. Am J Reprod Immun 2002;48:210–19.
[53] Murphy G, Reynolds JJ, Whitham SE, Docherty AJ, Angel P, Heath JK. Transforming growth factor beta modulates the expression of collagenase and metalloproteinase inhibitor. EMBO J 1987;6:1899–904.

[54] Iles RK. Ectopic hCGβ expression by epithelial cancer: malignant behavior metastasis and inhibition of tumor cell apoptosis. Mol Cell Endocrinol 2007;264–70.
[55] Carter WB, Sekharem M, Coppola D. Purified hCG induces apoptosis in breast cancer. Breast Cancer Res Treatm 2006;100:S243–4.
[56] Li D, Wen X, Ghali L, Al-Shalabi FM, Docherty SM, Purkis P, et al. hCG beta expression by cervical squamous carcinoma—*in vivo* histological association with tumour invasion and apoptosis. Histopathology 2008;53:147–55.
[57] Jankowska A, Andrusiewicz M, Grabowski J, Nowak-Markwitz E, Warchol JB. Coexpression of human chorionic gonadotropin beta subunit and its receptor in nontrophoblastic gynecological cancer. Intl J Gynecol Cancer 2008;18:1102–7.
[58] Cole LA. Biological functions of hCG and hCG-related molecules. Reprod Biol Endocrinol 2010;8:102–16.
[59] Cole LA. hCG, five independent molecules. Clin Chim Acta 2012;413:48–65.
[60] Cole LA, Muller Y. hCG in the management of quiescent and chemorefractory gestational trophoblastic diseases. Gynecol Oncol 2010;116:3–9.
[61] Cole LA. Minimally-aggressive gestational trophoblastic neoplasms. Gynecol Oncol 2012;126:145–50.
[62] Cole LA, Butlers SA. B152 anti-hyperglycosylated hCG free β-subunit A new possible treatment for cancer. J Reprod Med 2014.
[63] Kovalevskaya G, Genbacev O, Fisher SJ, Cacere E, O'Connor JF. Trophoblast origin of hCG isoforms: cytotrophoblasts are the primary source of choriocarcinoma-like hCG. Mol Cell Endocrinol 2002;194:147–55.
[64] Cole LA. Chapter 3: Introduction to pregnancy implantation, villous formation, and hemochorial placentation. In: Cole LA, editor. Human chorionic gonadotropin (hCG). Burlington, MA: Elsevier; 2010. p. 23–34.
[65] Cole LA, Khanlian SA, Muller CY, Giddings A, Kohorn EI, Berkowitz R. Gestational trophoblastic diseases: 3. Human chorionic gonadotropin free β-subunit a reliable marker of placental site trophoblastic tumors. Gyn Oncol 2006;102:159–63.
[66] Harvey RA, Pursglove HD, Schmid P, Savage PM, Mitchell HD, Seckl MJ. Human chorionic gonadotropin free beta-subunit measurement as a marker of placental site trophoblastic tumors. J Reprod Med 2008;53:643–8.
[67] Beebe JS, Huth JR, Ruddon RW. Combination of the chorionic gonadotropin free beta-subunit with alpha. Endocrinology 1990;126:384–91.
[68] Ruddon RW, Hanson CA, Bryan AH, Putterman GJ, White EL. Synthesis and secretion of human chorionic gonadotropin subunits by cultured human malignant cells. J Biol Chem 1980;255:1000–7.
[69] Kohorn EI. Hydatidiform mole and gestational trophoblastic disease in Southern Connecticut. Obstet Gynecol 1982;59:78–84.
[70] Jacobs PA, Hunt PA, Matsuura JS, Wilson CC, Szulman AE. Complete and partial hydatidiform mole in Hawaii: cytogenetics, morphology and epidemiology. Br J Obstet Gynaecol 1982;89:258–66.

Hyperglycosylated hCG and free β-subunit markers of gestational trophoblastic diseases

34

Laurence A. Cole
USA hCG Reference Service, Angel Fire, NM, USA

Two glycoproteins are made from the hCG α-subunit and β-subunit amino acid sequence by trophoblast cells. Root cytotrophoblast, the placental propagators, make hyperglycosylated hCG; fused cytotrophoblast cells, or syncytiotrophoblast cells, make the hormone hCG. Hyperglycosylated hCG distinguishes itself by having eight large sugar side chains and molecular weight of 42,800. Hyperglycosylated hCG is the most glycosylated protein known. hCG has a molecular weight of 37,180, and it is the longest circulating hormone (circulating half-life 36 h) [1].

hCG was discovered by Otfried Fellner, a gynecologist in Vienna, in 1913. He induced ovulation in immature rabbits with extracts of human placenta [2]. Toyoichi Hirose of the Osaka Medical College stimulated ovulation and normal luteal function in immature rabbits by repeated injection of human placental tissue in 1919 [3]. In 1927, Selamar Aschheim of the University of Berlin in Germany and Bernhard Zondek of the Berlin-Spandau Hospital demonstrated that the placentas of pregnant women produce a gonad-stimulating substance: a molecule called chorionic gonadotropin [4].

I discovered in 1987 that there was a major structural difference between the O-linked oligosaccharides attached to choriocarcinoma hCG and the sugar side chains on pregnancy hCG. The two molecules, choriocarcinoma and pregnancy hCG, had identical amino acid sequence [1]. This was the discovery of choriocarcinoma hCG, the molecule I named hyperglycosylated hCG [5,6].

Studies showed that hyperglycosylated hCG is a molecule independent from the hormone hCG with totally different functions. Hyperglycosylated hCG is an autocrine and not a hormone; it acts on cytotrophoblast cells as an autocrine functioning through a transforming growth factor-β (TGFβ) receptor [7–9]. Hyperglycosylated hCG functions during pregnancy to promote growth of cytotrophoblast cells, growth and invasion of placental tissue, and invasion of uterine tissues by the placenta or implantation of the placenta [10,11].

Examining cytotrophoblast cells and choriocarcinoma cells (cytotrophoblast cell) cultured on Matrigel basement membrane, one can determine their invasion capacity (Table 34.1) [10]. As shown in Table 34.1, trophoblast cells are promoted to invade by hyperglycosylated hCG. The hormone hCG, however, does not promote invasion and seemingly partially inhibits invasion (Table 34.1). This is when we started measuring hyperglycosylated hCG as a proportion of total hCG (hyperglycosylated hCG ÷ hCG plus hyperglycosylated hCG). The

Human Chorionic Gonadotropin (hCG). DOI: http://dx.doi.org/10.1016/B978-0-12-800749-5.00034-1

Table 34.1 Action of hyperglycosylated hCG and regular hCG on cytotrophoblast invasion and choriocarcinoma cell invasion of Matrigel membranes

	Mean penetration ± standard deviation
(A) Pregnancy cytotrophoblasts	
Control cultures	40 ± 10%
Hyperglycosylated hCG, 10 ng/ml	66 ± 13%[a]
Regular hCG, 10 ng/ml	34 ± 9%[b]
(B) JEG-3 choriocarcinoma cells	
Control cultures	48 ± 11%
Hyperglycosylated hCG, 100 ng/ml	88 ± 6%[a]
Regular hCG, 100 ng/ml	38 ± 3%[b]

Isolated cytotrophoblast cells were prepared from term placenta primary cultures. Primary cytotrophoblast and JEG-3 choriocarcinoma cells were separately cultured 24 h (5000 cells) on Matrigel basement membranes and control inserts in triplicate. Concentrations of hCG and hyperglycosylated hCG used to promote invasion were approximately four times that (ng/ml per 1000 cells) normally produced by the cells, 10 ng/ml for term pregnancy cytotrophoblast and 100 ng/ml for JEG-3 choriocarcinoma cells. The underside of Matrigel basement membranes, containing penetrated or invaded cells, was stained and counted. Cell penetration was compared with that of control inserts. The percentage penetration or invasion was calculated using the formula described by the manufacturer.

proportion indicates the promotional activity of hyperglycosylated hCG and any inhibitory activity of hCG as a percentage of total concentration [10,11].

Proportion of hyperglycosylated hCG joins gestational trophoblastic diseases and trophoblastic invasion [10,11]. hCG and hyperglycosylated hCG β-subunit both evolved from LH β-subunit with primate evolution. hCGβ and hyperglycosylated hCGβ evolution started in lower simian primates producing a hormone and autocrine with minimal biological activity. They evolved with development of advanced primates, and advanced further with hominids to molecules with an extensive circulating half-life and extensive biological activity [12,13]. Hyperglycosylated hCG developed from a molecule with minimal growth and invasive capacity in lower simian primates into a supergrowth and invasion promoter in humans [12,13]. Hyperglycosylated hCG promotes invasion in hydatidiform mole and choriocarcinoma by instigating the mechanism developed for pregnancy implantation [10,11], the only invasive mechanism inherent in humans [10,11].

Proportion of hyperglycosylated hCG was a useful marker indicating spontaneous abortion and biochemical pregnancy in implantation of pregnancy [14,15]. Here, we show that proportion hyperglycosylated hCG is an invaluable indicator of disease status in gestational trophoblastic diseases.

34.1 Hydatidiform mole

USA hCG Reference has consulted on numerous gestational trophoblastic disease cases. These include hydatidiform mole, persistent hydatidiform mole, choriocarcinoma, and gestational trophoblastic (choriocarcinoma before obtaining histology)

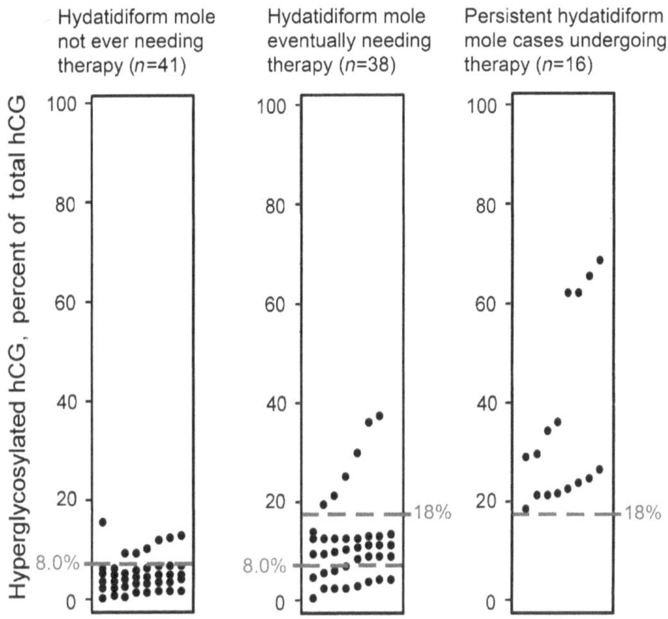

Figure 34.1 Proportion hyperglycosylated hCG in hydatidiform mole.

cases. In all incidences, proportion hyperglycosylated hCG was measured as an indicator of disease status.

Looking at hydatidiform mole cases at 1–4 weeks after diagnosis (Figure 34.1), a total of 41 cases that required no chemotherapy or hysterectomy at any time were consulted on. These were 9partial mole and 33 complete mole cases. The total hCG at the time of investigation had a median of 2006 mIU/ml. The range was 5.4–295,000 mIU/ml. The hyperglycosylated hCG had a median of 10.3 ng/ml and a range of <0.03–2656 ng/ml. The mean proportion hyperglycosylated hCG (hyperglycosylated hCG ÷ total hCG) was 5.6 ± 3.7% (standard deviation).

A total of 38 cases that eventually required chemotherapy for persistent gestational trophoblastic disease were consulted on (Figure 34.1). These were 6 partial mole and 32 complete mole cases. The total hCG had a median of 15,131 mIU/ml. The range was 700–652,300 mIU/ml. The median hyperglycosylated hCG was 183 ng/ml, with a range of 2.9–12,929 mg/ml. The mean proportion of hyperglycosylated hCG was 11.1% ± 8.0%, or virtually twice the proportion in cases not requiring therapy. Hyperglycosylated hCG is useful to identify hydatidiform mole cases that will require chemotherapy.

Sixteen cases with proven persistent hydatidiform mole needing chemotherapy for invasive disease were consulted on (Figure 34.1). These were 2 partial mole and 14 complete mole cases. The total hCG had a median at this time of 4325 mIU/ml. The range was 107–204,400 mIU/ml. The median hyperglycosylated hCG was 181 ng/ml, with a range of 2.0–4025 ng/ml. The mean proportion of hyperglycosylated hCG was 33% ± 18%. Clearly, hyperglycosylated hCG can identify cases needing therapy.

Looking for a cutoff to differentiate hydatidiform mole cases with no need for chemotherapy or hysterectomy at any time from those that eventually needed therapy, the cutoff of 8.0% was selected (Figure 34.1). At this cutoff, 68% of cases eventually needing therapy were detected, along with 17% false-positive cases not requiring chemotherapy. Proportion hyperglycosylated hCG can predict which cases will ultimately need chemotherapy (Figure 34.1). The cutoff of 18% was selected for persistent hydatidiform mole diagnosis. This cutoff was absolute in that it was exceeded by 16 of 16 otherwise proven persistent disease cases. Four of the 78 cases not having persistent hydatidiform mole were false-positive or needlessly selected (100% detection with 5.1% false-positive rate). Normally, it is common in hydatidiform mole cases to record three total hCG plateaued or increasing points before diagnosing persistent disease. A single measurement of proportion hyperglycosylated hCG >18% may be adequate.

A new test is described here that can forecast benign hydatidiform mole that becomes malignant (persistent hydatidiform mole) and requires chemotherapy, using a cutoff of 8.0%. Using this test, proportion of hyperglycosylated hCG, 68% of benign hydatidiform mole cases were correctly identified, with a 17% error rate. The proportion of hyperglycosylated hCG test is more absolute in identifying case with existing persistent hydatidiform mole. Using a cutoff of 18%, it correctly detected 16 of 16 women with existing disease needing chemotherapy, and it wrongly detected just 4 of 78 cases not needing chemotherapy (100% detection with 5.1% false-positive rate). If a woman with hydatidiform mole has a positive result on this test (>18% hyperglycosylated hCG), chemotherapy should commence.

34.2 Choriocarcinoma

A total of 50 cases have been investigated, 18 cases were diagnosed as gestational trophoblastic neoplasm (GTN) with no surgery or histology, and 32 cases were diagnosed as choriocarcinoma after pathology. All cases had received chemotherapy and were 1–20 weeks from diagnosis (Figure 34.2).

It is the opinion of the USA hCG Reference Service that there were clearly two classes of disease. There was aggressive disease or fast-growing disease (total hCG doubling rate before chemotherapy <2 weeks), of which we saw 34 cases. There was minimally aggressive disease (16 cases total). These were cases with very slowly progressing disease (hCG doubling rate 3.2–20 weeks). Nine of these 16 cases were chemorefractory and not successfully treated with methotrexate, actinomycin D, or EMA-CO, or with up to seven chemotherapy regimens.

The total hCG in the aggressive disease group ranged from 62 to 493,493 mIU/ml. In the minimally aggressive group, the total hCG ranged from 71 to 2362 mIU/ml. Proportion of hyperglycosylated hCG (Figure 34.2) distinguished the two groups. The 34 aggressive cases had proportions ranging from 49% to 100%, with a mean of 84.3 ± 16.6% (Figure 34.2). The 16 minimally aggressive cases had proportions ranging from 1.0% to 39% (Figure 34.2). The mean was 16.1 ± 11.9%. Using a cutoff of

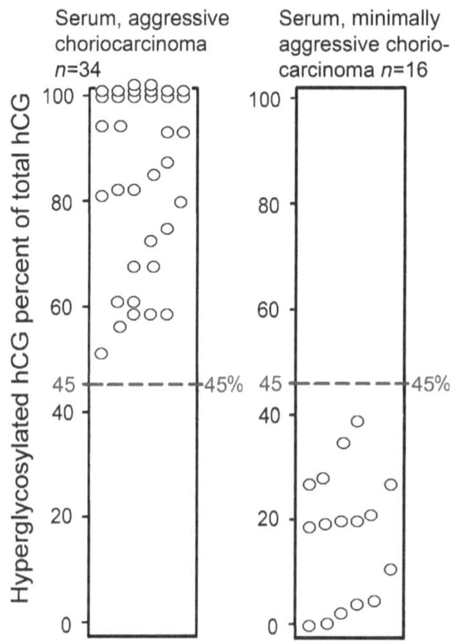

Figure 34.2 Proportion hyperglycosylated hCG in choriocarcinoma.

45% for the two groups, aggressive disease and minimally aggressive disease were completely distinguished (Figure 34.2). Proportion of hyperglycosylated hCG may be a useful test to distinguish these aggressive and minimally aggressive conditions.

The USA hCG Reference Service has found that proportion of hyperglycosylated hCG correlated with aggressiveness of disease. All cases of advanced disease (stage IV) with brain and lung metastases were producing >100% hyperglycosylated hCG. Hyperglycosylated hCG is clearly a useful test to categorize choriocarcinoma cases into aggressive and minimally aggressive disease. In cases of minimally aggressive or slow-growing disease, chemorefractory disease needs to be considered. The disease may be growing too slowly to respond to chemotherapy.

In experience of the USA hCG Reference Service with chemorefractory disease, it may be best to allow the disease to slowly advance to aggressive disease before administering chemotherapy. Disease advancing from 500 to 3000 mIU/ml total and raised hyperglycosylated hCG may achieve this objective.

Hyperglycosylated hCG assays are presently available from Quest Diagnostics and should be used to assess all gestational trophoblastic disease cases (Quest Diagnostics hyperglycosylated hCG test, orders #11303X and 90581). Hyperglycosylated hCG results are presented in nanograms and picograms per milliliter. Nanograms can be converted to total hCG units, milli-international unit equivalents, by multiplying by 11, where 1 ng/ml = 11 mIU/ml.

34.3 Placental site trophoblastic disease

Placental site trophoblastic tumor (PSTT) is a malignancy independent of choriocarcinoma and GTN derived from trophoblastic cells. Although choriocarcinoma and GTN may not be a true cancer involving cancer transformation, PSTT is a true cancer of placental trophoblastic cells. In that respect, it is very reminiscent of a non-trophoblastic cancers. Like a non-trophoblastic cancer, it produces primarily small quantities of the hCG free β-subunit or the hyperglycosylated free β-subunit. It does not produce a high concentration of total hCG and hyperglycosylated hCG like choriocarcinoma and GTN.

PSTT needs expert histology analyses to differentiate it from choriocarcinoma (expert such as Robert J. Kurman, MD, of Johns Hopkins Medical Center) or by a specialist at the Brewer Trophoblast Disease Center (Chicago and New England (Boston)) or Charing Cross (London) Trophoblast Disease Specialty Center. The only other method to specifically diagnose PSTT is to demonstrate that 30–100% of the total hCG immunoreactivity in blood is due to hCG free β-subunit and not total hCG. Unfortunately, laboratories in the United States only perform the hCG free β-subunit tests for Down syndrome screening. Under a patent agreement promoted by Johnson & Johnson, they are not allowed to use this test for cancer measurements. The only centers that can measure hCG free β-subunit in cancer are the USA hCG Reference Service and laboratories in Canada. I recommend the Hereditary Disease Laboratory at University of Alberta Hospital in Edmonton, Canada.

As shown in Table 34.2 listing 12 proven PSTT cases, disease is marked by increased hCG free β-subunit. Cases have free β-subunit accounting for 39–90% of total hCG. Cases lack significant hyperglycosylated hCG, which marks choriocarcinoma and GTN [15]. The use of hCG free β-subunit in identifying PSTT was confirmed recently in England [16].

Table 34.2 Twelve cases referred to the USA hCG Reference Service of PSTT proven by histology

Total hCG (mIU/ml)	Free β-subunit (% of total hCG)	Hyperglycosylated hCG (% of total hCG)
28	69%	16%
8.5	47%	34%
231	38%	37%
35	50%	0%
12.8	68%	0%
94	48%	0%
13	62%	0%
0.77	46%	0%
3.3	39%	0%
236	82%	0%
25	90%	0%
138	97%	5%
31	52%	0%
65 ± 84	60 ± 19%	7.1 ± 13%
3.3–236	38–97%	0–37%

References

[1] Cole LA, Dai D, Butler SA, Leslie KK, Kohorn EI. Gestational trophoblastic diseases: 1. Pathophysiology of hyperglycosylated hCG-regulated neoplasia. Gynecol Oncol 2006;102:144–9.

[2] Cole LA, Khanlian SA, Riley JM, Butler SA. Hyperglycosylated hCG (hCG-H) in gestational implantation, and in choriocarcinoma and testicular germ cell malignancy tumorigenesis. J Reprod Med 2006;51:919–29.

[3] Guibourdenche J, Handschuh K, Tsatsaris V, Gerbaud P, Leguy MC, Muller F, et al. Hyperglycosylated hCG is a marker of early human trophoblast invasion. J Clin Endocrinol Metab 2010;95:E240–4.

[4] Handschuh K, Guibourdenche J, Tsatsari V, Guesnon M, Laurendeau I, Evain-Brion D, et al. Human chorionic gonadotropin produced by the invasive trophoblast but not the villous trophoblast promotes cell invasion and is down-regulated by peroxisome proliferator-activated receptor-A. Endocrinology 2007;148:5011–19.

[5] Sasaki Y, Ladner DG, Cole LA. Hyperglycosylated hCG the source of pregnancy failures. Fertil Steril 2008;89:1781–6.

[6] Cole LA. Hyperglycosylated hCG and pregnancy failures. J Reprod Immunol 2012;93:119–22.

[7] Butler SA, Ikram MS, Mathieu S, Iles RK. The increase in bladder carcinoma cell population induced by the free beta subunit of hCG is a result of an anti-apoptosis effect and not cell proliferation. Br J Cancer 2000;82:1553–6.

[8] Berndt S, Blacher S, Munuat C, Detilleux J, Evain-Brion D, Noel A, et al. Hyperglycosylated human chorionic gonadotropin stimulates angiogenesis through TGF-β receptor activation. FASEB J 2013; Abstract 12-213686.

[9] Cole LA, Butler SA. Hyperglycosylated hCG hCGβ and hyperglycosylated hCGβ: interchangeable cancer promoters. Mol Cell Endocrinol 2012;349:232–8.

[10] Hamade AL, Nakabayashi K, Sato A, Kiyoshi K, Takamatsu Y, Laoag-Fernandez JB, et al. Transfection of antisense chorionic gonadotropin β gene into choriocarcinoma cells suppresses the cell proliferation and induces apoptosis. J Clin Endocrinol Metab 2005;90:4873–9.

[11] Khanlian SA, Smith HO, Cole LA. Persistent low levels of hCG: a pre-malignant gestational trophoblastic disease. Am J Obstet Gynecol 2003;188:1254–9.

[12] Cole LA, Sutton JM. Selecting an appropriate hCG test for management of gestational trophoblastic diseases and cancer cases. J Reprod Med 2004;49:545–53.

[13] Cole LA, Kohorn E, Smith HO. Gestational trophoblastic diseases: management of cases with persistent low human chorionic gonadotropin results. Obstet Gynecol Clin North Am 2005;32:615–26.

[14] Cole LA, Muller CY. New approach to treating gestational trophoblastic neoplasms. J Reprod Med 2009.

[15] Cole LA, Khanlian SA, Muller CY, Giddings A, Kohorn EI, Berkowitz R. Gestational trophoblastic diseases: 3. Human chorionic gonadotropin free β-subunit a reliable marker of placental site trophoblastic tumors. Gyn Oncol 2006;102:159–63.

[16] Schmid P, Nagai Y, Agarwal R, Hancock BW, Savage PM, Sebire NJ, et al. Prognostic markers and long-term outcome of placental-site trophoblastic tumours: a retrospective observational study. Lancet 2009;374:6–7.

Quiescent trophoblastic disease and minimally aggressive gestational trophoblastic neoplasm

35

Laurence A. Cole
USA hCG Reference Service, Angel Fire, NM, USA

35.1 Quiescent gestational trophoblastic disease

The presence and absence of invasive or malignant choriocarcinoma or gestational trophoblastic neoplasm (GTN) can be shown by the presence and absence of serum hyperglycosylated hCG. If a patient has a history of choriocarcinoma, or hydatidiform mole, and suddenly has no detectable hyperglycosylated hCG (<5% of total hCG) with plateauing total hCG concentration, then that patient probably no longer has invasive or malignant disease. This is quiescent gestational trophoblastic disease or inactive disease. A high proportion of patients with gestational trophoblastic disease history, whether hydatidiform mole, invasive mole, choriocarcinoma, or GTN, will go through a period of plateauing total hCG levels (1.1–246 mIU/ml) and no detectable hyperglycosylated hCG (Table 35.1).

Table 35.1 presents 174 cases of quiescent gestational trophoblastic disease observed by the USA hCG Reference Service. Cases have proven history of choriocarcinoma or GTN ($n = 30$), complete hydatidiform mole, or partial hydatidiform mole. Sixteen patients with quiescent gestational trophoblastic disease have a recent (6 months) history of ectopic pregnancy with no histology to identify hydatidiform mole, and 34 patients have a history of spontaneous abortion with no histology to identify hydatidiform mole.

Two patients with quiescent gestational trophoblastic disease had a hysterectomy. Histology revealed a small mass of tissue composed of only syncytiotrophoblast cells [1–3]. It is inferred from this finding and from the finding that quiescent gestational trophoblastic disease cases do not produce hyperglycosylated hCG (produced by cytotrophoblast cells) (Figure 35.1) and that quiescent gestational trophoblastic disease cases comprise only non-growing differentiated syncytiotrophoblast cells. They lack cytotrophoblast cells, the invasive and growing element in gestational trophoblastic diseases.

Quiescent gestational trophoblastic disease that appears to comprise only non-growing, non-invasive syncytiotrophoblast cells does not, in the experience of the USA hCG Reference Service, respond to any chemotherapy. The small mass of

Human Chorionic Gonadotropin (hCG). DOI: http://dx.doi.org/10.1016/B978-0-12-800749-5.00035-3

Table 35.1 **Quiescent gestational trophoblastic disease cases referred to the USA hCG Reference Service**

History	Total hCG mIU/ml	hCG-H mIU/ml	History	Total hCG mIU/ml	hCG-H mIU/ml	History	Total hCG mIU/ml	hCG-H mIU/ml
Choriocarcinoma	1.1	<0.33	Complete mole	6	<0.33	Partial mole	2.1	<0.33
Choriocarcinoma	2	<0.33	Complete mole	6	<0.33	Partial mole	2.4	<0.33
Choriocarcinoma	2.1	<0.33	Complete mole	7	<0.33	Partial mole	2.7	<0.33
Choriocarcinoma	5	<0.33	Complete mole	7.1	<0.33	Partial mole	3.3	<0.33
Choriocarcinoma	5.5	<0.33	Complete mole	8.2	<0.33	Partial mole	6.6	<0.33
Choriocarcinoma	8	<0.33	Complete mole	8.3	<0.33	Partial mole	7.6	<0.33
Choriocarcinoma	16	<0.33	Complete mole	9.2	<0.33	Partial mole	9.9	<0.33
Choriocarcinoma	17	<0.33	Complete mole	9.8	<0.33	Partial mole	9.9	<0.33
Choriocarcinoma	37	<0.33	Complete mole	11	<0.33	Partial mole	15	<0.33
Choriocarcinoma	19	<0.33	Complete mole	11	<0.33	Partial mole	11	<0.33
Choriocarcinoma	20	<0.33	Complete mole	11	<0.33	Partial mole	13	<0.33
Choriocarcinoma	29	<0.33	Complete mole	11	<0.33	Partial mole	13	<0.33
Choriocarcinoma	48	<0.33	Complete mole	11	<0.33	Partial mole	15	<0.33
Choriocarcinoma	97	<0.33	Complete mole	12	<0.33	Partial mole	18	<0.33
Choriocarcinoma	107	<0.33	Complete mole	12	<0.33	Partial mole	19	<0.33
GTN	3.1	<0.33	Complete mole	12	<0.33	Partial mole	21	<0.33
GTN	4.5	<0.33	Complete mole	13	<0.33	Partial mole	23	<0.33
GTN	3.2	<0.33	Complete mole	13	<0.33	Partial mole	23	<0.33
GTN	4.6	<0.33	Complete mole	13	<0.33	Partial mole	29	<0.33
GTN	4.9	<0.33	Complete mole	14	<0.33	Partial mole	38	<0.33
GTN	7.2	<0.33	Complete mole	14	<0.33	Partial mole	50	<0.33
GTN	7	<0.33	Complete mole	16	<0.33	Partial mole	61	<0.33
GTN	16	<0.33	Complete mole	21	<0.33	Partial mole	62	<0.33
GTN	30	<0.33	Complete mole	22	<0.33	Partial mole	93	<0.33
GTN	35	<0.33	Complete mole	30	<0.33	Miscarriage	<1	<0.33
GTN	51	<0.33	Complete mole	31	<0.33	Miscarriage	2.4	<0.33
GTN	50	<0.33	Complete mole	34	<0.33	Miscarriage	2.7	<0.33

GTN	115	< 0.33	Complete mole	35	< 0.33	Miscarriage	3.1	< 0.33
GTN	106	< 0.33	Complete mole	35	< 0.33	Miscarriage	3.4	< 0.33
GTN	144	< 0.33	Complete mole	38	< 0.33	Miscarriage	3.5	< 0.33
Ectopic	5.5	< 0.33	Complete mole	48	< 0.33	Miscarriage	3.9	< 0.33
Ectopic	5.5	< 0.33	Complete mole	51	< 0.33	Miscarriage	4.6	< 0.33
Ectopic	8.2	< 0.33	Complete mole	57	< 0.33	Miscarriage	5.7	< 0.33
Ectopic	13	< 0.33	Complete mole	42	< 0.33	Miscarriage	6.4	< 0.33
Ectopic	15	< 0.33	Complete mole	48	< 0.33	Miscarriage	6.6	< 0.33
Ectopic	16	< 0.33 <0.33	Complete mole	60	< 0.33	Miscarriage	6.8	< 0.33
Ectopic	17	< 0.33	Complete mole	61	< 0.33	Miscarriage	7.5	< 0.33
Ectopic	18	< 0.33	Complete mole	70	< 0.33	Miscarriage	7.6	< 0.33
Ectopic	20	< 0.33	Complete mole	79	< 0.33	Miscarriage	7.9	< 0.33
Ectopic	32	< 0.33	Complete mole	80	< 0.33	Miscarriage	8.6	< 0.33
Ectopic	33	< 0.33	Complete mole	90	< 0.33	Miscarriage	10	< 0.33
Ectopic	34	< 0.33	Complete mole	110	< 0.33	Miscarriage	10	< 0.33
Ectopic	41	< 0.33	Complete mole	119	< 0.33	Miscarriage	12	< 0.33
Ectopic	81	< 0.33	Complete mole	141	< 0.33	Miscarriage	13	< 0.33
Ectopic	94	< 0.33	Complete mole	161	< 0.33	Miscarriage	14	< 0.33
Ectopic	127	< 0.33 <0.33	Complete mole	208	< 0.33	Miscarriage	15	< 0.33
Biochemical	< 1	< 0.33	Complete mole	231	< 0.33	Miscarriage	18	< 0.33
Biochemical	2.4	< 0.33	Complete mole	246	< 0.33	Miscarriage	29	< 0.33
Biochemical	4.2	< 0.33	Complete mole	24	< 0.33	Miscarriage	33	< 0.33
Biochemical	84	< 0.33	Complete mole	24	< 0.33	Miscarriage	36	< 0.33
Complete mole	1.5	< 0.33	Complete mole	24	< 0.33	Miscarriage	40	< 0.33
Complete mole	1.7	< 0.33	Complete mole	26	< 0.33	Miscarriage	48	< 0.33
Complete mole	2.6	< 0.33	Complete mole	25	< 0.33	Miscarriage	52	< 0.33
Complete mole	3	< 0.33	Complete mole	26	< 0.33	Miscarriage	53	< 0.33
Complete mole	3	< 0.33	Complete mole	27	< 0.33	Miscarriage	58	< 0.33
Complete mole	3.6	< 0.33	Complete mole	27	< 0.33	Miscarriage	92	< 0.33
Complete mole	5	< 0.33	Complete mole	28	< 0.33	Miscarriage	99	< 0.33
Complete mole	5.8	< 0.33	Complete mole	29	< 0.33	Miscarriage	212	< 0.33

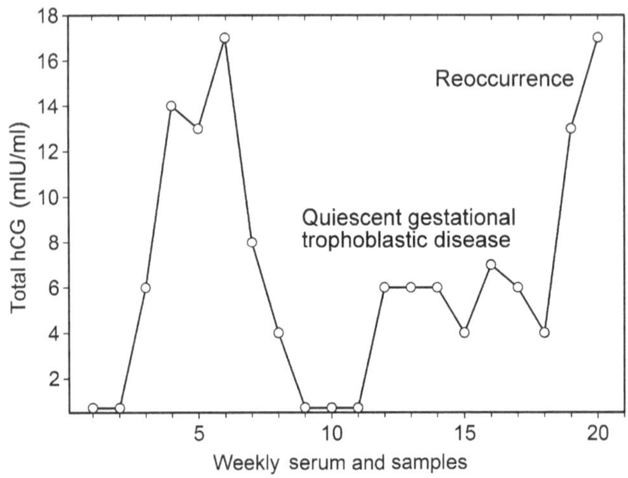

Figure 35.1 Example of quiescent gestational trophoblastic disease showing reoccurrence of malignant/persistent disease.

syncytiotrophoblast cells may be located in the uterus or outside of the uterus, so a hysterectomy does not necessarily resolve this disorder. This disorder is a completely benign non-invasive, non-growing condition. In the experience of the USA hCG Reference Service, most cases resolve by apoptosis or cell death, with no therapy, within 6 months. At that time, the total hCG concentration resolves to <1 mIU/ml. In the experience of the USA hCG Reference Service, ~10% of hydatidiform mole source cases lead to a recurrence of disease, with total hCG levels continuously increasing. In contrast, approximately 25% of GTN or choriocarcinoma cases reoccur. Quiescent gestational trophoblastic disease serum hCG (1.1–246 mIU/ml) may oscillate somewhat during weeks of detection. Serum hCG may naturally increase and decrease in weekly total hCG measurements by as much as 50%. In the case of a reoccurrence, three consecutive increasing total hCG values may be needed to confirm the persistence of disease.

Figure 35.1 illustrates a case of quiescent gestational trophoblastic disease and a reoccurrence of malignant/persistent disease. It is important to measure hyperglycosylated hCG to diagnose quiescent gestational trophoblastic disease. The proportion of hyperglycosylated hCG should be <5% of total hCG. When a patient has quiescent gestational trophoblastic disease diagnosed, it is important for her to use an oral contraceptive. The USA hCG Reference Service experience indicates that the hyperglycosylated hCG produced in pregnancy can evoke persistent trophoblastic disease.

35.2 Minimally aggressive GTN

Not surprisingly, unduly low levels of hyperglycosylated hCG are associated with minimally aggressive GTN, a slow-growing variant of invasive moles, choriocarcinoma,

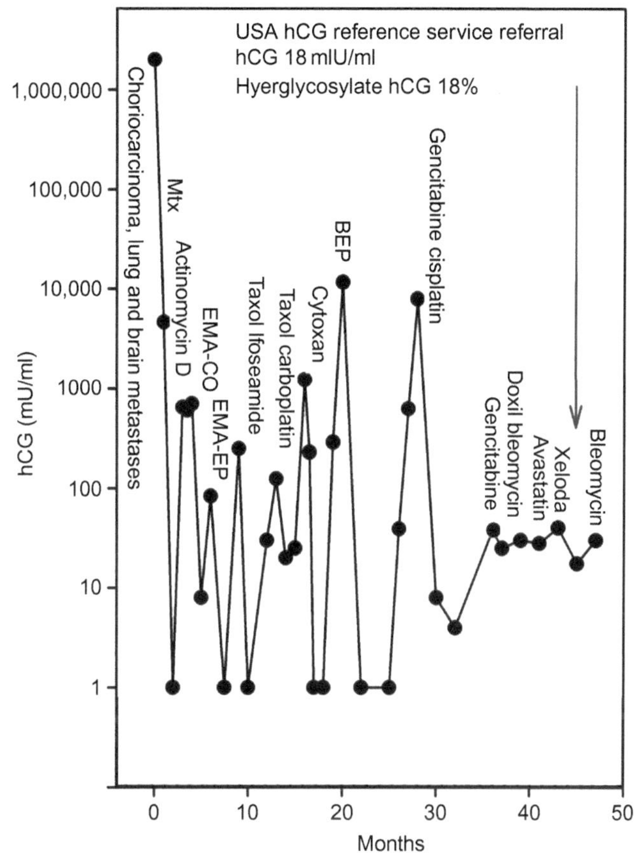

Figure 35.2 USA hCG Reference Service example case of minimally aggressive gestational trophoblastic disease. Mtx is methotrexate; EMA-CO is etoposide, methotrexate, and actinomycin D with cyclophosphamide and cisplatinin; EMA-EP is etoposide, methotrexate, and actinomycin D with etoposide and cusplatinin; and BEP is bleomycin, etoposide, and cisplatinin.

or GTN. This slow-growing variant of GTN, because it is slow-growing, often poorly responds to chemotherapy, causing chemorefractory disease. Here, I describe this minimally aggressive GTN and some suggestions developed by the USA hCG Reference Service regarding how to treat this slow-growing chemorefractory disease.

Figure 35.2 shows an extreme example of minimally aggressive gestational trophoblastic disease referred to the USA hCG Reference Service. Minimally aggressive GTN is identified either by low hyperglycosylated hCG as a proportion of total hCG (<1–40%) (Table 35.2) or by extremely slow total hCG doubling time (cancer growth) in the absence of chemotherapy (minimally aggressive GTN indicated by doubling time of 2.3–20 weeks) (Table 35.2).

A single patient with minimally aggressive GTN decided to turn to God and refused further chemotherapy (agreed to provide us with weekly serum samples), so

Table 35.2 **Twenty-nine cases of minimally aggressive GTN managed by USA hCG Reference Service June 2009 to April 2012**

Age	History	Total hCG mIU/ml	Hyperglycosylated hCG (% of total hCG)	Total hCG double time	Chemotherapy resistance	Notes
26	GTN	256	6.40%	6.1 weeks	Mtx	Lung nodule, protocol halted
33	CM	26	16%	4.0 weeks	Mtx, ActD	Lung nodule, protocol halted
29	CM	208	<1%	5.3 weeks	Mtx, Mtx	Successful >3000 protocol
34	CM	320	27%	4.6 weeks	Mtx, ActD, EMA-CO	Successful >3000 protocol
27	Chorio	445	38%	8.5 weeks	Mtx	Successful >3000 protocol
50	Chorio	1596	27%	9.0 weeks	EMA-CO, TI, Xeloda, Cytoxan	Successful >3000 protocol
34	Chorio	2362	3.90%	4.0 weeks	Mtx	Successful >3000 protocol
27	CM	60	3.80%	3.5 weeks	Mtx, ActD	Successful >3000 protocol
46	Chorio	71	1.40%	3.4 weeks	Mtx	Successful >3000 protocol
37	CM	735	12%	6.3 weeks	EMA-CO, EMA-EP, TI, BEP Xeloda	Successful >3000 protocol
54	Chorio	107	<1%	3.9 weeks	Mtx, AcD, CP, EMA-CO, ICE, VIP, BEP, CT, EMA-EP	Died from chemotherapy
26	PM	112	3.10%	20 weeks	Mtx	Lung nodule, protocol halted
25	GTN	23	21%	6.7 weeks	None	Protocol not followed
18	Chorio	175	19%	6.9 weeks	Mtx, ActD	Successful >3000 protocol
29	CM	208	<1%	7.1 weeks	Mtx	Successful >3000 protocol
29	Chorio	596	27%	5.0 weeks	Mtx, ActD	Successful >3000 protocol
23	CM	204	23%	26 weeks	Mtx	Successful >3000 protocol
19	GTN	97	4.70%	3.2 weeks	Mtx, EMA-CO, EMA-EP, BEP, CP	Successful >3000 protocol

27	Ectopic CM	154	5.10%	4.1 weeks	Mtx	Successful >3000 protocol
20	CM	78	<1%	4.4 weeks	Mtx	Successful >3000 protocol
18	GTN	642	2.40%	2.3 weeks	Mtx, ActD	Successful >3000 protocol
32	CM	101	14%	6.4 weeks	ActD, EMA-CO, CV	Successful >3000 protocol
34	GTN	150	<1%	2.4 weeks	TAH, EMA-CO	Successful >3000 protocol
36	Chorio	104	<1%	4.5 weeks	Mtx	Successful >3000 protocol
37	CM	824	5.10%	4.0 weeks	TVH, Mtx, ActD	Successful >3000 protocol
33	Chorio	97	4.70%	5.2 weeks	Mtx, EMA-CO, EMA-EP, BEP, CP	Successful >3000 protocol
19	CM	341	<1%	4.9 weeks	EMA-CO, TAH	Successful >3000 protocol
32	GTN	55	<1%	4.1 weeks	TAH, Mtx	Successful >3000 protocol
35	Chorio	79	11%	4.6 weeks	ActD, EMA-CO, EMA-EP, CV	Successful >3000 protocol

hCG doubling time estimated from recent hCG determinations while not receiving chemotherapy. Mtx, methotrexate; ActD, actinomycin D; EMA-CO, etoposide-methotrexate-actinomycin D, cycling with cyclophosphamide-vincristine; EMA-EP, etoposide-methotrexate-actinomycin D, cycling with etoposide-cisplatin; BEP, bleomycin-etoposide-cisplanin; CP, carboplatin with paditaxel; VIP, vinblastine-ifosfamide-cisplatin; ICE, ifosfamide-carboplatin-etoposide; CT, carboplatin-taxol; CV, cisplatin-VP16; TAH, trans-abdominal hysterectomy; TVH, trans-vaginal hysterectomy; CM, complete mole; PM, partial mole; Chorio, choriocarcinoma; GTN, gestational trophoblastic neoplasm.

we were able to examine the action of advancing disease on proportion of hyperglycosylated hCG. As shown (Figure 35.3), the proportion of hyperglycosylated hCG rapidly increases with advancing disease or increasing disease mass, making malignancy more aggressive and faster-growing.

We considered that if disease with a low proportion of hyperglycosylated hCG or slow growth were allowed to advance with no therapy so that disease advances, then the proportion of hyperglycosylated hCG would increase, as would the growth rate, and disease may start to respond to chemotherapy. On examining a second patient who, like this God-fearing patient, had increased hyperglycosylated hCG with advancing disease, it was decided that letting minimally aggressive gestational

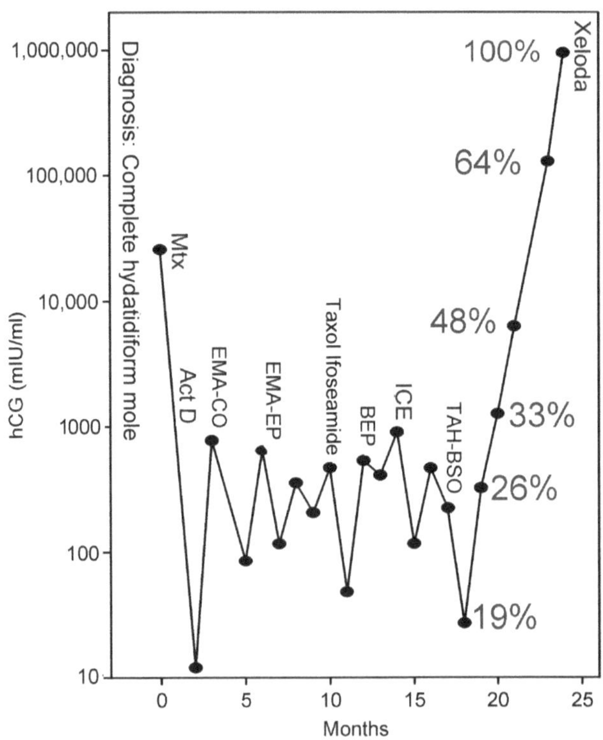

Figure 35.3 Chemorefractory choriocarcinoma patient refused further therapy after TAH-BSO (hysterectomy bilateral oophorectomy). USA hCG Reference Service followed the patient's hCG and hyperglycosylated hCG during the period that followed without chemotherapy. hCG and tumor mass slowly expanded with the eventual appearance of lung and brain metastases. The patient ultimately died from advancing choriocarcinoma. Mtx is methotrexate; Act D is actinomycin D; EMA-CO is etoposide-methotrexate-actinomycin D cycling with cyclophosphamide-vincristine; EMA-EP is etoposide-methotrexate-actinomycin D cycling with etoposide-cisplatin; BEP is bleomycin-etoposide-cisplatinin; and ICE is ifoseamide, carboplatinin, and etoposide. Large percentages are proportion of hyperglycosylated hCG (of total hCG).

trophoblastic disease cases advance in disease without therapy until total hCG was >3000 mIU/ml may facilitate a response to chemotherapy [4].

Cases with ineffective chemotherapy or chemorefractory disease were identified by the USA hCG Reference Service (Table 35.2). A total of 29 cases were identified (Table 35.2). As shown in Table 35.2, all cases had unduly low proportions of hyperglycosylated hCG of <1% to 38% of total hCG. All also had unduly slow hCG doubling times (in the absence of chemotherapy) of 2.3–20 weeks. In addition, all 29 cases had no success with chemotherapy. In all 29 cases it was recommended to the treating physician that therapy should be stopped until the total hCG increased to >3000 mIU/ml. During this process, in one case the protocol was not followed and in three cases protocol was abandoned because of appearance of metastases while hCG was increasing. One patient died. In the remaining 24 cases, the hCG was allowed to increase slowly to >3000 mIU/ml (in two cases it increased to >5000 mIU/ml), and then EMA-EP (methotrexate-actinomycin D, cycling with etoposide-cisplatin) chemotherapy was administered. In 24 of 24 cases, this was successful (Table 35.2). This protocol is recommended for chemorefractory minimally aggressive GTN cases.

References

[1] Khanlian SA, Smith HO, Cole LA. Persistent low levels of hCG: a pre-malignant gestational trophoblastic disease. Am J Obstet Gynecol 2003;188:1254–9.

[2] Cole LA, Sutton JM. Selecting an appropriate hCG test for management of gestational trophoblastic diseases and cancer cases. J Reprod Med 2004;49:545–53.

[3] Cole LA, Kohorn E, Smith HO. Gestational trophoblastic diseases: management of cases with persistent low human chorionic gonadotropin results. Obstet Gynecol Clin N Am 2005;32:615–26.

[4] Cole LA, Muller CY. New approach to treating gestational trophoblastic neoplasms. J Reprod Med 2009.

Background hCG

36

Laurence A. Cole
USA hCG Reference Service, Angel Fire, NM, USA

Background hCG is one of the poorest studied scopes of hCG research, but it is still important. Only by knowing what the true background levels of female hCG are can we fairly say what concentration of serum and urine hCG constitutes a pregnancy or the successful treatment of gestational trophoblastic disease or other disorder. We understand that hCG is primarily produced by trophoblast cells during pregnancy. However, it is understood from false-positive total hCG results that background hCG can be produced by the pituitary gland as part of the menstrual cycle. It is also understood from cancer studies that hCG and its variants are also produced by malignancies. Furthermore, recent studies show hCG variant production by genetic disorders like familial hCG syndrome [1,2]. Other sources of positive hCG tests include false-positive hCG tests, quiescent gestational trophoblastic disease, and gestational trophoblastic diseases (see Chapter 28) [3,4].

What is the background serum hCG due to pituitary menstrual cycle production? The definition of acceptable background according to the Clinical Chemistry Society is a concentration corresponding to 97.5th percentile. As shown by Snyder et al. [5], examining 240 serum samples, the 97.5th percentile is 2.5 mIU/ml hCG (Table 36.1). In our own study of 797 menstrual cycle serum samples, we calculated a similar 97.5th percentile of 2.7 mIU/ml [6] (Table 36.1). These 97.5th percentiles indicate that any total hCG more than 2.7 mIU/ml can be considered as positive or indicating pregnancy. Currently, 5 mIU/ml is generally accepted as the cutoff in a serum hCG pregnancy test.

More extensive studies have been performed with urine samples [7]. I have tested 9123 urine samples from the follicular phase, ovulatory phase, and luteal phase of the menstrual cycle (399 women followed-up daily for five menstrual cycles)[7]. As shown in Table 36.2, 2.8% or 1 in 259 of 9123 urine samples were elevated when tested with the Siemens Immulite total hCG pregnancy test or >1.0 mIU/ml. This created a 97.5th percentile for urine hCG of 1.0 mIU/ml. The 97.5th percentile is lower in urine than in serum. Over-the-counter pregnancy tests are available, such as First Response manual, with sensitivity as low as 3.5 mIU/ml. Examining the urine samples closer (Table 36.2), 31 of 3662 follicular phase urine samples (0.80%) and 206 of 2289 ovulatory phase urine samples are positive (9.0%), and 22 of 3172 luteal phase urine samples are positive (0.69%). Clearly, hCG is most commonly positive in the ovulatory phase.

Gonadotropin-releasing hormone (GnRH) pulses reach a peak with the LH peak. It is thought that pituitary hCG production is incidental to the GnRH peak [7]. On the single day of the LH peak in each case, a total of 312 urine samples were

Human Chorionic Gonadotropin (hCG). DOI: http://dx.doi.org/10.1016/B978-0-12-800749-5.00036-5

Table 36.1 Detection of hCG in serum during the menstrual cycle

	n	Clinical chemistry background limitation (97.5 centile)
Snyder et al., 2005 [5]	240	2.5 mIU/ml
Cole et al., 2009 [3]	797	2.7 mIU/ml

Table 36.2 Detection of hCG in urine during the menstrual period

	Follicular phase 1–10 days	Ovulatory phase 11–17 days	Luteal phase 18–28 days	Overall 1–28 days
Number of urines tested	3662	2289	3172	9123
Number with hCG >1 mIU/ml	31	206	22	259
Percentage positive	0.80%	9.0%	0.69%	2.8%
Range of hCG detection	1–4 mIU/ml	1–9.3 mIU/ml	1–2.7 mIU/ml	1–9.3 mIU/ml
97.5th percentile	<1 mIU/ml	1.6 mIU/ml	<1 mIU/ml	1.0 mIU/ml

collected. Of these, 107 (34%) were elevated (>1.0 mIU/ml). The elevated urines ranged from 1 to 9.3 mIU/ml. It would appear from this finding that one in three women may have a false-positive hCG pregnancy test on the day of LH peak. Women need to be cautious of hCG pregnancy tests at approximately day 13–16 of the menstrual cycle or at the time of the LH peak.

Considering pituitary hCG as the principal source of false-positive hCG tests, serum tests should has a pregnancy cutoff of >2.7 mIU/ml and urine pregnancy tests should have a cutoff of >1.0 mIU/ml. Other sources of false-positive tests need to be considered. These might include perimenopausal and postmenopausal women, cancer, and genetic disorders such as familial hCG syndrome [1,2,5–7].

Figure 36.1 shows the distribution of serum hCG in 103 perimenopause and postmenopause cases. As shown, hCG varies from <1 to 39 mIU/ml. This figure might be a little biased in that it only shows problem cases referred to the USA hCG Reference Service. How can we include perimenopausal and postmenopausal women in our sensitivity statistics? If we estimate that 5% of these women are tested for pregnancy, then we can estimate that it increases the 97.5th percentile to 9.6 mIU/ml, or dramatically takes away the sensitivity of pregnancy tests.

It is inferred that we cannot include perimenopausal and postmenopausal cases and, similarly, cancer cases and genetic disorder cases in our background hCG statistics. Perimenopausal and postmenopausal women need to be excluded from hCG pregnancy testing. This can be done by considering age older than 40 years or

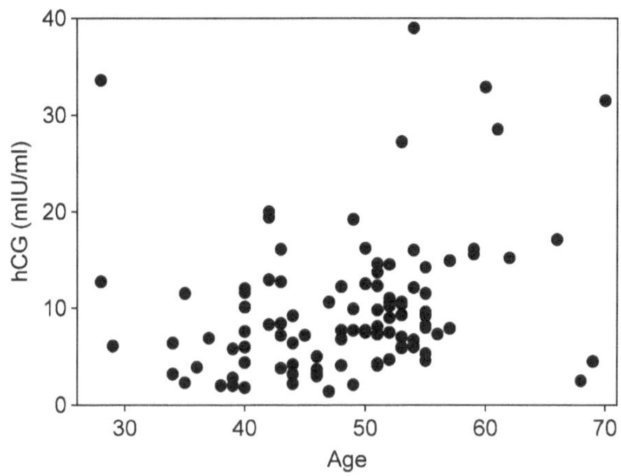

Figure 36.1 Postmenopausal hCG in 103 women referred to the USA hCG Reference Service.

considering the results of a follicle stimulating hormone test (>30 mIU/ml). Similarly, cancer cases need to be excluded from routine pregnancy screening. As shown in Chapter 20, 30% of individuals with cancer may have a positive result on a total hCG test. In cancer cases, the malignancy diagnosis should be enough reason to exclude them. In contrast, cases of familial hCG syndrome create a problem. This is a genetic disorder that may be difficult to diagnose [1,2]. However, we have no data yet regarding the incidence of this syndrome. We know it is a rare syndrome, and therefore we cannot calculate the effect it may have on background false-positives.

References

[1] Cole LA. Familial hCG. J Reprod Immunol 2012;93:52–7.
[2] Cole LA, Laidler LL. Inherited hCG. J Reprod Med 2010;55:99–102.
[3] Cole LA, Laidler L, Muller C. USA hCG reference service, 10 year report. Clin Biochem 2010;43:1013–22.
[4] Cole LA. Hyperglycosylated hCG, the most valuable test in the management of gestational trophoblastic neoplasms. Troph Dis Upd 2010;10:5–22.
[5] Snyder JA, Haymond S, Parvin CA, Gonowski AM, Grenache DG. Diagnostic considerations in the measurement of human chorionic gonadotropin in aging woman. Clin Chem 2005;51:1830–5.
[6] Cole LA. Background hCG in healthy non-pregnant women. Clin Chem 2005;51:1765–6.
[7] Cole LA, Ladner DG. Background hCG in non-pregnant individuals: need for more sensitive point-of-care and over-the-counter pregnancy tests. Clin Biochem 2009;42:168–75.

hCG standards

37

Ulf-Håkan Stenman
Helsinki University Central Hospital, Helsinki, Finland

Standardization of hormones became important when pharmaceutical preparations of human hormone preparations became available in the 1930s. Most of these were crude tissue or urine extracts, which were standardized with biological assays. The first international standard (first IS) for hCG was introduced in 1939 and was assigned its value in arbitrary international units (IU), with one tablet of the standard containing 10 mg of dry substance; its potency was thus 100 IU/mg [1]. The second IS was calibrated against the first IS, and the third and fourth were calibrated against the second [2,3]. Thus, the hCG standardization can be traced back to the first IS. The introduction of immunoassays for hormones into clinical use in the 1960s revealed problems caused by discrepancies between bioactivity and immunoreactivity. Consisting of partially purified urinary hCG preparations, the first and second standards contained degradation products and subunits of hCG, which lacked bioactivity but were immunoreactive. Thus, they were not suited for standardization of immunoassays that, to a variable degree, recognized hCG subunits and other gonadotropins. The third IS was a highly purified preparation of hCG, but because the second IS contained subunits, the value for the third IS had to be assigned on the basis of bioactivity [3,4]; however, the third IS also contained significant amounts of degraded forms of hCG (nicked hCG, hCGn), which were recognized by many hCG assays. With the development of better purification techniques, it became possible to prepare more pure preparations of hCG, its subunits, and partially degraded forms. The values of these were assigned in substance concentrations—the first time for a protein hormone. In 2006, these preparations were approved by the WHO as International Reference Reagents (IRRs) [5,6]. The properties of the standards are summarized in Table 37.1.

37.1 First IS

During the Third Conference on the Standardization of Hormones, held under the auspices of the Permanent Commission on Biological Standardization in Geneva in August of 1938, it was decided to arrange for the establishment of an IS for the gonadotropic substance of human urine of pregnancy. The first IS for hCG was established in 1939, under the auspices of the League of Nations. It consisted of a mixture of urinary hCG preparations donated by six companies. The dry preparations were diluted with lactose to a mass of 800 g and further dried over

Human Chorionic Gonadotropin (hCG). DOI: http://dx.doi.org/10.1016/B978-0-12-800749-5.00037-7

Table 37.1 Reference preparations for hCG and related substances

Standard	Source	Potency (IU/mg)	Established	Reference
First IS	Placental extract	10	1939	[1]
Second IS	Pregnancy urine	782	1964	[2]
IRP	Pregnancy urine	12,000	1976	[4]
Third and fourth IS	Pregnancy urine	9300	1980	[3]
IRR	Pregnancy urine	13,000	2005	[6]

IRP, international reference preparation.

P_2O_5 *in vacuo*. From this material, 10 mg tablets were prepared and batches of 25 tablets were packed in tubes that were each enclosed in 2854 ampoules. The definition of the unit was "the specific gonadotropic activity of 0.1 mm (=100 gamma) of the standard preparation shall be the IU for recording the activities of all gonadotropic preparations of human urine of pregnancy, *but only of such.*" Thus, the potency of this preparation was 10 IU/mg [1]. It should be noted that most of the dry substance in this preparation consisted of lactose, and the activity per milligram of protein is not known. This standard was successfully used to calibrate pharmaceutical preparations of hCG for the next few decades.

37.2 Second IS

Preparation of a new standard was initiated in 1960, when supplies of the first IS were running low. The Expert Committee on Biological Standardization of WHO authorized the National Institute for Medical Research in London to arrange establishment of a new standard. Because highly purified preparations of hCG were considered unstable, it was decided that material of medium potency should be used. The second IS was prepared from a single batch of 10 g of urinary hCG with a potency 2180 IU/ml donated by N. V. Organon (Oss, The Netherlands). This preparation was dissolved in 4 l of 0.5% lactose solution and 1 ml aliquots distributed into 3850 ampoules, which were freeze dried. Each ampoule contained 6.78 mg of dry substance consisting of lactose and the urinary hCG preparation. This standard was calibrated in an international collaborative study performed in nine laboratories using six different bioassays and one immunoassay. Based on this study, the standard was assigned a value of 5300 IU/ampoule, and the IU of hCG was defined as the activity contained in 0.001279 mg of the second IS for hCG [2]. This corresponds to a potency of 782 IU/mg, which means that most of the standard mass consisted of lactose.

Table 37.2 Comparison of WHO units (IU), mass units (μg), and substance concentrations (pmol)

Abbreviation	MW	IU	μg	pmol
hCG	37,500	1	0.11	2.9
hCGβ	23,500	1	1	42.5
hCGα	14,000	1	1	71.4
hCGβcf	12,000		1	83.3

The WHO IU for hCG is based on bioactivity, whereas the corresponding values for hCGα and hCGβ are based on mass.

37.3 Third IS

When the second IS was used for standardization, the presence of free subunits that, to variable degrees, were recognized by different immunoassays caused problems. In 1972, highly purified preparations of hCG and its subunits encoding CR119 were prepared by Canfield and Ross and were available from the National Institute of Child Health and Human Development in the United States. CR119 consisted of three highly pure preparations: CR119 hCG, hCGα, and hCGβ isolated from urine. CR119 hCG had a nominal activity of 13,450 IU/mg. A total of 550 mg of hCG and 280 mg of each subunit were made available to the WHO and used to prepare new standards [3,4]. These preparations were initially called reference preparations and, after thorough international evaluation organized by the WHO laboratory in the United Kingdom and National Institute of Biological Standards and Control (NIBSC), CR119 was adopted in 1980 as a WHO standard with a code [3]. It was assigned a value of 650 IU/ampoule based on its bioactivity and the ratio of bioactivity to mass (i.e., potency) was 9286 IU/mg. The same batch of hCG was used to prepare the fourth IS; therefore, the third IS 75/537 and fourth IS 75/589 are identical. The concentrations of hCG in IU of hCG can be converted to pmol by multiplying by 2.93 [7].

Standards for the free subunits of hCG were prepared by dissociating intact hCG and purifying the subunits. Because these lacked bioactivity, the standards for hCGα (75/551) and hCGβ (75/569) were assigned values based on mass with 1 μg corresponding to 1 IU [3]. Thus, their units are not comparable with those of hCG. Whereas 1 IU of hCG corresponds to 2.9 pmol, 1 IU of hCGβ is 42.5 pmol (Table 37.2). Many commercial hCG assays measured hCG and hCGβ together, and the results are based on the units for hCG and expressed in IU/l [8]. It should be recognized that this unit does not correspond to the WHO IU for hCGβ. Groups that measure subunits by specific assays usually express the concentrations of hCGβ and hCGα in substance concentrations as pmol/l [8]. In practice, the concentrations of hCG subunits are not expressed using the official WHO units.

Table 37.3 IFCC nomenclature and WHO designations for the first IRR for hCG and clinically important hCG-related molecules

Human chorionic gonadotropin, intact	hCG	hCG devoid of nicked forms and free subunits	99/688
Human chorionic gonadotropin, nicked	hCGn	Partially degraded hCG, missing peptide bonds in the hCGβ-40–50 region	99/642
Human chorionic gonadotropin, α-subunit	HCGα	Purified hCGα, dissociated from hCG	99/720
Human chorionic gonadotropin, β-subunit	hCGβ	Purified hCGβ, dissociated from hCG	99/650
Human chorionic gonadotropin, nicked β-subunit	hCGβn	Partially degraded hCGβ, missing peptide bonds in the hCGβ-40–50 region	99/692
Human chorionic gonadotropin, β-core fragment	hCGβcf	Residues hCGβ-6–40, joined by disulfide bonds to hCGβ-55–92	99/708

37.4 IRR for hCG and related substances

Methods to improve immunoassays standardization were discussed at two Bergmeyer conferences organized by the International Federation of Clinical Chemistry (IFCC) in 1990 and 1992. Participants were invited to suggest projects, and Ulf-H å kan Stenman's proposal to develop standardization of hCG immunoassays was approved. The project plan was elaborated by an expert group and published in 1993.

The main goals of the project were as follows: (i) establish a uniform nomenclature and abbreviations for the various molecular forms of hCG; (ii) prepare new calibrators for the various molecular forms of hCG and establish methods for determining the substance concentrations (mol/l) of these; (iii) improve quality-assessment materials and procedures; and (iv) define methods for characterization of hCG measurement procedures [8]. In 1995, IFCC appointed a working group (WG) for standardization of hCG determinations. The group consisted of Ulf-Håkan Stenman (Helsinki) (chairman), Steven Birken (New York), Jean–Michel Bidart (Paris), Bruce Nisula (Bethesda, MD), Peter Berger (Innsbruck), and Rob Norman (Adelaide). Nisula soon retired from the group and was replaced by Catharine Sturgeon, who acted as secretary until 2001. In 2001, she replaced Stenman as chairman when he was appointed member of the Scientific Division of the IFCC with the responsibility of supervising the WG.

The first goal of the WG, establishment of a nomenclature, was accomplished in 1993 (Table 37.3) [8]. The need to prepare new standards arose when the third IS was found to contain partially degraded hCG (i.e., hCGn) and some other impurities, causing problems when the preparations were used for immunoassay standardization. The decision to assign values to the new standards in substance concentrations was based on the reasoning that immunoassays reflect molar concentrations of protein rather than bioactivity [8].

Furthermore, the use of substance concentrations (i.e., mol/l) is the most appropriate way of comparing concentrations of two (or several) variants of the same molecule with different molecular weights (MW) and to express the sum of their concentrations. Molar concentrations can reliably be determined by amino acid analysis [5]. Financing for the process of creating new hCG standards became possible through support from the IFCC and 13 diagnostics companies: Abbott Diagnostics Division; Bayer PLC (Diagnostics Division); Beckman Coulter; Bioclone (Australia); BioMérieux; Chiron Diagnostics; CIS International; Dade Behring; Diagnostic Products Corporation; Ortho-Clinical Diagnostics; PerkinElmer Life Sciences; and Randox Laboratories. Together, they provided US$155,000 in exchange for early access to the hCG preparations [5].

Two lots of pharmaceutical-grade hCG with a biological activity of 4200 IU/mg of dry weight from Diosynth (Division of Akzo/Nobel) were used as the starting material for preparation of the standards. Steven Birken at Columbia University, New York, undertook purification of the four forms of hCG that were considered clinically important (i.e., hCG, hCGα, hCGβ, and core fragment of hCGβ (hCGβcf)). In addition, standards were also prepared for two partially degraded or nicked forms (i.e., hCGn and hCGβn) (Table 37.3). The nicked variants were prepared because they were needed for characterization of antibody and assay specificity. Certain antibodies that do not recognize hCGn will miss part of hCG that may form a substantial part of hCG in urine and also in serum [9–11]. Nicked hCGβ (hCGβn) is also useful for characterization of the epitope specificity of hCG antibodies [12]. A major part of the hCG immunoreactivity in urine of pregnant women consists of hCGβcf, and it is of potential clinical value as a urine marker for cancer [13]. Therefore, it is important for characterization of pregnancy tests and other immunoassays for detection of hCG immunoreactivity in urine.

The purification process and value assignment of the six new preparations have been described. Removal of the nicked forms of hCG was achieved by use of hydrophobic interaction chromatography and final purification of subunits by reverse-phase chromatography. With the exception of hCGβn, which contained some ribonuclease, the final preparations displayed a spectrum of isoelectric points reflecting the carbohydrate heterogeneity of the native forms. Contamination with other proteins and other forms of hCG was negligible [5].

The standards were assigned values in substance concentrations by amino acid analysis and the specific absorbance at 276 nm was determined. The activity of the hCG preparation determined by the seminal weight method in two laboratories was 10,300 and 15,400 IU/mg, whereas the Leydig cell testosterone production assay gave a value of 14,100 IU/l [5]. The mean value of all assays was 13,300 IU/mg; therefore, its potency was approximately 43% higher than that of the third and fourth IS, but there is a strong confidence interval.

Lyophilized preparations of the purified material were delivered to the National Institute of Standards and Control (Potters Bar, Herts, UK), where they were dissolved in a buffer with a carrier protein, lyophilized, and subjected to stability and recovery studies in a collaborative study. All preparations were found to be stable and the recovery after reconstitution was normal. The six preparations were approved by WHO as the

First International Reference Preparations for six hCG-related molecules (Table 37.3). The IRRs are primarily intended for characterization of hCG immunoassays [6].

The first IRRs have been used to characterize the reactivity of commonly used sandwich-type assays and two in-house radioimmunoassay (RIA) methods. In 14 assays, mean recovery of the fourth IS was 107% (range 93–126%), whereas that for IRR 99/680 hCG was 139% (range 109–164%). The higher values and the larger variation for the IRR reveal problems caused by the presence of hCGn in the fourth IS for assay calibration; hCGn is differently recognized by different assays. Recognition of hCGβ by assays designed to measure this form together with hCG was more variable, and all but one sandwich assay overestimated it (one by 150%). Four sandwich assays recognized hCGβcf, but they all underestimated it. RIAs, however, overestimated it [14]. The results of a similar study comprising eight assays gave very similar results [15]. These studies demonstrate the value of the new IRRs.

The results of these studies reveal problems that will be encountered when the presently used fourth standard has to be replaced. It is obvious that value assignment on the basis of the bioactivity of a new standard will result in a substantial change of approximately 40% in assay calibration. The original aim of the hCG standardization project was to prepare standards calibrated in substance concentrations. The metrologically correct way to accomplish this would be to use the IRRs to calibrate immunoassays for hCG, hCGβ, hCGα, and hCGβcf and to express the concentrations in mol/l [8]. When the IRRs were approved, WHO recommended that hCG immunoassays should still be calibrated against the fourth IS; however, because the fourth IS has to be replaced soon, standardization inevitably will change when a new standard is introduced. Based on preliminary information, the new standard will be based on IRR 99/680 (Chris Burns, personal communication).

Because of the heterogeneity of Chinese hamster ovary cells (CHO) moieties, the MW displays a spectrum of values. The average MW of hCG determined by MALDI-TOF mass spectrometry is 37,500; that of hCGα is 14,000; that of hCGβ is 23,500; and that of hCGβcf is 10,000 [5]. The calculated mass of the peptide moiety of hCGα is 10,206, and that of the glycated subunit containing two N-linked biantennary sialylated CHO chains is 14,165. The MW calculated on the basis of the amino acid composition of hCGβ is 15,532, and that containing two N-linked biantennary CHO chains and four type 1 *o*-core carbohydrates is 24,316. Thus, the calculated MW of a "typical" hCG molecule would be 38,931 [11]. The difference between the calculated and measured average MW indicates that the CHO chains are shorter than those used to calculate the theoretical MW. The MW of hCG produced by trophoblastic cancer is higher than that of pregnancy hCG [16], which is explained by the larger carbohydrate chains of hyperglycosylated hCG (hCG-H) [17,18].

37.5 Future prospects

hCG produced by choriocarcinoma often contains larger than normal carbohydrate moieties. This form of hCG, called hCG-H, can be determined by the use of a monoclonal antibody (B152) raised against hCG-H isolated from urine of a patient with choriocarcinoma [17,19]. Most hCG secreted during early pregnancy consists of hCG-H [20]. Several studies suggest that specific determination of hCG-H is useful for diagnosis of certain pregnancy complications [21,22] and for differentiation between benign and malignant trophoblastic disease [23]. Some hCG assays and pregnancy tests have been shown to underestimate hCG-H [24,25]. Final evaluation of this would require a standard for hCG-H.

When the goals of the hCG WG were defined, the potential utility of hCG-H was not recognized and a standard was not prepared; however, identification of a suitable source hCG-H is problematic. Although it occurs in serum and urine during early pregnancy, it is not yet known whether hCG-H from these sources is comparable with that occurring in patients with trophoblastic and other types of cancer. Studies of urine from pregnant women and patients with trophoblastic and testicular tumors have revealed large differences in the glycan structure of hCG [11]. Therefore, hCG-H is not a well-defined entity, and the form of hCG-H to be used as a standard remains to be determined.

37.6 Recombinant hCG

Recombinant hCG (rhCG) has been produced in CHO cells, and a pharmaceutical preparation is available from Serono. rhCG is also a potential new standard, and thus its biological and immunological properties are of interest. When rhCG is used for ovulation induction, a dose of 250 μg of rhCG is equivalent to 5000–10,000 IU of urinary hCG [26,27]. The hCG content of this preparation is expressed on the basis of the mass of the peptide moiety disregarding the carbohydrate content [27]. Thus, 250 μg of rhCG corresponds to approximately 360 μg of urinary hCG. It can thus be calculated to have a specific activity of 13,900 IU/mg [28], which is similar to the potency of approximately 12,000–15,000 IU/mg of the WHO hCG RR 98/688 [5]. These calculations are based on the typical carbohydrate structure of urinary hCG, but the carbohydrates of rhCG are slightly different; the N-linked carbohydrates are the same, but the O-linked carbohydrate chains on the CTP are slightly different [27]. rhCG is also available as a laboratory product, and it is a potential calibrator for immunoassay purposes; however, considerable differences in the immunoreactivity pituitary and recombinant growth hormone have been observed [29]. Therefore, introduction of a recombinant standard for hCG would require careful evaluation of its reactivity in various immunoassays.

References

[1] The Department of Biological Standards NifMR. Memorandum on the international standard for the gonadotropic substance of urine of pregnancy (chorionic gonadotropin). Bull World Health Org Lond 1939;8:884–6.

[2] Bangham DR, Grab B. The second international standard for chorionic gonadotropin. Bull World Health Org 1964;31:111–25.

[3] Storring PL, Gaines-Das RE, Bangham DR. International reference preparation of human chorionic gonadotropin for immunoassay: potency estimates in various bioassay and protein binding assay systems; and international reference preparations of the alpha and beta subunits of human chorionic gonadotropin for immunoassay. J Endocrinol 1980;84:295–310.
[4] Canfield RE, Ross GT. A new reference preparation of human chorionic gonadotropin and its subunits. Bull World Health Org 1976;54:463–72.
[5] Birken S, Berger P, Bidart JM, Weber M, Bristow A, Norman R, et al. Preparation and characterization of new WHO reference reagents for human chorionic gonadotropin and metabolites. Clin Chem 2003;49:144–54.
[6] Bristow A, Berger P, Bidart JM, Birken S, Norman R, Stenman UH, et al. Establishment, value assignment, and characterization of new WHO reference reagents for six molecular forms of human chorionic gonadotropin. Clin Chem 2005;51:177–82.
[7] Alfthan H, Haglund C, Dabek J, Stenman U-H. Concentrations of human chorionic gonadotropin, its b-subunit and the core fragment of the β-subunit in serum and urine of men and nonpregnant women. Clin Chem 1992;38:1981–7.
[8] Stenman U-H, Bidart JM, Birken S, Mann K, Nisula B, O'Connor J. Standardization of protein immunoprocedures. Choriogonadotropin (CG). Scand J Clin Lab Invest Suppl 1993;216:42–78.
[9] Cole LA, Kardana A, Ying FC, Birken S. The biological and clinical significance of nicks in human chorionic gonadotropin and its free beta-subunit. Yale J Biol Med 1991;64:627–37.
[10] Hoermann R, Berger P, Spoettl G, Gillesberger F, Kardana A, Cole LA, et al. Immunological recognition and clinical significance of nicked human chorionic gonadotropin in testicular cancer. Clin Chem 1994;40:2306–12.
[11] Valmu L, Alfthan H, Hotakainen K, Birken S, Stenman UH. Site-specific glycan analysis of human chorionic gonadotropin beta-subunit from malignancies and pregnancy by liquid chromatography—electrospray mass spectrometry. Glycobiology 2006;16:1207–18.
[12] Berger P, Sturgeon C, Bidart JM, Paus E, Gerth R, Niang M, et al. The ISOBM TD-7 Workshop on hCG and related molecules. Towards user-oriented standardization of pregnancy and tumor diagnosis: assignment of epitopes to the three-dimensional structure of diagnostically and commercially relevant monoclonal antibodies directed against human chorionic gonadotropin and derivatives. Tumour Biol 2002;23:1–38.
[13] Wehmann RE, Nisula BC. Characterization of a discrete degradation product of the human chorionic gonadotropin beta-subunit in humans. J Clin Endocrinol Metab 1980;51:101–5.
[14] Sturgeon CM, Berger P, Bidart JM, Birken S, Burns C, Norman RJ, et al. Differences in recognition of the first WHO International Reference Reagents for hCG-related isoforms by diagnostic immunoassays for human chorionic gonadotropin. Clin Chem 2009;55:1484–91.
[15] Whittington J, Fantz CR, Gronowski AM, McCudden C, Mullins R, Sokoll L, et al. The analytical specificity of human chorionic gonadotropin assays determined using WHO International Reference Reagents. Clin Chim Acta 2010;411:81–5.
[16] Mann K, Karl HJ. Molecular heterogeneity of human chorionic gonadotropin and its subunits in testicular cancer. Cancer 1983;52:654–60.

[17] Elliott MM, Kardana A, Lustbader JW, Cole LA. Carbohydrate and peptide structure of the alpha- and beta-subunits of human chorionic gonadotropin from normal and aberrant pregnancy and choriocarcinoma. Endocrine 1997;7:15–32.
[18] Birken S. Specific measurement of O-linked core 2 sugar-containing isoforms of hyperglycosylated human chorionic gonadotropin by antibody b152. Tumour Biol 2005;26:131–41.
[19] Birken S, Krichevsky A, O'Connor J, Schlatterer J, Cole L, Kardana A, et al. Development and characterization of antibodies to a nicked and hyperglycosylated form of hCG from a choriocarcinoma patient: generation of antibodies that differentiate between pregnancy hCG and choriocarcinoma hCG. Endocrine 1999;10:137–44.
[20] Kovalevskaya G, Birken S, Kakuma T, O'Connor JF. Early pregnancy human chorionic gonadotropin (hCG) isoforms measured by an immunometric assay for choriocarcinoma-like hCG. J Endocrinol 1999;161:99–106.
[21] Cole LA, Shahabi S, Oz UA, Bahado-Singh RO, Mahoney MJ. Hyperglycosylated human chorionic gonadotropin (invasive trophoblast antigen) immunoassay: a new basis for gestational down syndrome screening. Clin Chem 1999;45:2109–19.
[22] Kovalevskaya G, Birken S, Kakuma T, Ozaki N, Sauer M, Lindheim S, et al. Differential expression of human chorionic gonadotropin (hCG) glycosylation isoforms in failing and continuing pregnancies: preliminary characterization of the hyperglycosylated hCG epitope. J Endocrinol 2002;172:497–506.
[23] Cole LA, Butler S. Detection of hCG in trophoblastic disease. The USA hCG reference service experience. J Reprod Med 2002;47:433–44.
[24] Butler SA, Khanlian SA, Cole LA. Detection of early pregnancy forms of human chorionic gonadotropin by home pregnancy test devices. Clin Chem 2001;47:2131–6.
[25] Cole LA, Kohorn EI. The need for an hCG assay that appropriately detects trophoblastic disease and other hCG-producing cancers. J Reprod Med 2006;51:793–811.
[26] Chang P, Kenley S, Burns T, Denton G, Currie K, DeVane G, et al. Recombinant human chorionic gonadotropin (rhCG) in assisted reproductive technology: results of a clinical trial comparing two doses of rhCG (Ovidrel) to urinary hCG (Profasi) for induction of final follicular maturation in *in vitro* fertilization-embryo transfer. Fertil Steril 2001;76:67–74.
[27] Gervais A, Hammel YA, Pelloux S, Lepage P, Baer G, Carte N, et al. Glycosylation of human recombinant gonadotropins: characterization and batch-to-batch consistency. Glycobiology 2003;13:179–89.
[28] Stenman UH, Tiitinen A, Alfthan H, Valmu L. The classification, functions and clinical use of different isoforms of HCG. Hum Reprod Update 2006;12:769–84.
[29] Stenman UH. Detection of illicit use of growth hormone. Clin Chem 2009;55:387–8.

Part L

Methods

hCG and hyperglycosylated hCG purification and analysis from serum, urine, and culture fluids

38

Laurence A. Cole
USA hCG Reference Service, Angel Fire, NM, USA

Over the past 20 years, the USA hCG Reference Service has used a variety of methods to purify hCG and hyperglycosylated hCG from serum, urine, and culture fluids. Most of these methods are published [1–3]. These methods were used to obtain pure hCG and hyperglycosylated hCG for carbohydrate structure and peptide cleavage studies and for generating antibodies. Our basic purification protocol was derived from the purification methods of Birken et al. [4], which are used for hCG amino acid analysis and hCG standards. Over the past 20 years, we have purified approximately 40–45 hCG preparations from serum, urine, and culture fluids.

38.1 Urine hCG and hyperglycosylated hCG

The USA hCG Reference Service has purified more hCG and hyperglycosylated hCG preparations from large volumes of urine than any other source [1,2]. Generally, 4–10 mg of total hCG is required as starting material, which can be obtained from approximately 5–10 l of urine from pregnancy peak, advanced choriocarcinoma, or hydatidiform mole cases. It is essential to have at least 4 mg of hCG to start with, because lower concentrations of pure hCG are very sticky and often are lost to gels, albumin-like proteins, and plastic ware and glassware.

The first stage of purification is removal of lipids, salts, sugars, and small peptides. This is performed by precipitating the urine with acetone (one volume of urine to two volumes of acetone). The acetone is first cooled to 4°C before adding it to the urine. Once added to the urine, it is stirred continuously overnight in a refrigerator. In the morning, the mixture is carefully decanted down to a 1:l volume, transferred to 250- or 500-ml centrifuge bottles, and spun at 2000 rpm for 30 min. After aspirating all the liquid, the powder is dried overnight under a stream of desiccant-dried air. This is a simple method in which very little hCG or hyperglycosylated hCG is lost in the final powder. The total hCG precipitate is slowly dissolved in 100 ml of distilled water and centrifuged to remove any insoluble precipitate (hCG is extremely soluble in water).

The organic solvent precipitation is then repeated using 100% ethanol. Nine volumes of ethanol are added to the precipitate-dissolved hCG (nine volumes of

Human Chorionic Gonadotropin (hCG). DOI: http://dx.doi.org/10.1016/B978-0-12-800749-5.00038-9

ethanol added to one volume of hCG solution). Once again, the mixture is continuously stirred overnight and allowed to settle for 30 min. This liquid is then aspirated and centrifuged in the 250- or 500-ml centrifuge tubes. Small volumes of pure ethanol are used to extract the precipitate from the multiple centrifuge bottles. All the precipitates are then transferred into one bottle. The single bottle of precipitate is recentrifuged and allowed to dry overnight under a stream of desiccant-dried air. The single precipitate in the centrifuge vessel is taken up in 20 ml of distilled water and transferred to a smaller centrifugation vessel (25 or 50 ml). Any precipitate is removed by high-speed centrifugation and discarded.

The purification procedure is now limited to an hCG concentrate with maximum volume of 20 ml. A large column (2.6 cm × 100 cm or 500 ml) of Sephacryl S200 High-Resolution gel is prepared for the initial gel filtration. This gel should be equilibrated and washed slowly at peristaltic pump speed (20 ml/h or ~12 h per column volume) with two or three column volumes of ammonium bicarbonate 0.05 M (no pH adjustment needed). Ammonium bicarbonate must be the gel filtration wash because it is vaporized on lyophilization. The 20-ml hCG concentrate preparation is slowly passed through the 500-ml volume column and pumped at approximately 20 ml/h. Fractions of 100 × 5 ml are collected as the column is equilibrated with the sample and eluted. Fractions are assayed for total hCG (dilute 1/1000 for assay). A wide total hCG peak will be identified. All peak hCG tubes are collected (peak fraction to fractions containing 1/100th peak), pooled, and lyophilized. The lyophilizate is taken-up in a total of 10 ml of distilled water and centrifuged at high-speed to remove any precipitate.

The 10-ml lyophilizate free of salt (because ammonium bicarbonate vaporizes) is then placed on a 200-ml DEAE-Sepharose column. The column is equilibrated with two to three volumes of ammonium bicarbonate 0.05 M. After pumping the 10-ml concentrates onto the column, the column is eluted sequentially with 250 ml ammonium bicarbonate 0.05 M. It is then eluted with the same volume of buffer containing, sequentially, 0.05, 0.10, 0.15, 0.20, 0.25, and 1.0 M sodium chloride. Elution volumes (250 ml) are either collected in tubes or in beakers (changing beakers with each change in buffer concentration). Regular hCG normally elutes with 0.10 M sodium chloride and hyperglycosylated hCG with 0.15 M sodium chloride. The 250-ml concentrates are then dialyzed to remove sodium chloride (cannot be lyophilized), and after dialysis they are lyophilized.

A final chromatographic step is needed to remove impurities. Fine gel filtration is recommended, purifying the isolated regular hCG and hyperglycosylated hCG from DEAE separately. The lypohilizate material from DEAE is taken-up in a 5-ml volume of distilled water. A 100-cm × 1.3-cm gel filtration column is prepared. This should be Sephacryl S200 HR or S300 HR. This final gel filtration should be pumped slowly over 2 days and equilibrated with ammonium bicarbonate 0.05 M. The final pure hCG and hCG-H peaks should be collected and rapidly lyophilized, then tested for purity as described later. In the experience of the USA hCG Reference Service, these procedures are very efficient and yield approximately 60–70% of the starting material. The final gel filtration eluent should be 100% pure (check purity by SDS-PAGE and amino acid analysis). If, as occasionally can happen, hCG is not 100% pure, then either repeat final stage gel filtration or apply to reverse-phase HPLC (dissociates subunits).

38.2 Culture fluid hCG and hyperglycosylated hCG

It has been found that the methods used to purify urine (described earlier) also work to generate pure hCG and hyperglycosylated hCG from cell line culture fluids. It is noted that the culture fluids must be serum-free for purification to be possible (otherwise, see serum hCG purification described later). Culture fluids have to be such a volume to contain 2–10 mg of hCG or hyperglycosylated hCG in the starting material. The culture fluids might have to be as much as 400 l in volume to provide these concentrations. Volumes of hCG or hyperglycosylated hCG in culture fluids should be first reduced to 10 l before starting acetone precipitation by using a Millipore Minitan or other tangential flow dialysis protein concentration machine using cutoff molecular weight 10,000.

38.3 Serum hCG and hCG-H

Affinity purification should be considered for serum samples. They will not be effectively purified using the urine purification system described. The hCG Reference Service has had some success in purifying hCG from serum samples using this method. In our studies, we had 200 ml of serum from an individual with a Down syndrome pregnancy and a total hCG of 165,000 mIU/ml (approximately 15 μg/ml total hCG or 3.0 mg total hCG). In another study, we had 50 ml of serum from a hydatidiform mole case with a total hCG of 550,000 mIU/ml (approximately 50 μg/ml total hCG or 2.5 mg total). In both studies, we prepared 40 ml of affinity gel with a monoclonal antibody against the folded core of the β-subunit of hCG. The antibody was completely pure. Complete purity is essential for this purification process and thus must be checked. We bound the antibody to the gel at an approximate concentration of 1 mg/ml using cyanogen bromide – activated Sepharose.

A double-affinity chromatography procedure was required to purify regular hCG or hyperglycosylated hCG from serum. In the first procedure, we placed the serum in a beaker with gel and mixed it overnight very slowly (do not mix quickly or it will destroy the gel matrix). We then poured the gel into a small chromatography column and allowed it to drain. We then washed the column with 100 ml ammonium bicarbonate 0.05 M. We eluted the gel very slowly (10 ml/h) with 20 ml of guanidine hydrochloride 4 M. All solutions and gels should be at refrigerator temperature to avoid dissociation of hCG subunits. We then dialyzed the eluate overnight against ammonium bicarbonate 0.05 M. The dialyzate was then lyophilized. We found that the hCG in the dialyzate was only approximately 20% pure (mass of hCG versus mass of protein). We then repeated the affinity chromatograph procedure a second time, once again eluting with guanidine hydrochloride 4 M. This method gave an approximate yield of 40% regular hCG or hyperglycosylated hCG.

38.4 Analysis

All samples were analyzed for purity by reducing sodium dodecyl sulfate polyacrylamide gel electrophoresis with coomassie blue staining [2]. Only two bands corresponding to the α-subunits and β-subunits of hCG (elution at molecular sizes 18,000 and 32,000) will be observed. Alternatively, the observation of three bands could be consistent with nicked hCG β-subunit and an α-subunit, with two β-subunit bands corresponding to β1–β47 and β48–β145 (elution at molecular sizes 18,000 and 28,000).

References

[1] Kardana A, Elliott ME, Gawinowicz MA, Birken S, Cole LA. The heterogeneity of hCG: I. Characterization of peptide variations in 13 individual preparations of hCG. Endocrinology 1991;129:1541–50.

[2] Elliott MM, Kardana A, Lustbader JW, Cole LA. Carbohydrate and peptide structure of the α- and β-subunits of human chorionic gonadotropin from normal and aberrant pregnancy and choriocarcinoma. Endocrine 1997;7:15–32.

[3] Cole LA. Distribution of O-linked sugar units on hCG and its free alpha-subunit. Mol Cell Endocrinol 1987;50:45–57.

[4] Birken S, Canfield RE. Isolation and amino acid sequence of COOH-terminal fragments from the β-subunit of human chorionic gonadotropin. J Biol Chem 1977;252:5386–92.

Part M

Ethics, Comments and the Future

Ethics, Texas, and politics

39

Laurence A. Cole
USA hCG Reference Service, Angel Fire, NM, USA

Surprisingly, specializing in the field of hCG can expose one to some strange ethical issues and politics. In 2003, I ran into problems involving me and the police of Galveston, Texas.

I was phoned by a sergeant of the police in Galveston, Texas. The sergeant said that they urgently needed help from an expert on the hCG pregnancy test and that they had been given my name. As he explained, they had this acidic abdominal fluid that was the only remains of a young lady who was killed and wondered if I could test it for hCG and pregnancy. I then responded that the acid or digestive enzymes in the fluid may have chewed up or destroyed the hCG. I went on to explain that I had hCG tests that detected nicked or cleaved, dissociated, and further cleaved hCG molecules. The sergeant asked whether he could send me the fluids and if I could try and detect the pregnancy. I said that I would do it. I told him what address to send it to, and the phone conversation came to an end.

I related the contents of the phone call to Sarah A. Khanlian, my laboratory manager. Twenty minutes later she came into my office and related to me the full story of why the Galveston Police needed an urgent hCG pregnancy test. After hearing the story I phoned back the sergeant and told him that we could not do the testing. However, he forced us to do the testing. He got a court order from a local New Mexico judge to order us to complete the testing.

As Sarah showed to me, the whole Galveston story was news on the Internet. As stated, the hCG test was needed for a girl who hung herself multiple weeks previously. The hCG test needed to show definitively that she was pregnant.

As the Internet described, in Galveston, a young boy from the poor side of town was having a relationship with a young girl from the fancy houses in this town. Her parents told her specifically to stay away from this boy. They repeatedly said, "He is not good enough for you." The parents left the city on vacation for 3 weeks. The boy was invited to come to the fancy houses and visit the girl. As the girl shakingly related to him, while crying, she had just been given an hCG pregnancy test. The test was low-level positive. The test gave a result of 4–6 mIU/ml, so I was told. The testing doctor related to the girl that she was pregnant. While crying to the boy, the girl kept saying "how can I ever tell my parents when they return that I am pregnant? Especially that I am pregnant by you?" The boy wanted to help her. He stated that he did not have the money for her to have an abortion or anything like that. The girl kept crying. The boy grabbed the girl and hugged her, but it did not help much. The boy ended up leaving with the girl still crying.

The boy returned the next day. When he rang the door bell nobody answered. The boy entered, repeatedly shouting out the girl's name, but nobody replied.

Human Chorionic Gonadotropin (hCG). DOI: http://dx.doi.org/10.1016/B978-0-12-800749-5.00039-0

The boy searched the house and found the girl in her bedroom. She was standing on bed in her bedroom. She had constructed a noose out of electrical cable and had tied it around the beams in the ceiling. The girl had started to place the noose around her head. The young boy did not know what to do. The boy shouted "stop this, don't be silly, throw that wire away." The girl, however, proceeded to put the noose over her face. The boy shouted "stop," but the girl continued. He shouted "let's go to Jo's and have a drink and talk about it." She then said, "I cannot live with this pregnancy situation, or dare to tell anyone what has happened." Then she said, "If you have any love for me at all, help me move this wire over my chin." The boy felt that he was obliged to help her, and helped her manipulate the wire around her chin. The girl then jumped off the bed, quickly hanging herself.

The boy shouted "No!" as she hung herself, but he had no effect. The boy then looked at the scene with the girl hanging in the middle of the room. He did not know what to do; just the boy and the hung girl were in the house. He thought he should get away from the scene and not be found with the hung girl. He ran out from the house shouting "Please Jesus, no, don't take her!" and he then started crying as he walked slowly home.

The parents came home 2–3 weeks later and found their daughter hanging in the bedroom. They screamed and cried and could not understand why their special daughter had killed herself. They had a friend who was a city coroner perform an autopsy. He confirmed suicide and kept the autopsy simple. They then had her cremated. The autopsy did not reveal the extremely early pregnancy, a pinhead of cells. From my knowledge of reproductive biology, such a low hCG could be from a blastocyst communicating with the endometrium pre-implantation or from a hatched blastocyst producing hyperglycosylated hCG, going through the stages of implantation. All that was kept by the coroner before the girl was cremated were some rotting abdominal fluids. These were the fluids that Galveston Police were sending to me for pregnancy testing. Unfortunately, it would have been nicer if this story ended here, but it did not. The story drags on.

A few weeks later, the boy went with friends to see the movie *The Passion of Christ* starring Mel Gibson. The boy came out of the movie theater blown away by the movie, shouting out "Christ I love you!" He then said to himself that Christ would want him to tell that girl's parents and the police exactly what did happen, and why she was found dead.

The boy went to the police station near the movie theater and started to honestly relate to the police everything that had happened. In the state of Texas, there is no assisted suicide law, so that when the boy told the police the story, he was arrested and charged with murder. He related to the police, he helped the girl in positioning the wire noose and in getting it over her chin. He told them she was definitely pregnant. This is how I enter into the story.

It became very clear to me that the police wanted me to prove that she was pregnant from the rotting acidic abdominal fluid remains so that under Texas law they could charge this young boy with two murders, the girl's murder and her baby's murder. Under Texas law, two murders meant the death penalty. If my laboratory's tests proved that she was pregnant, then this poor boy could get executed. Under Texas law, the killing of any baby beyond fertilization or conception was considered murder. Based on the hCG concentration involved, this was seemingly an embryo pre-implantation or during implantation.

From my knowledge of reproductive biology, an embryo is a dying or failing being until after it has implanted. After fertilization and up to blastocyst implantation, it is small being encased by a zona pellucida or shell that stops it from growing or enlarging. It has no blood supply and no energy source. It could be called a dying being.

From my understanding, such a being has only a small chance of implanting properly and becoming a viable human being. How can killing such a blastocyst be considered murder? The Bible even says in Exodus Chapter 21 that killing a person is murder, which is punishable by death. It then says that killing a pregnant woman's baby is not murder. It is a crime punishable by a fine.

It was my personal feeling in this case that the police should have just given this boy a spank on the bottom for not stopping the girl from killing herself and then should have sent him away, not charge him with one or two murders.

I was now stuck testing this rotten abdominal fluid. All my staff refused to do anything. I was stuck doing it myself. Thinking about the acidic fluid and their nature, the test that I used to measure degraded fragments of hCG, hCG free β-subunit, degraded free β-subunit, and β-core fragment was the B204 immunoassay. This was a sensitive test, with sensitivity of 0.02 ng/ml (approximately 0.34 mIU/ml molar equivalents of hCG), which should detect most free β-subunit–β-core fragment degradation intermediates.

To my horror, the test was borderline positive, approximately 0.06 ng/ml (approximately 1.02 mIU/ml equivalents). I went a little crazy, saying this cannot be possible, and immediately started on the process of repeating the multititer plate assay. It once again was positive. The result was close to the same as the first result, confirming the assay. I felt terrible. I would have to report this low-positive result to the Galveston Police.

I went to see an Albuquerque lawyer for advice on what I had to write to Galveston Police. I first showed the lawyer copies of the New Mexico order to perform the test and report it to Galveston Police. I then showed the lawyer the report on the Internet. We carefully worded a letter together. The letter made two points: the concentrations detected were low and should be considered only as borderline or "maybe" results, and the hCG was measured in acidic abdominal fluid and not in blood, so that the acidity and the digestive enzymes in the fluid may interfere with the assay, which is standardized to serum hCG, and may cause false results. Third, I noted that the assay was standardized and compared with non-pregnant individual and pregnancy serum hCG-related molecules and that these could not be compared. Basically, the police received a very "iffy" report.

Galveston Police did not charge the boy with the murder of the fetus. He was, however, sent to prison for considerable time and is there currently. This story raises many ethical and legal questions. First, is helping someone to position a noose they had made over the chin considered murder, assisted suicide, some crime, or no crime at all? Second, if a fetus is involved in a claimed crime, possibly a fetus pre-implantation, is this murder, some crime, or no crime at all? I leave the story at this point.

Summary: hCG a remarkable molecule

40

Laurence A. Cole
USA hCG Reference Service, Angel Fire, NM, USA

hCG β-subunit is quite a remarkable molecule with an unusually flexible polypeptide and an extreme concentration of carbohydrate side chain consisting of 30% (hCG) to 42% (hyperglycosylated hCG) of the molecular weight. The peptide sequence of hCG is the source of multiple independent molecules, with each having separate biological functions. There is hCG the hormone, which is made by placental syncytiotrophoblast cells with multiple functions in controlling placental, uterine, and fetal development during pregnancy (Chapters 12–15). Hyperglycosylated hCG is a sugar variant of hCG made by placental cytotrophoblast cells. This is an autocrine that binds a TGFβ receptor and controls cytotrophoblast cell growth, apoptosis, and invasion and implantation of pregnancy (Chapters 5, 7, 10, 12, 18, and 37). hCG free β-subunit and hyperglycosylated hCG free β-subunit are also autocrines that antagonize the TGFβ receptor. These modulate most cancer cell growth and malignancy (Chapters 6, 20, 21, and 22). Pituitary hCG is a sulfated hormone that acts parallel to LH in controlling menstrual steroidogenesis and ovulation (Chapters 16 and 28). Then there is fetal hCG, made by the fetal kidney with limited structural information because of the limited fetal availability. It seemingly has roles in fetal organ growth and development (Chapter 13). hCG β-subunit is the only polypeptide known to be the root of multiple independent molecules like hCG, hyperglycosylated hCG, pituitary hCG and free β-subunit, hyperglycosylated hCG free β-subunit, and fetal hCG.

In some ways, hCG is a one-of-a-kind molecule. It is the most glycosylated glycoprotein known, with hCG accounting for 30%, hyperglycosylated hCG accounting for 39%, and hyperglycosylated hCG free β-subunit accounting for 42% glycosylation by molecular weight (Chapters 4 and 5). hCG β-subunit is strangely coded for by eight separate back-to-back genes on chromosome 18. Only four of these genes can be expressed in pregnancy by β1, β5, β8, and β9; the balance of hCG β-subunit genes can be expressed in cancer cases. Our question to hCG is, why such multiplicity?

hCG is one of the newest molecules in the human genome; it is unique to primates and humans. hCG was first evolved with early simian primates, introducing hemochorial placentation to a mammalian species (Chapter 17). The early simian hCG had only three N-linked oligosaccharides and two O-linked oligosaccharides with a pI of 6.3. This molecule has a short circulating half-life and a limited biologic activity. It was only with evolution to humans that hCG gained

Human Chorionic Gonadotropin (hCG). DOI: http://dx.doi.org/10.1016/B978-0-12-800749-5.00040-7

its fourth N-linked oligosaccharide and its third and fourth O-linked oligosaccharide, making a potent pI of 3.5 hCG with a long circulating half-life and extensive biological activity. These were needed to permit more extreme hemochorial placentation and the evolution of the human brain (Chapter 17).

To conclude, hCG is an extremely amazing molecule and is the most acidic glycoprotein. It exists as five independent structures with eight hCG β-subunit genes and is responsible for events ranging from human evolution to pregnancy to human cancers. We have a lot to learn.

hCG and the future

Laurence A. Cole
USA hCG Reference Service, Angel Fire, NM, USA

41

At the time when Robert Hussa wrote his comprehensive book on hCG in 1987, nobody could have predicted that the hCG field would have changed as much as it has over the past 33 years. As highlighted by Robert Hussa in Chapter 1, the number of citations regarding hCG has grown exponentially year by year. In 1987, hCG was one molecule, a hormone produced by the placenta with a single function in promoting progesterone production by corpus luteal cells. We now believe hCG exists as five separate molecules, each with independent hormonal, autocrine, or cytokine functions. Can we predict what will be known about hCG in a similar 36-year period by 2049? We might discover new variants with further functions. Of particular interest would be discovering an independent function for hCG free α-subunit.

In general, scientific knowledge expands twofold every 10 years. Therefore, it is inevitable that a great deal of new information regarding hCG will emerge over the next three decades, and it would be foolish to attempt to predict what these new discoveries might be. It is tantalizing, however, to speculate based on what we now know and to attempt to answer the numerous questions raised in this text. In the final chapter of this book, we take the liberty of looking into this future.

Recently, new receptor sites for hCG have been found in fetal organs, the placenta, and the uterus. Is there significance to these findings? If so, will they reveal new biological functions of this hormone? Is it involved in the growth and development of fetal organs and the fetal brain? Does hCG and hyperglycosylated hCG alone control placental growth and function? This information might be critical to our understanding of the hormone hCG. The future might see hCG measurement in assessment of pregnancy progress and administration of hCG to promote feto-placental development.

Today, we know that hyperglycosylated hCG promotes placental invasion, seemingly through its action on a TGFβ type II receptor. How is hyperglycosylated hCG action controlled and modulated? We also must confirm how and whether it promotes metalloproteinases and collagenases and other invasive enzymes. How does it work with associated cytokines? Are there methods for blocking hyperglycosylated hCG actions and preventing invasion in gestational trophoblastic diseases? In the future, we might see hyperglycosylated hCG or its derivatives being used to prevent biochemical pregnancies and miscarriages, and hyperglycosylated hCG administration during the course of pregnancy to prevent onset of preeclampsia and hypertensive disorders.

Free β-subunit and hyperglycosylated hCG free β-subunit seemingly, as shown by seven separate groups, drive malignancy in human cancer. Seemingly,

Human Chorionic Gonadotropin (hCG). DOI: http://dx.doi.org/10.1016/B978-0-12-800749-5.00041-9

carcinogenesis or cancer transformation involves the expression of the hCG gene pathways. Hopefully, we will learn how we can control free β-subunit expression to control human malignancies. Can we extend lives by inhibiting gene expression? There has been some success using hCG vaccines in treating advanced human malignancies. We hope the day will come when the treatment or cure of numerous human malignancies can be controlled by hCG vaccines or human hCG antibodies.

New total hCG tests that equally detect all hCG variants are needed. When Hussa's book on hCG was published in 1987, the only hCG test at that time was the radioimmunoassay. Today, the machine-automated immunometric assay is the current choice of tests. Over the next 33 years, we hope to see the development of new tests other than immunometric assays. New tests might equally recognize hCG, hyperglycosylated hCG, and other hCG-related molecules. New standards are urgently needed for hyperglycosylated hCG, pituitary sulfated hCG, and hCG and free β-subunit missing the β-subunit C-terminal peptide. These new standards will aid in the development of new assays.

The pituitary produces a sulfated form of hCG during the menstrual cycle and menopause. Could this have independent functions? This area of research has not yet been explored but might teach us new information regarding the roles of hCG during the menstrual cycle and menopause. We hope it will be initiated in the years to come.

Currently, the three-dimensional (3D) structure of hCG is limited by crystallography regulations. For example, only a deglycosylated hCG missing the β-subunit C-terminal peptide section, approximately 50% of the molecular weight of hCG or hyperglycosylated hCG, has been examined because of the limitations of N-linked and O-linked oligosaccharides in crystal formation. The oligosaccharides on hCG, and particularly on hyperglycosylated hCG, might dramatically affect our understanding of the structure of the molecule. With time, we hope new procedures like electron microscopy will evolve and allow us to examine the complete 3D structures of whole hCG, whole hyperglycosylated hCG, and whole hyperglycosylated free β-subunit. It is only with this information that we will truly understand each of the molecule's functions.

We also hope new antibodies and assays for hyperglycosylated hCG will be developed. These tests would be useful in pregnancy management and management of gestational trophoblastic diseases, and in accurately predicting biochemical pregnancies, pregnancy failures, and hypertensive disorders during pregnancy. We still do not fully understand how hCG gene expression is controlled. It is only with this understanding that the meaning of hCG expression can be fully appreciated. Over the past 50 years, the discoveries made in the hCG field have opened doors to other hCG-related discoveries. We hope this will continue for many years to come.

Kolibri 360 GmbH
Ferdinand-Jühlke-Straße 7
99095 Erfurt, Deutschland
produktsicherheit@kolibri360.de